Marine Biology

Marine Biology

An Ecological Approach

THIRD EDITION

James W. Nybakken

*California University at Hayward
and the Moss Landing Marine Laboratories*

HarperCollins*CollegePublishers*

Executive Editor: Glyn Davies
Project Editor: Ellen MacElree
Design Supervisor: Mary Archondes
Cover Design: Teresa A. Delgado
Cover Photo: Sea fan coral attracts a pipefish. (Photograph by Chris Newbert.)
Production Administrator: Jeffrey Taub
Compositor: Black Dot Graphics
Printer and Binder: R. R. Donnelley & Sons Company
Cover Printer: The Lehigh Press, Inc.

Marine Biology: An Ecological Approach, Third Edition

Library of Congress Cataloging-in-Publication Data

Nybakken, James Willard.
    Marine biology : an ecological approach / James W. Nybakken. —
3rd ed.
      p.   cm.
    Includes bibliographical references and index.
    ISBN 0-06-500822-7
    1. Marine biology. 2. Marine ecology. I. Title.
QH91.N9 1993
574.5′2636—dc20   92-20888
                                        CIP

92 93 94 95 9 8 7 6 5 4 3 2 1

Contents

3   OCEANIC NEKTON 90

4   DEEP-SEA BIOLOGY 125

7   MEIOFAUNA 283

8   ESTUARIES AND SALT MARSHES 302

Preface

*I*n recent years, there has been an increasing interest in the field of marine biology at the undergraduate level in our two-year community colleges as well as in four-year colleges and universities. That interest apparently derives from the growing awareness and concern at all levels of society of the importance of the world's oceans as sources of food, as reservoirs of minerals, as major suppliers of oxygen and regulators of climates, and as the ultimate dumping ground for the mounting burden of human waste materials. This concern has been popularized and brought into focus by the various international disputes on fishing rights, the whale problem, international law of the seas conferences, and numerous television programs and books addressed to the general public about marine life.

As a result of this heightened public awareness, marine biology courses have attracted a wide spectrum of students with varying backgrounds who desire a basic understanding of the biological processes that operate in the oceans.

It is important to understand that in the oceans, just as on land, there are scientific principles that govern the organization and perpetuation of organisms and associations. Although those principles operate somewhat differently in the ocean than on land because of the physical properties of water, they can be readily set forth and may be understood by those with a minimum background in other sciences and in mathematics.

The texts in marine biology available to serve these courses appeared to me to lack an ecological approach to the entire marine environment;

they tended to emphasize specific areas, habitats, and organisms to the exclusion of broader concepts and processes. I was thus stimulated to write this book as a basic introduction to marine biology, emphasizing ecological principles governing marine life throughout the world, not as a purely taxonomic or regional approach, so that it would be useful in all parts of the world. I have, furthermore, purposely downgraded any discussion of pure oceanography, except as it bears on organization of the associations or communities, in order to increase coverage of the biology. The sequence of topics in this book closely follows the sequence in my undergraduate course in marine ecology, which I have taught for many years at the Moss Landing Marine Laboratories.

This text is designed for the undergraduate student in marine biology. It presumes a certain minimum background in very basic concepts of chemistry, physics, and biology, but no more than would be obtained from a high school or introductory college course in each of these fields. Familiarity with the major invertebrate phyla is helpful, but not necessary. Some acquaintance with the basic ecological concepts is also helpful, but if lacking, may be obtained from Chapter 1. Although the text is generally aimed at the lower-division undergraduate student, I have tried to imbue this book with sufficient rigor and detail that it will also be useful in upper-division undergraduate courses in marine ecology and biological oceanography. To this latter end, I have included text citations to the primary literature.

This text stresses ecological processes and adaptations that act to structure marine associations and permit their persistence through time. It is thus not a guide to the local fauna and flora. It is presumed that familiarity with this, if necessary, will be obtained in laboratory and field trips in which the instructor will provide the taxonomic expertise.

Depending on the time available in the course, this book includes perhaps more material than can be adequately covered in a single quarter or semester. In that case, the instructor may choose to concentrate on those chapters or areas of most concern in his or her geographical region. Thus, for example, instructors in temperate zones may wish to de-emphasize coral reefs

and many may wish to spend less time on deep-sea biology and more on the intertidal, to which they have direct access. Suggestions for use of the book under varying course conditions are given in the instructor's manual. However, I hope that the comprehensive nature of this book means that whatever instructional choices are made, information will be available from the book.

CHANGES IN THE THIRD EDITIONS

The changes made in this edition maintain the rigorous treatment of all subjects established in the second edition and primarily reflect new knowledge and understanding about marine ecology.

Topics that have been substantially upgraded or changed from the second edition include the section on the differences between marine and terrestrial communities, the plankton microbial loop, vertical diel migration of plankton, hydrothermal vents, the roles of predation and competition in subtidal benthic communities, the ecology of subtidal rocky communities in various parts of the world, the role of sea urchins in kelp communities, comparisons of rocky intertidal shores in other parts of the world, a completely revised and expanded section on intertidal sand beaches, an expanded section on the role of biological interactions in structuring communities on mangrove roots, and additional differences in Atlantic and Pacific coral reefs. New subjects in this edition include an explanation of the differences in open ocean planktonic communities of the North Atlantic and North Pacific oceans, the role of geomagnetism in the orientation and navigation of marine mammals, the ecological significance of cetaceans, the role of recruitment in structuring benthic communities, switching within a single species from the role of predator to prey, a completely new section on the biology of polar seas and the biota of sea ice, tropical rocky intertidal ecology, the ecological significance of coral bleaching, the *Exxon Valdez* and Persian Gulf oil spills, drugs from the sea, and the phenomenon of global warming.

This edition also has an expanded set of color plates, which are organized into sections to complement the adjacent chapters. The new color photographs were graciously provided by Drs. Lovell and Libby Langstroth, Dr. George Matsumoto, Dr. Roger Seapy, and Mr. Hunter Lenihan.

In this edition I have again included additional information and examples that point out the controversies in marine ecology, because I believe that it is important for the student to learn that not only are there many unsolved problems in the field of marine ecology, but also that there are disagreements among scientists about how these various communities and systems are organized. In this latter connection, I think it is particularly important for the student to recognize that studies elucidating the mechanisms ordering a given community in a given geographical area likely cannot be extrapolated as a "universal" explanation for the organization of communities in similar environmental settings throughout the world.

The first edition was perceived by many as distinctly biased toward the Pacific coast, even though I felt that I had made an effort to avoid that bias. I tried to remedy that in the second edition and in this edition with more coverage of Atlantic and Gulf coast examples. I think that it is important to point out in this context that most of the East Coast of the United States south of New England as well as the Gulf Coast consists of sedimentary environments such as open sand beaches, protected sand beaches, muddy shores, and large estuaries and marshes. Rocky shores are rare. The Pacific coast, by contrast, consists mainly of rocky shores, with few estuaries and other sedimentary environments. I hope, therefore, that I will be forgiven for continuing to use Pacific coast examples in discussions of the rocky shores and commended for expanding a discussion of the dominant Atlantic and Gulf coast sedimentary environments. As part of an effort to make this book more useful globally, I have also included in this edition more examples and comparisons with communities in areas outside the North American continent.

Some who have used this book in more advanced classes have suggested that I remove the very basic introductory ecology material from Chapter 1 because their students have had it already. I have chosen not to do so in deference to those who teach more basic courses without major prerequisites and who feel that their students need it. I can suggest only that those whose classes are more advanced simply ignore it.

ACKNOWLEDGMENTS

As the author of this textbook, I am indebted to a large number of marine scientists whose research work is the foundation on which this book is based and who are too numerous to mention in entirety here. A few of this vast number are mentioned in the text and listed in the literature cited. I reiterate here my thanks to all those colleagues and friends mentioned in the prefaces to the first and second editions who helped me by reviewing chapters and contributed illustrations for the preceding editions.

In addition to those colleagues who helped me in the first editions of this book, I should like to thank the many who have helped with this edition by providing me with constructive criticisms of the second edition and pointing out errors. I am particularly indebted here to Nicholas Welschmeyer, James Harvey, Robert Higgins, and Robert Buddemeier. I am also grateful to the following reviewers for their contributions: Larry Brand, University of Miami; Al Bratt, Calvin College; Charles W. Holliday, Lafayette College; and Mary K. Wicksten, Texas A&M University.

Finally, I should like to thank the staff of HarperCollins, particularly Glyn Davies, Executive Editor for Life Sciences, for his encouragement in the process of this revision.

Because of the broad scope of the book and necessary limitations in length, I have often been forced to treat certain complex or poorly understood processes through generalizations that some may consider serious oversimplifications. Undoubtly some errors and omissions remain. I therefore close with a request to all who may use this book that they feel free to pass on to me any errors, omissions, mistakes in interpretations, comments, or suggestions that they may have, so that I might correct my ignorance and improve any future revisions.

James W. Nybakken

Introduction to the Marine Environment

About 71 percent of the surface of this planet is covered by salt water. Beneath this surface, the water depth averages 3.8 km, giving a volume of 1370×10^6 km^3. Since life exists throughout this immense volume, the oceans constitute the single largest repository of organisms on the planet. These organisms include representatives of virtually all phyla and are tremendously varied. All, however, are subject to the properties of the seawater that surrounds them, and many features common to these plants and animals are the result of adaptations to the watery medium and its movements. Before we can begin a consideration of the major associations or assemblages of organisms, we need to examine briefly the physical and chemical conditions of seawater and aspects of its motion (oceanography), and

some basic ecological principles and terms that will be central to an understanding of the processes discussed in the remainder of this text. Finally, we will make some comparisons between aquatic and terrestrial ecosystems to point out some fundamental differences in organization.

PROPERTIES OF WATER

Water is the substance that surrounds all marine organisms. It also composes the greater bulk of the bodies of marine plants and animals, and it is the medium in which various chemical reactions take place, both inside and outside living organisms.

Chemical Composition

Pure water is a very simple chemical compound composed of two atoms of hydrogen (H) joined to one oxygen (O) atom. Expressed symbolically, it is H_2O. The H atoms are bonded to the oxygen asymmetrically such that the two hydrogens are at one end of the molecule and the oxygen at the other. The bonding between the H atoms and the oxygen is via shared electrons, each H sharing its single electron with the oxygen. In this manner, oxygen receives the two electrons needed to complete its outer electron shell, and each hydrogen the one needed for its outer shell. However, the oxygen atom tends to draw the electrons furnished by the hydrogen atom closer to its nucleus. This creates a slight negative charge at the oxygen end of the molecule, while the removal of the electrons away from the hydrogens results in a slight positive charge at that end. This electrical separation creates a polar molecule.

The polar nature of the water molecule means that the hydrogen end, which is positive, will attract the negative, or oxygen, end of other water molecules. This gives rise to weak bonds, called *hydrogen bonds,* between adjacent water molecules. These bonds are only about 6 percent as strong as the bonds between the hydrogen and oxygen in the water molecule itself, and they are easily broken and re-formed (Fig. 1.1). It is this hydrogen bonding between adjacent water molecules and the polarity of the water molecule that is responsible for many of the unique chemical and physical properties of water. Indeed, if water were not polar and did not form such bonds with adjacent molecules, it would be a gas, not a liquid, at room temperature, and its freezing point would be lower than temperatures found over most of the earth's surface. Under such conditions, life would be impossible.

Physical and Chemical Properties

Because of the hydrogen bonding, water tends to stick firmly to itself, resisting external forces that would break these bonds. This is called *cohesion.* At air-water interfaces, the strength of cohesion forms a "skin" over the water surface, strong enough to support small objects. This phenomenon is called *surface tension.* The surface tension of water is the highest of all common liquids and permits the existence of associations of organisms either suspended below it or moving over the top. Cohesion is also responsible for the *viscosity* of water. Viscosity is a property of a material that measures the force necessary to separate the molecules and allow passage of an object through the liquid. Viscosity of seawater is altered by temperature and salinity. It increases with a decrease in temperature or an increase in salinity. This resistance to flow or movement is important in the sinking rate of objects (pp. 49–53) and in problems

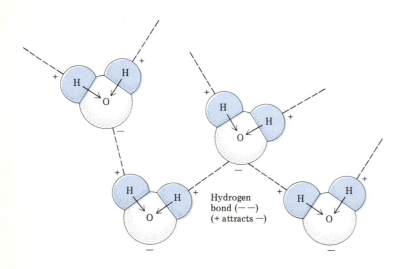

Hydrogen
bond (— —)
(+ attracts —)

Figure 1.1 Diagrammatic representation of a series of water molecules, indicating the polar nature of the molecules and the hydrogen bonding.

of movement in the water by animals (see pp. 96–98).

Another set of properties concerns the effects of heat. Water, when heated, evaporates slowly in comparison with other liquids of simple molecules. This means that the *heat of vaporization* of water is high—the highest of most common substances. This is a direct result of the strength of the hydrogen bonding between molecules, which must be broken in order to allow the escape of a molecule. Because of the high heat of vaporization, water evaporates slowly and, by so doing, absorbs considerable heat, leading to cooling. Similarly, this high heat of vaporization means a high boiling point (100°C), with the result that water is a liquid on earth rather than a gas. Related to this is the *latent heat of fusion,* which is the amount of heat gained or lost per unit mass when a substance changes from a solid to a liquid, or vice versa. Water, again, has the highest value of most common liquids. Ice, when melting, therefore takes up large quantities of heat. When ice is formed, large quantities of heat are given off.

The high values for both the heat of vaporization and latent heat of fusion mean that it takes more heat to effect a change in temperature in a given quantity of water than in virtually any other common substance. This high *heat capacity* means that water is a strong buffer against both rising and falling temperatures and therefore ameliorates the climate; also, it means that the range of temperatures experienced within any body of water is less than that in air.

Water has a peculiar *density-temperature* relationship. Most liquids become more dense as they are cooled. If cooled until they become solid, the solid phase of such liquids is more dense than the liquid phase. This is not true for water. Pure water becomes more dense as it is cooled until it reaches 4°C. Cooling below this temperature decreases the density, and when freezing occurs, there is a marked decrease in density. Therefore, ice is lighter than water and floats upon it. This property is of utmost importance to life in the oceans; otherwise, major volumes of the oceans would be uninhabitable as large blocks of ice. Different water masses in the

world's oceans have different densities due to their different temperatures and salinities. Water masses may sink or rise through other water masses depending on their relative densities and may also spread out and flow horizontally. Such behavior is common in the world's oceans, and such density differences are responsible for deep ocean currents.

The chemical properties of water that are of importance are those concerned with the solvent capacities (Fig. 1.2). Water is almost a universal solvent, with the ability to dissolve more substances than any other liquid. This is because the solvent action is of two types: one depends on the polar character of the molecule and the other on the hydrogen bonding. Various nonpolar organic and inorganic compounds containing oxygen atoms or hydrogen atoms bonded to either oxygen or nitrogen atoms are held in solution by hydrogen bonding. The polar dissolving action of water for various salts depends on the interaction of the salt ions with the charges on the water molecule. When a salt is dissolved in water, it breaks into its component ions. Thus, common table salt, NaCl, when put into water, breaks into Na^+ and Cl^- ions. It is through interaction of the charges on the salt ions and the water molecules that salts are held in solution.

Still other chemical properties of water depend on the breaking of the strong bonds between the hydrogen and the oxygen in the water molecule. In any volume of water, a few of the water molecules have separated the charges so completely that the molecule breaks into two charged parts, H^+ (hydrogen ion) and OH^- (hydroxyl ion). Here, one hydrogen atom moves away, leaving its electron with the bonding electrons on the oxygen atom, therefore making the free hydrogen positive and leaving the OH negative. Certain substances, when they dissolve in water, do so by reacting with these ions. An example is carbon dioxide (CO_2), which reacts as follows:

$$CO_2 + HOH \rightleftharpoons H_2CO_3$$

These properties are all summarized, together with some additional ones, in Table 1.1.

(A)

(B)

Figure 1.2 *(A)* Solvent action of water through creation of hydrogen bonds. *(B)* Solvent action of water through interaction of salt ions with charged parts of water molecules.

Seawater

Seawater is water in which are dissolved a variety of solids and gases. A 1000-g sample of seawater will contain about 35 g of dissolved compounds, collectively called salts. In other words, 96.5 percent of seawater is water and 3.5 percent dissolved substances. The total amount of dissolved material is termed the *salinity*. Marine biologists and oceanographers, however, generally prefer to deal with salinity in terms of parts per thousand, abbreviated as $^0/_{00}$. Thus, if a typical seawater sample has 35 g of dis-

solved compounds in 1000 g, it has a salinity of 35 $^0/_{00}$.

Dissolved substances include inorganic salts, organic compounds derived from living organisms, and dissolved gases. By far the greatest fraction of the dissolved material is composed of inorganic salts present as ions. Six inorganic ions comprise 99.28 percent by weight of the solid matter. They are chlorine, sodium, sulfur (as sulfate), magnesium, calcium, and potassium (Table 1.2). These can be considered the major ions. An additional four minor ions add an

Table 1.1 SOME PROPERTIES OF WATER

Property	Compared to other substances
Surface tension	The highest of all common liquids
Conduction of heat	The highest of all common liquids except mercury
Viscosity	Relatively low viscosity for a liquid (decreases with increasing temperature)
Latent heat of vaporization: The quantity of heat gained or lost per unit mass by a substance changing from a liquid to a gas or gas to liquid phase without an increase in temperature (cal/g)	The highest of all common substances
Latent heat of fusion: The quantity of heat gained or lost per unit mass by a substance changing from a solid to a liquid or liquid to solid phase without an accompanying rise in temperature (cal/g)	The highest of all common liquids and most solids
Heat capacity: The quantity of heat required to raise the temperature of 1 g of a substance 1°C (cal/g/°C)	The highest of all common solids and liquids
Density: Mass per unit volume (g/cm^3 or g/ml)	Density determined by (1) temperature, (2) salinity, (3) pressure, in that order; the temperature of maximum density for pure water is 4°C; for seawater the freezing point decreases with increasing salinity
Dissolving ability	Dissolves more substances in greater quantities than any other common liquid

Source: After Ingmanson and Wallace, 1973.

additional 0.71 percent by weight so that ten ions together make up 99.99 percent by weight of the dissolved substances.

The salinity of various parts of the open ocean away from coastal areas varies within a narrow range, usually from 34 to 37 °/$_{00}$, and averages 35 °/$_{00}$. The differences in salinity are due to differences in evaporation and precipitation. Higher values occur in subtropical and tropical oceans, where there is a high evaporation rate, and lower values in temperate oceans, where there is less evaporation. In inshore areas and partially enclosed seas, the salinity is more variable and may be near 0 °/$_{00}$ where large rivers discharge fresh water, to near 40 °/$_{00}$ in the Red Sea and Persian Gulf.

Even though it has been found that variations in salinity from near 0 to 40 °/$_{00}$ exist, the ratios among the most abundant ions remain virtually constant. This is important; it means that concentrations of nearly all ions in a given sample of water can be determined by measuring only one. This fundamental relationship is the basis for the measurement of salinity in seawater. Salinity can be established by measuring a single aspect, which may be the chlorinity (= chlorine concentration), for example, or the electrical conductivity or refractive index, which also depend on the salt content.

Among the remaining 0.01 percent of dissolved substances in seawater are several inorganic salts that are of crucial importance to

Table 1.2   MAJOR AND MINOR
            CONSTITUENTS OF 34.8 ‰
            SEAWATER

Ion	Percent by weight
A. Major	
Chloride (Cl^-)	55.04
Sodium (Na^+)	30.61
Sulfate (SO_4^{2-})	7.68
Magnesium (Mg^{2+})	3.69
Calcium (Ca^{2+})	1.16
Potassium (K^+)	1.10
Subtotal	99.28
B. Minor	
Bicarbonate (HCO_3^-)	0.41
Bromide (Br^-)	0.19
Boric acid (H_3BO_3)	0.07
Strontium (Sr^{2+})	0.04
Subtotal	0.71
Total	99.99

organisms in seawater. Included here are the inorganic nutrients, the phosphates and nitrates, which are required by plants to synthesize organic material in photosynthesis, and silicon dioxide, which is required by diatoms and radiolarians to construct their skeletons. In contrast to most ions, nitrates ($NO_3 =$) and phosphates ($PO_4 \equiv$) do not exist in a constant ratio with other elements or ions and tend to be in short supply in surface waters, varying in abundance as a result of biological activity (see Chapter 2 for discussion). Supplies of these essential nutrients may often become limiting to plant production in some cases (see pp. 60–63).

Other substances existing as trace amounts include elements essential to life processes, such as iron, manganese, cobalt, and copper. Though present in minute amounts, they are usually not limiting to the existence of life in the oceans. Certain organic compounds such as vitamins are also present in minute amounts, but little is known of their variation.

The salt content of seawater has a definite effect on its properties. The maximum density of

pure water occurs at 4°C, but with seawater the density continues to increase to the freezing point. Because of the salt content, the freezing point is also reduced from 0°C, the amount being a function of the salinity. For seawater of 35 ‰, the freezing point is −1.9°C. Upon freezing, however, the salts are excluded and the density decreases, and thus ice floats on the surface. The importance of this density increase below 4°C is that very cold and very dense surface water can be formed and sink to the bottom of the ocean basins where it forms the environment for deep-sea organisms.

Two gases dissolved in seawater are of metabolic importance: oxygen and carbon dioxide. The solubility of gases in seawater is a function of temperature; the lower the temperature, the greater the solubility. Therefore, the colder the water, the more oxygen it can hold. Even so, the solubility of gases in water is not great. At 0°C, 35 ‰ seawater contains about 8 ml of O_2 per liter, whereas air has 210 ml/liter. At 20°C, 35 ‰ seawater contains only 5.4 ml/liter. The reason that deep ocean water does not become anoxic (devoid of oxygen) through biological activity as it flows across the ocean floor is that, when water that is to become the deep ocean water sinks from the surface, it is so cold that it has a maximum amount of oxygen, more than can be consumed by the limited populations of animals in deep water. Oxygen is not distributed uniformly with depth in the ocean. A typical vertical profile of the oxygen content shows a maximum amount in the upper 10–20 m, where photosynthetic activity by plants and diffusion from the atmosphere often leads to supersaturation (Fig. 1.3). With increasing depth, the oxygen content declines, reaching a minimum value somewhere between 500 and 1000 m in open-ocean waters. This *oxygen-minimum zone* may have oxygen values that approach zero in some areas. Below this zone, oxygen values increase somewhat with depth, but they usually do not reach surface values except perhaps in the tropics (Fig. 1.3). The occurrence of the oxygen-minimum zone is usually attributed to biological activity depleting the oxygen coupled with the absence of photosynthetic activity and contact with the atmosphere to permit renewal. The oxygen increase below the oxygen-minimum

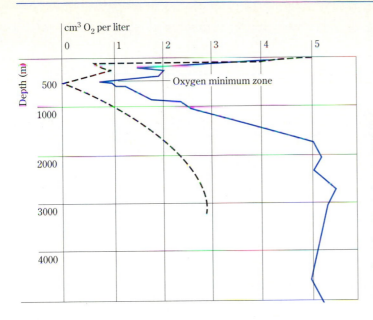

cm^3 O$_2$ per liter

Depth (m)

Oxygen minimum zone

Figure 1.3 Change in dissolved oxygen with depth in the eastern tropical Pacific Ocean (dashed line) and the tropical Atlantic Ocean (solid line). (Modified from D. E. Ingmanson and W. J. Wallace, Oceanology: An introduction, copyright © 1973 by Wadsworth Publishing Company, Inc. Reprinted by permission of Wadsworth Publishing Company, Belmont, Calif. 94002.)

zone is thought to be the result of the influx of the cold, oxygen-rich waters that originally sank in the polar regions.

The solubility of carbon dioxide is somewhat different from that of oxygen, since it reacts chemically in the water. Carbon dioxide is abundant in seawater, which has a considerable capacity to absorb the gas. This is because carbon dioxide, upon entering seawater, reacts with the water to produce carbonic acid, as follows:

$$CO_2 + HOH \rightleftharpoons H_2CO_3$$

Carbonic acid further dissociates into a hydrogen ion and a bicarbonate ion:

$$H_2CO_3 \rightleftharpoons H^+ + HCO_3^-$$

Bicarbonate may further dissociate into another hydrogen ion and a carbonate ion:

$$HCO_3^- \rightleftharpoons H^+ + CO_3^{2-}$$

The major reservoir of carbon dioxide in the ocean is the bicarbonate ion, as can be seen from its place among the major ions in seawater (Table 1.2). In contrast to oxygen, it is more abundant in seawater than in air. Therefore, carbon dioxide is rarely limiting to plants in seawater.

The carbon dioxide–carbonic acid–bicarbonate system is a complex chemical system that tends to stay in equilibrium. Thus, if CO_2 gas is removed from seawater, the equilibrium will be disturbed and carbonic acid and bicarbonate will shift to the left in the above equations, until more CO_2 is produced and a new equilibrium is set up.

The above reactions result in the production or absorption of free hydrogen ions (H^+). The abundance of hydrogen ions in solution is a measure of *acidity*. More H^+ ions mean a more acid solution, and fewer H^+ ions a more alkaline solution. *Alkaline* solutions are those that have large numbers of OH^- ions and few H^+ ions. Acidity and alkalinity are measured on a logarithmic scale of 1 to 14 units. These units are called pH units and the scale the *pH scale*. The higher the concentration of H^+ ions on this scale, the lower the pH value. Hence, low pH values indicate acid conditions. Conversely, high pH values indicate low H^+ concentrations and high OH^- concentration. The neutral point is pH 7, where equal numbers of both ions occur (Fig. 1.4).

Although pure water is neutral in pH because dissociation of the water molecule produces equal numbers of H^+ and OH^- ions, the presence of CO_2 and the strongly alkaline ions sodium, potassium, and calcium in ocean waters tends to change this, so that seawater is slightly alkaline, usually ranging from pH 7.5 to 8.4. The

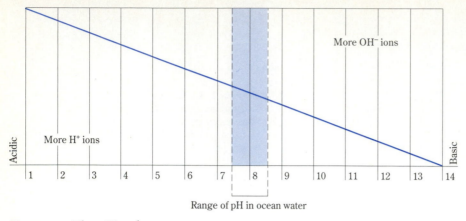

Figure 1.4 The pH scale.

carbon dioxide–carbonic acid–bicarbonate system functions as a *buffer* to keep the pH of seawater within a narrow range. It does this by absorbing H^+ ions in the water when they are in excess and producing more when they are in short supply. This is accomplished by shifting the above reactions to the right when there are too few H^+, thereby producing more bicarbonate ion and carbonate ion, and to the left when there are too many, producing more undissociated carbonic acid and bicarbonate ion.

BASIC OCEANOGRAPHY

This is not a book on *oceanography,* which is the discipline devoted to the study of all aspects of the physics, chemistry, geology, and biology of the sea. We are concerned here only with giving a basic understanding of life in the sea and how it is organized and persists. However, in order to understand the ecology of various marine associations, it is necessary to know something about the structure and motion of the ocean water masses.

Geography and Geomorphology of the Oceans

Whereas all the major oceans are connected to each other, converging in the area around the Antarctic continent, the world ocean has been separated for convenience into four major divisions—Pacific, Atlantic, Indian, and Arctic—in order of decreasing size. Projecting from, or partially cut off from, these larger oceans are smaller marginal seas such as the Mediterranean, Caribbean, Baltic, Bering, South China, and Okhotsk. The locations of the major oceans and seas are given in Fig. 1.5.

The oceans are not equally distributed over the earth. Oceans cover more than 80 percent of the Southern Hemisphere but only 61 percent of the Northern Hemisphere, where most of the earth's landmasses occur.

On the margins of the major landmasses, the ocean is very shallow, overlying an underwater extension of the continent called the *continental shelf.* Forming only 7 to 8 percent of the total ocean area, the continental shelf slopes gently from shore to depths of one to several hundred meters. The shelf extends offshore for up to 400 km off eastern Canada, but extends only a few kilometers offshore along most of the Pacific coast of North America. At the outer edge of the shelf, there is an abrupt steepening of the bottom to become the *continental slope.* The continental slope descends precipitously to depths of 3–5 km. At these depths, the bottom becomes the flat, extensive, sediment-covered *abyssal plain.* Such plains monotonously cover the floors of vast areas of the oceans at depths between 3 and 5 km (Fig. 1.6).

Abyssal plains are broken in several places by various *submarine ridges.* These ridges are extensive, contiguous submarine mountain chains

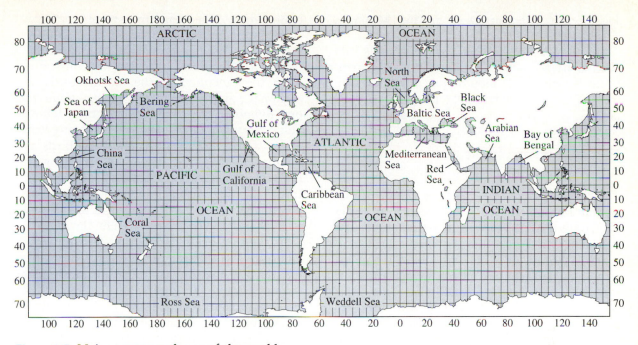

Figure 1.5 Major oceans and seas of the world.

that have been found in all oceans. The best known is the mid-Atlantic ridge, which bisects the Atlantic Ocean into east and west basins and runs from Iceland into the south Atlantic, where it links with a similar ridge in the Indian Ocean. Occasionally, the ridges break the surface to form islands. The Azores, Ascension, and Tristan da Cunha are islands formed by the mid-Atlantic ridge. These extensive ridge systems mark the boundaries of the various crustal plates of the earth and are often the sites of volcanic activity.

In certain areas, the abyssal plains are cut by deep, narrow troughs called *trenches*. Most of these trenches lie in an arc bordering the islands and continents in the Pacific Ocean. The trenches have depths from 7000 to more than 11,000 m. The deepest area known is the

Figure 1.6 Diagrammatic cross section of an ocean basin, showing the various geographic features. (Not to scale; large vertical exaggeration.)

11,022-m Challenger Deep in the Marianas Trench.

Finally, there may be isolated islands and submarine *seamounts* formed by isolated volcanic action. Such mountains, as opposed to the ridges, rise individually from the abyssal plain.

Temperature and Vertical Stratification

Temperature is a measure of the energy of molecular motion. In the world oceans, it varies horizontally with changes in latitude and also vertically with depth. Temperature is a singularly important factor in governing the life processes and the distribution of organisms. Vital life processes, collectively termed *metabolism,* function only within a relatively narrow range of temperatures, usually between 0 and 40°C. However, a few organisms are able to tolerate temperatures somewhat above and below these limits. These include the blue-green algae living in 85°C hot springs. Within the temperature range in which life processes operate, metabolism is temperature dependent. In general, for organisms that do not regulate their internal temperatures, metabolic processes are increased by a factor of 2 for each 10°C rise in temperature. With the exception of the marine birds and mammals, marine organisms are *poikilothermic* or *ectothermic,* meaning that their body temperatures vary with that of the surrounding water mass. *Homeothermic* or *endothermic* birds and mammals have the ability to regulate their own internal temperature, regardless of the temperature of the water mass. Most marine organisms are adapted to live and reproduce within more narrow temperature ranges than the total 0–40°C range. Since most are also poikilothermic and since the seawater temperatures vary latitudinally, the distribution of marine organisms follows closely the geographical differences in ocean temperatures.

On the basis of surface ocean temperatures and overall organism distribution, four major biogeographical zones can be established: *polar, cold temperate, warm temperate* (= subtropical), and *tropical* (= equatorial) (Fig. 1.7). Transition zones between these areas also exist and bound-

Warm temperate

Figure 1.7 Major biogeographical regions of the world's oceans, based on temperature.

aries may vary somewhat with season so that these zones are not absolute.

Temperature in the oceans also has a marked variation with depth. Surface waters in the tropical regions are warm, 20–30°C, throughout the year and temperate-zone surface waters are warm in the summer.

Below the warm surface water, the temperature begins to fall, and over a narrow depth range of 50–300 m undergoes a very rapid decline. The depth zone of most rapid temperature decline is the *thermocline.* Below the thermocline the temperature continues to fall with depth, but at a very much slower rate, so that the water mass below the thermocline is nearly isothermal all the way to the bottom. The thermocline is a persistent feature of tropical waters and occurs in temperate waters in the summer months. Thermoclines are absent in polar waters. Thermoclines extend through a more narrow depth range over the relatively shallow waters of the continental shelf than those formed in the open ocean. The importance of this zone is discussed in the next chapter.

Temperature also has an effect on the density of seawater. Warm seawater is less dense than cold seawater of the same salinity. Density is also a function of salinity, increasing salinity causing increasing density. However, the range of temperatures found throughout the world oceans is greater than the range of salinities, and temperature is therefore more important in affecting the density. The rapid change in temperature that produces the thermocline also means the density of seawater changes rapidly over the same depth range. Such a zone of rapid density change is called a *pycnocline.*

Water Masses and Circulation

As a result of different temperatures and salinities and their effect on density, the seawater of the world's oceans can be separated into different water masses. The *upper water mass* (surface water mass) of the oceans includes all well-mixed water above the thermocline. Below the thermocline is a *deep-water mass* extending to the bottom.

The upper water mass of the oceans is in constant motion. The motion is produced primarily by the action of winds blowing across the surface of the water. These winds produce two kinds of motion, *waves* and *currents.* Waves range in size from ripples only a few centimeters in height to storm waves, which may tower as high as 30 m. Other than height, waves are further characterized by *wavelength,* which is the horizontal distance between the tops or crests of successive waves (Fig. 1.8). The *period* of a wave is the time required for two successive wave crests to pass a fixed point. In addition to wind, waves may be generated by earthquakes, volcanic explosions, and underwater landslides, which create the destructive waves known as *tsunamis.* The attraction of the moon and sun produces the standing waves known as *tides* (see Chapter 6).

Wind wave height in the open ocean is dependent on the force of the wind, the distance or *fetch* over which the wind blows, and the length of time the wind blows.

All waves behave similarly. Once generated, the waves move outward and away from the center of origin. This horizontal progression does not result in significant horizontal transport of the individual water molecules. The water molecules transcribe a circle, moving upward and to one side as a wave crest approaches and to the other side and down to near their original position as the wave crest passes by. While the wave form and the energy are transported horizontally, the water remains stationary (Fig. 1.8). The passage of waves generates movement, not only in the surface water molecules, but also in water down to a depth approximately equal to one-half the wavelength. With each depth interval below the surface equal to one-ninth the wavelength, the orbits followed by the water particles diminish by one-half. Therefore, by the time a depth of half a wavelength is reached, the movement is almost imperceptible.

As waves enter shallow water and begin to encounter the frictional resistance of the bottom, they slow their forward motion and the wavelength decreases. As a result, they begin to increase in height and become steeper. At a point where the water depth is 1.3 times the height of the wave, it will "break," releasing the energy onto the shore. Waves are important biologically only in shallow water, as discussed in Chapter 6.

Currents are water movements that result in the horizontal transport of water masses. The

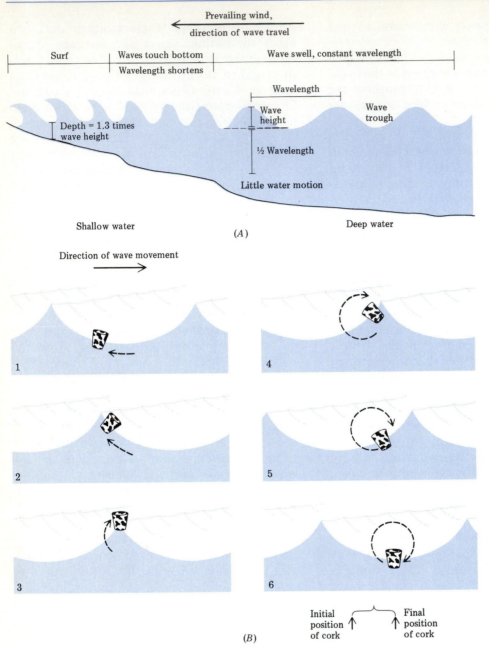

Figure 1.8 *(A)* Features of waves and change of wave form as it enters shallow water. *(B)* Cork float in the water demonstrates the passage of a wave, indicating that the water itself does not move. As the wave passes, the cork transcribes an imaginary circle (dashed line).

major ocean current systems are produced by a few major wind belts that succeed each other latitudinally around the world and where the winds are steady and persistent in direction. These winds are caused by differential heating of the atmospheric air masses aided by the *Coriolis effect*. The backbone of the system is the north-

east trade wind blowing from northeast to southwest between the equator and 30°N latitude and the southeast trades in similar position south of the equator, moving air from southeast to northwest. Between 30° and 60°N and S latitudes, the westerlies blow from the southwest to the northeast in the Northern Hemisphere and to the southeast in the Southern Hemisphere. In the polar regions cold air flows toward the equator as the polar easterlies (Fig. 1.9).

These winds set the surface waters into motion, producing the slow, horizontally moving currents that are capable of transporting huge volumes of water across vast distances in the oceans. Such currents influence the distribution of marine organisms and also lead to the displacement of biogeographical zones through the transport of warm water into colder regions, and vice versa. The major currents of the world's oceans are shown in Fig. 1.10.

A comparison of Figs. 1.9 and 1.10 will show that the ocean currents do not flow parallel to wind direction. They are deflected into roughly circular gyres that move clockwise in the Northern Hemisphere and counterclockwise in the Southern Hemisphere. The deflections and gyres are the result of the Coriolis effect. The Coriolis effect, in turn, is the result of the rotation of the earth on its axis. The spinning of the planet imparts a deflection to moving water, displacing it to the right in the Northern Hemisphere and to the left in the Southern. Because the rotation of the earth is from west to east and because the deflection of the currents created by the trade winds imparts a water movement at the equator parallel to the equator moving from east to west,

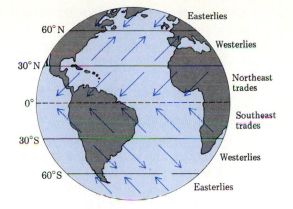

Figure 1.9 The main wind belts of the earth and their prevailing direction of motion (arrows).

the net water movement is from east to west, piling up water on the western side of ocean basins. As water builds up on the western sides, it meets the continental or island chain landmasses and is deflected north or south as continental boundary currents. These boundary currents, in turn, moving poleward, fall under the influence of the westerly winds. The westerlies impart more energy to the currents and drive them in an easterly direction, eventually crossing the ocean basins to return water to the eastern side of the basin. Continental landmass-

Figure 1.10 The major ocean surface current systems.

es on the eastern side deflect the moving water toward the equator, which completes the pattern. These huge circular current patterns, called *gyres,* are found in all major ocean basins (Fig. 1.10).

These global currents are surface currents. How deeply do they affect the water column? This can be estimated by noting what happens to deeper layers of water when wind stress causes the surface water to move. As noted above, because of the Coriolis effect, the direction of the water movement is deflected from that of the initiating wind. The energy of the wind is passed down through the water column and it sets each successively deeper water layer into motion. Each layer, however, receives a decreasing amount of energy, and therefore its velocity is lower. At the same time, due to the Coriolis effect, each layer set in motion is deflected with respect to the one immediately above it. The result is the *Ekman spiral* of current directions and velocities from the surface downward (Fig. 1.11). The depth at which wind stress fails to impart motion varies, but it is roughly limited to the upper hundred meters.

In certain areas and under certain conditions, the wind-induced lateral movements of the water may also bring about a vertical circulation or *upwelling* of water. Along the eastern margins of

ocean basins, for example, the wind-driven surface currents along the continental margins flow toward the equator. At the same time, the Coriolis effect tends to push these surface waters offshore. This water is then replaced by deeper water transported vertically to the surface. Similarly, along the equator the two equatorial currents flowing west are deflected away to the right north of the equator and left to the south. To replace this water, subsurface water upwells to the surface (Fig. 1.12).

Movement in the deep-water mass is quite different from that in the surface mass. The deep-water mass is isolated from the wind; therefore, its motion cannot be dependent on it. Movement in the deep-water mass does, however, result from changes occurring in water at the surface. Seawater increases in density with a decrease in temperature and an increase in salinity. When seawater increases in density, it sinks. Therefore, in order to move water into the deep basin of the oceans, it is necessary to increase its density at the surface. This is accomplished in two ways. Warm water from the tropics is high in salinity due to evaporation. In the northern Atlantic Ocean, such warm saline water is transported out of the tropics by the Gulf Stream. In the region of Iceland and Greenland, it meets the very cold waters of the Labrador Current mov-

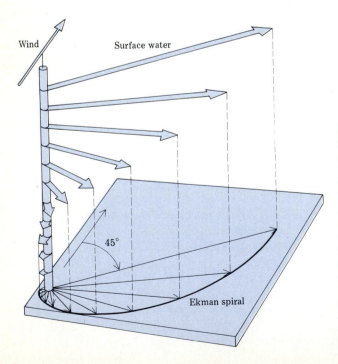

Figure 1.11 The Ekman spiral. The wind-driven surface current moves at an angle of 45° to the direction of the wind, to the right in the Northern Hemisphere, to the left in the Southern. Successively deeper water layers are deflected even further with respect to those immediately above them and move at slower speeds. Net water movement is at 90° to the wind. (Modified from Arthur Strahler, The earth sciences, 2nd ed., 1971, Harper & Row.)

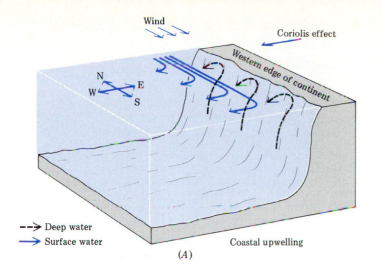

Deep water ->
Surface water →

(A) Coastal upwelling

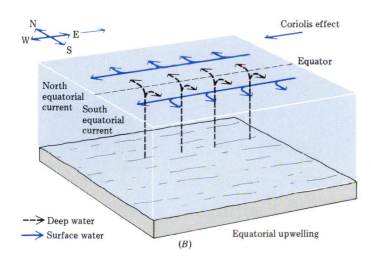

Deep water ->
Surface water →

(B) Equatorial upwelling

Figure 1.12 *(A)* Coastal upwelling in the Northern Hemisphere. Along the western margins of continents, the wind and Coriolis effect act to move water offshore as indicated by the blue arrows. This water is replaced by water moving up from the depths (dashed arrows). *(B)* Equatorial upwelling. Along the equator, the Coriolis effect acts on the westward-flowing currents, pulling water north in the Northern Hemisphere and south in the Southern Hemisphere (blue arrows). This is replaced by cool water moving up from the depths (dashed arrows).

ing south. Mixing then cools this highly saline water, increasing its density, and it sinks to form North Atlantic deep water. Similarly, the meeting of warm water moving south with cold water in Antarctic seas also causes the sinking of water masses. A final, very-high-density water mass is produced in the Weddell Sea in Antarctica, where very cold water becomes more saline when winter freezing occurs. This mass sinks to become the bottom water of most ocean basins.

Since these dense waters are all cold and produced at the surface, they have large amounts of oxygen, which is then transported to the depths. Without this oxygen, deep water would be anoxic.

These masses move very slowly north and south to form the deep water of all ocean basins. Many hundreds of years are required for these masses to move through the ocean basins. North Atlantic deep water, for example, may have been away from surface contact for hundreds of years from the time it sank near Iceland until the time it again surfaced in the Antarctic region. It is, however, a major contributor to the productivity of Antarctic seas, because when it surfaces, it brings with it large amounts of inorganic nutrients (NO_3^-, PO_4^{3-}) accumulated during the many years beneath the photosynthetic zone. A general outline of these deep-water masses and movement is given in Fig. 1.13.

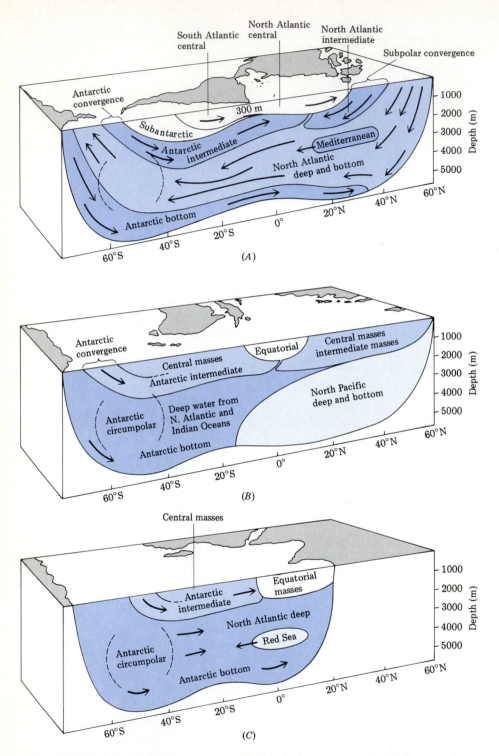

Figure 1.13 Subsurface water masses and circulation patterns in the three major oceans. *(A)* Atlantic Ocean. *(B)* Pacific Ocean. *(C)* Indian Ocean. (From D. E. Ingmanson and W. J. Wallace, Oceanology: An introduction, copyright © 1973 by Wadsworth Publishing Company, Inc. Reprinted by permission of Wadsworth Publishing Company, Belmont, Calif. 94002.)

SOME ECOLOGICAL PRINCIPLES

Because this is a text that emphasizes habitats and the ecology of marine organisms and because certain users of this book may lack a background in basic ecology, it is necessary at this point to cover briefly some basic ecological concepts and terms.

Terms and Definitions

Ecology is the science that treats the spectrum of interrelationships existing between organisms and their environments and among groups of organisms. It is important to realize initially that living things do not exist as isolated individuals or groups of individuals. All organisms interact with others of their own species, with other species, and with the physical and chemical environments that surround them. In this interactive process, the organisms have an effect on each other and on the surrounding environment. Similarly, the different factors of the environment affect the activities of the organisms. The various organisms and the environmental factors may be further organized into various levels, each of which is somewhat broader than the preceding. A *species* is a natural group of actually or potentially interbreeding individuals reproductively isolated from other such groups. All the individuals of a given species in an area constitute a *population.* Several different species populations that tend to occur together in a particular geographical area constitute an ecological *community.* A community or a series of communities and the surrounding physical and chemical environment together constitute an *ecosystem.* Ecosystems are the most complex entities in the above series and have many interactive components. They are so large and complicated that ecologists tend to study them by concentrating on their component parts, such as various communities or populations. Ecosystems, however, may be considered on a large or small scale, depending on the number of communities included and the dimensions of the surrounding nonliving environment. On the largest scale, it is possible to consider the earth as a single ecosystem comprising all the various terrestrial, freshwater, and marine communities. At the other end of the scale, it is possible to consider a tidepool or freshwater pond as an ecosystem. Both have the biological and nonbiological components the definition requires.

The oceans of the world may also be considered a single ecosystem in which a series of communities is influenced by and, in turn, influences the physical and chemical factors of the surrounding seawater. This large ecosystem may be subdivided into smaller sections or areas in which the physical and chemical parameters have differential effects on the populations of organisms, thus imposing changes in the composition and adaptations of organisms within the area subject to those effects. For example, the physical factors acting on a rocky seashore, such as waves and movements of the bottom, are not those that affect small organisms floating in the open ocean. Therefore, both areas contain different organisms and different adaptations. It is the aim of this book to give an understanding of the functioning of the total marine ecosystem through a consideration of the functioning of these major subsections, which form the bases of the subsequent chapters.

Ecosystem Components

An ecosystem is a functional unit of variable size composed of living and nonliving parts, which interact. The component parts and the whole system function through a sequence of operations involving energy and the transfer of energy. With few exceptions, the original energy source is the sun. Energy from the sun is captured by the *autotrophic* component, the green plants. Energy captured is stored in the chemical bonds of organic materials in the plants, which are the "food" that drives the *heterotrophic* component of the system. Heterotrophic organisms include all other life forms, which obtain their energy through consumption of the autotrophic plants, through consumption of organisms that have ingested plants, or through the absorption of dissolved organic matter from the environment. Such an arrangement of autotrophs and succeeding levels of heterotrophs is called a *trophic structure,* in which each successive level is called a *trophic level.* Trophic structure is a characteristic feature of all ecosystems. The first trophic level is the autotrophic or producer level, where the energy is initially captured and stored in organic compounds. As the energy is passed from level to level in such a system, the majority of it is lost through heat and metabolic use by the organisms. The amount lost at each transfer is variable, but substantial, usually ranging between 80 and 95 percent. The system thus becomes self-limiting in that, at a certain point, not enough energy remains to be passed on to sustain another level. This is visualized as either an energy pyramid or trophic pyramid (Fig. 1.14). In such a structure, the trophic level that consumes the plants (autotrophs, first level) includes animals called *herbivores*. Herbivores, in turn, are consumed by *carnivores,* which, in turn, are eaten by still other, often larger, carnivores. All levels above the second therefore consist of carnivores or *omnivores* (animals that eat both plant and animal material). Within each trophic level or population, the amount of living material at any instant in time is the *standing crop.*

The final component of the trophic structure of an ecosystem consists of the *decomposers.* These are organisms, chiefly bacteria, that break down the complex organic molecules of dead organisms, releasing simple molecules usable again by the autotrophs or by those heterotrophs that can absorb organic molecules from the environment. Decomposers act on every trophic level.

The necessary abiotic components of the trophic structure of an ecosystem are an energy source, a nutrient source, and a water source. Plants cannot fix energy and produce complex organic molecules without sunlight for energy and without a series of inorganic nutrients, of which nitrates and phosphates are the most important.

Since ecosystems are composed of from one to many communities, each, in turn, composed of many populations of producers, consumers, and decomposers, the transfer of energy through an ecosystem may follow several pathways. Each pathway that transfers energy from a given source plant or plants through a given series of consumers is called a *food chain.* The combination of all food chains in a given community or ecosystem is called a *food web.* The food web is thus a summary of all the pathways by which energy moves from one level to another through a community or ecosystem.

Decomposer
5 g/m^2

Second-level carnivore
1.5 g/m^2

First-level carnivore
11 g/m^2

Herbivore 37 g/m^2

Phytoplankton 809 g/m^2

(A)

First-level carnivore
48 kcal

Herbivore 596 kcal

(Bacteria)
3890 kcal

Net production: 8,763 kcal/m^2/year
Gross production: 36,380 kcal/m^2/year

(B)

Figure 1.14 Trophic pyramids. *(A)* Biomass pyramid. *(B)* Energy pyramid. The thickness of the bars indicates the relative amounts. (From R. L. Smith, Ecology and field biology, 3rd ed., Harper & Row, 1980.)

Biogeochemical Cycles

Among many of the chemical elements and compounds in ecosystems there is a cycling back and forth between organisms and the physical environment. Such repeated transfers are called *biogeochemical cycles.* A few of these cycles involve chemical compounds vital to the continued maintenance of life in the ecosystem and are therefore extremely important. Perhaps the most significant are the cycles involving carbon, nitrogen, and phosphorus (Fig. 1.15).

In all these cycles there is a major reservoir or pool of the element, from which the element is continually moving in and out as it passes through organisms. Each cycle also contains a sink, into which a certain amount of the chemical passes and from which it is not recycled in the normal course of events. Over long periods of time the loss to the sink may become limiting, unless the sink can be tapped again. This latter situation usually develops through geological action that acts to move the sink into an area where organisms, erosion, or other factors release the elements. Biogeochemical cycles tend, finally, to have self-regulating feedback mechanisms that keep the cycles in equilibrium.

In the carbon cycle, the reservoir is in the form of CO_2, which exists in water in the carbonic acid–bicarbonate–carbonate system (Fig. 1.15). It is fixed into organic compounds by plants, transferred to animals through herbivory, absorption, and predation, and returned to the reservoir via respiration and bacterial action. A nonrenewable loss occurs when heavy carbonate materials such as shells are deposited in deep oceans as calcareous sediments.

In the phosphorus cycle, the major reservoir is in phosphate rock. Here, erosion brings the chemical into the water, where it cycles through animals and plants and is returned to the general circulation through decay and excretion. Phosphorus is lost when it sinks to the bottom in the form of various skeletal materials. These accumulate in deep sediments.

Air is the major reservoir in the nitrogen cycle. Nitrogen gas is not used by most organisms and must be converted into a nitrogen compound before it can enter the cycle. This is normally done by bacteria and/or algae. Useful nitrogen compounds may also be produced through volcanic activity. Once in compounds, the nitrogen cycles through organisms with some loss to deep sediments (Fig. 1.15).

Biotic Structure of Ecosystems

Communities and ecosystems have the same trophic levels throughout the world. However, the species constituting each level differ among geographical areas and, in some areas, each level may have more or fewer species than another. This implies that ecologically equivalent species replace one another geographically and that the species structure of communities is variable, even if the trophic structure is not. For example, seagrass beds in the tropics may consist of several species of grasses, whereas those in the temperate zone but one. They are different species, but ecologically fulfill the same role, as autotrophs. Species that are ecological equivalents are those that perform the same function or role in a given community. They are therefore said to occupy the same ecological niche.

A *niche* in ecology is usually defined today as the role of an organism in a community. The fact that the term "niche" has had other meanings in the past has often given rise to confusion. A niche can be established for a species for virtually every environmental and biological variable affecting a species. For any given variable the range within which the species can exist and reproduce is a part of its niche. The potential distribution of a species along all such niche axes or ranges would then be the *fundamental niche* for the species. This theoretical concept of a fundamental niche is difficult to comprehend because it requires us to think in a large number of dimensions simultaneously, whereas we are used to only three, and because we cannot completely determine the fundamental niche of any organism. We therefore tend to restrict our discussion of the niche to only a few variables. We speak then of niches with respect to food, space, breeding sites, and so forth. These are *realized niches* and represent the actual distribution of the species in the real world. They will be the niches to which the term is applied in this book.

Ecological niches may be broad or narrow. Narrow niches mean that the role or function in the community or ecosystem has been more finely subdivided and the species is therefore more specialized. Broad niches, on the other hand, mean the function is more generalized. At

(A)

(B)

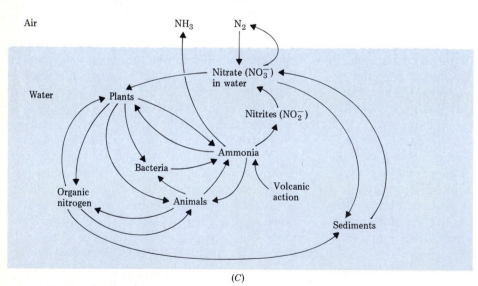

(C)

Figure 1.15 Biogeochemical cycles in the sea. *(A)* The carbon cycle. *(B)* The phosphorus cycle. *(C)* The nitrogen cycle.

the extremes, species may therefore be either *specialists* or *generalists* with reference to their niches. In our preceding example, the autotroph niche is broad in the temperate zone and therefore occupied by a single generalist grass; in the tropics, several more specialized grasses subdivide the niche.

In contrast to niche, an ecological *habitat* refers to the place where an organism is found. Therefore, the small copepod crustacean *Calanus finmarchicus* (see Fig. 2.5) has as a habitat the surface waters of the North Atlantic Ocean, but its niche is that of a herbivore feeding on the microscopic plants in the area.

Most communities have a characteristic species structure that consists of a few species that are abundant and a larger number of species that are uncommon. The numerically abundant species are usually called *dominants* and are often used to characterize a community. Thus, we can speak of a "mussel bed" on the rocky shore reflecting the dominance by the mussels *Mytilus edulis* or *Mytilus californianus.* This pattern of few common and many uncommon or rare species seems to be true at each trophic level. It is also true whether the community has many or few species in total. Species structure in ecological communities may be measured in several ways. *Species richness* is a simple listing of the total number of species in a community or trophic level. *Species diversity* is a measure that combines into a single figure both the number of species (richness) and the distribution of the total number of individuals among the species (evenness). It is expressed through various mathematical diversity indices, which need not concern us here. High species diversity has been used in the past as an indicator of stable environments and communities, but current ecological thought recognizes many exceptions to this dogma.

There are currently two major schools of thought in ecology that address how high species diversity is produced and maintained in communities. One school, the "equilibrium" school, suggests that the species composition of communities is usually in a state of equilibrium and that high diversity is due to a large number of habitats and/or finely divided narrow niches that are maintained by various feedback mechanisms (= niche diversification) and the stable nature of the physical environment that allows the equilibrium to be achieved. The other school, the nonequilibrium or intermediate disturbance school of Connell (1978) and others, says that communities and species are rarely in a state of equilibrium and that high species diversity is maintained through continual or gradual environmental change and periodic disturbance. This promotes a changing species composition through the presence of many species that are not highly specialized. In this model, communities that experience very frequent disturbances that kill most or all the resident species would tend to be colonized by only those few species that happen to have the ability to attain maturity and reproduce quickly before the next catastrophic event. At the other end of the spectrum, those areas that experience infrequent, small disturbances and only at long time intervals would also tend to have low diversity. In this case it is because the most efficient competitors have the time to eliminate most or all of the other species and take over the community. Only when the disturbances are intermediate does diversity increase. This is because enough time is available to permit establishment of a variety of slower-growing and reproducing species, and the interval between disturbance events is short enough to avoid competitive exclusion. It is probably not possible at our current level of knowledge to decide whether one or the other or both of these theories is correct for all the various marine communities, but the evidence that has been gathered thus far would seem to favor the intermediate disturbance theory for at least some communities.

Each species in a community has certain tolerances with respect to all environmental factors. If the limit of tolerance is exceeded in some area by a given factor (say, temperature), the species will be absent. Similarly, each species requires a certain minimum amount of various materials. If the concentrations of these necessities, such as nitrate, fall below the minimum, the species disappears. More important, if any *single factor* exceeds the tolerance level or if any single necessary substance is reduced below the minimum, the species will be eliminated. This is true even if all the other factors and substances are favorable. This is known as *Leibig's law of the minimum.*

Since these *limiting factors* vary for each species and since communities are defined on the basis of recurrent groups of species, it follows that the boundaries between communities, representing changes in various environmental factors, are also not sharp. The result is areas of transition between adjacent communities where species gradually drop out and others come in. Such boundary zones are called *ecotones*. Compared to communities that usually change rather slowly in time and space, ecotones change rapidly.

Communities are not static units. They change in structure and composition with season and over longer periods of time. Some communities tend to change in an orderly fashion over periods of many years until they reach a stage that perpetuates itself indefinitely as long as the climate does not change or there is no disturbance. In such a sequence, each community modifies the environment, in turn making it suitable for the next community. This orderly process of community change controlled through modification of the physical environment is called *ecological succession* and the terminal, persistent community the *climax*. The intermediate communities are called *seres*.

The above scenario is the classical or *facilitation* model of succession as originally developed for terrestrial plant communities by Clements (1936). A second model for succession, termed the *inhibition* model, was proposed by Egler (1954) and it differs from the classical in that no species is competitively superior, and whichever species gets to a site first holds it against later settlers. In this model, succession is not an orderly predictable process and there is no climatic climax. Here succession proceeds from short-lived to long-lived species. A third model, proposed by Connell and Slayter (1977), called the *tolerance* model, is intermediate between the other two. In this model early colonizing species are not necessary, and any species can start succession. Community change then occurs as species that are more tolerant, or more competitively superior, prevail.

In the marine environment ecological succession seems to follow the second or third model. In bottom communities, following a disturbance the area is usually reoccupied by organisms that happen to have their propagules in the water and available to settle. The first occupants may therefore change, depending on the time of year that the disturbance occurs. Furthermore, the initial occupiers often do not modify the environment to make it unsuitable for them and more suitable for subsequent organisms. In fact, they may even prevent the occupation of the area by other, later successional forms by various means and thus delay the climax community development. Some examples of these kinds of succession are given in Chapters 5 and 6. Finally, succession probably does not occur in the open waters among pelagic communities.

Ecological Control and Regulation

Populations, communities, and ecosystems are all regulated by various factors. The major controlling factors for ecosystems and communities are energy, the physical factors collectively termed climate or environment, and the interaction among various species that compose the systems. Ultimately, nearly all systems on earth are limited by the amount of energy available from the sun. However, tolerance limits of various species to abiotic factors such as temperature, light, nutrients, and salinity also limit the extent of populations and communities in the oceans. The final aspects of regulation and control considered here are those dealing with the interaction among populations that act to keep populations within limits.

Biologists have long known that all species possess the reproductive potential to produce much larger populations than are observed under natural conditions and that if population explosions occur, they are quickly reduced. What are the biological factors that exert control? They are grouped under competition, predation, and parasitism and disease.

Competition is an ecological term referring to the interaction among organisms for a necessary resource that exists in short supply. Competition may be *intraspecific* (among individuals of the same species) or *interspecific* (among individuals of different species). Competition may be for any number of things, but usually is limited to items such as light, food, nutrients, water, and space. In a competitive interaction, either the competitors manage to share the limited resource or one excludes the other. In the first case both individuals are hampered, which inhibits their growth, development, and reproduction, thus limiting the numbers. In the second case, one individual is eliminated, again controlling population. Interspecific competition is usually between two closely related species and has led to the *competitive exclusion principle,* which says no two species with exactly the same requirements can coexist in the same place at the same time; that is, complete competitors cannot coexist. As population numbers increase, competition usually increases, because the limited resource becomes scarcer. This increased competition increases the stress on animals and plants and absorbs energy otherwise used for reproduction, thus limiting populations.

Predation can be defined as the consumption of one species by another. The animal that is the consumer is the *predator;* the victim, the *prey.* A special case concerns those animals that consume plants. They are called *herbivores.* A *grazer* is an animal that feeds on plants or sessile animals. Predators and herbivores vary considerably in their ability to regulate the numbers of organisms they consume. In some cases, the predator may be the most important factor in regulating numbers of a prey species. In other cases, a predator may have little effect on a prey population. In the first case, the removal of the predator will have a marked effect on the prey population, causing it to increase dramatically. In the latter case, predator removal has little effect on the prey population. We shall see examples of both these situations in the following chapters. A special case of predation concerns those predators that have a profound effect, not only on their prey population, but also on the entire community of which they are a part. In these cases, the removal or depletion of the numbers of the predator has the effect of causing great changes in the presence and abundance of many species in the community, most of which are not the prey of the predator. As a result, the entire community structure may be changed. Such predators are called *key industry* or *keystone* species.

Parasitism and disease are the final biological controls on populations. As with predation, they may exert strong control or they may have little effect. *Parasites* are organisms living in or upon other organisms from which they derive nourish-

ment and shelter. Many, perhaps all, marine organisms have parasites, but we know considerably less about the role of parasites and disease in regulating populations of marine organisms than we do for terrestrial organisms. Perhaps the best-documented change induced in a marine community by a disease is the loss of the eelgrass beds in the Atlantic Ocean in the 1930s (see pp. 202–203).

LARVAE AND LARVAL ECOLOGY

A very large number of marine organisms, both fish and various invertebrate phyla, inhabiting both pelagic and benthic systems, produce larvae in their life cycles. *Larvae* are independent, morphologically different stages that develop from fertilized eggs and that must undergo a profound change before assuming adult features. They are almost always considerably smaller than the adult stage of the species. Larvae and their ecological characteristics are of considerable significance in the establishment and maintenance of many marine communities and associations. This is because, as Chia (1988) notes, in any given marine habitat the distribution and abundance of benthic invertebrates are maintained by four factors: larval recruitment, migration, asexual reproduction, and mortality. Most benthic communities are composed of species that reproduce by producing various larval types that undergo a free-swimming stage in the water column before metamorphosing to produce the benthic adults. Even many organisms that pass their adult lives in the water column also produce larvae. Therefore, an understanding of larval ecology is central to understanding how such communities persist. We discuss the basic features of larval ecology here, leaving the specific role for larvae in each larger community to be covered in the appropriate chapter.

Larval Types and Strategies

As noted by Vance (1973), there are three possible paths that development may take in a benthic invertebrate with a given amount of energy available with which to produce young. One way would be to produce a great many very small eggs. Such eggs would hatch quickly into larvae that would be free swimming in the plankton. Because so little yolk was put into each egg, the larvae would be dependent for nutrition on the plankton. Such larvae are called *planktotrophic*. A second course would be to produce fewer eggs and endow each with somewhat more energy in the form of yolk. Such eggs would hatch into larvae that, because of their yolk reserve, would not feed in the plankton and spend less time in the plankton before settling. Such larvae would use the plankton phase mainly for dispersal. Larvae of this type are called *lecithotrophic* larvae (Fig. 1.16).

A final option would be to dispense with the larval phase altogether. In this course, the adult produces a very few eggs, each with a very large amount of yolk. Such eggs undergo a long development without additional energy sources. They pass through the larval stages in the egg and hatch as juveniles. There is no free-swimming larval stage. This is *nonpelagic* or *direct development*. Such hatchlings are called *nonpelagic larvae* or *juveniles*.

There are two major assumptions of the Vance model. One is that mortality in the plankton is constant over time and is not related to size of the larvae, and the other is that egg size is inversely related to time spent in the plankton.

Each of the above developmental pathways has both advantages and disadvantages. In the case of the planktotrophic larvae, the advantages are that a very large number of young can be produced for a given amount of available energy, and wide dispersal is assured through the long time spent in the plankton. The disadvantages are that the larvae depend for nutrition on the plankton, a notoriously unpredictable situation, and that long residence in the plankton increases the chances that a predator will consume the larvae. The chances of being consumed by a predator in the plankton are extremely high and the pelagic zone is thus a very dangerous habitat.

Lecithotrophic larvae have the advantage of spending less time in the plankton and thus have less chance of being consumed. They also are not dependent on the vagaries of the plankton for food. The disadvantage of this strategy is that, because of the greater amount of energy placed into each egg, fewer eggs and larvae can be produced per unit of available energy. A shorter

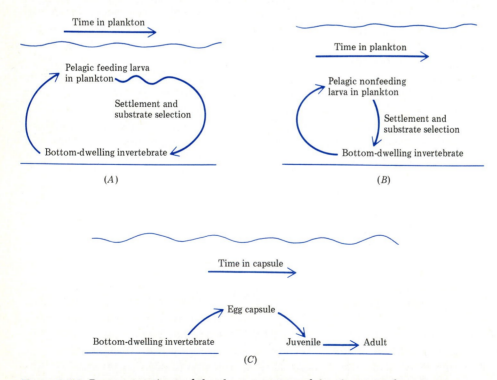

Figure 1.16 Representations of the three patterns of development found among benthic invertebrates. *(A)* Planktotrophic. *(B)* Lecithotrophic. *(C)* Nonpelagic.

time in the plankton also means less dispersal ability. A lecithotrophic larva also presents a larger target for visual predators in the plankton.

The nonpelagic development has the advantage of reducing the planktonic mortality to zero. In this case, however, the disadvantages are that only a few eggs can be produced because of the great amount of energy each takes, and that dispersal is nil.

Since all three types of development exist, it can be assumed that under certain conditions one type or another is favored. Even though we are currently ignorant of all the reasons why some areas and conditions favor one larval type over another, it is possible to suggest some that are valid.

In polar waters, where the productivity is reduced and confined to a narrow summer peak (pp. 66–67) and where it takes a long time for benthic animals to obtain sufficient energy to reproduce, nonpelagic development is common. Such development avoids the problem of precise timing to hit the very short polar bloom period, where failure would mean the death of most, if not all, larvae, and it ensures the highest survival rate of the few eggs that can be produced.

Lecithotrophic larvae are common under conditions somewhat similar to those above. Lecithotrophy is preferred over nonpelagic development if dispersal is of prime concern to the survival of the organism in question. Lecithotrophy is also more energy efficient than planktotrophy when planktonic mortality is high. Both lecithotrophy and nonpelagic development appear to be common in deep-sea organisms.

The advantages of planktotrophy are that eggs are cheap to produce, can be produced in large numbers, and have high dispersal rates. The disadvantages are primarily the great attrition rate suffered while in the plankton and the unpredictable plankton food source. Planktotrophic larvae therefore should be common where the plankton food is predictable over long periods, where planktonic mortality is low, where dispersal is of importance, and where development time is short. The area that fulfills most of these criteria is the tropics, and it is there that we find most forms with planktotrophic larvae.

Indeed, it was Thorson (1950) who first suggested that there was a latitudinal gradient in the proportions of the three larval types. From his studies, primarily in cold temperate and Arctic waters, Thorson suggested that planktotrophy was most common among tropical marine organisms, and lecithotrophy and nonpelagic development became more prevalent as one approached polar latitudes. Planktotrophy also decreased and lecithotrophy and nonpelagic development increased in moving from shallow waters to the deep sea.

The problem with the above simple models is that there is no nice fit to it if one studies a variety of marine invertebrates. Strathmann (1977, 1978) has elucidated the problems by demonstrating that the predicted effect of various hypotheses about larvae and larval development is confounded when looking at the real world. For example, species with a feeding larval form (= planktotrophic) very often have low survival and recruitment most years, but occasionally they have a very successful year. One would tend to expect that an organism that produces such larvae would have to be long-lived as an adult and breed many times (= iteroparous) simply because in most years the young would not survive, and it would take several years of producing young to have the probability of a successful recruitment. However, many short-lived animals that breed only once (= semelparous) in fact have feeding planktotrophic larvae.

Similarly, it has been suggested that small animals more often are brooders, producing a few eggs, because they simply are not big enough to allocate a large amount of energy to produce numerous eggs. To ensure the survival of the few, they brood them through to a more advanced state. This seems inherently unsatisfactory in a logical sense because if brooding ensures a higher percentage of survival of young, then why shouldn't all animals, large and small, do it? Certainly, the larger animals would gain by getting a higher return of juveniles per energy unit input. In other words, there must be some other compensating advantage to sticking with the "risky" planktotrophic development other than simple survival of the young. One such advantage might be dispersal, since large animals are less likely to be moved over long distances by water currents and rafting than are small animals. This is particularly true if the large adults are also sedentary. However, both animals that are permanent members of the plankton and those that as adults are sedentary or sessile on the bottom have planktotrophic larvae. Since adults in the water column are as dispersible as the larvae, this would tend to negate the argument for dispersibility. One reason why larger animals may not brood is that they may simply lack the physical space to brood a large set of eggs or be unable to aerate the very large egg mass properly. Strathmann (1985) has even suggested that, since eggs must have oxygen and since there are physical constraints on how egg masses are constructed so as to allow access by oxygen to the interior eggs, it may be that planktotrophy is an adaptation to prevent suffocation of eggs in large egg masses.

In summary, whereas we have several hypotheses as to why organisms may have a particular larval developmental type, we can find many exceptions. Together these suggest that there are many factors involved in selection for a given larval type, that they do not act similarly for all the animals in a given habitat, and that there may be morphological or phylogenetic constraints on the development of a particular type.

Obrebski (1979) has also offered a model for the existence of different larval types. This model suggests that the type of larva—planktotrophic, lecithotrophic, or nonpelagic—should be predicted by the relative abundance of suitable substrates for settlement. In this model, those benthic organisms colonizing abundant habitat patches would have less need for long-lived planktotrophic larvae, whereas those inhabiting rare patches would benefit from having long-lived planktotrophic larvae in order to search out the rare habitat type.

Vance's models have also been challenged by Underwood (1974), who has shown that surveys of the existing data reveal many discrepancies to the predicted modes of reproduction. This has led Underwood to conclude that other factors are of importance and that energy considerations alone are inadequate to explain the different strategies.

We find ourselves in the somewhat awkward position of having several different hypotheses for the existence and perpetuation of the different larval types, but with the real data not seeming to fit any model completely and with

several larval types coexisting in the same area. This must strongly suggest that other factors are also at work in determining larval development type. Why should this be? We can only suggest that for each species there is a necessity to balance all the factors, including energy allocation, dispersal, relative abundance of microhabitat space, longevity of adults, competitive ability of adults, and undoubtedly others, against the need for larger numbers of young. The type of reproduction that has evolved is that which balances all of these factors on the one hand with survival on the other. This is reinforced by the finding of certain benthic infaunal species that are able to alter their reproductive strategy among the three developmental patterns depending on prevailing conditions. An example is the common polychaete worm *Capitella capitata,* which Grassle and Grassle (1974) have shown can produce either planktonic or benthic larvae depending on the conditions.

Larval Ecology and Community Establishment

Many, if not most, marine benthic communities are composed primarily of species that have a free-swimming larval stage. The exception to this rule appears to be the aforementioned polar waters. Since over larger areas the benthic communities seem to persist over time (see Chapter 5 for examples) but are dependent for this continued existence on the settlement of larvae from the plankton, and since the adults live only a short time, how is this stability maintained? In other words, what mechanisms are there that will act to ensure that the "right" larvae will settle and mature to keep the community going? A partial answer to this is to be found in the characteristics of larvae.

Since the pioneering work of D. P. Wilson (1952), many laboratory studies on larvae have disclosed that larval settlement and metamorphosis can be regarded as responses on the part of the larvae to complex and highly specific environmental stimuli, including light, gravity, and fluid movement, but the most influential clues appear to be chemical, emanating from biological sources. Most larvae have been shown to have preferences as to the areas in which they will settle. In other words, larvae do not just settle out of the water column on whatever substrate is available when the time comes for them to metamorphose into adults. Larvae have the ability to "test" the substrate, probably responding to certain physiochemical factors; if the substrate is unsuitable, they do not settle out or metamorphose. This means that certain substrates will always be attractive to certain larvae and repellent to others. Larvae also respond to the presence or absence of adults of their own species. Many larvae preferentially settle where the adults are living. They are, as Crisp and Meadows (1962) have shown, attracted to the area by a chemical or *pheromone* secreted by the adult. Since the presence of the adult means that the area is suitable for habitation, this mechanism is one that acts to ensure survival of the young. It also ensures that the community will persist.

The larvae of many invertebrates also have the ability to delay their own metamorphosis for a certain period of time if they cannot find suitable substrates at the time they are supposed to settle. This delay of metamorphosis is a finite period, however, and after a certain time, metamorphosis will take place even over unsuitable substrates. The ability to delay settlement, however, is yet another factor that can be employed to ensure that the larvae will settle in the proper place.

Larvae, as Thorson (1966) has noted, also respond to other physicochemical factors such as light, pressure, and salinity. Many free-floating larvae are positively phototactic in the early stages of their larval life. This condition ensures that they will reside in the upper, faster-moving waters where dispersal is greatest. Later, as time for settlement approaches, they become negatively phototactic and migrate downward toward the bottom. Some larvae are very sensitive to light and pressure such that they inhabit only certain levels in the water column, areas with a definite light regime and pressure. Segregation of larvae into various layers in the water column also means that wherever a given layer impinges on the bottom, it will have only a certain complement of larvae available for settlement, a further way of ensuring a nonrandom bottom community (Fig. 1.17).

These various studies designed to determine the mechanisms that control larval settlement and metamorphosis have tended to strongly favor the hypothesis of active habitat selection by the larvae. However, most of these studies have been done in the laboratory in still water and have not considered the alternative hypothesis that larvae may be passively deposited on the sea floor by physical processes. Presently Butman (1987) has suggested that various hydrodynamic processes are also active in determining where larvae settle.

Finally, different species have different times of reproduction and hence different times of the year during which larvae are in the water column. As we shall see subsequently, this may also have an effect on the associations that develop.

Despite the considerable abilities of larvae to discriminate among substrates and select places to settle, there is often considerable variability from year to year in both species abundances and composition within bottom infaunal communities. Samples taken close to each other often show considerable differences in species composition and abundance. If the larvae have such selective powers, why then is there such variability? The answer to this lies in the life histories of the various invertebrates composing the community, their interaction with each other and with the physical environment, and the effects of predators and competition.

Life History Strategies

According to MacArthur (1960), it is possible to recognize two quite different types of life history patterns among organisms in any habitat. These are introduced here and will be referred to in subsequent chapters. One type can be called *opportunistic* or *"r* selected." Opportunistic marine species are characterized by having short life spans, rapid development to reproductive maturity, many reproductive periods per year, larvae present in the water during much or all of the year, and high death rates. Usually, they are small animals and often are sedentary or *sessile* (fixed in place). The second type of life history

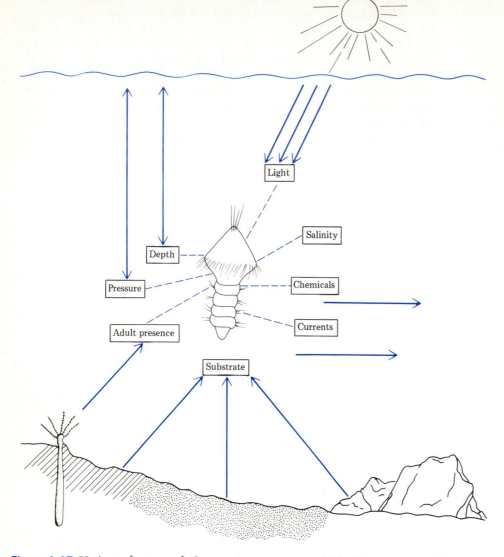

Figure 1.17 Various factors of the environment to which larvae respond in selecting a suitable site for settling.

can be termed *equilibrium* or *"K* selected." Equilibrium species are those that have long life spans, relatively long development time to reach reproductive maturity, one or more reproductive periods per year, and low death rates. Usually, these species are somewhat larger in size than opportunists and are often mobile (Table 1.3).

The above characteristics define the two ends of a continuum. It must be remembered that although some species will have all characteristics of an opportunistic species and others

those of equilibrium species, still others will fall somewhere in the middle and have varying mixtures of these features. For purposes of discussion, however, it is easier to consider the ends of the spectrum.

Opportunistic species tend to be favored where the substrate is subjected to frequent disturbance from some particular agent. In the case of shallow water soft bottoms, the common disturbance agent is waves or other water motion that stirs up the bottom or carries away the upper layer of sediment. Another disturbance

Table 1.3   SUMMARY OF THE CHARACTERISTICS OF
OPPORTUNISTIC AND EQUILIBRIUM SPECIES

Features	Equilibrium (K)	Opportunistic (r)
1. Reproduction periods	Few per year	Many per year
2. Development	Slow	Rapid
3. Death rate	Low	High
4. Recruitment	Low	High
5. Colonizing time	Late	Early
6. Adult size	Generally large	Generally small
7. Mobility	High	Low

may be the rapid deposition of sediment on the bottom. In certain cases, biological activity may also disturb the bottom, for instance when large fish such as rays dig into the bottom. When such a disturbance occurs, it usually causes the mortality of the resident organisms. The result is an open area, which can be resettled by organisms. Because of their attribute of having larvae in the water most of the time, opportunists quickly settle these open areas. Furthermore, if disturbances occur frequently, the opportunists will have the advantage because they can quickly mature and reproduce again before being destroyed by the next disturbance.

Equilibrium species, on the other hand, tend to inhabit areas that are not subject to frequent disturbance, thus allowing them to complete their life cycles. Too frequent disturbance would mean destruction before they could reach reproductive maturity, and hence no larvae would be present in the water column to settle the area again.

Whenever the bottom is subject to frequent disturbance, the benthic communities are usually composed primarily of opportunists, whereas in areas where disturbance is less frequent, more equilibrium species are found. This is because equilibrium species are generally better competitors than opportunistic species and, given sufficient time, will outcompete the opportunists.

Because of their different characteristics, there are differences in abundances between these two groups. Communities in which the number of opportunists is high show great fluctuations in numbers of individuals over time, whereas communities composed of equilibrium species are more constant in abundances.

If the above situation prevails, it would appear that communities would be composed of either opportunists or equilibrium species, depending on frequency of disturbance. In reality, samples of the bottom usually contain representatives of both groups. How does this happen? Apparently, the equilibrium species are slower to colonize disturbed areas than are opportunists, but they are better competitors once they are there. That is to say, once there, they tend to push out (outcompete) the opportunists. Over time, then, there is a sequence in which, immediately following a disturbance, the area is settled by opportunists. As time goes on, the equilibrium species settle in, and finally they outcompete the opportunists. When sampling, then, differing numbers of both groups would result, depending on the time since the last disturbance. Therefore, varying numbers of both groups can be seen in bottom samples (Fig. 1.18).

The reason for different numbers of individuals and species from different areas in the same bottom is at least partially due to the fact that disturbances do not affect the whole bottom equally, but rather act selectively. The result is to produce a patchy array of communities, each in varying degrees of recovery from disturbance and with differing numbers of opportunists and equilibrium species.

Since wave-induced disturbance decreases with depth, it is not surprising that deeper communities tend to be inhabited by more equilibrium species than shallow water, unless another type of disturbance creates unstable conditions. Such a disturbance could be biological— for example, the reworking of sediment by fishes or burrowing deposit feeders.

COMPARISON OF TERRESTRIAL AND MARINE ECOSYSTEMS

It is quite true that ecological principles apply equally whether one is dealing with terrestrial, freshwater, or marine ecosystems. However, the special conditions prevailing in salt water that are the result of the physical and chemical properties of water have, directly or indirectly, channeled evolution and adaptation of marine organisms. The result has been some striking

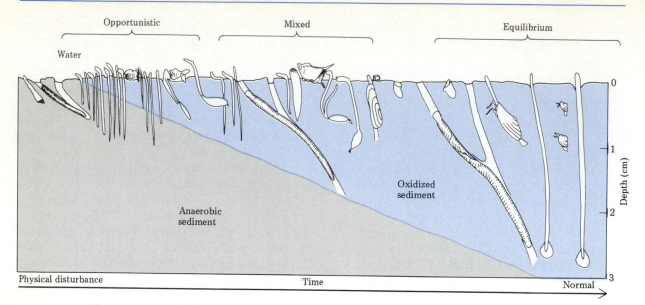

Figure 1.18 Changes in a soft bottom community with time after a physical disturbance. (From D. C. Rhoads, P. L. McCall, and J. Y. Yingst, Disturbance and production on the estuarine seafloor, *American Scientist*, September–October, 1978. Reprinted by permission of American Scientist, Journal of Sigma Xi, The Scientific Research Society.)

differences in the organization of marine communities when compared with terrestrial communities. These will be considered next.

Physical and Chemical Differences

Seawater has several physical features that have a profound effect on the organization of marine communities. They are its greater density with respect to air (830 times as dense as air), its greater viscosity (60 times that of air), its ability to transmit sound four times faster than air, its high electrical resistivity (10^{16} greater than air), and its ability to absorb light. Of these factors, the density and viscosity differences have the most widespread effects on marine organisms. The greater density of seawater means that relatively large organisms and particles can remain suspended in it. This is not possible in air. One significant result of this is that the marine ecosystem has evolved a whole community of small organisms that are perpetually afloat, the plankton. No such comparable community exists terrestrially floating in the air. As a result, all communities in the sea are bathed in a medium that itself contains a community! The presence of this suspended community has, in turn, led to

the evolution of animals that are adapted to filter organisms and particles out of seawater. These *filter feeders* are unique to aquatic systems. There is nothing really comparable on land. The closest analog would be spiders, which capture flying insects in their webs. Furthermore, since the waters of the oceans are in perpetual motion, many of these filter feeders are sessile, extending their filtering nets up into the moving water mass. Sessile filtering animals are not possible in the terrestrial environment because the bathing medium flowing by contains no perpetually suspended food organisms as particles. Finally, marine organisms, especially the sessile species, have evolved motile larval forms. Such motile larvae, when put into the plankton, permit dispersal of the species.

As a result of the suspension of a whole community in the bathing medium, it is more difficult to have isolated communities and isolated species distributions in the sea.

An interesting sidelight of the viscosity difference between air and water is that water becomes less viscous as it heats up, whereas air becomes more viscous. Therefore, tropical marine animals encounter less resistance to movement than do temperate or polar animals, whereas tropical birds flying experience more drag than temperate or polar birds do.

Another physical factor indirectly responsible for differences between marine and terrestrial organisms is *gravity*. Because marine plants and animals are buoyed up by the greater density and viscosity of water, they do not need to have a significant amount of their biomass invested in structural material such as skeletons or cellulose in order to hold themselves erect against the force of gravity. Similarly, where movement is concerned, terrestrial animals must raise their mass against the force of gravity for each step. Such movement requires significantly more energy than swimming movements do in aquatic organisms and, hence, more energy storage.

This absence of large amounts of structural material and energy storage in the bodies of marine organisms is reflected in the differences in the dominant biochemical compounds found in terrestrial and marine organisms. Among terrestrial organisms, the dominant compounds are carbohydrates; among marine organisms, the predominant material is protein. Living organisms that are composed primarily of carbohydrate are long lived, slow growing, and rich in stored energy. Protein-dominated organisms, on the other hand, are rapid growing and without significant energy storage.

In contrast to air, water strongly absorbs light. As a result, light entering water can only penetrate to a certain depth before it is completely absorbed. Although this depth varies, it is sufficiently shallow (200–400 m) that the great majority of the volume of water in the oceans is without light. This means that plant life and primary productivity are limited to an extremely narrow band near the surface. Whereas terrestrial communities, with the exception of caves, all possess sufficient light for plant growth, many marine communities exist without the benefit of an autotrophic component.

Not only does water absorb light, it does so differentially, depending on the particular wavelength. The result is that certain wavelengths penetrate deeper than others. This is not true on land, where the same spectral composition of light impinges at all levels in any community. The importance of this to marine communities is discussed on pp. 57–60.

The speed of sound at 20°C in air is 346

m/sec, whereas at the same temperature in water it is 1518 m/sec, a fourfold increase. In addition, the wavelength in water increases by a factor of four. The biological consequences of this concern the use of sound by animals to monitor their environment, find food, and avoid predators. The faster movement of sound in water coupled with the more limited range of vision due to the absorption of light by water means that sound is more useful and has a greater range for orienting an animal to its surroundings in water than does vision. Whereas both terrestrial and aquatic animals employ sound to orient and detect prey, the process is more widespread in the sea. It is interesting to note that to detect prey items with sound, the sound waves must be several wavelengths smaller than the potential prey. Since in water the wavelengths of sound increase fourfold, marine animals must use shorter wavelengths (= higher frequency) than terrestrial animals to detect the same size prey (see Chapter 3 for further discussion).

The physical factor with the greatest difference between air and water, as Denny (1990) points out, is electrical resistivity. Water has a resistivity 10^{16} times that of air! The consequences of this are that some aquatic animals have developed an ability to detect electrical activity in other animals at a distance, whereas this sense is absent in terrestrial animals.

Because of its greater density, water experiences greater changes in pressure with depth than does air. This, as we are now beginning to learn, has profound effects on the metabolism of marine organisms and their distribution. This is discussed more fully in Chapter 4.

It is also worth pointing out that, unlike air, seawater contains energy-rich organic compounds in the form of dissolved organic matter (DOM) that can itself be a potential food source.

A final physical difference between terrestrial and marine systems concerns oxygen. In air, oxygen constitutes a nearly constant 21 percent of the volume throughout the world. However, water holds less oxygen, and furthermore its concentration can vary with the temperature and salinity (pp. 6–7). Finally, oxygen diffuses through water at a rate one-ten thousandth of the rate in air—meaning that, once reduced, it regenerates slowly.

These facts concerning oxygen in seawater

together create a situation for marine organisms in which not only can the oxygen level be reduced in the external environment, due to limited solubility or biological activity, but also in the immediate boundary layers around the organism due to the slowness of diffusion. These conditions have led marine organisms to develop rather extensive morphological, physiological, and behavioral adaptations to provide adequate oxygen in the face of variable availability. Some of the more important include extensive respiratory surfaces, particularly in the form of integumental respiration, and small size, therefore giving a greater surface-to-volume ratio.

Life History Differences

There are several life history traits common to many marine organisms that are not seen in terrestrial organisms. In contrast to terrestrial organisms, marine plants and animals may shed both male and female gametes into the water and thus have external fertilization. Terrestrial plants may shed male gametes (= pollen) to be moved by air currents, but female gametes are never shed. Furthermore, there are no marine equivalents of the many terrestrial pollinators, animals that transfer male gametes in return for a reward. Finally, for animals, parental care (brood protection) and energy investment in offspring are much lower in aquatic systems than in terrestrial systems. Strathmann (1990) suggests that this is probably because the abundance of food in water permits feeding at a very small size, because desiccation is not a factor, and perhaps because the lower diffusion constant of oxygen in water and higher viscosity of water may simply preclude furnishing enough oxygen to a brooded clutch of young.

Structural and Functional Differences

One striking difference between terrestrial and marine communities is the insignificance in the latter of large macroscopic plants. Terrestrial communities are universally dominated by large flowering plants that are persistent and long lived. With the exception of certain large kelp plants, the communities of the sea have no large plants and the dominant autotrophs are single-celled microscopic plants of various groups of algae. In turn, this means that the dominant herbivores of the sea are also small, often microscopic, animals, in contrast to the large-bodied herbivores common to terrestrial communities. There are, for example, few herbivores in the sea to compare in size with the antelopes of the African plains or the bison of the North American grasslands. The dominant marine herbivores are microscopic crustaceans called *copepods* (see Fig. 2.5).

Another characteristic of terrestrial plants is that significant portions are composed of rigid structural materials such as wood and fiber, which themselves are relatively indigestible by most herbivores. As a result, herbivore grazing in terrestrial communities rarely removes significant amounts of the entire plant community. By contrast, the small herbivores of the seas usually consume the entire plant, and a major difference between terrestrial and marine ecosystems is the ability of the herbivores to remove the plants completely (see pp. 68–70).

Terrestrial communities are characterized by a matrix of long-lived plants and a fauna that is generally shorter lived. In the seas, at least among those communities associated with the sea bottom, the plants, if present, are short lived, and the matrix of the community is of relatively long-lived animals. Thus, one speaks of redwood forests, oak forests, and tall grass prairies on land, but of coral reefs, clam beds, oyster reefs, and mussel beds in the seas.

As a result of the microscopic nature of most marine plants and herbivores, the majority of large animals in the sea are carnivores. Large marine animals are therefore on higher trophic levels than large land animals. From the standpoint of human beings, harvestable food in the form of animal protein is available primarily from large herbivores on land (second trophic level), such as cattle and sheep, whereas harvestable food from the sea is usually in the form of carnivores from the third or fourth trophic levels, such as salmon, tuna, and halibut. A corollary of this is that most marine food chains have about five links (steps) to reach the top carnivore, whereas terrestrial food chains tend to be shorter, averaging three links to reach the top carnivore (Fig. 1.19).

There is also a question about the efficiency of energy production and transfer between terrestrial and marine ecosystems. Generally, the production of organic material is higher in terrestrial ecosystems than in marine, but the efficiency of transfer from first to second trophic levels is higher in marine ecosystems. Thus, although starting out with less energy fixed, marine systems appear to lose less with the first transfer. Attempts at further assessments of the efficiency of energy transfer in marine food chains are complicated because of the various feedback mechanisms that tend to "blur" the trophic level to which to assign the animal. The primary group responsible for this blurring of the trophic level is the filter feeders. These animals tend to feed upon suspended particles, and the distinction as to what is retained on their filters is made on the basis of size, not whether the particle is animal or plant or living or dead. This means they often take in both plant and animal material and feed on several trophic levels. A second

reason for blurred trophic levels is the tendency of marine species to switch positions in the trophic hierarchy. Thus juvenile fishes, for example, may feed on herbivorous copepods, but when adult change to feed upon other fishes higher in the trophic spectrum. An additional problem is the ability of various marine organisms to switch trophic levels as adults. Thus, for example, salmon may feed on squid at one time, but switch to lantern fish at another, thus shortening the food chain since lantern fishes are themselves the food of squid. Similarly, euphausiid crustaceans ("krill") may be herbivores in shallow water and carnivores in the open ocean.

The above-mentioned differences between terrestrial and marine ecosystems should be kept in mind as we explore in the remainder of this book the various habitats and communities of the sea.

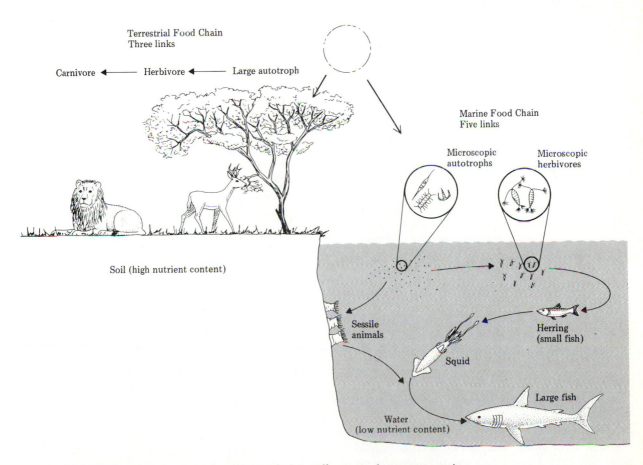

Figure 1.19 Diagrammatic representation of the differences between marine and terrestrial food chains.

DIVISION OF THE MARINE ENVIRONMENT

The marine ecosystem is the largest aquatic system on the planet. Its size and complexity make it difficult to deal with as a whole. As a result, it is convenient to divide it into more manageable, albeit somewhat arbitrary, subdivisions, each of which can then be discussed in terms of the ecological principles that govern the adaptations of the organisms and the organization of the communities. No universally acceptable scheme of subdivision of the marine environment has yet been proposed. The one followed here is modified from Hedgpeth (1957) and has enjoyed widespread use among biologists for more than 30 years. It therefore probably comes as close as any to a mutually acceptable division.

Major Subdivisions of the World Ocean

Beginning with the waters of the open ocean, there are subdivisions that can be made in both the vertical and horizontal directions. The entire area of the open water is termed the *pelagic* realm; pelagic organisms are those that live in the open sea away from the bottom. This is in contrast to the *benthic* realm, which is a general term referring to organisms and zones of the sea bottom. Horizontally, the pelagic realm can be divided into two zones. The *neritic* zone encompasses the water mass that overlies the continental shelves. The *oceanic* zone includes all other open waters (Fig. 1.20). Progressing vertically, the pelagic realm can be further subdivided. Two schemes are possible. The first is based on light penetration. The *photic* or *euphotic* zone is that part of the pelagic realm that is lighted. Its lower boundary is the limit of light penetration and varies in depth with clarity of the water. Generally, the lower boundary is between 100 and 200 m. A synonym for this zone is the *epipelagic* zone. Because it is the zone of primary production in the ocean, it is of major importance and the subject of Chapters 2 and 3. The permanently dark water mass below the photic zone is termed the *aphotic* zone. Some scientists prefer to recognize a transition zone between the photic and aphotic called the *disphotic* zone. This transition area has enough light for vision but not enough for photosynthesis and extends down to about 1000 m. Disphotic is synonymous here with mesopelagic.

The pelagic part of the aphotic zone can itself be subdivided into zones that succeed each other vertically. The *mesopelagic* is the uppermost of the aphotic areas. Its lower boundary in the tropics is taken to be the 10°C isotherm, which may be at 700 to 1000 m, depending on the area. Next is the *bathypelagic,* lying between 10 and 4°C, or in depth between 700 and 1000 m and between 2000 and 4000 m. Overlying the plains of the major ocean basins is the *abyssalpelagic,* which has its lower boundary at about 6000 m. The open water of the deep oceanic trenches between 6000 and 10,000 m is called the *hadalpelagic* (Fig. 1.20).

Corresponding to the last three pelagic zones are three bottom or *benthic* zones. The *bathyal* zone is that area of bottom encompassing the continental slope and down to about 4000 m. The *abyssal* zone includes the broad abyssal plains of the ocean basins between 4000 and 6000 m. The *hadal* is the benthic zone of the trenches between 6000 and 10,000 m. This entire aphotic area, though encompassing by far the largest volume and area of the world's oceans, is little known and forms the subject of Chapter 4.

The benthic zone underlying the neritic pelagic zone on the continental shelf is termed the *sublittoral* or *shelf* zone. It is illuminated and is generally populated with an abundance of organisms constituting several different communities, including seagrass beds, kelp forests, and coral reefs. The majority of this area is covered in Chapter 5, but the specialized coral reef communities are discussed in Chapter 9.

Two transitional areas exist, one between the marine environment and the terrestrial, the other between marine and fresh water. The *intertidal* zone or *littoral* zone is that shore area lying between the extremes of high and low tide; it represents the transitional area from marine to terrestrial conditions. It is a zone of abundant life and is well studied. It is the subject of Chapter 6. *Estuaries* represent the transition area where fresh and salt water meet and mix. They form the subject of Chapter 8.

It is well to remember that the above classifi-

Figure 1.20 Divisions of the oceans (not to scale). (Modified from J. Hedgpeth, ed., The treatise on marine ecology and paleoecology, Vol. 1, Ecology, The Geological Society of America, 1957.)

cation scheme is not absolute but is used for convenience, and that organisms and communities may extend into two or more zones. This is particularly true for those organisms that are powerful swimmers and divers. These organisms are discussed in Chapter 3.

BIBLIOGRAPHY

Bascom, Willard. 1980. Waves and beaches, the dynamics of the ocean surface. Anchor Books, N.Y. 366 pp.

Briggs, J. C. 1974. Marine zoogeography. McGraw-Hill, N.Y. 475 pp.

Butman, Cheryl. 1987. Larval settlement of soft sediment invertebrates: The spatial scales of pattern explained by active habitat selection and the emerging role of hydrodynamical processes. Oceanog. Mar. Biol. Ann. Rev. 25:113–165.

Caswell, H. 1981. The evolution of "mixed" life histories in marine invertebrates and elsewhere. Amer. Nat. 117:529–536.

Chia, Fu-shang. 1988. Differential larval settlement of benthic invertebrates, pp. 3–12. *In:* Ryland, J. S., and P. A. Tyler (eds.). Reproduction, genetics and distributions of marine organisms. Olsen & Olsen, Fredensborg, Denmark.

Clements, F. E. 1936. Nature and structure of the climax. J. Ecol. 24:252–284.

Connell, J. H. 1975. Some mechanisms producing structure in natural communities: A model and evidence from field experiments, pp. 460–490. *In:* Cody, M. L., and J. M. Diamond (eds.). Ecology and evolution of communities. Belknap Press, Cambridge, Mass.

Connell, J. H. 1978. Diversity in tropical rain forests and coral reefs. Science 199(4335):1302–1310.

Connell, J. H., and R. O. Slayter. 1977. Mechanisms of succession in natural communities and their role in community stability and organization. Amer. Nat. 111:1119–1144.

Crisp, D. J., and P. S. Meadows. 1962. The chemical bases of gregariousness in cirripedes. Proc. Roy. Soc. London (Biol.) 150:500–520.

Denny, Mark. 1990. Terrestrial versus aquatic biology: The medium and the message. Amer. Zoo. 30(1):111–122.

Dietrich, G., K. Kalle, W. Krauss, and G. Siedler. 1980. General oceanography: An introduction, 2nd ed. Wiley, N.Y. 626 pp.

Egler, F. E. 1954. Vegetation science concepts. I. Initial floristic composition, a factor in old field vegetation development. Vegetation 14:412–417.

Grassle, J. F., and J. P. Grassle. 1974. Opportunistic life histories and genetic systems in marine benthic polychaetes. J. Mar. Res. 32:253–284.

Gross, M. G. 1982. Oceanography: A view of the earth, 3rd ed. Prentice-Hall, Englewood Cliffs, N.J. 498 pp.

Hedgpeth, J. 1957. Classification of marine environments, pp. 17–28, and Concepts of marine ecology, pp. 29–52. *In:* Hedgpeth, J. E. (ed.). The treatise on marine ecology and paleoecology. Vol. I, Ecology. Memoir 67, Geol. Soc. Amer.

Ingmanson, D. E., and W. J. Wallace. 1973. Oceanology: An introduction. Wadsworth, Belmont, Calif. 325 pp.

Krebs, C. J. 1985. Ecology, the experimental analysis of distribution and abundance. Harper & Row, N.Y. 800 pp.

MacArthur, R. 1960. On the relative abundance of species. Amer. Nat. 94 (874):25–34.

Obrebski, S. 1979. Larval colonizing strategies in marine benthic invertebrates. Mar. Ecol. Progr. Ser. 1:293–300.

Palmer, A. R., and R. R. Strathman. 1981. Scale of dispersal in varying environments and its implication for life histories of marine invertebrates. Oecologia 48:308–318.

Smith, R. L. 1980. Ecology and field biology, 3rd ed. Harper & Row, N.Y. 800 pp.

Smith, R. L. 1986. Elements of ecology. Harper & Row, N.Y. 677 pp.

Strahler, Arthur. 1971. The earth sciences, 2nd ed. Harper & Row, N.Y. 824 pp.

Strathmann, R. R. 1974. The spread of sibling larvae of sedentary marine invertebrates. Amer. Nat. 108:29–44.

Strathmann, R. R. 1977. Egg size, larval development, and juvenile size in benthic marine invertebrates. Amer. Nat. 111:373–376.

Strathmann, R. R. 1978. The evolution and loss of feeding larval stages of marine invertebrates. Evolution 32:899–906.

Strathmann, R. R. 1985. Feeding and nonfeeding larval development and life history evolution in marine invertebrates. Ann. Rev. Ecol. Syst. 16:339–361.

Strathmann, R. R. 1990. Why life histories evolve differently in the sea. Amer. Zoo. 30(1):197–208.

Strathmann, R. R., and M. F. Strathmann. 1982. The relation between adult size and brooding in marine invertebrates. Amer. Nat. 119:91–101.

Thorson, G. 1950. Reproductive and larval ecology of marine bottom invertebrates. Biol. Rev. 25:1–45.

Thorson, G. 1966. Some factors influencing the recruitment and establishment of marine benthic communities. Neth. J. Sea Res. 3(2):267–293.

Underwood, A. J. 1974. On models for reproductive strategy in marine benthic invertebrates. Amer. Nat. 108:874–878.

Vance, R. 1973. Reproductive strategies in marine benthic invertebrates. Amer. Nat. 107(955):339–352.

Whittaker, R. H. 1975. Communities and ecosystems, 2nd ed. Macmillan, N.Y. 385 pp.

Wilson, D. P. 1952. The influence of the nature of the substratum on the metamorphosis of the larvae of marine animals, especially the larvae of *Ophelia bicornis* Savigny. Anna. l'Insti. Oceanogr. Monaco 27:49–156.

Chapter 2

Plankton and Plankton Communities

*A*s we have seen in the previous chapter, there are certain fundamental physical differences between terrestrial and marine environments, which contribute to the differences that we observe in the organization of the communities in the two areas. Perhaps nowhere are these differences more dramatic and easier to see than in a consideration of the free-floating and weakly swimming associations of organisms that we collectively call the *plankton.* It is appropriate to begin a consideration of the ecology of the oceans with the plankton because the plants of the plankton contribute by far the greatest amount of photosynthesis in the oceans. Thus, it is in the plankton that the major amount of the energy of the sun is trapped, which may subsequently be transferred to the many other communities of the ocean. This vital role of initial

fixation of energy makes plankton important in the economy of the oceans. Just as there could be no life on land without the energy-fixing grasses, trees, and shrubs, similarly there could be no life in the oceans without the energy-fixing minute planktonic plants.

Terms and Definitions

The term plankton is a general term as noted above. Planktonic organisms are animals and plants that have such limited powers of locomotion that they are at the mercy of the prevailing water movements. This is in contrast to the *nekton,* which comprise the strong swimming animals of the open sea, capable of exercising movement against the prevailing water flow. Plankton may be further subdivided. *Phyto-*

plankton comprise the free-floating plants of the sea that are capable of photosynthesis. Zooplankton are the various free-floating animals. Bacterioplankton include the various minute bacteria and blue-green algae.

Because planktonic organisms are traditionally captured with the use of nets of various mesh sizes, they have also been classified on the basis of size. This size classification does not distinguish between plant and animal. The most recent, probably most acceptable, though arbitrary, scheme of classification for plankton has seven size classes (Table 2.1). Megaplankton are all those organisms above 20 cm. Macroplankton comprise the organisms from 2 to 20 cm in size. Mesoplankton fall between 0.2 and 20 mm. Microplankton are those plankton that fall between about 20 and 200 μm. These first groups are those usually captured in standard plankton nets and may be collectively called net plankton. The nanoplankton are very small organisms ranging in size from 2 to 20 μm. The picoplankton are minute organisms from 0.2 to 2.0 μm in size and consist primarily of bacteria and cyanobacteria. Picoplankton and nanoplankton are sometimes called microplankton. The smallest class is the femtoplankton, from 0.02 to 0.2 μm. Little is known of this last group. The nanoplankton, picoplankton, and femtoplankton cannot be captured in plankton nets because in order to filter them out of the water columns, the mesh would have to be so fine that when the nets were pulled behind a boat, the water would not pass through. As a result, these organisms can be obtained only by centrifuging samples of seawater, filtering water samples on fine filters such as millipore filters, or allowing them to settle out from water samples.

A final set of terms concerns the life history characteristics of the zooplankton organisms. Holoplankton are those organisms that spend their entire lives in the plankton. Meroplankton, on the other hand, are those species that spend but a part of their lives in the plankton. Meroplanktonic organisms include a large number of larvae of animals that as adults either live on the bottom or swim as nekton.

THE PHYTOPLANKTON

Although this text is designed to cover the principles of organization of life in the sea, not to detail the various taxonomic groups, it is worthwhile to include here a brief description of the common phytoplankton. The phytoplankton include a wide range of photosynthetic organisms. The larger phytoplankton—that is, those normally captured in nets—consist predominantly of only two groups, diatoms and dinoflagellates, which dominate the net phytoplankton throughout the world. The smaller phytoplankton, those in the nanoplankton and picoplankton size ranges that pass through nets, are more variable. It is the nanoplankton group that now appears to be more important in productivity (see pp. 79–81).

Diatoms

The diatoms (class Bacillariophyceae) are easily differentiated from the dinoflagellates because

Table 2.1 SIZE CLASSIFICATION SCHEME FOR PLANKTON ORGANISMS

Class name	Femtoplankton (0.02–0.2 μm)	Picoplankton (0.2–2.0 μm)	Nanoplankton (2.0–20 μm)	Microplankton (20–200 μm)	Mesoplankton (0.2–20 mm)	Macroplankton (2–20 cm)	Megaplankton (20–200 cm)
Virioplankton							
Bacterioplankton							
Mycoplankton							
Phytoplankton							
Protozooplankton							
Metazooplankton							
Size (m)	10^{-7}	10^{-6}	10^{-5}	10^{-4}	10^{-3} 10^{-2}	10^{-1}	10^{0}

Source: From Sieburth, 1979.

they are enclosed within a unique glass "pillbox" and have no visible means of locomotion. Each box is composed of two parts or valves, one valve fitting over another (Fig. 2.1). The living part of the diatom is within the box. The box is constructed of silicon dioxide, the same material that is a major constituent of glass. Each box is highly ornamented with species-specific designs, pits, and perforations, and this feature has made these organisms very popular with microscopists and, more recently, with scanning electron microscopists (Fig. 2.2).

Diatoms may occur singly, each individual occupying a single box, or they may occur in chains of various kinds, which themselves add to the ornamentation seen in the individual boxes (Figs. 2.2, 2.4).

In reproduction, each diatom divides into two halves; one half then occupies the top valve of the box, while the other takes the bottom. Each then secretes a valve so that the typical box is recreated. Each of these new valves is secreted *within* the old valve and hence, as the process continues through several generations, the average size of the diatom population decreases. Thus, individuals of species of diatom vary in size. Obviously, there must be a limit to the process of reduction. This occurs after a certain number of generations, when the diatom casts off both valves and becomes a structure called an *auxospore*. Within this spore, new valves are secreted that reestablish the original size of the diatom species (Fig. 2.1). Although this is the general mechanism of diatom reproduction and

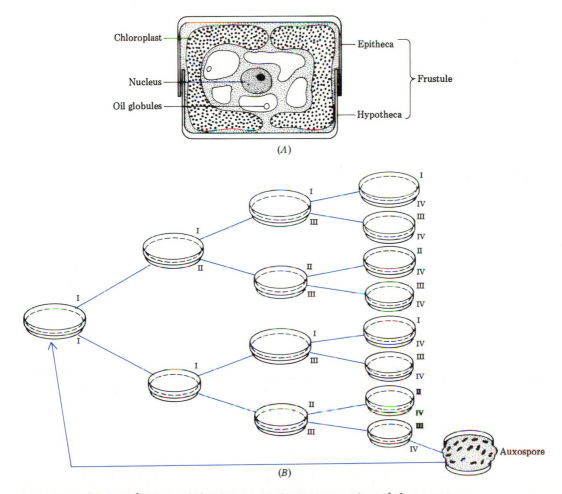

(A)

(B)

Figure 2.1 Diatom features. *(A)* Diagrammatic representation of the structure of a diatom. *(B)* Diagrammatic representation of the diminution of size in successive divisions and the restoration of size following auxospore formation.

(A) (B) (C)

Figure 2.2 Scanning electron micrographs of diatoms. *(A) Chaetoceros* sp. (700×). *(B) Thalassiosira decipiens* (1000×). *(C) Pseudoeunotia doliolus* (1400×).

reestablishment of size, there are some diatoms that undergo replication without valve decrease and do not require auxospore formation to reestablish size.

Diatoms are abundant as species and individuals. Although it is not possible to illustrate all here, some of the more common planktonic genera are shown in Figs. 2.2 and 2.4.

Some diatoms are not planktonic but benthic. We shall return to these in Chapter 6.

Dinoflagellates

The second major group, the *dinoflagellates* (class Dinophyceae), is recognizable by their possession of two flagella, which they use to move themselves through the water. They lack an external skeleton of silicon but are often armored with plates of the carbohydrate cellulose (Figs. 2.3, 2.4). They are also characterized by having chromosomes that are condensed throughout the cell cycle. Dinoflagellates are generally fairly small organisms and are solitary, rarely forming chains. They reproduce by simple fission, as do the diatoms, but in this case each of the daughter cells retains half of the original cellulose armor and forms a new part to replace the missing half without any diminution in size. Hence, successive generations do not change in

size. Some dinoflagellates also are capable of producing toxins that are released into seawater. If dinoflagellates become extremely abundant (2–8 million cells per liter), the cumulative effect of all the toxins released may affect other organisms, causing mass mortality. Such extreme concentrations, or blooms, of dinoflagellates are called *"red tides"* and are responsible for the massive localized mortality seen in fish and invertebrates in various places.

Some dinoflagellates have nonmotile stages called *zooxanthellae,* which are symbionts in the tissues of many invertebrates such as corals, sea anemones, and giant clams (see discussion in Chapter 10). Other dinoflagellates, such as the common *Noctiluca* (Fig. 2.11), are also highly bioluminescent (see Chapter 5 for discussion) and, when present in large numbers, can actually light up the wakes of boats and the breaking waves on a beach. Many dinoflagellates, such as *Noctiluca,* are not photosynthetic, but feed as do amoebas by engulfing food particles (phagocytosis), thus playing the role of a grazer.

The Smaller Phytoplankton

Constituents of the nanoplankton and picoplankton size classes (sometimes lumped as "nanoplankton") include a number of photosyn-

(A)   (B)   (C)

Figure 2.3 Scanning electron micrographs of dinoflagellates. *(A) Dinophysis* sp. *(700×). (B) Ceratium* sp. *(500×). (C) Dinophysis* sp. *(500×).*

thetic organisms. These organisms are of considerable importance in primary productivity and in oceanic food webs, which has only been realized in recent years (see discussion on pp. 79–82). The more important groups are the *haptophytes* (Coccolithophoridae, Haptophyceae) and the *blue-green algae,* also called the *cyanobacteria* (Cyanophyceae) (Fig. 2.4). The cyanobacteria are prokaryotic cells that possess chlorophyll *a,* but this is not in plastids and occurs in single cells, in filaments, or in chains. Cyanobacteria are abundant in the tropics,

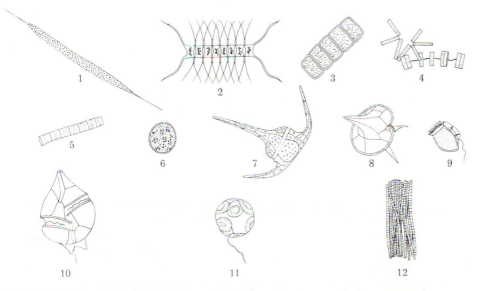

Figure 2.4 Diagrammatic representation of some characteristic genera of marine phytoplankton. Diatoms: (1) *Rhizosolenia,* (2) *Chaetoceros,* (3) *Navicula,* (4) *Thalassiosira,* (5) *Skeletonema,* (6) *Coscinodiscus.* Dinoflagellates: (7) *Ceratium,* (8) *Peridinium,* (9) *Dinophysis,* (10) *Gonyaulax.* Coccolithophores: (11) *Coccolithus.* Blue-green alga: (12) *Trichodesmium.* (Not to scale.)

where they occasionally form dense mats of filaments and color the water. The Red Sea was named from the red color of the cyanobacterium *Trichodesmium (Oscillatoria) erythraeum.*

Haptophytes are tiny eukaryotic flagellates (less than 20 μm) with one or two chromatophores; a unique threadlike appendage, the haptonema; and distinctively marked external calcareous or organic plates. They are abundant in all oceanic waters. The most widespread and abundant haptophytes are the coccolithophores, easily distinguished by the tiny calcareous plates (= coccoliths) on their outer surface. They have complex life histories, with several morphologically different cells present in the same species, and several modes of reproduction. Coccolithophores are now recognized as a major source of primary production in many ocean areas. Other, less abundant microalgae include the silicoflagellates (Chrysophyceae), the cryptomonads (Cryptophyceae), and certain motile green algae (Chlorophyceae).

Pelagic bacteria, or bacterioplankton, are also found in all oceans. They seem to be most abundant near the sea surface, where they are usually found in association with organic particles in the water column, collectively called *particulate organic carbon* (POC), or on various gelatinous zooplankton pieces known as *marine snow* (see pp. 84–85). They are usually motile and gram-negative and occur in coccoid, rod, and spiral forms. They decrease markedly with depth, and their role in the "microbial loop" of the oceanic food web is now clearly recognized (see pp. 79–82).

THE ZOOPLANKTON

In contrast to the phytoplankton, which consist of a relatively small variety of plants, the zooplankton are extremely diverse, consisting of a host of larval and adult forms representing most of the animal and many of the protistan phyla. As with the phytoplankton, two groups can be identified: the larger "net" zooplankton and the smaller microzooplankton (nanozooplankton and picozooplankton). We know much more about the larger "net" zooplankton because of their size and ease of collection. Marine biolo-

gists have long considered these larger net zooplankton to be the most important grazers of the phytoplankton. In recent years, however, the importance of the smaller microzooplankton as grazers has been established (see pp. 79–82). Among the "net" zooplankton one group stands out as being more important than the others. The class Copepoda (subphylum Crustacea, phylum Arthropoda) are small holoplanktonic crustaceans that dominate the net zooplankton throughout the world's oceans (Fig. 2.5). These small animals are of vital importance in the economy of the ecosystems in the oceans, because they are one of the primary herbivorous animals in the sea. It is they who graze upon the aquatic pastures constituted by the larger phytoplankton, especially in coastal waters, and provide one of the links between the primary production of the plants and the numerous large and small carnivores.

Copepods

Free-living planktonic copepods are generally small, between one and several millimeters in length. They swim weakly, using their jointed thoracic limbs, and have a characteristic jerky movement. They employ their very large antennae to slow their rate of sinking. Most of the free-living planktonic copepods have a characteristic body shape and hence are readily recognizable (Fig. 2.5).

Copepods graze on phytoplankton either by means of a filtering mechanism that works by removing algal cells from the water or by use of the anterior appendages. As Koehl and Strickler (1981) have demonstrated, copepods produce a stream of water past their bodies by flapping several anterior appendages (2nd antennae, mandibular palps, 1st maxillae and maxillipeds). When an algal cell enters this stream of water, the anterior appendages respond by beating asymmetrically, redirecting the water so the entrained algal cell moves toward the animal. The copepod then rapidly flings apart its 2nd maxillae, which then close over the water and the algal cell, forcing water out through the setae while retaining the algal cell (Fig. 2.6). Some copepods are carnivorous, seizing prey with their appendages.

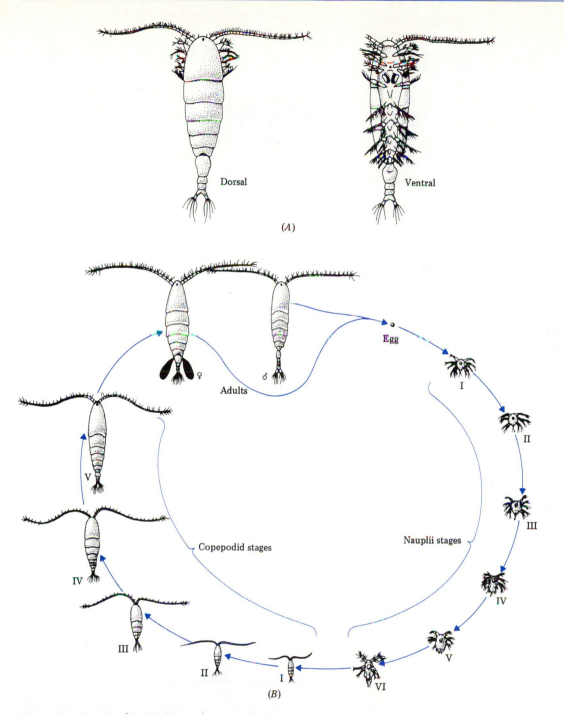

Dorsal

Ventral

(A)

Egg

I

II

III

IV

V

VI

Nauplii stages

Adults

♀   ♂

Copepodid stages

V

IV

III

II

I

(B)

Figure 2.5 Copepods. *(A)* Typical copepod showing major anatomical features. *(B)* Outline of the typical life cycle of a copepod.

In copepods, the sexes are separate, and sperm are transferred to the female packaged in spermatophores. After fertilization, the eggs are enclosed in a sac carried by the female, attached to her body. They hatch as *nauplius* larvae and progress through several naupliar stages and then several more *copepodid* stages before becoming adult (Fig. 2.5). As we shall see, the life

Figure 2.6 *(A)* Diagram of a copepod in feeding position as viewed from the left side. Only the left appendage of each pair is shown. Abbreviations are: A1 = 1st antennae; A2 = 2nd antennae; m = mouth; MP = mandibular palp; M1 = 1st maxillae; M2 = 2nd maxillae; MXP = maxilliped; S = swimming appendage. *(B–D)* Diagrams of feeding appendage movement (darker blue arrows) and water current movement (light blue arrows). Arrows with narrow shaft and wide head indicate movement laterally toward the reader; arrows with a wide shaft and narrow head indicate movement away from the reader. *(E–G)* Diagrams indicating mechanism of food capture in calanoids. *(E)* shows movement of water bypassing 2nd maxillae when no cells are present; *(F)* shows the 2nd maxillae flung open when an algae cell is present (small arrow); *(G)* shows final capture of algal cell by an inward sweep of 2nd maxillae. (From M. A. R. Koehl and J. R. Strickler, Copepod feeding currents: Food capture at low Reynolds numbers. Limn. Oceanog. 26:1062–1072, 1981.)

history of the adult and larval forms has an effect on the phytoplankton cycles observed in the sea.

Other Zooplankton

Because of the great diversity of zooplankton organisms, it is not practical to cover all in detail in a book such as this. Space permits us to consider only the larger and more significant taxonomic groups. Within each of these taxa are often many different species.

Among the holoplankton, the kingdom Protista is the most important, approaching or even exceeding the copepods as grazers, but is usually overlooked because of the small size of the individuals and their difficulty of capture (Fig. 2.7). The dominant groups of the "net" zooplankton are in the phylum Sarcomastigophora, particularly the order Foraminiferida ("for-

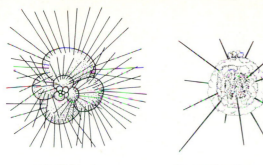

(A) *Globigerina*   (B) *Acanthometron pellucidum*   (C) *Parafavella*   (D) *Stephanoeca*   (E) *Bicoeca*

Figure 2.7 Holoplankton. Examples of some typical plankton members of the kingdom Protista. *(A)* Phylum Sarcomastigophora, order Foraminiferida. *(B)* Phylum Sarcomastigophora, order Radiolaria. *(C)* Phylum Ciliophora. *(D)* Phylum Sarcomastigophora. *(E)* Phylum Sarcomastigophora. (Not to scale.) *(A to C* after De Boyd L. Smith, A guide to marine coastal plankton and marine invertebrate larvae. Copyright © 1977 by Kendall/Hunt Publishing Co. Reprinted with permission.)

ams") and the order Radiolaria (see Fig. 2.7*B*). Both radiolarians and foraminiferans are single-celled organisms that produce skeletons, calcium carbonate in the case of forams and glass (SiO_2) in the case of radiolarians. Radiolarians are exclusively marine, and forams primarily so. Members of these two orders are so abundant and widespread that their skeletons have formed thick layers of foraminiferan and radiolarian *ooze* extending over vast areas of the deep-sea floor. Among the nanoplankton many small protistans are the most important primary consumers. The dominant groups are various ciliates of the phylum Ciliophora and small nonphotosynthetic flagellates of the phylum Sarcomastigophora. Both of these groups are exceedingly abundant and are the major grazers of the nanophytoplankton (Fig. 2.7).

Holoplanktonic members of the phylum Cnidaria include the various jellyfishes of the classes Hydrozoa and Scyphozoa (Fig. 2.8*A,B*; Plates 1,3) and the curious, complex hydrozoan colonies known as siphonophores (see Fig. 2.12*A,B*; Plate 2). Scyphozoan jellyfishes are among the largest plankton organisms, and may occasionally be found in large numbers (Plate 1).

Closely related to the Cnidaria is the phylum Ctenophora (Plate 5). With few exceptions, this phylum is entirely planktonic (Figs. 2.8*C,D*; 2.11*C*; 2.13*A*). All are voracious carnivores, capturing food with sticky tentacles or engulfing it with an oversized mouth. Locomotion is via rows of fused, large cilia called *ctenes.*

The phyla Nemertinea and Annelida are represented in the holoplankton by a few highly specialized forms, which are not abundant (Plate 4). The annelids include polychaete worms of the families Tomopteridae and Alciopidae, the latter with the largest and best developed eyes in the Annelida. Most planktonic nemerteans live in deep water.

The phylum Mollusca is the second largest phylum in the animal kingdom. It is usually considered to be composed of slow-moving benthic animals. However, it includes a considerable variety of specially adapted holoplanktonic forms. Perhaps the most highly modified planktonic mollusks are the *pteropods* and *heteropods* (Figs. 2.8*E,F;* 2.11*D*). Both groups are probably derived from benthic snails and slugs and are classified in the class Gastropoda. Pteropods are of two types, shelled (O. Thecosomata) and naked (O. Gymnosomata) (Plate 7). Shelled pteropods have fragile shells and swim using their winglike foot. They are particle feeders employing mucous nets and are preyed upon by the faster-swimming naked pteropods. Heteropods are large carnivores with transparent, jellylike bodies (Plates 8, 9). A special case is the

(A) *Muggiaea*

(B) *Pelagia*

(C) *Pleurobrachia*

(D) *Beröe*

(E) *Clione limacina*

Side view   Ventral view

(F) *Euclio pyramidata*

(G) *Sagitta setosa*

Figure 2.8 Holoplankton. Examples of some typical plankton members of the phyla Cnidaria *(A, B)*; Ctenophora *(C, D)*; Mollusca *(E, F)*; and Chaetognatha *(G)*. (Not to scale.) *(A, E, F, G* after G. E. Newell and R. C. Newell, Marine plankton, a practical guide, Century Hutchinson, 1973. *B, C, D* after De Boyd L. Smith, A guide to marine coastal plankton and marine invertebrate larvae. Copyright © 1977 by Kendall/Hunt Publishing Co. Reprinted with permission.)

pelagic shelled snail, *Janthina,* which maintains itself at the surface by clinging to the underside of its own bubble raft (Fig. 2.12*C;* Plate 10). A final group of planktonic mollusks are the squids. Many, perhaps most, squids are powerful, fast swimmers that must be considered nekton. There are, however, a significant number of very small squids that are not strong swimmers and must be considered planktonic (see Fig. 2.11*B;* Plate 6).

It is among the phylum Arthropoda that the greatest number of plankton organisms occur. In the seas of the world, virtually all belong to the subphylum Crustacea. In addition to the domi-

nant copepods, the holoplanktonic crustaceans include members from the order Cladocera (Fig. 2.9*A,B*), class Ostracoda (Fig. 2.9*D*), order Mysidacea, order Amphipoda, order Euphausiacea (Fig. 2.9*C;* Plate 11), and order Decapoda. Most of these taxa are small filter feeders straining plants and/or small animals out of the water.

The phylum Chaetognatha is a very small phylum consisting of about 100 species, which, with the exception of one genus, are planktonic. Chaetognaths, or "arrow worms" as they are called, are abundant members of the plankton throughout the world (Figs. 2.8*G,* 2.13*D*). All are

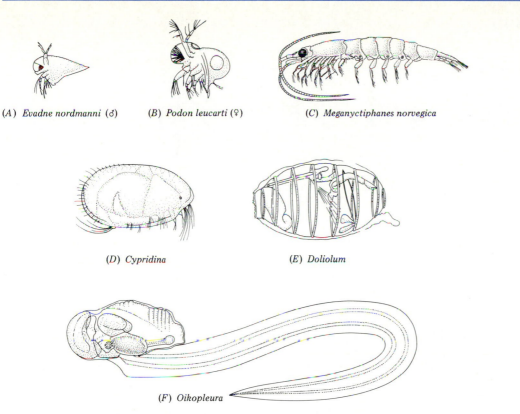

(A) *Evadne nordmanni* (♂)   (B) *Podon leucarti* (♀)   (C) *Meganyctiphanes norvegica*

(D) *Cypridina*   (E) *Doliolum*

(F) *Oikopleura*

Figure 2.9 Holoplankton. Typical planktonic forms of the phyla Arthropoda (*A, B, C, D*) and Chordata (*E, F*). (*A, B*) Order Cladocera, subphylum Crustacea. (*C*) Order Euphausiacea, subphylum Crustacea. (*D*) Class Ostracoda, subphylum Crustacea. (*E*) Salp of the class Thaliacea. (*F*) Class Larvacea. (Not to scale.) (*A, B, C* from G. E. Newell and R. C. Newell, Marine plankton, a practical guide, Century Hutchinson, 1973. *D, E, F* from De Boyd L. Smith, A guide to marine coastal plankton and marine invertebrate larvae. Copyright © 1977 by Kendall/ Hunt Publishing Co. Reprinted with permission.)

voracious predators on copepods and other planktonic organisms.

The final phylum represented in the holoplankton is the Chordata. Planktonic chordates belong to the classes Thaliacea ("salps") and Larvacea (Fig. 2.9*E,F;* Plate 12). These gelatinous-bodied animals are filter feeders. Larvaceans construct a "house" around themselves and pump water through a screen in the house to filter out their food. "Houses" are continually built and shed.

Compared to the holoplankton, meroplankton as a group are even more diverse. A bewildering array of larval forms constitute the meroplankton. These larvae are derived from virtually all animal phyla and from all different marine habitats. The number of larval forms is greater than the number of species in the sea that produce them because many species go through a series of larval stages before becoming adult. Thus, each of these species may be represented in the plankton by several different stages. This is particularly true in the Crustacea, where some decapods may have as many as 18 different larval stages, and most have more than one. Phyla that are not represented at all in the holoplankton, such as Bryozoa, Phoronida, Echinodermata, and Porifera, have larval forms (Fig. 2.10*A,B,C,D,E,F*). Other abundant phyla, such as Nemertinea, Mollusca, and Annelida, which have but a few highly adapted holoplankton groups, have planktonic larvae (Fig.

(*A*) *Echniocardium cordatum*
echinopluteus

(*B*) *Asterias rubens*
bipinnaria

(*C*) *Asterias rubens*
brachiolaria

(*D*) *Labidoplax digitata*
auricularia

(*E*) *Electra pilosa*
cyphonautes

(*F*) *Phoronis* sp.
actinotrocha

(*G*) Serpulid trochophore
(side view)

(*H*) *Balcis alba*
(early veliger)

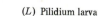

(*I*) *P. puber* megalopa

(*J*) *Carcinus maenas*
(3rd zoea)

(*K*) Nauplii of barnacles

(*L*) Pilidium larva

Figure 2.10 Meroplankton. Examples from several phyla. (*A*) Echinodermata, sea urchin. (*B, C*) Echinodermata, starfish. (*D*) Echinodermata, sea cucumber. (*E*) Bryozoa. (*F*) Phoronida. (*G*) Annelida, Polychaeta. (*H*) Mollusca, gastropod. (*I, J*) Arthropoda, Crustacea, Decapoda. (*K*) Crustacea, barnacle. (*L*) Nemertinea. (*A–J* from G. E. Newell and R. C. Newell, Marine plankton, a practical guide, Century Hutchinson, 1973. *K–L* after De Boyd L. Smith, A guide to marine coastal plankton and marine invertebrate larvae. Copyright © 1977 by Kendall/ Hunt Publishing Co. Reprinted with permission.)

2.10*G,H,L*). The Crustacea have by far the greatest number of larval types, and these include larvae of the abundant holoplankton organisms also. Even the nekton may be represented here, since many fishes have eggs and larvae in the plankton. Although space does not allow treatment of all the types, we will discuss next some adaptations of the plankton.

FLOTATION MECHANISMS

Plankton tend to have a density (mass per unit volume) that is somewhat greater than that of seawater. This means that any plankton organism will eventually tend to sink in the water column. This can be deleterious for both phytoplankton and zooplankton; phytoplankton would sink below the lighted areas of the sea and hence be unable to photosynthesize, and zooplankton would sink out of the area where their phytoplankton food occurs. On the other hand, there may be some advantage to a selective sinking in the water column for phytoplankton because there are more nutrients deeper in the water (see pp. 60–63). However, in general, uncontrolled or continued sinking is undesirable for both groups. Since they are weak swimmers unable to cope with winds and currents, and since living flesh tends to sink, how then do these organisms cope with the problem of staying in the upper layers of the sea?

Principles

In order to answer the above question, we must consider some physical and chemical characteristics that will bear on this problem. Since the density of seawater is a function of two factors, its temperature and its salinity, seawater becomes more dense as its salinity increases and less dense as its temperature increases. Thus, wherever there are temperature changes over the course of a year, such as in temperate seas, the density also varies. Another physical characteristic is the viscosity of seawater. Viscosity is related to temperature and salinity in that the more saline water is and the lower its temperature, the more viscous it is and the less rapidly things sink in it.

A second set of principles has to do with the effect of shapes on the rate of sinking in dense liquids. Objects of different shape but similar weights fall at differing rates in both gases, such as air, and liquids, such as water. The rate of fall is proportional to the amount of resistance the body offers to the gas or liquid through which it moves. Objects with a great amount of surface for a given weight tend to fall at slower rates than objects of similar weight with less surface area. Included here is the physical law that surface area increases as the square of the linear dimensions of the object, but the volume increases as the cube of the same dimensions. This means that for a given or fixed shape the smaller the body, the greater the surface area relative to the mass.

From the above physical laws governing changes in density and viscosity of water and those governing resistance, it is possible to derive a very simple equation that will relate sinking rates of organisms to these parameters. The equation is

$$SR = \frac{W_1 - W_2}{(R)(V_w)}$$

where SR = sinking rate; W_1 = density of the organism; W_2 = density of seawater; $W_1 - W_2$ = amount of overweight (overweight = the amount by which the flesh of the organism exceeds the weight of a similar volume of water); R = surface of resistance; V_w = viscosity of water.

The organisms can do nothing about the density or the viscosity of water. Hence, all adaptations of the organism to reduce the sinking rate must be concerned with either reducing the amount of overweight or else increasing the surface of resistance.

Let us consider first those adaptations designed to reduce the amount of overweight.

Reduction of Overweight

One mechanism that may be employed to reduce overweight is simply to alter the composition of the body fluids such that they are less dense than an equal volume of seawater. It is necessary, however, to maintain the same number of solute particles in the body fluid as in the surrounding seawater in order to avoid osmotic problems. *Osmosis* is the passage of water across semipermeable membrane, such as cell walls, to equalize dilution on both sides of the membrane. How is this change in composition accomplished? Perhaps the most common way is to replace heavy chemical ions in the body fluids with lighter ones. This allows the animal to maintain the same osmotic condition (same number of particles) while becoming lighter with respect to seawater. An example is the dinoflagellate *Noctiluca* (Fig. 2.11*A*), whose internal fluid contains ammonium chloride (NH_4Cl) and which is iso-osmotic with seawater but is less dense, having a

(A) *Noctiluca*   (B) Cranchid squid   (C) *Pleurobrachia*   (D) Heteropod (*Firoloida*)

Figure 2.11 Flotation mechanisms. Examples of organisms that exclude heavy ions, forming body fluids less dense than seawater. (Not to scale.) (Redrawn from J. H. Wickstead, An introduction to the study of tropical plankton, Century Hutchinson, 1965.)

specific gravity of 1.01 versus the 1.025 of seawater. Similarly, the cranchid squids, which, for squid, have very fat, bulbous bodies, are filled with NH_4Cl, making them less dense than seawater (Fig. 2.11B).

In a similar fashion, certain planktonic forms such as salps, ctenophores, and heteropods (Fig. 2.11C,D) actively exclude heavy ions such as SO_4^{2-} from their bodies and replace them with osmotically similar but lighter chloride ions.

Another mechanism is the development of special gas-filled floats. Since gas is much less dense than a similar volume of water, buoyancy is assured. Perhaps the most familiar examples of this approach are the floats of the Portuguese man-of-war (*Physalia*) and the swim bladders of fishes (Fig. 2.12A,B). If the animal is able to regulate the gas pressure in the float or bladder, this mechanism can be used to regulate the position of the animal in the water column so that it can move up or down at will. This is what some fishes do.

Similar to the use of gas-filled floats is the employment of liquids that are less dense than water. Prominent among these liquids are oils and fats. These can serve a dual purpose, as fats and oils also act as food reserves for the organisms. Copepods, for example, often store excess food in the form of oil droplets under the carapace, and these droplets aid in buoying the animal. Diatoms also store food as oils.

Whereas some planktonic organisms have employed the above mechanisms to alter their density and hence remain afloat, many others have not. If we return to our original equation,

we can see that, if overweight cannot be reduced, the only other option is to increase the surface of resistance to the water and thereby at least slow the rate of sinking.

Changes in Surface of Resistance

Surface of resistance may be increased in a number of ways, and all may be observed in planktonic organisms.

One of the most common observations made about plankton is that the general body size is small. This is particularly true for tropical plankton. As we noted earlier, surface area increases as the square of the linear dimension, whereas volume increases as the cube. This means that the smaller the organism, the greater the surface area relative to volume. Thus, by remaining small, plankton organisms offer far more surface area of resistance to sinking per unit volume of living material than if they were large. Small size is particularly important for tropical plankton, because they live in warmer and hence less viscous water, where sinking would be faster.

The other way of increasing the surface of resistance is to change the shape of the body. If one were to drop a round ball and a flat coin of equal weight into water, the ball would sink more quickly because it has less surface of resistance to the fluid. In a like manner, planktonic organisms have evolved various flattened body shapes or appendages. Such examples (Fig. 2.13) include a host of different species from many different phyla.

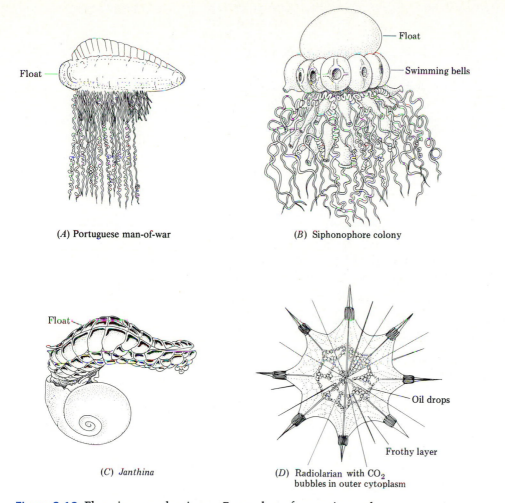

(A) Portuguese man-of-war

(B) Siphonophore colony

(C) *Janthina*

(D) Radiolarian with CO_2 bubbles in outer cytoplasm

Figure 2.12 Flotation mechanisms. Examples of organisms that use gas- or fluid-filled floats. *A* and *B* are both siphonophore colonies. (*A*, from G. E. Newell and R. C. Newell, Marine plankton, a practical guide, Century Hutchinson, 1973. *B, D,* from L. Hyman, The invertebrates. Vol. 1, Protozoa to Ctenophora, McGraw-Hill, 1940. By permission.)

Even more common than changes in body shape is the development of various spines and body projections. These structures add considerable resistance, but little to the weight. Such adaptations are common in the various diatoms (Fig. 2.2*A*), radiolarians, foraminiferans, and crustaceans.

Water Movements

A final mechanism of buoyancy has to do not with the organisms, but with the nature of water movements in the ocean. In the ocean, the surface water heats up during the day and cools at night. This alternating heating and cooling changes the density and leads to the creation of *convection cells,* which are small units of water that are either sinking or rising according to their density. These are gentle movements, and plankton may be moved by them.

On a small scale, the action of the wind blowing over the water can produce localized vertical movement of plankton due to the production of *Langmuir convection cells.* Langmuir cells are produced when wind speeds are above 3 meters per second. Each Langmuir cell is a few meters wide and hundreds of meters long. In each cell, wind causes the water to move away

(A) Velamen

(B) Phyllirhoe

(C) Phyllosoma larvae
(Scyllaridae)

(D) Sagitta enflata

(E) Ceratium

(F) Porcellanid larvae

(G) Late ophiopluteus
larva

(H) Acanthochiasma
serrulatum

(I) Calocalanus plumulosus

Figure 2.13 Flotation mechanisms. Examples of organisms that are flat or that have spines: *(A)* Phylum Ctenophora. *(B)* Phylum Mollusca, class Gastropoda. *(C)* Phylum Arthropoda, class Crustacea. *(D)* Phylum Chaetognatha. *(E)* Order Dinoflagellata. *(F)* Phylum Arthropoda, order Decapoda. *(G)* Phylum Echinodermata, class Ophiuroidea. *(H)* Phylum Protozoa, class Sarcodina. *(I)* Phylum Arthropoda, class Crustacea. (Not to scale.) *(A–G* from J. H. Wickstead, An introduction to the study of tropical plankton, Century Hutchinson, 1965. *H* from G. E. Newell and R. C. Newell, Marine plankton, a practical guide, Century Hutchinson, 1973. *I*, from *The Cambridge Natural History.* Vol. 4, Crustacea, reprinted in 1968 by Wheldon & Wesley.)

from the center (divergence) toward the outside, where it meets the water from the adjacent cell. These two convection currents converge and downwell along the line of convergence for a short distance. They then move horizontally, where they again meet water moving in a similar direction from the adjacent cells, whereupon they rise up to the surface to complete the cycle (Fig. 2.14). Plankton organisms in such cells will tend to be carried downward at the margins of

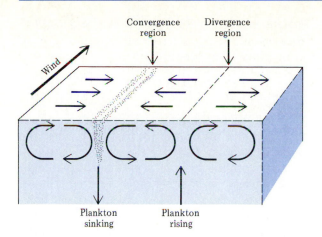

Figure 2.14 Langmuir convection cells induced by wind.

the cell (convergence) and upward in the centers (divergence). On a larger scale, upwelling (pp. 62–63) can carry plankton organisms up toward the water surface.

PRIMARY PRODUCTION

The basis for almost all life on the earth resides in the ability of green plants to use the energy of sunlight to synthesize energy-rich organic molecules from inorganic materials. This is the process of *photosynthesis*. The general equation for this process is

$$6CO_2 + 6H_2O \xrightarrow[\text{nutrients}]{\text{sunlight}} \underset{\substack{\text{energy-rich}\\\text{organic}\\\text{compounds}}}{C_6H_{12}O_6} + 6O_2$$

The basis for nearly all life in the sea is the photosynthetic activity of the aquatic plants. There are, however, significant differences in the form of the plants and in the location and extent of maximum photosynthetic activity imposed by the particular physical and chemical conditions of the oceans (see pp. 57–64). These will be covered subsequently. Photosynthetic organisms are diverse and widespread in the world's oceans. They include the pelagic phytoplankton and cyanobacteria, benthic microalgae, benthic macroalgae (kelps, seaweeds), seed plants (sea-

grasses, mangroves, salt marsh plants), and symbiotic algal cells in corals and other animals. However, the pelagic phytoplankton (including cyanobacteria) are responsible for perhaps 95 percent of all marine primary productivity. It is for this reason we consider primary productivity in detail here.

Primary Productivity

Primary productivity is defined as the rate of formation of energy-rich organic compounds from inorganic materials. Primary productivity is thus usually considered synonymous with photosynthesis, but this is not quite correct, since a minor amount of primary productivity may be produced by chemosynthetic bacteria (see Chapter 4). For all intents and purposes here, however, primary productivity is confined to photosynthesis.

The total amount of organic material fixed in the primary productivity process is termed the *gross primary production*. Since some of this total production must be used by the plants themselves to operate their own life processes, collectively called respiration, a lesser amount is available for transfer or use by the other organisms of the sea. *Net production* is the term given to the amount of total production that is left after losses from plant respiration and that is available to support other trophic levels.

For either gross or net primary production, the rates are usually expressed in terms of grams of carbon fixed per unit area or volume of seawater per time interval. For example, production may be reported in grams of carbon per square meter per year (g $C/m^2/yr$), which represents the production integrated over all depths from the surface to the base of the euphotic zone.

A few other terms require definition at this point. The *standing crop*, as applied to plants, is the total amount of plant biomass present in a given volume of water at a given time. Primary productivity and standing crop can vary considerably on a time scale of days to a year, and this variation is the result of a large number of factors that act both directly and indirectly on the photosynthetic process within plants and on the plants themselves. These will be discussed later in this chapter.

As we saw in Chapter 1, water has a profound effect on the light that penetrates it. Because water absorbs light, there is less light energy available as one goes deeper, and eventually, light disappears. For plants, this means that at some depth, the light energy is just sufficient for the plant to fix energy at a rate equal to the rate at which the plant uses energy in its own metabolic processes. In other words, there is a depth at which the respiratory usage of energy by the plant equals the ability of the photosynthetic mechanism to produce energy. If the plant goes deeper, the respiratory needs continue at the same rate, but the decreasing light is insufficient to allow the photosynthetic process to keep up and so there is a net loss of energy. The depth at which the rate of respiration of a plant is just equaled by the rate of photosynthesis is called the *compensation depth.* Above this depth, the rate of photosynthesis exceeds that of respiration and there is a net production of carbon, or a net primary production.

The compensation depth varies over the world's oceans because it is dependent on the clarity of the water. The clearer the water, the greater the light penetration and the deeper the compensation depth. In general, the compensation depth is deeper in clear, open ocean waters and shallower in inshore waters where large amounts of particulate matter are found in the water.

Since the compensation depth varies with depth of light penetration, it follows that it should also be possible to define the compensation depth in terms of light intensity. In fact, this is true, and one can speak of a *compensation intensity,* which is the light intensity at which photosynthesis equals respiration. Although this intensity is somewhat different for different phytoplankton species, it is approximately that depth at which the intensity has been reduced to 1 percent of the surface light intensity. The compensation depth can then be approximated as the depth to which 1 percent of the incident light penetrates.

It is well to remember that the compensation depth varies, not only in different geographic areas of the world's oceans as noted previously, but also from day to day and season to season in any one geographical area, as a result of changes in light with season (see pp. 57–60), as a result of changes in water clarity, and as a result of availability of nutrients.

Measurement of Primary Productivity

If we return to the photosynthetic equation on p. 53, we can see that it should theoretically be possible to measure the rate of production of organic compounds by measuring the rate of appearance or disappearance of some component of that equation. Thus, if one could measure the rate of disappearance of CO_2 or the appearance of O_2, this would be a measure of the rate of photosynthesis. In fact, this is done, but for various reasons it is practical only for the two components mentioned.

The classical method that has been used for many years to measure primary productivity is called the *light-dark bottle method.* In this method, two identical bottles are employed. One bottle is completely transparent, while the other is made completely opaque by painting it black or covering it with something like aluminum foil. Into each is placed the same volume of seawater, taken from the body of water for which you wish to estimate the productivity. The water thus contains the naturally occurring phytoplankton and zooplankton organisms. The oxygen content of the water to be added to the two bottles is determined on a separate sample to establish the initial concentration. The bottles are then stoppered, attached to a line, and returned to the appropriate depth in the sea. In the bottle that is dark, no photosynthesis can occur, but the plants and animals within continue to respire and use up oxygen. In this bottle, then, the original oxygen content will decrease as the respiring plants and animals use up the dissolved oxygen. In the light bottle, on the other hand, photosynthesis continues in excess of respiration and, hence, oxygen builds up over the initial values, since it cannot escape. After the two bottles have incubated for a period of time determined by the investigator, the bottles are brought up, unstoppered, and the water is analyzed for oxygen content. Usually, the oxygen content is determined by the Winkler method. Once the oxygen content is known for each bottle after incubation, it is possible to calculate the rate of photosynthesis from these results and the initial concentration of oxygen of the water when put into the bottles. In order to understand how this is done, it is necessary to consider what has happened in each bottle. When the original seawater was added to each bottle, it contained a certain amount of oxygen already present (initial oxygen concentration = O_1). In the light bottle, this oxygen was used by the organisms for respiration, but at the same time, the phytoplankton organisms were producing more oxygen in photosynthesis. The final oxygen concentration in the light bottle, then, is the result of oxygen added by photosynthesis plus initial oxygen concentration, less the oxygen used in respiration. Since we know the initial amount of O_2, in order to estimate the total amount of oxygen produced by photosynthesis, we need to know how much is consumed in respiration. If there were no respiration, only photosynthesis, in the light bottle, the increase in oxygen would be greater than that actually observed. This is where the dark bottle comes in. In the dark bottle, only respiration occurred and the oxygen concentration decreased. Since both bottles have the same volume and were incubated the same length of time, the decrease in oxygen in the dark bottle is a measure of the total respiration in the light bottle. To find the total amount of oxygen produced in photosynthesis, then, it is necessary to make a few simple additions and subtractions. First, the initial oxygen concentration of the water (before incubation) is subtracted from the final oxygen concentration in the light bottle, and in the dark bottle the final oxygen concentration is subtracted from the initial concentration. The value of the subtraction in the case of the light bottle is the *net photosynthesis,* or the amount of photosynthesis in excess of respiration. The value from the dark bottle is simply the respiration of the organisms. *Gross photosynthesis* is then obtained by adding the amount of oxygen respired to the net photosynthesis or subtracting the dark bottle from the light bottle. It is a measure of the total amount of photosynthesis that occurred during incubation. It is possible to convert the rates of oxygen production to rates of carbon dioxide assimilated by using a conversion factor. These factors are different depending on the organic compound produced (carbohydrate, lipid, or protein) but an average conversion factor would be 1.2 to 1.4.

Of course, the values for both gross and net photosynthesis will differ with different incubation depths, because light values change with depth and affect the photosynthetic process. This traditional method is only applicable in waters where there is considerable photosynthetic activity because the sensitivity of the Winkler method is too low to pick up oxygen changes under sparse phytoplankton populations and short 2- to 4-hour incubation times. Longer incubation times would permit use of the Winkler method but cannot be done because various "bottle effect" errors such as the increased growth of oxygen-consuming bacteria on the walls of the container become too large. More recently, oxygen production has been measured directly using electronic devices, which are sensitive to lower oxygen values.

Another method for estimating primary productivity is the *^{14}C method.* This has been the preferred technique for calculating primary productivity for the past 25 to 30 years. Here, radioactive ^{14}C is introduced into a bottle containing a volume of seawater populated by phytoplankton. The ^{14}C is usually introduced as $H^{14}CO_3$ (bicarbonate), since that is the major reservoir (90 percent) for CO_2 in the ocean. A known quantity of $H^{14}CO_3$ is added and the bottle is then incubated for a period at the appropriate depth in the sea (or in special racks on deck). At the end of the incubation period, the bottles are brought up and the water filtered onto fine membrane filters or glass fibers, which catch all phytoplankton organisms. The amount of radioactivity on the filters is measured with a counter. The amount of ^{14}C that appears on the membrane (e.g., in the particles) is proportional to the rate of production. In order to correct for possible nonphotosynthetic uptake of ^{14}C, a dark bottle is often run along with the light bottle. One calculates the carbon uptake from the following equation:

$$C\ uptake = \frac{^{14}C\ in\ particles\ on\ the\ filter\ \times\ available\ inorganic\ carbon\ \times\ 1.05}{Total\ ^{14}C\ added}$$

The available inorganic carbon in the sample is usually measured by some chemical procedure. The 1.05 factor is included in the equation to correct for the fact that phytoplankton discriminate between ^{14}C and ^{12}C, absorbing the former more slowly.

Both of the above methods have errors and problems associated with them and both rest upon certain assumptions. In the case of the light-dark bottle method, the assumption is that conditions in both bottles are similar except for light. In fact, this may not be the case and will introduce error. Bacterial growth tends to be accelerated by the introduction of surfaces, and hence the increased bacterial numbers may affect the oxygen levels in the bottles. This problem is common to both ^{14}C and light-dark bottle methods. Another problem common to both methods is that phytoplankton organisms enclosed in a bottle undoubtedly do not act normally, and this may be reflected in their photosynthetic rate. For the light-dark bottle technique, a serious problem is comparative insensitivity when concentrations of phytoplankton are low or when waters are polluted or contain high bacterial concentrations. The chief additional problems with the ^{14}C method are that there may be cell breakage on the filters, allowing some ^{14}C to leak out, and that the method does not account for ^{14}C fixed in photosynthesis and then respired out as $^{14}CO_2$, which then is redissolved in seawater. Since we cannot account for this latter problem, it is not always clear whether the method measures gross or net photosynthesis or something in between.

Recently, various independent methods used to assess carbon dioxide uptake or oxygen release have suggested that the ^{14}C method underestimates primary productivity, particularly in waters with low nutrient contents and low phytoplankton densities. Part of this problem may lie in the technique used since different-sized incubation bottles have been shown by Peterson (1980) to give widely varying productivity estimates. Part of the problem may also result from inadvertent contamination from metals associated with bottle samples and ^{14}C stock solutions.

Because of these various shortcomings, newer methods of assessing primary productivity have been and are being attempted. Included here are the use of electronic particle counters to measure directly the increase in numbers of particles of particular size classes—those known to be phytoplankton—over time, the change in ATP (adenosine triphosphate) content of cells,

and large-scale optical assessment via satellite imaging.

Standing Crop

Standing crop may be plants, animals, or both. The standing crop (at any point in time) is the result of the difference between the factors tending to increase the numbers of individuals—reproduction and growth—and those factors tending to decrease biomass or numbers—death and sinking or lateral transport out of the area. If reproduction and growth rates are, or have been, high and death or removal low, then standing crop will be high, and vice versa.

Standing crop of phytoplankton in the oceans is a difficult factor to measure accurately. Partly, this has to do with the patchy distribution of plankton organisms and the problems of sampling (see pp. 82–84 for further discussion of this) and partly with the problems inherent in the methods. The usual method of measuring standing crop is to measure some component common to all plants, usually the chlorophyll *a* content of a given volume of seawater. Since all plants must have chlorophyll to photosynthesize, the total amount of chlorophyll in a given volume of seawater should be a direct measure of the total biomass of plants present. Chlorophyll can be measured by taking advantage of its ability to fluoresce when excited by an appropriate wavelength of light or by directly extracting the chlorophyll from the plants with a chemical such as acetone and then measuring the amount of color in a colorimeter. The problems with this method are numerous. The method assumes, for example, that cellular chlorophyll content is constant, but it is not; it varies among the phytoplankton species as well as among cells of the same species, depending on their condition. It also varies with time of day and light intensity. Finally, in the method that extracts chlorophyll before measurement, the very method of extraction may alter the chlorophyll.

More recently, newer methods of assessing biomass have been used. Of these the one that has received some considerable attention is the measurement of adenosine triphosphate (ATP). ATP is found only in living cells and disappears quite rapidly on death. It is a useful indicator of standing crop. The method devised by Holme-Hansen and Booth (1966) depends on the use of the protein luciferin and the enzyme luciferase. These two compounds in the presence of ATP produce light (= bioluminescence), which is then measured quantitatively by a photocell. Drawbacks of the method are that ATP content is variable, depending on several biological and physical factors, making extrapolation of ATP levels to units of living cells difficult; the problem of interference by microzooplankton; and the inability of biologists to reach agreement on the percentage of ATP measured in a unit volume of water that should be attributable to animals or bacteria. A potential advantage of the ATP method is that once sufficient data are available to assign values accurately to all components of the plankton, this method will give biomass values for zooplankton, phytoplankton, and bacterioplankton collectively.

FACTORS AFFECTING PRIMARY PRODUCTIVITY

Since the functioning of the entire ecosystem of the world's oceans is dependent for energy almost exclusively on the photosynthetic activity of the small phytoplankton confined to the thin layer of lighted surface waters of the oceans, it is important to understand the conditions under which their productivity is either enhanced or inhibited. This is the first step toward reaching an understanding of how our oceans function.

Physical and Chemical Factors

If we consider terrestrial plants, there is a series of physical and chemical factors that affect their growth, survival, and productivity. Critically important factors would include light, temperature, nutrient concentration, soil, and water. Considering the freely floating phytoplankton, we find that the number of factors from the aforementioned list that are important is considerably reduced. Obviously, for plants suspended in water, neither water nor soil is of any importance. Temperature, which can vary in the terrestrial biosphere to such an extent as to influence productivity, has a considerably nar-

rower range in the marine environment. Because this range almost never encompasses lethal limits for life and because the changes are always gradual due to the inherent physical properties of water, temperature is of lesser importance to productivity in the ocean. Thus, of all the physicochemical factors affecting terrestrial plant production, only two, *light* and *nutrient supply,* are of significance in limiting productivity in the ocean. Since, however, the phytoplankton are suspended in water and therefore affected by forces that act to move water around, and because both light and nutrients are also affected by the water masses, a new, very important factor not seen on land enters into consideration. This factor is a composite one, which we may call *hydrography,* and comprises all those factors that act to move water masses around in the oceans, such as currents, upwelling, and diffusion. It is the interplay of these three factors—light, nutrients, and hydrography—that provides the limits to phytoplankton productivity in the oceans and the geographical differences that we observe. It is now necessary to consider how each factor acts separately to limit or enhance production and then to consider how all three act in concert to produce the observed productivity patterns of the oceans.

Light   Photosynthesis is possible only when the light reaching the algal cell is above a certain intensity. This means that the phytoplankton are limited to the uppermost layers of the ocean where light intensity is sufficient for photosynthesis to occur. The depth to which light will penetrate into the ocean, and hence the depth at which production can occur, is dependent on a number of factors; these include absorption of light by the water, the wavelength of light, transparency of the water, reflection from the surface of the water, reflection from suspended particles, latitude, and season of the year.

A number of meteorological features influence light before it even impinges on the surface of the water (Fig. 2.15). Factors such as clouds and dust act to interfere with light in such a way that lesser amounts survive the passage through the atmosphere to impinge on the surface of the water. This reduces the available light initially, without reference to water conditions.

Figure 2.15 The fate of incident light in the ocean.

When light strikes the surface of the water, a certain amount of light is reflected back; the amount depends on the angle at which the light strikes the surface of the water. If the angle from the horizontal is low, a large amount will be reflected. Conversely, the nearer the angle is to 90° (that is, perpendicular to the horizontal surface of the water), the greater will be the penetration and the lesser will be the reflection. Light that is reflected is lost to the system, so from the standpoint of phytoplankton productivity, maximum penetration is the most desirable.

The angle at which the light strikes the surface of the water is directly related to the maximum height of the sun above the horizon. In the tropical regions of the earth, the sun is directly overhead at midday, or virtually perpendicular to the sea surface, giving an angle for maximum penetration of light into the water column. Furthermore, this position changes little with sea-sons of the year, so light conditions are maximal all year long. As one progresses toward the poles from the equator, the sun may be directly overhead during the summer months, but may be far from this position at other times of the year. The closer one approaches the poles, the greater becomes the difference in the height of the sun above the horizon among the seasons of the year. This reaches its maximum in the Arctic and Antarctic regions, where the sun is absent during the winter or is so low to the horizon that no light can penetrate the water. The presence of ice in these areas also reduces light. This means that as one moves away from the equator, north or south, the amount of light penetrating the surface of the ocean, and hence available for use, changes significantly with season. As a result, the amount of photosynthesis that can occur also changes, being maximal in the summer and minimal in the winter. Only in the tropics is the light optimal at all seasons (Fig. 2.16).

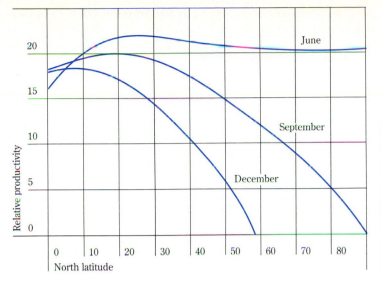

Figure 2.16 Relative amount of photosynthesis as a function of latitude for three seasons of the year. (From J. Ryther, Geographical variations in productivity. *In:* The sea, Vol. 2, edited by M. N. Hill, Wiley. Copyright © 1963. Reprinted by permission of John Wiley & Sons, Inc.)

The portion of light that enters the water column is subject to further reduction from two additional processes acting on it within the water. The first is scattering from various suspended particles in the water column. Suspended living or dead particles intercept the light and either absorb it or scatter it. Scattered light experiences a longer path length, which in turn results in further attenuation. Second, water itself absorbs light, making it unavailable for the plants. The amount absorbed is a function of wavelength and the path length or water depth in perfectly clear, clean water. This absorption of light by water is the reason that the vast majority of the water masses of the ocean are dark below a certain level. Because of this absorption, photosynthesis is automatically restricted to the thin, uppermost lighted layer. In the clearest ocean water this may correspond to the upper 100 m. Water, however, does not absorb all wavelengths of light equally. Sunlight, as it arrives at the surface of the ocean, is composed of radiation in a spectrum of wavelengths measured in nanometers (1 nanometer = 10^{-9} m). This spectrum includes all the visible colors ranging from violet to red (Fig. 2.15), or wavelengths from about 400–700 nm. As these wavelengths enter into seawater, the violet and red components are very quickly absorbed by the water. The green and blue components are absorbed less rapidly and hence penetrate most deeply. Eventually, they too are absorbed by the water. All the red and violet light is absorbed within the first few meters of even the clearest seawater, whereas 10 percent of the blue light may penetrate to more than 100 m under similar circumstances. It is important here to realize that, for any given wavelength of light, a certain fraction of its remaining intensity is lost by absorption with each additional increment of water depth. Thus, even though blue and green light penetrate deeper into the water column, the intensity decreases with depth, and it is intensity that is needed by plants. The intensity, relative to the surface, can be estimated from the *extinction coefficient:* $I_z = I_0 e^{-kz}$, where I_0 and I_z are the intensities at the surface and at depth; k is the extinction coefficient (for pure water it is 0.035), and z is the depth or path length in meters.

The depth to which a given intensity of light penetrates is thus a function of the transparency of the water and the differential absorption by the water. Since differential absorption by water is constant, the changes in depth of effective light penetration are due primarily to particle concentration. Where there are large numbers of particles in the water, such as in coastal waters, the depth of light penetration may be severely reduced and the amount of light insufficient for photosynthesis below a few meters. On the other hand, in the clearest tropical water, where few interfering particles exist, light intensity may be sufficient for photosynthesis down to 100–150 m.

Phytoplankton photosynthesis is light dependent. The rate of photosynthesis is high at intermediate light levels and decreases as the light intensity either decreases or increases (Fig. 2.17). On the other hand, the rate of respiration of the phytoplankton cells does not change much with depth. This means that, as the algal cells go deeper in the water column, the rate of photosynthesis declines as the light intensity decreases until at some point the photosynthetic rate equals the respiration rate. At this point there is no net production of organic material. As defined earlier, this depth is called the compensation depth, and it is approximated by the depth to which 1 percent of the incident radiation penetrates. The compensation depth marks the lower limit of the photic zone and varies geographically from a few meters in very turbid inshore waters to depths of 120–150 m or more in the open waters of tropical oceans. It also varies on a seasonal basis in temperate areas where high turbidity during certain seasons (the plankton bloom) reduces it to a few meters (see p. 61), whereas at other times the sparse populations of organisms increase it.

The compensation depth also changes with season, due to the change in the position of the sun, and it may be virtually absent during the winter months in high latitudes.

For most phytoplankton, the photosynthetic rate varies with light intensity. Near the surface of the water column, where the light intensities are the highest, most species show a leveling off or a decrease in photosynthesis (Figs. 2.17, 2.18). This is due either to an inhibition at high light levels or else to saturation of the photosynthetic apparatus such that it is not possible to increase rates. Different species have different curves of photosynthetic rate plotted against light intensity, giving different optimal light intensities for maximum photosynthesis (Fig. 2.17). This may be of considerable significance in seasonal succession (see p. 77).

Nutrients   The major inorganic nutrients that are required by phytoplankton for growth and reproduction are nitrogen (as nitrate, NO_3^-; nitrite, NO_2^{2-}; or ammonium, NH_4^+) and phosphorus (as phosphate, PO_4^{3-}). Diatoms and silicoflagellates also require silicate (SiO_2) in significant amounts. Other inorganic and organic nutrients may be required in small or minute amounts. All of these nutrients are of great importance partly because they occur in such small amounts in seawater. They are thus the limiting factors for phytoplankton productivity under most conditions, and the oceans of the world can be considered as nutrient-poor deserts when compared to terrestrial counterparts. For example, fairly rich agricultural land contains about 0.5 percent nitrogen in the upper meter of soil. In one cubic meter of soil, this amount of nitrogen is sufficient to permit the production of 50 kg of dry organic matter. Terrestrial plants, under ideal conditions, can produce several kilograms of dry organic matter in excess of their own need per year per square meter. Thus, the reservoir of nitrogen in that one cubic meter is sufficient to allow a plant to grow for many years. As a result, it is possible in terrestrial systems to have long-lived plants that continue to grow for many years. Forests are a good example, for they represent the accumulation of years of organic production. Of course, eventually, unless the

Figure 2.17 Photosynthesis versus light intensity curves for three groups of phytoplankton. (From T. R. Parsons, M. Takahashi, and B. Hargrave, Biological oceanographic processes, 3rd ed., Pergamon, 1984.)

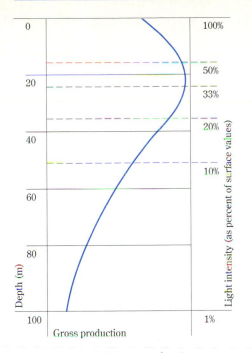

Figure 2.18 Inhibition of photosynthesis in the surface layers of water as measured by gross production at different levels in the water column. (After E. Steeman-Nielsen, Productivity, definition and measurement. *In: The sea*, Vol. 2, edited by M. N. Hill, Wiley, copyright © 1963. Reprinted by permission of John Wiley & Sons, Inc.)

nitrogen is renewed in the soil, nutrient exhaustion will limit growth.

On the other hand, the richest ocean water contains only about 0.00005 percent nitrogen, or 1/10,000 of the amount in soil. This means that a cubic meter of such water could permit a production of only 5 g of dry organic matter. In contrast to soil, however, where plant roots may penetrate only 1 m, oceanic phytoplankton should, theoretically, have access to the nitrogen in a column of water that extends as deep as the plants can exist. What is this depth and does it compensate for the greatly reduced nitrogen concentration? As we learned in the previous section, phytoplankton are limited in depth by light. If we assume the most ideal light conditions, we might have plant production extending down to as much as 100–120 m. This would mean that the plants would potentially have access to a volume of 100 m³ (1 m² × 100 m

deep) and hence a potential production of 500 g of dry weight (5 × 100), or about 1/100 of the amount on dry land. In fact, production levels of that magnitude never occur in the ocean, and the maximum amount of organic production that can accrue under a square meter of seawater under ideal conditions is only about 25 g. Why the discrepancy?

Several factors act in concert to reduce the maximum productivity. In the first place, the above example assumed that the nitrogen content of seawater was constant throughout the water column. Such is usually not the case. The upper layers of water usually have a reduced concentration compared to lower waters, as will be shown subsequently. Also, due only to light absorption by water, the production at 100 m would be less than at 10 m. More important, the increasing numbers of phytoplankton cells have a profound effect. As the phytoplankton population grows in the upper 100 m of water, the plants themselves absorb more and more of the light. As a result, there is less and less light penetrating to the deeper levels. Less light means that the compensation depth begins to move upward and becomes shallower (Fig. 2.19). Thus, the original 100-m reservoir of nutrients is reduced and, as plants increase in numbers, more and more of the water column and nutrient supply become inaccessible to the phytoplankton, thus reducing total potential productivity. Finally, as the plants grow, they absorb the nutrients, and absorbed nutrients are not available to other plants.

The result of all these factors acting together is to reduce the theoretical production. The reduction in the compensation depth and the absorption of nutrients is so great that it has been estimated that by the time a phytoplankton population reaches a density of 2 g/m³, an original 100-m compensation depth would have been decreased to as little as 3.5 m, and all nitrogen would be transferred into plant bodies. Hence, in terms of production and of standing crop of plant material, the oceans appear to be deserts in comparison with fertile land. The rate of production between land and sea, on a global scale, may not be significantly different, but the differences in nutrient concentration preclude that rate from continuing in the sea for long periods of time.

No plants   Few plants   Large numbers of plants

Figure 2.19 Change of compensation depth with increasing phytoplankton numbers. Dashed line indicates compensation depth.

If, however, the nutrients are used up in the upper, lighted zone, an untapped reservoir remains in the water mass below the photic zone. Since this mass is several orders of magnitude larger in volume than that of the lighted (photic) zone, it represents a considerable reservoir of nutrients that could greatly enhance production. If this vast reservoir could be tapped, then high rates of production and large standing crops could be sustained on a long-term basis.

Unfortunately, certain physical factors prevent general access to this reservoir over most of the world's oceans. Water, as we learned in Chapter 1, has different densities, depending on its temperature and its salinity. Cold, saline water is more dense than warm, less saline water. Below the photic zone throughout the world's oceans the water is cold, and therefore dense. In the photic zone the water varies in temperature. In the tropics it is warm, and therefore less dense, all year round. In the temperate zone this water is warm in summer and cold in winter. In polar regions it is cold and dense year round. In the tropics the difference in density between the warmer upper and cold lower layers is of such magnitude that the two do

not mix. Hence the nutrients cannot reach the photic zone. In the temperate seas the same situation prevails in the summer, but in the winter the temperature of the two water masses becomes similar and mixing can occur. In polar regions there are no significant differences in temperature, and mixing can occur year round.

Another component is required to effect the tapping of the reservoir—and that is a mechanism to accomplish mixing. Even though the water masses approach the same density, some force must be exerted to mix the masses together. The force that is available over most of the world's oceans is the wind. Strong winds blowing over the surface of the water create the mixing force (Fig. 2.20). The wind is of sufficient strength to cause mixing of the water masses and to transport nutrient-rich water into the euphotic zone, when the two masses are similar in density and temperature. In polar seas this is possible at all times (when not covered by ice), and in temperate seas only in the winter. The great differences in density between the photic and aphotic zones all year round in the tropics and in the temperate zones in the summer are of such magnitude, however, that no wind, even

Figure 2.20 Thermal structure of temperate ocean water masses in two seasons, winter and summer, showing the relationships between nutrient concentration, temperature, density, and wind-induced mixing.

from storms such as hurricanes, is of sufficient strength to mix the layers. This means that the upper layer remains nutrient-poor in the tropics at all times and in the temperate zone in summer, unless certain specific hydrographic conditions act to overcome the stability induced by density differences.

The special hydrographic conditions that act to bring nutrient-rich deep water up into the photic zone are upwelling, divergence of currents, and special currents. Upwelling occurs where the surface water moves away from shore and is replaced by nutrient-rich deep water brought to the surface. Persistent upwelling along the west coast of North America is responsible for high productivity in that region. Divergences are the result of transverse ocean currents flowing away from each other, bringing up deeper nutrient-rich water. Such a divergence occurs between the North Equatorial Current and the Equatorial Countercurrent in the Pacific; this is one area in the open tropical ocean where productivity is high, due to upwelled nutrients (Fig. 1.12). It should be noted here that it now seems

that some fraction of the nutrients, particularly nitrogen and perhaps also some trace metal nutrients, come from dust settling out of the atmosphere.

Turbulence and Critical Depth   Vertical mixing not only brings up nutrients, it is also responsible for carrying phytoplankton cells down into the depths. As long as vertical mixing is confined only to the upper illuminated zone, the plant cells can be carried downward only a short distance and will remain where there is sufficient light for photosynthesis. When mixing includes the lower water mass, however, it is possible for the plant cells to be carried well below the compensation depth. If the mixing is especially vigorous, the phytoplankton may spend most of their time below the compensation depth, and there will be no net production, because the time spent in the lighted zone in active photosynthesis is insufficient to fix as much organic matter as is used when they are below the compensation depth. This effect of turbulence has led to the development of the critical-depth concept (Fig.

2.21). The *critical depth* is that depth at which total gross photosynthesis of the phytoplankton in the water column equals total respiration. It is different from the compensation depth, which is the depth at which the rate of photosynthesis equals the respiration rate. The critical depth is always greater than the compensation depth because we are dealing here with a vertical mixing process in which the algal population is circulated between lighted and unlighted areas. When it is in the light, it can photosynthesize in excess of respiration and build up organic matter. The decisive point is the length of time spent in each area; this in turn depends on how deep the mixing occurs. Deeper mixing means more time spent below the photic zone, where respiration uses up organic material more rapidly than the short time in the lighted zone can compensate for. Whenever the wind-driven vertical mixing is less than the critical depth, photosynthesis is greater than respiration and a net production occurs.

Geographical Variations in Productivity

Having now considered the role of light, transparency, and nutrients separately, it is possible to consider how these factors interact with each other to produce the geographical or latitudinal variations in productivity that we observe in the oceans.

Temperate Seas   In the temperate-zone seas, the amount of light varies seasonally. As a result, the amount of solar energy entering the water varies, which in turn alters the temperature in the upper water layers. The thermal structure of the water column thus changes seasonally (Fig. 2.22). In the summer months the sun is high, days are long, and the upper layers heat up and become less dense than underlying layers. In other words, the water column is thermally stratified and no mixing occurs. In the fall the amount of solar energy entering the water col-

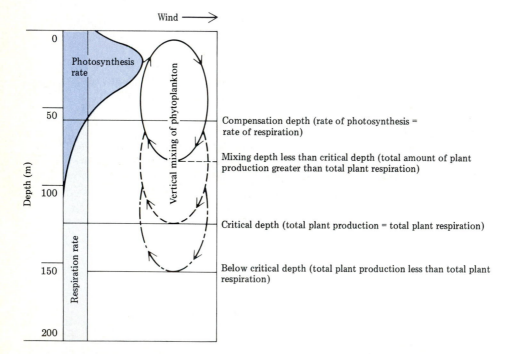

Figure 2.21 The relationship between the compensation depth and the critical depth. Critical depth is the depth to which the total phytoplankton biomass may be circulated and still spend enough time above the compensation depth to have a total production equal to its total respiration during the same time period. (Modified from D. H. Cushing, Productivity of the sea, Oxford biology reader no. 78, Oxford University Press, 1975. Reprinted with permission from Carolina Biological Supply Co.)

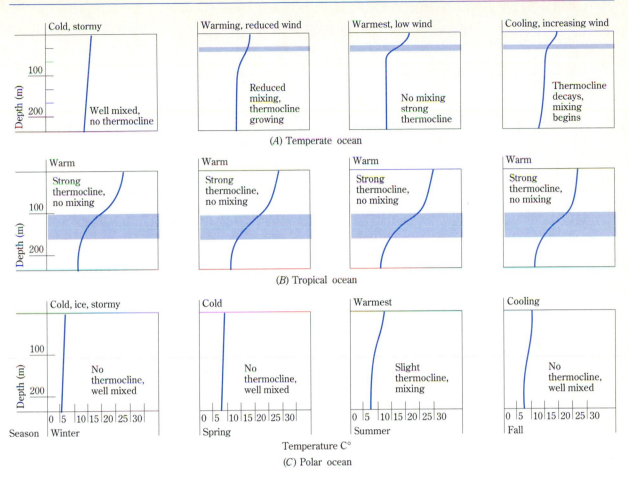

Figure 2.22 The thermal structure and extent of mixing in temperate, tropical, and polar seas during the four seasons of the year.

umn decreases, days become shorter, upper layers cool, and thermal stratification decreases. Finally, a point is reached where the temperature of the surface layers has been reduced to such an extent that the density of the layer is little different from that of the underlying mass. At this point, mixing can occur whenever sufficient wind is available. In winter, usually the storm season in the temperate zone, the sun is lowest on the horizon, solar energy input to the water is at a minimum, thermal stratification is at a minimum or absent, and mixing occurs. With the onset of spring, the days become longer, the solar energy increases, the upper layers begin to rise in temperature, and the system moves toward reestablishment of thermal stratification.

In contrast to the tropics, all the major factors that affect productivity change seasonally in

temperate seas. This is reflected in the change in production over the year (Fig. 2.23), with a major peak in spring, a lesser peak in the fall, and low productivity in winter and summer. We may explain this as follows: The low winter productivity is because of low light levels, due to the low position of the sun on the horizon and because the winter storms mix the isothermal water column and carry plant cells below the critical depth. In the spring the increased light and solar energy increase the temperature of the upper layers. With increasing temperature come increasing differences in density between upper and lower layers. Under such conditions the wind cannot mix the water to as great a depth as in winter, and at some point algal cells are no longer carried below the critical depth. Since nutrients in upper layers have been replenished during the winter mixing, conditions are good

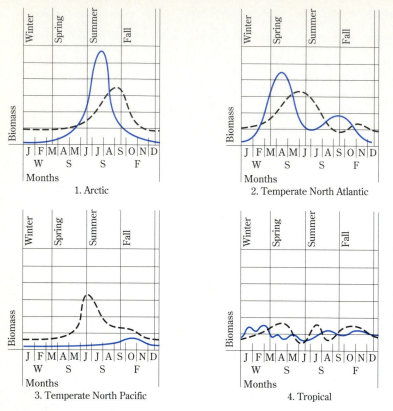

1. Arctic

2. Temperate North Atlantic

3. Temperate North Pacific

4. Tropical

Figure 2.23 Summary of the seasonal cycles of "net" phytoplankton and zooplankton in four different geographical areas. The solid line represents phytoplankton; the dotted one zooplankton. (Modified from T. R. Parsons, M. Takahashi, and B. Hargrave, Biological oceanographic processes, 3rd ed., Pergamon, 1984.)

for phytoplankton growth, and we observe the *spring bloom* (Fig. 2.23). As spring passes into summer, the water column becomes more thermally stratified, mixing with lower levels ceases, and light conditions reach optimal levels. Because mixing ceases due to stratification, nutrient replenishment ceases and production falls, even though light levels are optimal. With the advent of fall, the thermal stratification begins to break up and nutrients are returned to upper levels. If, in the fall, the mixing alternates with calm weather such that the plants spend more of their time in the upper layers and are not carried below the critical depth, a small bloom will occur because of the increased nutrients. This bloom declines in late fall, due to decreasing light and increased mixing. In the winter, low light levels and deep mixing of the water column keep productivity low. You will note that there are differences in the seasonal cycle curves between the temperate North Atlantic and North Pacific in Fig. 2.23. These differences are the result of somewhat different hydrographic conditions, coupled with different life cycles

of the dominant grazing animals (see pp. 72–74).

Tropical Seas   In the tropical seas, the upper waters are well lighted throughout the year because the sun does not show marked changes in height above the horizon. Light conditions are therefore optimal for phytoplankton production. At the same time, the continual input of energy from the sun maintains the surface layers of water at temperatures much higher than those in deeper waters. This means that there is a great difference in density between surface and deep waters and hence mixing does not occur. We say that such waters are *thermally stratified.* This thermal stratification extends throughout the year (Fig. 2.22). In the tropical seas, the sun and hence light conditions are optimal for high productivity, but because the sun's energy creates a thermal stratification in the water column that prevents mixing and the upward transport of nutrients, the productivity is low but constant throughout the year (Fig. 2.23). Tropical seas are very clear and have the deepest compensation depths, but they are that way because there are few phytoplankton in the water column due to the low nutrient content.

Polar Seas   In polar areas, productivity is restricted to a single short period in the polar summer, usually July or August in the Arctic (Fig. 2.23). At this time, the snow cover on the ice has disappeared, allowing sufficient light to enter the water through the ice to permit phytoplankton growth. In areas outside of the permanent ice pack, breakup of the ice at this time opens the leads, allowing sufficient light to enter the water, permitting phytoplankton growth. Following the single burst of production, the production quickly declines. Nutrients are not limiting and the water column is never strongly stratified (Fig. 2.22). The reason for the lack of production at other times is due primarily to lack of light. Light intensity is insufficient for a fall bloom, and during the long winter, light is either absent or else prevented from reaching the water column by a layer of snow over the ice pack. (See a more detailed analysis of polar seas in Chapter 5.)

Productivity in Inshore and Coastal Waters

The previously discussed latitudinal variations in phytoplankton productivity apply to open-ocean areas away from the influence of landmasses. The situation in the water masses adjacent to land is somewhat different. There are several factors that contribute to this difference.

In the first place, inshore waters tend to receive a considerable input of the critical nutrients, PO_4 and NO_3, due to runoff from the adjacent land (where, as we have seen, the nutrients are far more abundant). Because of this input, inshore waters usually do not show nutrient depletion. A second factor contributing to the difference is the water depth. Most inshore waters are shallower than the aforementioned "critical depth." Thus, the phytoplankton cannot be carried below the critical depth in any kind of weather. Given sufficient light, production can occur at any time, even in the winter. A third factor is that shallow inshore waters rarely have a persistent thermocline, and hence no nutrients are locked up in bottom waters. A final influencing factor is the presence of large amounts of terrigenous debris in the water, which may act to restrict depth of the photic zone and thus counteract the high nutrient concentration and shallow depth.

Interaction of these factors on a latitudinal basis produces changes, both in the cycle of productivity and in the total production when compared to offshore areas (Table 2.2). In the temperate regions, instead of a bimodal production cycle, as seen offshore, production remains high all through the summer, because nutrients are not limiting due to runoff from land and lack of a permanent thermocline. Average production in inshore temperate waters on a yearly basis is higher than in offshore waters due to the greater nutrient concentrations and lack of critical-depth problems. That production is not even higher inshore is probably due to the presence of large amounts of light-absorbing debris in shallow water, and the fact that in offshore water, production can occur to a greater depth. In other words, in shallow waters production is limited to the upper 5–10 m, whereas offshore it may go as deep as 50 m.

Table 2.2   PRODUCTION FOR SEVERAL DIFFERENT GEOGRAPHICAL AREAS

Location	Productivity in g C/m^2/year
Long Island Sound (temperate inshore)	380
Continental shelf	100–160
Tropical oceans	18–50
Temperate oceans	70–120
Antarctic oceans	100
Arctic Ocean	<1

Source: After Raymont, 1980.

In tropical waters the difference between inshore and offshore waters is particularly dramatic. Inshore tropical waters have a productivity as much as ten times that of offshore waters. This must be attributed in large part to the increased nutrient concentration inshore compared with offshore areas.

Inshore production is further enhanced through contributions of benthic plants, a component not present in offshore areas. The production of benthic algae and seagrasses is considered in Chapter 5 but is mentioned here for completeness.

Whereas we may have higher productivity in inshore or neritic waters, we also have much more variability over time and space due to the effects of local geography, river and stream discharge, storms, and tidally induced water movements. As a result, it is more difficult to predict the productivity on a year to year basis.

THE OCEAN ECOSYSTEM: THE CLASSIC MODEL

At this juncture we begin a discussion of the interactions among the various planktonic organisms beginning with the larger or net plankton. The sections that follow will detail first the classical or earlier concepts that describe the interactions among the larger phytoplankton, diatoms and dinoflagellates, and the larger zooplankton, primarily copepods. These ecological interactions still dominate coastal, upwelling, and polar seas. This classical model is followed by a discussion of the importance of the microplankton.

Grazing

As on land, the plant production is transferred into the food chain of the pelagic community through the grazing activity of herbivores. Because the phytoplankton are small, the herbivores are also small. Many protistan and invertebrate planktonic species are herbivores, but by far the dominant larger herbivores in all oceans are various species of copepods (Fig. 2.6). The dominance of the herbivorous larger or net zooplankton by copepods is so great (70–90 percent of biomass) that it is possible to consider only the copepods in discussing the effects of grazing on the larger or net phytoplankton populations. Furthermore, in most temperate seas, only from one to three species of copepod are numerically dominant.

Thus far, we have considered the changes in net phytoplankton production as a function only of interacting physical factors. In the oceans, however, phytoplankton exist together with various herbivores. What effect do they have on the phytoplankton populations? Evidence suggests a great effect.

Careful inspection of the numbers of phytoplankton cells per unit volume of water has shown that for the temperate North Atlantic Ocean the rate of decline of phytoplankton numbers following the spring bloom is steeper than would be predicted based on the nutrient decrease. Furthermore, nutrients never disappear completely from the upper waters, as would be expected if they were limiting. It can also be shown that the curve of zooplankton abundance in the temperate North Atlantic is similar in form to that of the phytoplankton, only delayed slightly in time (Fig. 2.23). All of the above have suggested to some workers (Riley, 1946; Cushing, 1959; Harvey et al., 1935) that the copepods are responsible for the regulation of the larger phytoplankton populations. This is reinforced by data from the North Pacific Ocean where there is no spring bloom of phytoplankton (Fig. 2.23). This is presumably the result of the presence of feeding copepods in the North Pacific at that time as opposed to the North Atlantic (see following section).

The rapidity with which copepods can remove phytoplankton cells is really remarkable (Table 2.3). In experimental conditions, Fleming (1939) established a diatom population that initially had 1 million cells per liter and divided once each day. It was grazed by a copepod population adjusted to remove 1 million cells per day. Under such conditions, the diatom population would remain in a steady state of 1 million cells per liter. When the copepod density was doubled (double the grazing rate), the effect was to reduce the diatom population to 27,000 cells per liter in five days. If the copepod numbers and hence grazing rate were increased five times, the diatoms were eliminated in five days. Since the rate of diatom reproduction remained

Table 2.3   CHANGES IN THE NUMBERS OF INDIVIDUALS OF A
PHYTOPLANKTON POPULATION WITH A CONSTANT RATE OF
REPRODUCTION WHEN SUBJECTED TO TWO DIFFERENT GRAZING
INTENSITIES

Time in days	Phytoplankton population density (cells/liter)		
	Initial grazing intensity	Grazing intensity doubled	Grazing intensity increased fivefold
0	1,000,000	1,000,000	1,000,000
1	1,000,000	487,000	62,000
2	1,000,000	237,000	3,900
3	1,000,000	106,000	240
4	1,000,000	56,000	15
5	1,000,000	27,000	1

Source: Adapted from Fleming, 1939.

unchanged in the above scenarios, it means that any increase in copepod numbers in the plankton could have a potentially significant effect on the numbers of diatom individuals. It is important here to understand that this decrease in numbers is independent of any change in the photosynthetic rate and, hence, the rate of primary production. It is quite possible to have a diatom population decreasing in numbers due to grazing, while actually increasing the rate of primary production. Of course, with fewer plant cells, the total amount of carbon fixed would decline with declining algal cell populations, but it would still be possible for the *rate* of fixation of carbon in each cell to increase. It is for this reason that the population density or standing crop of phytoplankton may be a poor or misleading indicator of primary productivity.

It is thus possible to observe a situation in which the standing crop of phytoplankton is declining due to grazing, while the rate of primary production is increasing or remaining steady. Under such circumstances, the diatom population is dividing rapidly because the photosynthetic rate allows the buildup of protoplasm for cell division, but as fast as the diatoms reproduce, they are grazed off by the copepods and other herbivores. In this case, the major portion of the carbon fixed in photosynthesis appears not in the standing crop of phytoplankton, but in the standing crop of zooplankton. Does such a situation occur in the oceans and, if so, what is the magnitude of the carbon fixed

that appears in the zooplankton versus phytoplankton? Plankton ecologists have used various mathematical equations employing the concentrations of nutrients, water transparency, light availability, and photosynthetic rates to calculate what the productivity of a given body of water should be and then comparing it with what is observed. The results for many areas have been remarkable. For example, Hart (1942) reported that, in the Antarctic seas around South Georgia Island and in the English Channel, the standing crop of phytoplankton was 2 percent of calculated production. In oceanic areas of the Antarctic seas where grazing was more intense, Mare (1940) found that the standing crop of algae was only 0.5 percent of calculated production. In other words, the great majority of primary productivity resides not in the bodies of the plant cells but in the zooplankton. These data also show that copepods are extremely efficient grazing animals and are, in fact, capable of reducing the populations of phytoplankton drastically in patches of the ocean. This is in contrast to the situation in natural terrestrial communities, where the grazers rarely remove all the vegetation and the standing crop of plants is high.

The ability of copepods to decrease phytoplankton populations to very low levels means that it is likely that the sudden decrease in the phytoplankton curve after the spring bloom in temperate waters could be caused by the grazing of increased numbers of copepods rather than being a result of nutrient depletion. This concept is supported by the finding that in the Antarctic seas, where nutrients are not limiting due to upwelling, the phytoplankton population also drops precipitously after the spring bloom. Additional evidence comes from the fact that grazing, and hence breakup of phytoplankton cells during passage through the copepod guts, releases the nutrients, which are then excreted back into the water. Under these circumstances, nutrients would be constantly regenerated in the upper layers and productivity never halted. Additional support for grazing as a control mechanism for phytoplankton is furnished by the patchiness of plankton. In the oceans, there are dense patches of phytoplankton and clear patches with copepods. One explanation is that copepods overgraze the phytoplankton.

If the copepods are so efficient at reducing the plant population, how does the spring bloom in the North Atlantic ever get started? The answer to this is that the copepods reproduce more slowly than the phytoplankton. Thus, in early spring, following the winter low period, the phytoplankton are able to respond more rapidly to the improving light conditions and lowered turbulence. They reproduce rapidly and build up to high levels in absence of significant grazing. The copepods, also at low levels, begin to breed as well, but their cycle is longer (weeks versus days for phytoplankton) and they cannot build up large enough populations until after the phytoplankton have bloomed (see subsequent section on copepod cycles). After that delay, however, when their populations are high, they quickly graze down the plant population to the low summer level. Copepod levels also then decline because of less food and predation, and because copepod egg production is dependent on food available to the breeding adult.

The above situation prevails in the temperate and subpolar North Atlantic. In the tropics, where nutrients limit population levels of plants to low levels at all times, the situation is different. In these areas, there are no real pulses in either the plant or copepod cycles, and instead there is a steady but inconspicuous consumption by copepods of the small phytoplankton crop.

Copepod Cycles Because the copepods dominate the larger zooplankton in all world oceans as the major grazers of diatoms and dinoflagellates and can determine the form of the phytoplankton population curve, it is worthwhile to review the major aspects of their life cycle, which contributes to our understanding of the aforementioned grazing patterns. This section will concentrate on copepod cycles in temperate and polar waters.

Relatively few of the numerous copepod species have been thoroughly studied, but in most temperate and polar waters only a few species predominate, and for some of these, notably *Calanus finmarchicus* of the Atlantic Ocean, the life cycle is well known. Since other species similar to *Calanus* also dominate Arctic and North Pacific waters and have somewhat similar life cycles, a general scheme may be established.

All copepods have a similar pattern of development. The eggs hatch as naupliar larvae and pass through a series of six naupliar stages. They then enter another larval stage called the *copepodid,* passing through five of these before becoming adult.

For the large dominant Atlantic copepods of the species *Calanus finmarchicus,* the early winter months are spent in the last or fifth copepodid stage in the deeper water of the open ocean. Here they persist without feeding, living on fat or oil reserves in their bodies. They do not migrate at this time (see p. 74). Those copepodids that survive the winter begin their final molt to the adult stage in late winter or early spring. At this time, the total numbers of individuals are at a minimum for the year. After molting to the adult condition, the copepods migrate to the surface waters, and begin feeding. They do not begin breeding, however, until after the phytoplankton numbers have begun to increase. This delay in the onset of breeding permits the phytoplankton to escalate their numbers to produce the high population numbers of the spring bloom. This first brood of eggs then goes through the naupliar and copepodid stages and itself matures and reproduces. The resulting increase in numbers of copepods then begins to reduce the phytoplankton numbers, creating the drop in the phytoplankton curve and the increase in the zooplankton curve seen in Fig. 2.23. This cycle of generations continues through the spring and summer, and the number of generations per year is dependent on the environmental conditions. Off Scotland and Maine, for example, there may be three complete generations. Further north, the number is reduced to two off Norway, and in Arctic seas, such as off Greenland, to one. Further south, off the Middle Atlantic states and English Channel, four generations may be produced. Each cycle, from egg production through adult to egg production again, takes about two months under favorable conditions. The life span of these large calanoid copepods is about two months for those individuals born at this time (Fig. 2.24), which means that adults do not live long after breeding.

The last brood produced in late summer does not complete the developmental cycle. Instead, when they reach Stage V copepodids, these animals migrate down into deeper water and remain there through the winter, subsisting on stored food reserves. They return to the surface waters in the spring to feed before being able to lay eggs. For this particular generation, the life span is then much longer, perhaps five to seven months, depending on the area.

Egg production is influenced by the abundance of phytoplankton available as food. Thus, successive generations of copepods build up large numbers in the spring, feeding on the spring bloom of phytoplankton. Similarly, the decrease in copepod numbers in the summer reflects the decrease in phytoplankton due to excessive grazing by the large numbers built up in the spring.

In summary, the life cycle of *Calanus finmarchicus* is about two months from egg through

Figure 2.24 *(A)* The life cycle of *Calanus finmarchicus* in the North Atlantic Ocean. *(B)* The life cycle of *Neocalanus plumchrus* in the North Pacific Ocean.

larval stages to adult, and successive generations are produced through the spring and summer in temperate and polar waters. The number of generations decreases as one moves poleward because of the increasing shortness of the season. Copepod numbers reflect phytoplankton densities.

The dominant calanoid copepods of the North Pacific Ocean, *Neocalanus plumchrus* and *Neocalanus cristatus,* have a somewhat different life history cycle than *C. finmarchicus.* This, coupled with certain environmental differences between the two oceans, leads to the different form of the seasonal curves for zooplankton and phytoplankton seen in Figure 2.23. To begin with, the two *Neocalanus* species are larger than *C. finmarchicus* and feed on both phytoplankton and zooplankton. *Calanus finmarchicus,* on the other hand, is primarily a herbivore. The *Neocalanus* species are also less efficient in their grazing activity. Finally, the seasonal cycles are different. Both *N. plumchrus* and *N. cristatus* have but one generation per year. Both species migrate to depth (300 m) in the summer and overwinter there as adults with sufficient energy reserves to mate and lay eggs. The eggs hatch and the larval copepods are back in the surface waters in the late winter ready to graze before the onset of the spring phytoplankton bloom.

While the above is the situation for the dominant large *Calanus*-type copepods, there are other smaller copepods in both the Atlantic and Pacific oceans such as *Acartia, Pseudocalanus,* and *Microcalanus,* which co-occur with *Neocalanus* and *Calanus* and which also differ slightly in life cycle. In these forms, the generation time is shorter and they tend to have more generations per year. Although they build up Stage V copepodids in the fall, there is no conclusive evidence of overwintering and it may be that these small forms breed throughout the year, albeit at a reduced rate in the winter and at other periods of low phytoplankton availability.

This latter type of life cycle with continuous breeding year round is found in many tropical copepods. Because of their small size, the constancy of the conditions, and the continuous phytoplankton growth, development is rapid and many broods are produced per year. This spreads out the grazing pressure more or less equally over the year to coincide with unfluctuating phytoplankton populations.

In general, then, the copepod cycles are maintained such that there is maximum utilization of almost all larger or net phytoplankton production. However, a certain, but as yet undetermined, amount of energy is lost through fecal pellets and exoskeletons sinking out of the photic zone.

Copepod Influence on Atlantic and Pacific Plankton Community Ecology

At this juncture it is useful to consider the differences in the life cycles of the dominant copepods in the North Atlantic and North Pacific alongside what we know about the slight differences in environmental factors between the two oceans to demonstrate how these interactions can lead to the very different seasonal cycles of phytoplankton and zooplankton in the two areas.

The environmental differences between the two oceans are slight, but significant. At the same latitude the Atlantic is somewhat warmer, 8.5°C at depth in the winter compared to the Pacific, which is 3.8°C. The winter mixing extends to beyond 200 m in the Atlantic, but only to 100 m in the Pacific. Finally, the nutrient levels in the Pacific are always above 5 μm/ liter, whereas in the Atlantic they are reduced below 1 μm/liter in the summer. Other physical and chemical factors are similar between the two oceans. There are, however, some differences in organisms. Among the photosynthetic organisms the Atlantic has larger numbers of diatoms and dinoflagellates, whereas the Pacific has larger numbers of small flagellates. The growing season is also shorter in the Atlantic than the Pacific.

Putting all these factors together with the life history differences among the copepod species produces surprisingly different results, as Parsons and Lalli (1988) have noted. Beginning in the winter we find that in both oceans the dominant copepods overwinter at depth, but the difference in temperature means that the Atlantic copepods have a higher metabolic rate such that by early spring they have insufficient reserves left to lay eggs. In contrast, the larger Pacific copepods that overwinter at a lower temperature and hence lower metabolic rate retain sufficient reserves to lay eggs without resuming feeding (Fig. 2.25).

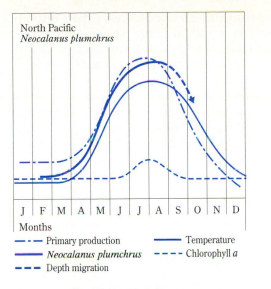

North Pacific
Neocalanus plumchrus

J F M A M J J A S O N D
Months

–·– Primary production — Temperature
— *Neocalanus plumchrus* – – – Chlorophyll *a*
– – – Depth migration

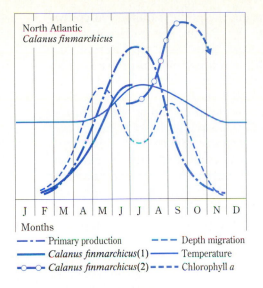

North Atlantic
Calanus finmarchicus

J F M A M J J A S O N D
Months

–·– Primary production – – – Depth migration
— *Calanus finmarchicus*(1) — Temperature
—○— *Calanus finmarchicus*(2) – – – Chlorophyll *a*

Neocalanus plumchrus		*Calanus finmarchicus*
Overwinters with low metabolic loss at 3.8°C. *Result*: Copepod has sufficient reserves to lay eggs without feeding in the spring.	*Winter*	Overwinters with high metabolic loss at 8.5°C. Copepod has insufficient reserves to lay eggs without feeding in the spring.
Phytoplankton growth starts slowly but is limited by light and temperature. *Result*: Zooplankton can track the spring bloom and suppress any increase in chlorophyll *a*; efficient phasing.	*Early Spring*	Phytoplankton growth cannot start in mixed layer > 200 m. When growth starts, it is rapid due to higher temperatures. Zooplankton cannot keep up with phytoplankton growth and chlorophyll *a* increases; inefficient phasing.
Phytoplankton providing most of the primary production are all small (< 200 mμ). *Result*: Low phytoplankton cell density and small size favors alternate strategy of carnivorous feeding on microzooplankton resulting in loss of food chain efficiency. Relatively slow gowth of zooplankton and slow development rate (i.e., stages) leads to large size.	*Late Spring*	A much larger proportion of the phytoplankton is made up of large diatoms. High phytoplankton stocks and large phytoplankton size favor herbivory with rapid zooplankton growth rate and development rate to next generation, resulting in small size of adults.
N. plumchrus gradually reaches stage V and migrates to depth to avoid high-surface temperatures of summer. *Result*: Large size acheived of aestivating *N. plumchrus* assures winter survival at low temperature.	*Early Summer*	*C. finmarchicus* grows first new generation (1). Growth rate is rapid, but phytoplankton exhaust winter nutrient supply. *C. finmarchicus* is ready to produce a second generation (2), which begins by growing faster than the first due to higher temperatures; phytoplankton production must now be based on efficient recycling of nutrients.
Annual cycle of *N. plumichrus* in surface layers entirely replaced by late summer opportunistic copepods and carnivores. *Result*: Phytoplankton shows small increase after *N. plumchrus* departure but this is again suppressed by summer herbivores.	*Late Summer*	Second generation of *C. finmarchicus* grows to provide immature copepodites which must overwinter. Other herbivores together with carnivorous zooplankton develop. Second generation is more capable of surviving winter than first generation since cooling trend favors larger body size of late development stages.

Figure 2.25 A comparison of the differences in the community ecology of the larger plankton organisms of the North Atlantic and North Pacific oceans. (Modified from T. R. Parsons, and C. M. Lalli. Comparative oceanic ecology of the plankton communities of the subarctic Atlantic and Pacific oceans, Oceanog. Mar. Bio. Ann. Rev. 26:317–359, 1988.)

In early spring the phytoplankton growth begins early in the Pacific but proceeds slowly because of reduced light and temperature. In the Atlantic phytoplankton growth is delayed due to the depth of the mixed layer, but once it begins, the light is better and so it proceeds more rapidly, aided by the then higher temperatures. As a result of the slow phytoplankton growth and the continual presence of smaller herbivores in the surface waters, combined with the entrance into the surface waters of the recently hatched copepodids of *Neocalanus,* the grazers keep pace with the plants, preventing any sudden bloom. In the Atlantic, on the other hand, the zooplankton cannot keep up with the plant growth and reproduction that are aided by the high nutrient content and the lack of a dominant copepod grazer. A grazer is absent because *Calanus finmarchicus* must feed first before building up enough reserves to produce eggs, which in turn must have a finite development time before hatching. As a result there is a delay in the grazer population, and a phytoplankton bloom ensues. In other words, there is an efficient coupling of the phytoplankton and zooplankton populations in the Pacific but not in the Atlantic.

By late spring the phytoplankton providing the bulk of the primary productivity in the Pacific are small in size and low in density. This means that the larger *Neocalanus* cannot efficiently graze on them, and they turn to carnivorous feeding on the smaller zooplankton, thus suppressing their numbers. In the Atlantic the majority of the phytoplankton consists of large diatoms in high density, and they are grazed efficiently by *C. finmarchicus.* This large food supply leads to rapid growth and maturity of *Calanus,* and they begin to produce the next generation.

In early summer the *Neocalanus* have reached maturity and migrate to depth to avoid higher surface temperatures. Their large size and large food reserves ensure that they will survive and be able to produce eggs. Meanwhile, the grazing in the surface layers is taken over by smaller zooplankton now released from the feeding pressure of the *Neocalanus.* These smaller zooplankton keep the grazing pressure on the phytoplankton throughout the fall and winter, when *Neocalanus* are at depth. The higher level of nutrients permits some phytoplankton growth even during the low light levels of the winter. In the Atlantic by summer the phytoplankton have exhausted the nutrients at about the time that *C. finmarchicus* begin to reproduce the second generation. This second generation of *Calanus* begins growth at a faster rate than the parent generation because of high temperatures and large food supply. However, the phytoplankton numbers now begin to decrease due to nutrient exhaustion and grazing by the high numbers of *Calanus.* Phytoplankton production now becomes dependent on nutrient regeneration in the euphotic zone. Finally, the second generation of *C. finmarchicus* grows to the last copepodid stage but with little reserve energy, thanks to the declining phytoplankton population, and migrates to depth to overwinter.

A probable, but as yet unproven, result of the poor coupling of the phytoplankton and zooplankton cycles in the Atlantic is that much of the carbon fixed in the spring bloom ends up as detritus that sinks and probably sustains large numbers of benthic fishes there. By the same token the closer coupling of the phytoplankton and zooplankton cycles in the Pacific may be responsible for the greater numbers of pelagic fishes in the Pacific.

Vertical Migration   One of the more puzzling phenomena in the sea is that of the vertical migration of zooplankton. *Vertical migration* is the name given to the daily migration of certain zooplankton organisms down into the depths during daylight hours and up into the surface waters at night. It is entirely separate from the seasonal migration noted above, where copepods descended in the fall to deeper waters to overwinter. Vertical migration may be just a special case of the more general phenomenon of the regular vertical migration of many open-ocean pelagic animals from one depth zone to another and back. Movements of these other so-called "deep scattering layers" are discussed in Chapter 4. We cover here the special case of the vertical migration of epipelagic plankton.

Daily (diel) vertical migrations have been known for more than 100 years. What makes this phenomenon so puzzling is that it is difficult to offer a satisfactory explanation as to why such small organisms should expend significant amounts of their limited energy resources for this purpose. The distances traversed in these daily movements may be from 100 to 400 m. Since we are considering animals of only a few millimeters in size, this is the equivalent of humans walking 25 miles to work and then 25 miles back at night.

Vertical migration has been observed in all zooplankton taxa, but not all zooplankton migrate. With respect to taxonomic group, size, or feeding habits, there are few generalities that can be drawn from a comparison between those that do migrate and those that do not. This inability to find correlations makes its occurrence even more difficult to understand.

It is now generally agreed that the major stimulus that initiates and controls vertical diel migrations is light. Vertical migrators respond negatively to light, moving themselves deeper as the surface light intensity increases. Conversely, they move toward the surface as the surface light intensity decreases. In a typical pattern, the migrators are found at the surface during the night; with the approach of dawn and advent of light, they begin their descent. As the light intensity increases through the morning, the animals move deeper, usually keeping themselves at a certain light intensity. At noon or whenever the intensity is maximal, they will be at the deepest position. Then, as the sun sinks through the afternoon and light intensity decreases, they begin their journey toward the surface again, arriving there around sunset and

remaining until just before dawn. As might be expected, *reverse migrators* proceed in exactly the opposite way, descending as the light decreases and rising as the light increases. Since the advent of sonar or echo-sounding gear in World War II, it has been possible actually to observe this phenomenon by watching the trace formed on the recorder by sound reflected from the bodies of these animals as they move up and down. Their bodies form a layer in the water column (Fig. 2.26).

Since the ocean is not uniform in its physical parameters in a vertical direction, other factors may modify the light-regulated scheme. For example, temperature may also influence migration. Some migrators are limited by temperature change in the depth to which they go. In such cases, the maximum depth to which they descend is set by temperature, which overrides the stimulus of light. Yet another constraint is the presence of the bottom. Hence, in open ocean areas where submarine ridges or mountains rise to within daytime depths of the vertical migrators, the migrators may congregate at the bottom. Such concentrations of animals are a prime food source and may be responsible for the observation that both submarine banks and seamounts are often densely populated with predators.

Although light is the main stimulus to which the animals respond and the guide to their positioning in the water column, it does not

Figure 2.26 Sonogram record of movement of vertical migrating plankton. (After G. B. Farquhar, ed., Proceedings of an international symposium on biological sound scattering in the ocean, 1970, U.S. Government MC Report 005.)

explain why the animals make the journey. What advantage can there be for undertaking such energy-demanding movements? Over the years, a number of explanations have been offered by marine ecologists. At present, four have been elucidated, but there is no universal agreement among ecologists as to which, if any, of the explanations is correct.

Perhaps the most popular explanation, which cannot be attributed to any single person, but has been reviewed by McLaren (1963) and Haney (1988), is that the zooplankton move from the lighted waters into dimly lit, deeper waters to avoid predation by visual, diurnal predators such as various fishes, cephalopods, and birds that forage at the surface. Similarly, reverse diurnal migrations are a response to avoid nocturnal surface-dwelling predators or diurnally migrating nonvisual predators. The problems with this explanation are that (1) the daytime depths of residence for many zooplankters still have sufficient light for visual predators to operate; (2) other zooplankton migrate deeper than would be required by this hypothesis; (3) many of the migrators possess and use bioluminescent organs that seem to counteract the effect of the darkness; and (4) many potential predators are also migratory. Support for this hypothesis has come mainly from investigations in freshwater lakes where studies have demonstrated that in systems with few or no visual predators vertical migration is lacking, whereas in those systems with such predators vertical migration is common. Recently, however, Ohman (1990) has demonstrated that the marine copepod *Pseudocalanus newmani* may undergo vertical migration, reverse migration, or no migration depending on whether the primary predators are visually hunting planktivorous fish, nocturnally feeding nonvisual zooplankton, or are absent. This study also suggests that diel vertical migration is not a fixed trait in zooplankton species but rather a dynamic behavioral trait that may be modified by prevailing conditions.

A second hypothesis, somewhat related to the above, is that of light damage avoidance. This hypothesis suggests that zooplankton organisms migrate to avoid damage from solar radiation. Since damaging ultraviolet light does penetrate several meters deep in clear oceans, this seems a

reasonable hypothesis. It is also known that certain pigments, primarily carotenoids, protect organisms from damage by ultraviolet light. Therefore a dilemma: Whereas the presence of pigment may protect the zooplankton from light, enabling it to inhabit lighted waters, it also makes the animal more visible to visual predators. Thus it seems we must assume that the greater risk of predation due to the high visibility of pigmented animals is offset by either the escape response of the zooplankton (= vertical migration) or the absence of visual predators. At present, no experimental or field studies have shown a direct causal relationship between light damage and vertical migration.

A third hypothesis for the occurrence of vertical migration, suggested by A. C. Hardy (1953), is that it allows the weakly swimming zooplankton to effectively change their horizontal position. In order to understand how this is possible, it is necessary to return to the previously described principle known as the *Ekman spiral.* As noted in Chapter 1, as one moves downward from the surface waters, the speed and the direction of a surface current also change. The speed decreases and the direction of its movement changes to the right of the surface-most current (Fig. 1.11). This means that a zooplankton organism, as it descends, will encounter currents that move not only more slowly but in slightly different directions. Once the zooplankter has descended into the deeper, slower-moving water, the surface currents will bring different water across and above the animal. If the zooplankter then ascends again, it will find itself in a different water mass than that from which it descended. What is the value of this? Since zooplankton organisms lack the ability to swim against currents, they would automatically be confined to a given water mass in the ocean. Given their great ability to graze out the phytoplankton from an area, they would soon be without food and starve. However, by simply sinking into the depths and then rising again, they would come up in a different surface water mass, potentially ungrazed. This explanation is supported by the universal observation that phytoplankton and zooplankton are very patchy in occurrence. This mechanism might also be employed by zooplankton to avoid or leave very dense phytoplankton patches. Hardy and Gunther (1935) have suggested that a toxic substance is produced by the phytoplankton, which has deleterious effects on zooplankton.

A final reason for diel migration, suggested by McLaren (1963) and McAllister (1969) and refined by Enright (1977) and McLaren (1974), concerns production and energetics. This two-part hypothesis suggests that (1) the rate of phytoplankton production is greater when subjected to discontinuous nocturnal grazing as opposed to a population subjected to continuous grazing; and (2) the zooplankton obtain an energetic advantage by spending a portion of their time in cold, deep waters rather than maintaining themselves constantly in the warm, upper waters. The second part of this hypothesis rests on the ability to demonstrate that more energy is saved by remaining in deeper, cold waters during the day than is expended in making the migration. It has been demonstrated that a neutrally buoyant zooplankter requires remarkably small amounts of energy to migrate, but it remains to be proved that the amount saved by remaining at depth is in excess of this amount. Implicit in this energetics theory is that feeding in warm water increases the efficiency of feeding and of assimilation of food, while residence at depth improves efficiency of growth.

This hypothesis is supported by the observation that in polar regions, where the temperature differential between deeper and surface waters is lowest, vertical migration is also least often observed. Conversely, the strongest vertical migration patterns are observed in the tropics, where the greatest differential exists between surface water and depth. However, attempts to verify this hypothesis both in the laboratory and field have thus far failed.

The above argument may also be used to explain the overwintering of copepods at depths where phytoplankton is scarce. Such overwintering at low temperatures conserves energy.

In summary, it can be stated that vertical migration is a worldwide phenomenon of certain zooplankton. The stimulus is primarily light, although it may be modified by other factors such as temperature. It occurs most commonly in strongly thermally stratified seas and becomes suppressed or disappears where seas approach isothermal conditions. It is also absent in temperate seas during winter months.

The hypotheses for its occurrence include avoidance of predation, avoidance of damaging ultraviolet light, change of position in the water column, and as a mechanism to increase production and conserve energy. It is probable, as Haney (1988) suggests, that no single explanation for occurrence will be found true in all systems, and that there are multiple causes for vertical migration in different systems as well as within a single system.

Seasonal Succession in Phytoplankton

In addition to the marked changes in abundance observed in the phytoplankton over the course of a year, there is also a marked change in species composition. This change in the dominant species from season to season is called *seasonal succession.* Under seasonal succession, one or more species of diatom, dinoflagellate, or coccolithophore dominate the plankton for a shorter or longer period of time and then are replaced by another set of species. This pattern is repeated yearly. This "succession" is thus different from the typical terrestrial ecological succession in which various plant associations replace each other until finally a so-called "climax" community develops, which persists through time.

Figure 2.27 displays the dominant phytoplankton species at several geographic locations to emphasize the widespread occurrence of this successional phenomenon. Some successional changes occur even in polar (East Greenland) and tropical areas (Bermuda).

What are the causative factors of this phenomenon? Margalef (1963) and Smayda (1980) have suggested several. Considering that the seasonal succession is most often and clearly seen in temperate seas, which have a marked change in temperature over the course of a year, temperature has been suggested as a cause. Certainly, this may be one of the factors, but it is unlikely to be solely responsible because certain dominant species recur at different temperatures. Furthermore, temperature changes rather slowly in seawater, but the replacement of dominant species often is very much more rapid.

Another suggested reason is the change in nutrient level over the year, with differing concentrations favoring different phytoplankton

Figure 2.27 Seasonal succession of phytoplankton in four different geographical areas. With the exceptions of coccolithophores and *Ceratium*, which is a dinoflagellate, all genera are diatoms.

species. While this factor may also contribute, observations suggest population changes are not closely correlated, that phytoplankton populations rise and fall much more quickly than nutrient concentrations change.

Yet another explanation for succession is that it is the result of *biological conditioning* of the seawater by the organisms in it. By this we mean that one group of organisms, when dominant in the water column, secretes or excretes very small quantities of organic compounds or *metabolites* that have a definite effect on other organisms, either inhibiting or promoting their growth. At the same time, these organic compounds could affect the very organisms producing them, mak-

ing the water either more or less suitable for their own existence.

These organic metabolites could, and probably do, include a number of different classes of organic compounds. Some are likely toxins, such as those released by the dinoflagellates during *"red tides,"* which inhibit growth of other plants. In such cases, the population explosion of dinoflagellates is so great that the water becomes brownish red in color from the billions of dinoflagellate cells. At the same time, although each secretes only a very minute amount of a toxin, the massive dinoflagellate numbers cause the toxin to reach concentrations where many creatures are killed. It is this same toxin that can be concentrated in certain filter-feeding organisms such as clams and mussels and that renders them toxic to humans at certain times and places.

Yet another class of metabolite is the *vitamins*. It is now known that certain phytoplankton species have requirements for certain vitamins and that there are considerable differences between species as to requirements. The B vitamins, especially vitamin B_{12}, thiamine, and biotin, seem to be the most generally required. It is possible that some species may be unable to thrive until a particular vitamin, or group of vitamins, is present in the water, and these are produced only by another species. Hence, a succession of species could occur whereby first the vitamin-producing phytoplankter is present and then is followed by the vitamin-requiring species.

Other organic compounds that may act as inhibitors or promoters of various species include amino acids, carbohydrates, and fatty acids. Although it is currently suspected that these organic metabolites may have an important role in the phenomenon of species succession and it has been demonstrated in the laboratory that phytoplankton species vary both in their ability to produce necessary vitamins and in their requirements for such in order to grow, evidence is still sketchy as to their real role in the sea.

There is also evidence to suggest that grazers, particularly selective grazers, can influence the phytoplankton species composition. Many copepods and invertebrate larvae have been shown to pick out selected phytoplankton species from mixed groups, in this way changing the species composition.

A growing body of evidence now suggests that all of the factors considered here are operating simultaneously to produce species succession. The importance of any one factor will vary with the particular phytoplankton species and the environmental conditions.

To summarize, there is a general seasonal pattern of succession, progressing from diatom- to coccolithophorid- or dinoflagellate-dominated systems (Fig. 2.27). In temperate waters small-celled, rapidly dividing diatom species initiate the spring bloom. They are then succeeded first by larger diatoms in the late spring and then by a summer warm-water dinoflagellate community. In tropical and subtropical waters coccolithophores and dinoflagellates are generally dominant all year, although they may be interrupted by periodic diatom blooms.

THE OCEAN ECOSYSTEM: A CHANGING MODEL

Up to this point we have been considering the usual explanations for the organization and functioning of the marine plankton system. Much of this classical description, while still valid as a basic framework and still true in polar seas, coastal seas, and upwelling areas, has been considerably modified in the past 20 years, primarily due to the advent of new technology. This new technology includes such techniques as satellite oceanography, remote sensing, epifluorescence and differential interference microscopy, scanning and transmission electron microscopy, ATP assays, nucleopore filters, and new methods for measuring minute amounts of various particles and chemicals. We now take up these new developments to build the final picture of the plankton ecosystems of the world as we understand them today.

Microorganisms and Productivity

The classical methods for studying phytoplankton productivity from net samples employing the ^{14}C method had the larger net phytoplankton, the diatoms and dinoflagellates, as the dominant primary producers. In an attempt to understand the relative significance of nanophytoplankton to the total phytoplankton productivity, a number of workers using various new methods have now demonstrated that the phytoplankton of the epipelagic zones of both temperate and tropical seas are dominated by nanoplankton. A similar situation may prevail in polar seas, but there are currently few data to support it. The domination is not only by numbers of individuals, but by the amount of photosynthesis as well. In most cases, Malone (1980) suggests that the nanoplankton account for 80 percent or more of all photosynthetic activity and 75 percent of the phytoplankton biomass in open ocean waters, especially in the tropics and subtropics (Table 2.4). Pomeroy (1974) even suggests that much nanophytoplankton photo-

Table 2.4 SUMMARY OF NANOPHYTOPLANKTON AND NET PHYTOPLANKTON PHOTOSYNTHESIS (P) AND CHLOROPHYLL BIOMASS (chl) IN DIFFERENT OCEANIC WATERS[a]

Region	Nanophyto-plankton		Net phyto-plankton		Percen-tage net plankton	
	P	chl	P	chl	P	chl
A. Tropical						
Eastern Equatorial Pacific	0.6	20.7	0.1	3.0	14	13
North Equatorial Pacific		21.1		1.0		4
North Equatorial Atlantic					10	22
Sargasso Sea	22.5	4.3	3.6	0.4	14	9
Caribbean	0.3	12.2	0.03	1.2	9	9
Indian Ocean					7	10
B. Upwelling						
Peru Current	2.7	18.2	0.2	1.6	7	8
California Current	0.7	15.6	0.1	2.6	12	14
Costa Rica Dome	1.8	—	1.2	—	41	—
Equatorial Upwelling (91°W)	0.4	—	2.0	—	83	—
C. Temperate						
Northeast Pacific					<25	<25

[a]Numbers are means. Photosynthesis is in micrograms of carbon per liter per hour. Chlorophyll is given in milligrams per square meter of surface.

Source: Modified from Malone, 1980.

synthesis is probably by organisms that are less than 30 micrometers in size. In the past the contribution of these tiny photosynthetic organisms was ignored because they passed through nets, were easily destroyed on filters, were sparsely distributed, and only could be studied alive.

Nanoplankton seem to be less dominant in coastal (neritic) and upwelling areas where net phytoplankton play a proportionately greater role, but where productivity and biomass are also more variable.

Not only are small nanophytoplankton abundant, they also tend to show less seasonal variability in biomass than the larger net phytoplankton. That is to say, they do not show the characteristic curves shown in Fig. 2.23. This lack of variability is particularly noticeable in tropical and warm temperate zones such as the central gyre of the North Pacific and the Sargasso Sea area in the Atlantic. These seem to be steady-state systems with very little seasonal or geographic change, according to McGowan

(1974). Marine biologists attribute this steady-state situation to the stable hydrographic regime, which gives rise to the continued low nutrient concentration. Low nutrient concentration seems to favor smaller photosynthetic organisms that have a proportionately greater surface area to absorb the nutrients but a lower relative need. Small organisms also sink more slowly.

The dominant groups of nanophytoplankton appear to be coccolithophores and very small flagellates of various kinds (Fig. 2.4).

A final consideration, heretofore neglected in analyses of plankton communities, is that of symbiosis. It now appears that many of the larger planktonic protozoa, particularly the Foraminiferida and Radiolaria, contain symbiotic intra- and extracellular photosynthetic symbionts of different taxa. These symbionts are probably responsible for a significant, but unknown, fraction of primary productivity, according to Fenchel (1988). Even more interesting is the recent finding by Stoecker et al. (1987) that many planktonic ciliates retain the functioning chloroplasts of their algal food in a short-term "chloroplast symbiosis" that has previously been described only for some marine mollusks (see Chapter 10). The contribution of this type of symbiosis to the pelagic planktonic productivity is presently unknown.

Nanoplankton, Respiration, and Grazing

If the majority of photosynthetic organisms are minute, it would seem logical that they must be consumed by heterotrophic organisms that are also of small size. Just as our phytoplankton nets are too large to capture nanophytoplankton, so it appears are the filtering systems of the dominant adult copepods. What then are the primary consumers of the nanophytoplankton? As yet we are not certain, but two groups of the kingdom Protista appear to be most significant. They are the nonphotosynthetic flagellates (phylum Sarcomastigophora) and various ciliates (phylum Ciliophora; Fig. 2.7). Both of these groups can ingest the tiny nanophytoplankton. The flagellates appear to do so by acting like amoebae and engulfing the individual cells, while the ciliates actually filter the nanophytoplankton cells from the water column.

How significant are these heterotrophs? Their importance can be estimated by measuring the respiration rate in the epipelagic zone of the ocean. The respiration rate is more difficult to measure than the photosynthetic rate because the size spectrum of organisms is larger than producers, because the total respiratory rate is very low, requiring methods of concentration of organisms, and because of inherent problems in techniques. Despite these problems, the respiration rate of net plankton versus nanoplankton has been estimated by investigators such as Pomeroy and Johannes (1968) and Hobbie et al. (1972) and has resulted in a rather close agreement among them that the respiration rate of nanoplankton exceeds that of net plankton by a factor of 10. What this, in turn, suggests is that nanozooplankton are consuming most of the energy fixed by the autotrophs. In other words, contrary to our classical model of diatoms and dinoflagellates being grazed by copepods as the major first step in the epipelagic food chains, we now believe the major initial link is from nano-

phytoplankton to ciliates and small flagellates. This is not to say that the classical chain is not valid, only that it is probably of less importance, at least in the open ocean.

Before we can conclude our discussion of this new concept of how the food web is organized in the open ocean, we must consider a few final components.

Bacteria, Particulate, and Dissolved Organic Matter

Early studies on the epipelagic zone determined that the water had a rather stable and static amount of dissolved organic matter, which was generally perceived to be too low to support any significant numbers of bacteria except on particles where organic matter was concentrated. More recently, various more sophisticated analyses for dissolved and particulate carbon and detection of living material have established quite a different picture.

The amount of dissolved organic carbon (DOC) in the waters of the world's oceans constitutes one of the largest reserves of organic carbon on our planet. The total amount has long been considered stable, with residence times of thousands of years. At the same time we know that perhaps one-fourth of the carbon that is fixed by photosynthesis is lost to seawater as dissolved organic matter "leaked" from cells. How can we reconcile this continual production with generally stable amounts measured in the water? Where does it go and why does it not build up? The answer appears to be that microorganisms, primarily bacteria, may account for this discrepancy through direct uptake of the dissolved organic matter. Andrews and Williams (1971) have estimated that such direct uptake may account for up to 50 percent of the total annual production of DOC.

The above suggests further that bacteria are abundant in the water column, more abundant than we had thought. Indeed they are. The introduction of direct plate counts with fluorescent microscopy has now demonstrated population numbers of from 10^5 cells/ml in oligotrophic waters to 5×10^6 cells/ml in coastal waters. According to Hobbie and Williams (1984) most are free living as *bacterioplankton,* are heterotrophic, and are about 0.4 μm in diameter. Larger bacteria are associated with planktonic particulate matter. These bacteria and their importance were missed by earlier scientists because of their very small size, which permitted them to pass through most filters then in use. Only the new nucleopore filters retain them. Indeed, according to Sieburth (1982), it appears that bacteria passing a 1 μm filter may account for 90 percent of the uptake of dissolved organic carbon in the water column.

Bacterioplankton also have a very high assimilation efficiency according to Pomeroy (1974), up to 60 to 65 percent as compared to the classic 10 to 20 percent efficiency for larger organisms. Added to their large numbers, this means that these bacteria are converting a substantial amount of DOC into particulate organic carbon (POC) in the form of bacterial cells. It also suggests how what was formerly considered "lost" production (i.e., the carbon lost by "leakage" from the phytoplankton) is not really lost but is recycled back into the food chains. The bacteria, in turn, are consumed by different nanozooplankton organisms and thus can channel a significant amount of energy into planktonic food webs. This has been termed the *microbial loop* by Pomeroy (1974; Fig. 2.28).

The bacteria are also partially responsible for the regeneration of nutrients in the photic zone, thereby permitting the continued productivity of the phytoplankton even in the absence of an influx of nutrient-rich water. Such regeneration is particularly important in the tropics and subtropics.

In addition to these heterotrophic bacteria, the photosynthetic cyanobacteria have been demonstrated by Johnson and Sieburth (1979) to be ubiquitous in seawater. They are responsible for a significant fraction of the photosynthetic activity, particularly in oligotrophic waters. Together with the heterotrophic bacteria, these two groups are the major components of the picoplankton-size class.

All of the above means that we have had to change our view of the planktonic food web.

The Marine Planktonic Food Web

In the classical marine planktonic food web we had the net phytoplankton—the diatoms and larger dinoflagellates—as the dominant produc-ers. They were consumed, in turn, by copepods and other crustaceans. Copepods were preyed upon by various carnivores, both fishes and other invertebrates. A final trophic level in this progression would include the top carnivores such as large fishes, cephalopods, and marine birds and mammals. With our new evidence we must now conclude that the dominant primary producers are various nanophytoplankton, such as coccolithophores, and picoplankton, such as cyanobacteria, supplemented by the bacterioplankton, which make available and recover potentially "lost" DOC. These, in turn, are consumed by various small protistans, mainly tiny flagellates and ciliates, which then are fed upon by larger net zooplankton. These various interactions are summarized in Fig. 2.28. A comprehensive view of both the classical planktonic food web and the microbial loop together is given in Fig. 2.29.

Spatial Distribution of Plankton

A characteristic feature of plankton populations is that they tend to occur in *patches.* These patches vary spatially on a scale of a few meters to hundreds of kilometers. They also vary in time—daily, seasonally, and annually—as well as vertically in the water column. It is this patchiness and its constant change in time and in space that has made it so difficult for plankton biologists to learn about the ecology of plankton. At its most extreme, because the water in which the plankton is suspended is constantly moving, it means that each sample taken by a plankton biologist removes a different volume of water, so each sample is unique and no such things as replicates exist. This makes life difficult at best in attempting to generate applicable principles and is one reason for our still unsatisfactory understanding of plankton ecology.

Patches may be generated by both physical and biological effects. Physical effects are particularly important. Large-scale patchiness can be due to the aforementioned changes in the stability of the water column that result in changes in nutrient availability (see pp. 60–63). Another cause of patchiness is the large-scale *advection,* or movement of water masses in which the plankton are embedded. On the largest scales, the various ocean current systems produce con-

Plate 1 Large planktonic scyphozoan jellyfish
Pelagia colorata with juvenile cancer crabs. *(Photo
courtesy of Drs. Lovell and Libby Langstroth.)*

Plate 2 The siphonophore *Nanomia bijuga.* Only
the upper portion of this planktonic colony is
visible in the photo. *(Photo courtesy of Dr. George
Matsumoto.)*

Plate 3 A hydrozoan jellyfish, *Crassota alba.* *(Photo courtesy of Dr. George Matsumoto.)*

Plate 5 A ctenophore of the genus *Pleurobrachia.* Note the long tentacles used to capture prey. *(Photo courtesy of Dr. George Matsumoto.)*

Plate 6 A small pelagic squid of the genus *Chiroteuthis.* Whereas the larger squids are considered nektonic, there are many small ones such as this that are primarily planktonic. *(Photo courtesy of Dr. George Matsumoto.)*

Plate 4 One of the few holopelagic members of the annelid class Polychaeta, the genus *Tomopteris.* *(Photo courtesy of Dr. George Matsumoto.)*

Plate 7 The largest pelagic pteropod mollusk genus, *Corolla.* These animals feed using a large mucous net, not shown here, that may exceed 2 m in diameter. *(Photo courtesy of Dr. George Matsumoto.)*

Plate 8 A small heteropod mollusk of the genus *Oxygyrus* with a reduced transparent shell. *(Photo courtesy of Dr. Roger Seapy.)*

Plate 10 *Janthina janthina*, a shelled gastropod living at the air-water interface suspended from a bubble raft. *(Photo courtesy of Dr. Roger Seapy.)*

Plate 9 The heteropod *Pterosoma planum,* a visual planktonic predator. *(Photo courtesy of Dr. Roger Seapy.)*

Plate 11 *Euphausia*, an individual of one of the most abundant groups of zooplankton in the cold oceans of the world. *(Photo courtesy of Dr. Roger Seapy.)*

Plate 12 The salp *Pegea coenfoederata. (Photo courtesy of Dr. George Matsumoto.)*

Plate 13 A school of pelagic fishes. *(Photo courtesy of Drs. Lovell and Libby Langstroth.)*

Plate 15 Female northern fur seals, *Callorhinus ursinus,* and single black pup on the breeding grounds of St. Paul Island in the Bering Sea. *(Photo by the author.)*

Plate 14 A killer whale, *Orcinus orca,* surfacing through the pack ice in Antarctica. *(Photo courtesy of Dr. John Oliver.)*

Plate 16 Harbor seal, *Phoca vitulina,* resting on an underwater rock. *(Photo courtesy of Drs. Lovell and Libby Langstroth.)*

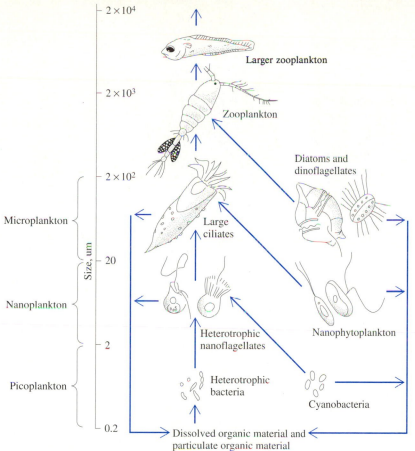

Size, um

2×10^4

2×10^3

2×10^2

20

2

0.2

Larger zooplankton

Zooplankton

Diatoms and dinoflagellates

Microplankton

Large ciliates

Nanoplankton

Heterotrophic nanoflagellates

Nanophytoplankton

Picoplankton

Heterotrophic bacteria

Cyanobacteria

Dissolved organic material and particulate organic material

Figure 2.28 Diagrammatic representation of the microbial loop and its relationship to the "classical" plankton food web. Autotrophs are on the right and heterotrophs on the left. (Modified from T. Fenchel, Marine planktonic food chains, Am. Rev. Ecol. Sys. 19:19–38, 1988.)

vergent *gyres* in some areas, such as the North Pacific and Sargasso Sea, and divergent systems in others, such as the upwelling areas at the equator (see Fig. 2.31). Such areas cover very large parts of the globe and, as we shall see subsequently, actually define major plankton biomes.

On a somewhat smaller scale, on the order of tens to hundreds of kilometers, another source of patchiness comes from ocean *eddies* or *rings.* Rings are masses of water rotating either clockwise (anticyclonic) or counterclockwise (cyclonic) that have broken off from larger ocean current systems and have moved into different water masses. Because they have water of different physical and chemical characteristics from the waters into which they move, and because they carry plankton characteristic of the current system in which they originated, they are usually biologically different from the surrounding water. They are, in other words, large moving rings

of water. Such rings are best known from the North Atlantic, where they are spawned from the Gulf Stream (Fig. 2.30).

Another source of large-scale patchiness is the presence of *coastal fronts* in neritic waters. As water flows over the continental shelf, or other regions where the water is shallow enough, the friction with the bottom causes vertical turbulence in the water column. This turbulence usually transfers nutrients and changes the water chemistry, and thus the plankton populations are also changed.

Physical processes responsible for small-scale patchiness include the aforementioned Langmuir cells (see pp. 51–53). These cells usually give rise to patches that are several hundreds of meters long because the plankton organisms accumulate along the lines of convergence (see Fig. 2.14). At the smallest scale, simple local turbulence is responsible for the observed plankton patches.

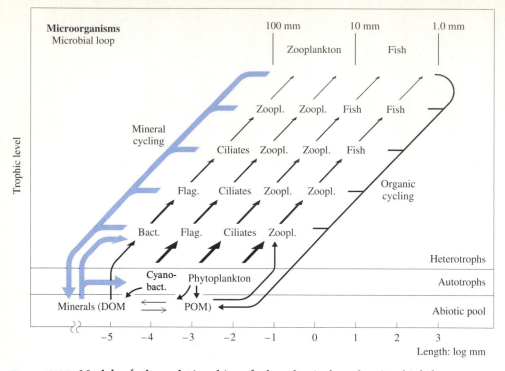

Figure 2.29 Model of the relationship of the classical and microbial loop planktonic food chains. Black arrows represent flow of energy and materials; color arrows, flow of materials alone. The food chain base represents a size range 3 orders of magnitude (smallest bacteria 0.2 μm, largest diatoms 200 μm; therefore, any trophic level will have a size-range factor of 10^3) and conversely each size class of organisms (100 μm) will represent at least 3 trophic levels. The thickness of color arrows (left) represents the approximate relative magnitude of minerals released in excretion at each trophic level; corresponding organic losses (feces, mucus, etc.) are shown on the right-hand side. (Modified from F. Azam et al., The ecological role of water column microbes in the sea, Mar. Ecol. Prog. Ser. 10:257–263, 1983.)

Plankton may also exhibit patchiness in a vertical direction. Physical factors contributing to this type of patchiness include light intensity, nutrients, and density gradients in the water column. Phytoplankton in particular tend to be unevenly distributed vertically, leading to the existence of a *chlorophyll maximum*. In the chlorophyll maximum area the chlorophyll concentration increases by anywhere from 2 to 20 times what it is in other areas of the photic zone. Such chlorophyll maxima are found in all temperate and tropical seas. They occur at greater depths (100–150 m) in the tropics but ·are a persistent feature of these waters. In temperate waters the maximum usually occurs only during the summer. The chlorophyll maxima usually occur at depths where the light intensity is 1 percent of what it is at the surface (= compensation depth). Although many reasons have been suggested for the occurrence of such maxima, there is at present no universally acceptable explanation for its occurrence.

Biological factors causing patchiness usually create small-scale anomalies. A host of biological phenomena may be involved, but a few stand out. The ability of zooplankton to migrate vertically and to graze out phytoplankton cells at a rapid rate (see pp. 68–70) certainly can create patches where phytoplankton organisms are rare and zooplankton abundant. Similarly, active swimming by certain zooplankton organisms can cause them to aggregate in dense groups.

One particular biological feature that needs to be emphasized is the phenomenon of marine

Figure 2.30 A schematic representation of the path of the Gulf Stream and the distribution and movement of rings. (From Oceanus, vol. 17, no. 3. Copyright © Woods Hole Oceanographic Institution, 1973.)

snow. *Marine snow* is the name given to the various kinds of amorphous particulate material derived from living organisms found in the water column. Marine snow consists of cast-off mucous nets produced by planktonic filter feeders, such as pteropod mollusks (Fig. 2.8) and tunicates (Fig. 2.9), and also the remains of dead gelatinous larger zooplankton, such as salps (Fig. 2.9) and jellyfishes (Fig. 2.8). The size of marine snow is usually 0.5 cm or less. These floating particles attract or accumulate living organisms as well as other nonliving floating and sinking particles, such as diatom frustules and fecal pellets. They therefore become floating microcosms with populations of organisms that are not found in the surrounding open water. Marine snow produces a solid substrate for the attachment of organisms that require a surface for continued existence. Since these gelatinous particles are also a locus for bacteria and small flagellates, there may be the regeneration of a certain amount of nutrients from the entrapped fecal pellets and other particulate matter. This,

in turn, may lead to the enhancement of the numbers of photosynthetic organisms such as dinoflagellates. This whole microcosm is therefore qualitatively and quantitatively different with respect to its component organisms than the organisms living free in the surrounding water. Each piece of marine snow is, in fact, a patch.

Marine snow is of widespread existence in the world's oceans and is one means of creating heterogeneous or patchy conditions in an otherwise homogeneous environment. Because marine snow is destroyed by plankton nets, the existence of these particles and their associated fauna and flora was only recognized when divers began looking at plankton in the natural environment.

The Major Plankton Biomes

Given all the uncertainties so far discussed with regard to the study and understanding of plankton, it may be surprising to learn that we now have enough data to recognize certain wide-spread patterns of the distribution and abundance of plankton. It is possible to establish broad associations, communities, or biotic provinces, and to make some reasonable statements concerning the composition of these associations and their change in time. It is fitting to close this chapter with such a summary.

As a result of the oceanic circulation that was outlined in Chapter 1, both the Atlantic and Pacific oceans have the water circulated in several large "gyres." The two largest are the central gyres north and south of the equator in both the Atlantic and Pacific oceans. They are largely warm-water tropical or subtropical systems. Between them lies an area of water flowing east to west, the tropical equatorial region. Poleward

from each central gyre lies a temperate subpolar gyre (Fig. 2.31). Transition zones exist between the gyre systems acting much as ecotones between terrestrial communities.

Extensive work by McGowan (1974), Reid et al. (1978), and Hayward and McGowan (1979) has established that each of these various gyre systems in the Atlantic and Pacific have characteristic associations of plankton species that are distinct enough to qualify as biotic provinces or communities. They are also partially closed systems showing little mixing with adjacent gyres in the transition zones.

Some of the differences between these biotic provinces are due to the different physical and chemical characteristics of the water masses. The large central North and South Atlantic and

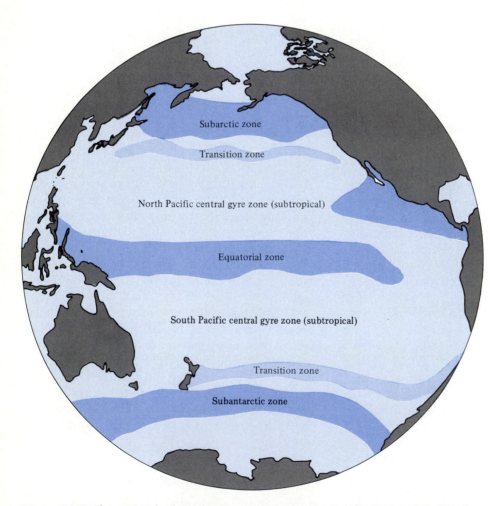

Figure 2.31 The main plankton biotic provinces in the Pacific Ocean. (Modified from C. B. Miller, The biology of the oceanic Pacific, Oregon State University Press, 1972.)

Pacific gyres are tropical and subtropical with permanent thermoclines, warm upper waters, and low nutrient concentrations, whereas the subpolar gyres have cold water, high nutrients, and undergo seasonal changes (see pp. 64–67). Productivity is low in the central gyres, a condition termed *oligotrophic*. The subpolar gyres, by contrast, are high in productivity, a condition termed *eutrophic*.

A number of features characterize all of these biotic provinces. In the first place, despite what we know about the general patchiness of plankton, the number and kinds of species within each province tend to be very uniform, on spatial scales ranging from a few meters to hundreds of kilometers. As McGowan (1974) has said, they are "monotonous." Furthermore, at least for the very large central gyres, the species composition and relative abundances are also uniform on time scales up to 16 years. They appear to be very stable systems.

These systems are also apparently old, at least in the Pacific Ocean. Evidence there suggests that these same sets of species or similar ones have been in existence for about 10 million years.

With respect to the features of the plankton in the central subtropical gyres, the individual zooplankton organisms tend to be small and often transparent or blue, whereas subpolar zooplankton organisms are larger. The biomass is also smallest in the central gyres and greatest in subpolar and equatorial provinces. The central subtropical gyres and equatorial zones have the greatest numbers of zooplankton species, and the subpolar zones have the fewest. Carnivores are also relatively more abundant in the subtropical gyres and equatorial regions.

Phytoplankton organisms tend to be more broadly distributed throughout the various provinces. No phytoplankton organisms appear to be endemic to any given province.

Perhaps the most puzzling question still to be answered for these large systems is the ecological mechanisms that regulate them and keep them so uniform and stable. One of the best ways to answer this question would be to set up controlled experiments. This, however, is not possible in the open ocean, so biologists have had to resort to other, more indirect methods. At present, it appears that the North Pacific Gyre, at least, is food limited due to the low nutrient levels, but there is no evidence of competition among the zooplankton filter feeders for the limited phytoplankton. Neither is there any evidence for niche partitioning and food specialization among more than 200 species of copepods. One might expect such partitioning in a system as stable as this. What then does regulate these systems? We simply do not know.

BIBLIOGRAPHY

Andrews, P., and P. J. LeB. Williams. 1971. Heterotrophic utilization of dissolved organic compounds in the sea. III. Measurements of the oxidation rates and concentrations of glucose and amino acids in sea water. J. Mar. Biol. Assoc. U.K. 51:111–125.

Anonymous. 1982. Marine pelagic protozoa and microzooplankton ecology. Anna. Inst. Oceanog. Suppl. Tome 58. 350 pp.

Azam, F., T. Fenchel, J. G. Field, J. S. Gray, L. A. Meyer-Reil, and F. Thingstad. 1983. The ecological role of water column microbes in the sea. Mar. Ecol. Prog. Ser. 10:257–263.

Bougis, P. 1976. Marine plankton ecology. Elsevier, N.Y. 355 pp.

Capriulo, G. M., and D. V. Nirivaggi. 1982. A comparison of the feeding activities of field collected tintinnids and copepods fed identical natural particle assemblages, pp. 325–334. *In:* Marine pelagic protozoa and microzooplankton ecology. Ann. Inst. Oceanog. Suppl. Tome 58.

Cushing, D. H. 1959. On the nature of production in the sea. Fish. Invest. London, Ser. II. 22(6):1–40.

Cushing, D. H. 1975. The productivity of the sea. Oxford biology reader no. 78. Oxford University Press, London. 16 pp.

Cushing, D. H., and J. J. Welsh (eds.). 1976. The ecology of the seas. Saunders, Philadelphia. 467 pp.

Denton, E. J. 1974. Buoyancy in marine animals. Oxford biology reader no. 54. Oxford University Press, London. 16 pp.

Enright, J. T. 1977. Diurnal vertical migration: Adaptive significance and timing. Part 1, Selective advantage: A metabolic model. Limnol. Oceanog. 22:856–872.

Fenchel, T. 1988. Marine planktonic food chains. Ann. Rev. Ecol. Sys. 19:19–38.

Fleming, R. H. 1939. The control of diatom populations by grazing. Journal du Conseil Permanent International pour l'Exploration de la Mer. 14:210–227.

Fraser, J. 1962. Nature adrift, the story of marine plankton. G. T. Foulis and Co., Ltd., London. 178 pp.

Haney, J. F. 1988. Diel patterns of zooplankton behavior. Bull. Mar. Sci. 43(3):583–603.

Hardy, A. 1956. The open sea: Its natural history. Vol. I, The world of plankton. Houghton Mifflin, Boston. 334 pp.

Hardy, A. C. 1953. Some problems of pelagic life, pp. 101–121. *In:* Essays in marine biology (Richard Elmhurst Memorial Lectures). Oliver and Boyd, Edinburgh. 144 pp.

Hardy, A. C., and E. R. Gunther. 1935. The plankton of the south Georgia whaling grounds and adjacent waters, 1926–27. Discov. Rep. 11:511–538.

Hart, T. J. 1942. Phytoplankton periodicity in Antarctic surface waters. Discov. Rep. 21:263–348.

Harvey, H. W., L. H. N. Cooper, M. V. Lebour, and F. S. Russell. 1935. Plankton production and its control. J. Mar. Biol. Assoc. 20:407–441.

Hayward, T. L., and J. A. McGowan. 1979. Pattern and structure in an oceanic zooplankton community. Amer. Zool. 19:1045–1055.

Hill, M. N. (ed.). 1963. The sea. Vol. 2, The composition of sea water, comparative and descriptive oceanography, IV, Biological oceanography. Wiley, N.Y. pp. 347–485.

Hobbie, J. E., and P. J. LeB. Williams (eds.). 1984. Heterotrophic activity in the sea. Plenum, N.Y. 569 pp.

Hobbie, J. E., O. Holme-Hansen, T. T. Packard, L. R. Pomeroy, R. W. Sheldon, P. J. Thomas, and W. J. Wiebe. 1972. A study of the distribution and activity of microorganisms in ocean water. Limn. Oceanog. 17:544–555.

Holme-Hansen, O., and C. R. Booth. 1966. The measurement of adenosine triphosphate in the ocean and its ecological significance. Limn. Oceanog. 11:510–519.

Johnson, P. W., and J. McN. Sieburth. 1979. Chroococcid cyanobacteria in the sea: A ubiquitous and diverse phototropic mass. Limn. Oceanog. 24:928–935.

Koehl, M. A. R., and J. R. Strickler. 1981. Copepod feeding currents: Food capture at low Reynolds numbers. Limn. Oceanog. 26:1062–1073.

Malone, T. C. 1980. Algal size, pp. 433–463. *In:* Morris, I. (ed.). The physiological ecology of phytoplankton. University of California Press, Berkeley.

Mare, M. F. 1940. Plankton production off Plymouth and the mouth of the English Channel in 1939. J. Mar. Biol. Assoc. 24:461–482.

Margalef, R. 1963. Succession in marine populations, pp. 137–188. *In:* Vira, R. (ed.). Advancing frontiers of plant sciences, Vol. 2.

McAllister, C. D. 1969. Aspects of estimating zooplankton production from phytoplankton production. J. Fish. Res. Bd. Can. 26:199–220.

McGowan, J. A. 1974. The nature of oceanic ecosystems, pp. 9–28. *In:* Miller, C. B. (ed.). The biology of the Pacific Ocean. Oregon State University Press, Corvallis.

McGowan, J. A., and P. W. Walker. 1985. Dominance and diversity maintenance in an oceanic ecosystem. Ecol. Monogr. 55:103–118.

McLaren, I. A. 1963. Effects of temperature on growth of zooplankton and the adaptive value of vertical migration. J. Fish. Res. Bd. Can. 20:685–727.

McLaren, I. A. 1974. Demographic strategy of vertical migration by a marine copepod. Amer. Nat. 108:91–102.

Morris, I. (ed.). 1980. Studies in ecology. Vol. 7, The physiological ecology of phytoplankton. University of California Press, Berkeley. 625 pp.

Newell, G. E., and R. C. Newell. 1973. Marine plankton, a practical guide. Hutchinson, London. 244 pp.

Ohman, M. D. 1990. The demographic benefits of diel vertical migration by zooplankton. Ecol. Monogr. 60(3):257–281.

Parsons, T. R., and C. M. Lalli. 1988. Comparative oceanic ecology of the plankton communities of the subarctic Atlantic and Pacific oceans. Oceanog. Mar. Biol. Ann. Rev. 26:317–359.

Parsons, T. R., M. Takahashi, and B. Hargrave. 1984. Biological oceanographic processes, 3rd ed. Pergamon, N.Y. 330 pp.

Peterson, B. J. 1980. Aquatic primary productivity and the ^{14}C-CO_2 method: A history of the productivity problem. Ann. Rev. Ecol. Sys. 11:359–385.

Platt, T. (ed.). 1981. Physiological bases of phytoplankton ecology. Can. Bull. Fish. Aquatic Sci. 210. 346 pp.

Pomeroy, L. R. 1974. The ocean's food web, a changing paradigm. Bioscience 24:499–504.

Pomeroy, L. R., and R. E. Johannes. 1968. Respiration of ultraplankton in the upper 500 meters of the ocean. Deep Sea Res. 15:381–391.

Raymont, J. E. G. 1980. Plankton and productivity in the oceans, 2nd ed. Macmillan, N.Y. Vol. 1, Phytoplankton, 489 pp. Vol. 2, Zooplankton, 824 pp.

Reid, J. L., E. R. Brinton, A. Fleminger, E. L. Venrick, and J. A. McGowan. 1978. Ocean circulation and marine life, pp. 65–130. *In:* Charnock, H., and G.

Deacon (eds.). Advances in oceanography. Plenum, N.Y.

Riley, G. A. 1946. Factors controlling phytoplankton populations on Georges Bank. J. Mar. Res. 6:54–73.

Russell-Hunter, W. D. 1970. Aquatic productivity. Macmillan, N.Y. 306 pp.

Sherr, E. B., and B. F. Scherr. 1991. Planktonic microbes: Tiny cells at the base of the ocean's food webs. Trends Ecol. Evol. 6(2):50–54.

Sieburth, J. M. 1979. Sea microbes. Oxford University Press, N.Y. 491 pp.

Sieburth, J. M. 1982. The role of heterotrophic nanoplankton in the grazing and nuturing of planktonic bacteria in the Sargasso & Caribbean Seas. Ann. Inst. Ocean. Suppl. Tome 58 pp. 285–296.

Smayda, T. J. 1980. Phytoplankton species succession, pp. 493–570. *In:* Morris, I. (ed.). The physiological ecology of phytoplankton. University of California Press, Berkeley.

Smith, DeB. L. 1977. A guide to marine coastal plankton and marine invertebrate larvae. Kendall/Hunt Publishing Co., Dubuque, Iowa. 161 pp.

Stoecker, D. K., A. E. Michaels, and L. H. Davis. 1987. A large proportion of marine planktonic ciliates contain functional chloroplasts. Nature 326:790–792.

Waterbury, J. B., S. W. Watson, R. R. L. Guillard, and L. E. Brand. 1979. Widespread occurrence of a unicellular, marine planktonic cyanobacterium. Nature 277:293–294.

Wickstead, J. H. 1965. An introduction to the study of tropical plankton. Hutchinson, London. 160 pp.

Wimpenny, R. S. 1966. The plankton of the sea. Elsevier, N.Y. 426 pp.

Chapter 3

Oceanic Nekton

In contrast to the plankton, the *nekton* comprises those organisms that have developed powers of locomotion so that they are not at the mercy of the prevailing ocean currents or wind-induced water motion. They are, in fact, capable of moving at will through the water. Most are large animals and include the largest and fastest-moving organisms in the sea. Whereas plankton are dominated by a host of invertebrate animals, nekton are predominantly vertebrates; among these the fishes are the most numerous, both in species and individuals, but representatives of every vertebrate class except amphibians are found here.

Whereas nekton in the larger sense includes all those organisms capable of sustained locomotion against the water motion without consideration as to habitat, in this chapter we will consider only those nektonic animals that are distributed in the epipelagic zone of the open ocean, the *oceanic nekton.* The ecology and adaptations of nekton associated with inshore waters, such as kelp beds and coral reefs, and with the deep sea will be considered in subsequent chapters. The special adaptations of the oceanic nekton are different from those of nekton living in deeper waters and inshore waters, and hence we justify this division of the nekton.

Because these animals are fast swimming and often wide ranging in the immense open seas of the world, they are difficult to study at sea and are virtually impossible to keep captive under natural conditions. We know very little about most aspects of the ecology or life history of these forms. In the dearth of such field data and virtual absence of laboratory information, we have been forced to infer much of our understanding of their ecology from the study of anatomical and physiological characteristics of captured specimens.

Most students will have little opportunity for direct contact with living oceanic nekton forms except for fishes and occasional seal, sea lion, or porpoise sightings or strandings; however, the immensity of the ocean area inhabited by this fauna and its important role in the ecology of the upper layers of the seas of the world make it important to study. Some nektonic fishes, such as the tunas, sustain a major world fishery. This fauna also includes the whales, perhaps the most popular group of marine organisms in the public eye, over which considerable controversy has raged in recent years with respect to their continued survival. To perpetuate the existence of the whales requires some knowledge of the habitat in which they live.

COMPOSITION OF THE OCEANIC NEKTON

The oceanic nekton is composed of a wide variety of various bony fishes, sharks, and rays, as well as lesser numbers of marine mammals, reptiles, and birds. The only invertebrates that can be considered nekton are the various cephalopod mollusks.

Several different groups of fishes may be recognized in the nekton. These are, first of all, those fishes that spend their entire lives in the epipelagic. These fishes may be termed *holoepipelagic*. Included here are certain sharks (thresher shark, mackerel shark, blue shark), most flying fishes, tunas, marlins, swordfish, saury, oarfish, and others (Fig. 3.1). These fishes often lay floating eggs and have epipelagic larvae; they are most abundant in the surface waters of the tropics and subtropics.

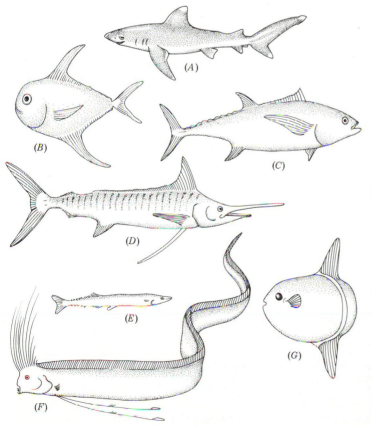

Figure 3.1 Some characteristic holoepipelagic fishes. *(A)* White tip shark, *Pterolamiops longimanus.* (B) Spiny marine bream, *Taractichthys longipinnus.* (C) Yellowfin tuna, *Thunnus albacares.* (D) Striped marlin, *Tetrapterus audax.* (E) Saury, *Cololabis saira.* (F) Oarfish, *Regalecus glesne.* (G) Ocean sunfish, *Mola mola.* (Redrawn from N. V. Parin, Ichthyofauna of the epipelagic zone, trans. from the Russian by the Israel Program for Scientific Translations, 1970.) (Not to scale.)

A second group of oceanic fishes is termed *meroepipelagic*. These fishes spend only part of their life cycle in the epipelagic. This is a more diverse group and includes those fishes that spend their adult lives in the epipelagic but spawn in inshore waters (herring, whale shark, dolphin, halfbeaks) or in fresh water (salmon) (Fig. 3.2). Still other fishes enter the epipelagic only at certain times, such as at night, when certain of the deeper-water fishes such as lantern fish migrate to the surface layers to feed. This latter group, together with all other deeper-living fishes (collectively called "midwater fishes"), is considered in the next chapter.

Most fishes spend their early life in the epipelagic but their adult lives in other areas. These juveniles form a persistent part of the epipelagic fauna but can best be considered meroplankton since they are restricted in their ability to move. They will not be considered further here.

The second component of the oceanic nekton is the marine mammals. Nektonic marine mammals include the whales (order Cetacea) and the seals and sea lions (order Carnivora, suborder Pinnipedia). Other marine mammals exist, such as manatees and dugongs (order Sirenia) and sea otters (order Carnivora), but these animals are not pelagic because they occupy inshore waters at all times. Little is known of the role of manatees and dugongs in any community, and they will not be considered here. The role of sea otters in the economy of inshore waters is discussed in Chapter 5.

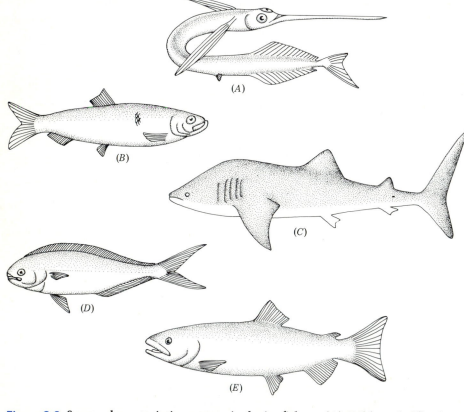

Figure 3.2 Some characteristic meroepipelagic fishes. *(A)* Ribbon halfbeak, *Euleptorhamphus viridis. (B)* Herring, *Clupea harengus. (C)* Whale shark, *Rhincodon typus. (D)* Dolphin, *Coryphaena hippurus. (E)* Salmon, *Oncorhynchus keta.* (Redrawn from N. V. Parin, Ichthyofauna of the epipelagic zone, trans. from the Russian by the Israel Program for Scientific Translations, 1970.) (Not to scale.)

Nektonic reptiles are almost exclusively turtles and sea snakes. Marine iguanas exist in the Galápagos Islands, and saltwater crocodiles inhabit many island areas of the Indonesian and Australian region, but these again are littoral animals, only rarely venturing out of sight of land. The fossil record indicates that, during the Cretaceous period some 60 million years ago, marine reptiles were much more common and varied than today. At that time, large plesiosaurs, ichthyosaurs, and mosasaurs roamed the warm seas.

Technically, most seabirds are not nektonic, since they fly over the open ocean or swim on the surface. But they do feed in the upper layers and some may dive to depths near 100 m to feed. They also occur in large numbers in some areas where it is likely that they have a significant effect on marine prey populations. Therefore they enter into the economy of these waters and are considered here. Perhaps the only group of truly nektonic birds are the flightless penguins of the Southern Hemisphere. But cormorants and other seabirds do dive for food and spend a good deal of time as swimmers.

ENVIRONMENTAL CONDITIONS

The environmental factors acting in the epipelagic zone inhabited by the nekton are, of course, the same as those reported in the preceding chapter, for the plankton (see pp. 57–64), and include light, temperature, density, and currents. However, the relative importance of different factors in selecting adaptations and life strategies of the nekton may well be different. It is important to note that the perception of this environment is very different for a large, fast-swimming fish or mammal than it is for a small copepod.

Several environmental conditions should be emphasized as being of considerable significance to the nekton and for which adaptations have evolved. In the first place, this is an area with a profound "three-dimensionality." Second, there is no solid substrate anywhere, and thus the animals are always suspended in a transparent medium with no shelter that may be employed as a refuge from potential predators. Because of the general lack of structures in this area, there is nothing visible for animals to take bearings on when moving from place to place horizontally. Finally, the lack of a substrate means that there is no firm support for the animals, most of which have flesh somewhat denser than the surrounding seawater.

Three-dimensionality, combined with the lack of any obstacles, facilitates the evolution of adaptations for great mobility. Great mobility and ability to cover large distances, in turn, selects for the development of nervous and sensory systems that will provide and process the information necessary to navigate in the area, to find and capture food, and to avoid predation.

Similarly, the lack of any shelter, coupled with the large size of most nekton forms, leads to development of faster swimming speeds as one of the few options left to escape predators and, in turn, to enable them to capture food. This lack of shelter also leads to development of camouflage as another option. Continuous suspension of the more dense body of nektonic animals in the less dense water leads to the progressive development of various adaptations toward keeping afloat.

ADAPTATIONS OF OCEANIC NEKTON

Buoyancy

Perhaps the most significant adaptations of nektonic animals are those that serve to keep the animals suspended in the water and propel them through the water at a fast rate of speed. It is these adaptations that have facilitated the strong convergence that we observe in morphology of otherwise very different animals.

The first order of business in living in the epipelagic is to stay afloat. This, as we saw in the last chapter, is also the primary concern of the plankton. Most nektonic animals have densities very close to that of seawater. Since living tissues are generally heavier than seawater, the fact that many of these large animals are nearly neutrally buoyant means that there must be lower-density areas in their bodies that counteract the higher density of most tissues.

Most fishes have a *gas* or *swim bladder* in their bodies. This structure, which forms perhaps 5 to 10 percent of the fish by volume, serves to counteract the denser flesh of the fish and thus give the fish a neutral buoyancy (Fig. 3.3). Most fishes can regulate the amount of gas in the gas bladder and thus change their buoyancy state. Two types of gas bladder systems are known: the *physostome,* in which there is an open duct between the gas bladder and the esophagus, and the *physoclist,* in which there is no duct. Physostome fishes move gases in and out of the bladder via the duct by gulping air at the surface, but usually fill the swim bladder through the gas gland and a *rete mirabile* system. The rete mirabile is a network of small blood vessels that branch off a large vessel. Physoclist fishes also secrete gas into the swim bladder through the gas gland and rete mirabile, but in

order to void the gas, they must do so through a special gas absorptive organ called the *oval.*

In very fast-moving fishes, such as bonito *(Sarda)* and mackerel *(Scomber),* which also move vertically in the water column, the gas bladder cannot adjust quickly enough to compensate for pressure changes and maintain neutral buoyancy, and it thus becomes more of a liability. Hence, Denton and Marshall (1958) noted that the fastest fishes tend to lack gas bladders and maintain buoyancy by other means.

Gas-filled cavities in the form of lungs also help sustain the neutral buoyancy of all the air-breathing nektonic animals. Some marine mammals have accessory air sacs (Fig. 3.3). In the case of these animals, they can regulate their buoyancy through the amount of air they hold in

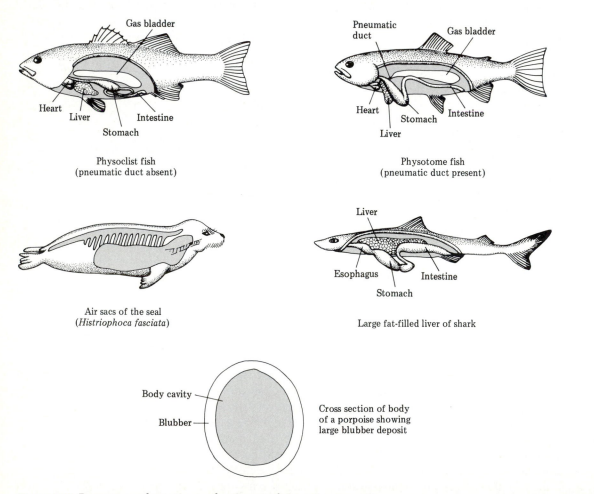

Physoclist fish
(pneumatic duct absent)

Physotome fish
(pneumatic duct present)

Air sacs of the seal
(*Histriophoca fasciata*)

Large fat-filled liver of shark

Cross section of body
of a porpoise showing
large blubber deposit

Figure 3.3 Buoyancy adaptations of nektonic fish and mammals.

their lungs. Birds also have additional air sacs. In most diving seabirds (excluding penguins), air trapped under the feathers provides the greatest amount of buoyancy. Among marine mammals, the sea otter and fur seals also use air trapped in the dense wool undercoat for buoyancy.

Another mechanism of ensuring neutral buoyancy is replacement of the heavy chemical ions in the body fluids with lighter ones. We observed this in plankton as well (see pp. 49–50). The only nektonic animals in which this occurs are the squids. Squids tend to have body cavities in which heavy sodium ions are replaced with lighter ammonium. As a result, an equal volume of body fluid is less dense than the same volume of seawater. Whereas this is a common and often-used mechanism among plankton, it is rare among nekton simply because for it to be really effective the amount of ammonium-dominated fluid must be large. Large, fluid-filled spaces give the animal a rotund appearance, which thus markedly decreases its ability to move rapidly.

Increasing buoyancy through reduction of the amount of bone or other hard parts is not a viable option for these animals, since strong and rigid skeletal frameworks are needed for the muscle system to operate effectively to propel the animals through the water. This is, of course, a marked contrast to the plankton.

Still another mechanism to increase buoyancy is to lay down lipid (fat or oil) in the body. Lipid is less dense than seawater and thus can be important in contributing to buoyancy. Large amounts of lipid are present in many nektonic fishes, primarily those that lack swim bladders, such as sharks, mackerels, bluefish *(Pomatomus)*, and bonito. Presumably, the lipid, at least partially, makes up for the absence of the swim bladder. Lipid may be deposited in various body parts such as muscles, internal organs, and the body cavity, or else be localized in one organ. In pelagic sharks, for example, the lipid is localized in the much enlarged liver, and in many shark species fat deposition in the liver is part of their development; thus, some young sharks start out negatively buoyant and gradually, during growth, become neutral or positively buoyant, as fat stores build up in the liver. In marine mammals, the lipid usually is deposited as a layer of fat just below the skin (= "blubber"), where it

serves not only to aid buoyancy but to insulate against heat loss, store energy, and add to streamlining.

In addition to these static means of maintaining or increasing buoyancy, certain nektonic animals also show some hydrodynamic mechanisms for producing additional buoyancy during movement. Perhaps the most common are the formation of lifting surfaces in the anterior region, usually represented by pectoral fins or flippers, and the presence of a heterocercal tail. A heterocercal tail is one in which the upper lobe is the larger and better developed. In this system, the fin or flippers act as movable ailerons, just as on an airplane, and when canted at the appropriate positive angle, cause the individual to rise in the water column as the tail provides the propulsive thrust (Fig. 3.4). If the tail is heterocercal, its motion also provides a portion of the upward thrust as well. In some forms, the lift provided by the fins or flippers is aided by the whole anterior portion of the body, which can also be

Figure 3.4 Various tail and fin shapes in fishes showing the lift provided. *(A)* Shark *(Squalus acanthias)* with slightly heterocercal tail. *(B)* Sturgeon *(Acipenser stellatus)* with heterocercal tail. *(C)* Mackerel *(Scomber scombrus)* with homocercal tail. *(a)* Force created by the pectoral fins or analogs; *(b)* force due to residual weight; *(c)* vertical component of propulsive force furnished by the caudal fin. (Modified from Yu. G. Aleyev, Nekton, Dr. W. Junk BV., 1977. Reproduced by permission of Kluwer Academic Publishers.)

inclined at an angle in order to provide lift. The best development of these dynamic buoyancy forces occurs in those forms that are negatively buoyant.

In general, there is a tendency among the more primitive fishes to have hydrodynamic (water movement) adaptations to create lift, whereas the more advanced forms seem to evolve static or passive means to achieve neutral buoyancy. This is because less energy is expended to obtain neutral buoyancy than to have to move constantly to achieve the lift necessary to keep a body suspended in the water. Because of air-filled lungs, air-breathing mammals tend to be nearly neutrally buoyant.

Locomotion

A second, and related, group of adaptations in nektonic animals are those having to do with movement of the animals through the water. These adaptations can be divided into two groups: those necessary to create the propulsive force and those that reduce the resistance of the body to passage through the water.

The force necessary to propel a nektonic animal through the dense water is created by some part or parts of the animal's body. The most common means of producing forward movement is by undulatory motions of the body or of the fins. Virtually all nektonic fishes show these types of motion. In the undulatory mechanism, the animal moves itself forward by sweeping the posterior part of the body and fins from side to side. This throws the body into a series of short curves that start at the head and move down the body. This side-to-side motion is cre-ated by alternate contractions of the body mus-culature, first on one side and then on the other, in interaction with forces from the reacting water. Elastic properties of body and fins and the lateral shape of the fish determine the phase relationships between the muscle contraction pattern and the undulating movements of the body. In most fishes the wave of muscle activity along the body on each side is much faster than the wave of curvature. In the saithe (*Gadus virens*), for example, the lateral muscles behind the head and near the tail are instantaneously active.

The amplitude of lateral movements of fishes with a slender body increases gradually toward the tail. Propulsive forces are generated because the lateral movements accelerate a mass of water around the fish. The head enters undisturbed standing water. Just behind the head accelera-tion of water begins and continues along the body until the end of the tail blade releases the jet of water in a direction on the average oppo-site to the swimming direction.

In short, stubby fishes the propulsive wave is less apparent because the amplitude on the body is low, but it shows a sharp increase near the caudal peduncle. This peduncle is usually nar-row or even dorsoventrally flattened. The prime characteristic feature of this group is the posses-sion of a lunate tail blade (Fig. 3.5). Propulsive forces are generated exclusively by the tail blade acting as an undulating hydrofoil. A similar propulsive system is used by the whales, but the movements of the tail are up and down rather than side to side. The speed of a fish is related to the speed with which the muscle contraction waves pass down the body; it also depends on other aspects of the body shape. In general,

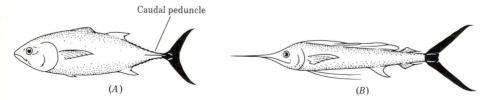

Caudal peduncle

(A)

(B)

Figure 3.5 Fast-swimming fishes with the characteristic lunate-shaped tail and narrow caudal peduncle. *(A)* Tuna *(Thunnus thynnus).* *(B)* Sailfish *(Istiophorus platypterus).*

short, stubby fishes are faster than long, narrow fishes (Fig. 3.6).

Another mode of propulsion is to employ undulatory movements of fins. In this mode of locomotion the body remains stationary and the fins are moved in one of several ways to cause forward motion. This form of locomotion is slower than the previous ones. Examples of this type are found in the rays such as the manta ray (*Manta hamiltoni*), certain squids (*Todarodes*), and ocean sunfish (*Mola mola;* Fig. 3.7). In most fishes, however, the lateral fins are used for maneuvering or lift, as in sharks with a heterocercal tail.

Except for cetaceans and sea snakes, the common form of propulsion among marine air-breathing vertebrates is through paddling movements of either the forelimbs or hindlimbs, or both. The limbs of marine turtles, seals and sea lions, and penguins are all modified into flat, paddle-shaped appendages, which the animals use to move themselves through the water, as we would employ oars (Fig. 3.8). The speed of movement through the water using paddles depends on the frequency of the propulsive stroke. In organisms that employ few strokes, such as turtles, speed is low; in others, such as penguins—where up to 200 strokes per minute have been recorded by Brooks (1917) in *Pygoscelis papua*—the rate of movement may be very fast (10 meters per second for the 200 strokes per minute).

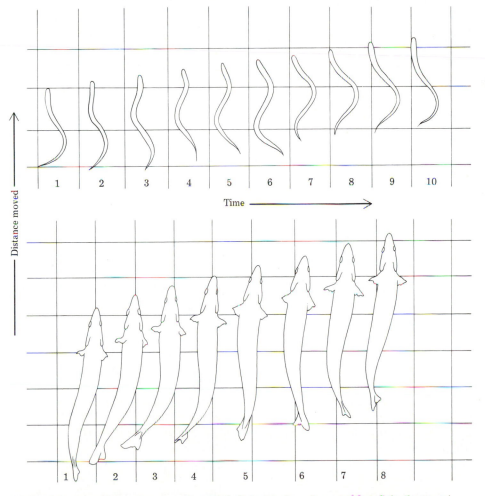

Figure 3.6 Propulsion in an elongated fish (top) and a stubby fish (bottom). (Adapted from Yu. G. Aleyev, Nekton, Dr. W. Junk BV., 1977. Reproduced by permission of Dr. W. Junk BV.)

Figure 3.7 Fishes using undulatory fins for locomotion. (A) Ocean sunfish (Mola mola). (B) Manta ray (Manta hamiltoni). (Not to scale.) (From Yu. G. Aleyev, Nekton, Dr. W. Junk BV., 1977. Reproduced by permission of Dr. W. Junk BV.)

The seals and sea lions, when swimming, are not able to reach the speeds seen in whales. Eared seals (sea lions) swim using their front flippers as paddles, but earless seals use their webbed hind feet, spread vertically like the double caudal fin of fishes. The final type of propulsion is "jet propulsion" using water. This form of propulsion is the domain of the oceanic squids, who are capable of producing very rapid movement, up to 16 m/sec.

Surface of Resistance and Body Shape

Because water is a very dense medium, it is difficult to move an object through it, and it is even more difficult to move an object at a fast rate of speed. To propel a body through water requires considerably more energy than to move the same body through air. It requires less energy to move an object through water if it has a shape that reduces the surface of resistance to the water to a minimum. Since nektonic animals must move and since they have limited, at least finite, energy resources, it follows that a major set of adaptations must be made toward reducing the surface of resistance and hence the drag of the body.

There are several types of drag or resistance to movement that must be overcome. *Frictional resistance* is proportional to the amount of surface area in contact with the water (Fig. 3.9). Minimum frictional resistance is produced by a spherical object, which, of geometric shapes, has the minimum surface area for a given volume. If, however, one wishes to move a nektonic object through water, another type of resistance to movement becomes of great importance. This is

form resistance, where the drag is proportional to the cross-sectional area of the object in contact with the water. In such a case, the spherical object presents a very large area and hence is unsuitable as a shape for nektonic animals. To reduce form resistance to a minimum, the shape should be relatively long and thin, such as a thin cylinder or wire. The final type of resistance that must be overcome is the so-called *induced drag* or *turbulence* due to changes in speed and direction of the flow. The flow around small animals or slow, large ones is usually smooth and called *laminar.* With increasing speed or size the laminar flow tends to be disrupted and thrown into vortices or eddies. This will greatly increase the drag. The flow pattern along large, fast animals is not laminar but turbulent. The water particles vibrate and have less tendency to be thrown into eddies. The turbulent flow follows the contour of a streamlined body, even at high speeds where a laminar flow would have been disrupted. Swordfish generate the turbulent flow with the rough sandpaperlike skin on their sword. This greatly reduces drag and makes high speeds possible. The best streamlined bodies have a teardrop shape, somewhat blunt in front and tapered to the posterior end. Such a shape gives the lowest resistance for the largest volume if the ratio between the largest diameter and the length is about 0.22. Whales, dolphins, and tunas closely approach this ideal shape (Fig. 3.10).

In addition to body shape, nektonic animals have other adaptations to reduce drag. These adaptations generally fall under the category of streamlining the external surface of the body so that there are no protuberances that may break up the smooth flow of water over the body and hence increase drag. To achieve this, species that

(A)   (B)   (C)

Figure 3.8 Swimming by paddle motions in three classes of marine vertebrates. *(A)* The green turtle *(Chelonia mydas)*. *(B)* The sea lion *(Arctocephalus pusillus)*. *(C)* The penguin *(Pygoscelis papua)*. (Redrawn from Yu. G. Aleyev, Nekton, Dr. W. Junk BV., 1977. Reproduced by permission of Dr. W. Junk BV.)

move rapidly, such as tunas, are structured so that almost all normally protruding body structures are recessed into depressions or grooves from which they may be elevated only when needed. Thus, in fast-moving fishes, eyes, though large, do not protrude beyond the sides of the body. Pectoral and pelvic fins fit into grooves except when in use, and body scales are reduced

Cross section

Side view

(A)   (B)   (C)

Figure 3.9 Frictional forces on variously shaped objects moving through water. *(A)* A flat disc. *(B)* A cylinder. *(C)* A teardrop-shaped object. All have the same cross-sectional shape and area but differ in their three-dimensional form. The teardrop shape is best for a fast-moving fish and has about one-fourteenth the resistance of that on disc *A*.

or absent (Fig. 3.10). Similarly, in marine mammals, the hair is lost or reduced in length, for it produces more drag than bare skin. Mammary glands are flattened, and genitalia of the males do not protrude beyond the skin, except when in use.

Defense and Camouflage

By far the strongest adaptations seen in nektonic animals are those that relate to the ability to achieve rapid forward movement in the water column. These adaptations are of such importance that they take precedence over, or preclude, any adaptations with respect to defense against potential predators, if such adaptations would necessitate modifications requiring de-

creased locomotory ability. This immediately limits the number of possible defensive adaptations. At the same time, most of these nektonic animals are large as adults, which means that most have few potential predators; the very largest (whales) oftentimes have few predators other than humans and killer whales (Plate 14). Thus, the need for elaborate defensive mechanisms is somewhat reduced.

Certain defensive mechanisms remain possible within the framework of fast locomotion. Of these, the most common and widely used is that of *camouflage*. Since we are now dealing with an environment that has no physical hiding places, visible in all three dimensions, camouflage mechanisms could conceivably take one of three directions: transparency of the body, cryptic

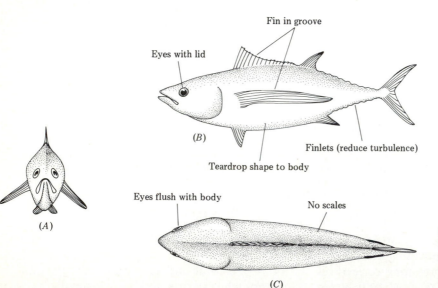

Fin in groove

Eyes with lid

(B)

Teardrop shape to body

Finlets (reduce turbulence)

(A)

Eyes flush with body

No scales

(C)

Figure 3.10 Three views of a tuna showing the adaptations necessary for fast movement. *(A)* Front view. *(B)* Side view. *(C)* Top view.

coloring, and alteration of body shape (= cryptic body shape).

If the body of an organism is transparent and suspended in the transparent surface water of the ocean, the animal will become invisible when viewed in the water. As we saw in the plankton, such transparency is a common defensive adaptation of many species. It is, however, not found among the nekton. The reason is that, as the size and thickness of nektonic animals increases, it becomes much more difficult to keep a body transparent, particularly when the body is highly muscular for propulsion. Thus, camouflage by transparency is not an option for nekton.

Camouflage through alteration of body shape is possible so long as the shape does not interfere with that necessary for fast locomotion. Among nektonic vertebrates, the commonest manifestation is to develop a ventral keel to the body to eliminate a conspicuous shadow on the belly of

the animal when viewed from below. What is the origin of this shadow and how does body shape help to reduce it? To answer this, we must know something about how light appears in the surface waters of the ocean. When light enters the water, the rays penetrate down in a conelike path. At the same time, some of the light is reflected or scattered back in all directions by the particles in the water. This scattered light is available to illuminate objects in the water from different angles other than the surface, but its intensity is very much less than the downwelling surface light. If an animal is suspended in this water column, it is illuminated most intensely from above, while the scattered light illuminates from the sides and below. Because light intensity from the side and below is so low, a shadow appears under the animal where its body has cut out the strongly downwelling surface light (Fig. 3.11). If the body is now extended ventrally as a keel to form a sharp ventral edge rather than a

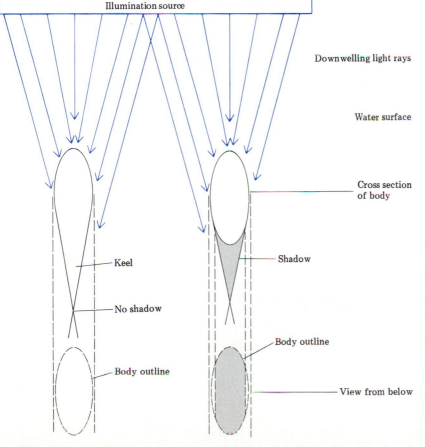

Downwelling light rays

Water surface

Cross section of body

Keel

Shadow

No shadow

Body outline

Body outline

View from below

No shadow, body invisible   Shadow, part of body shadowed visible

Figure 3.11 Diagram showing how a keel on the ventral surface of an animal eliminates the dark shadow normally cast downward by an unkeeled animal. The presence of the shadow means that an animal living deeper and looking upward would see the unkeeled nektonic animal due to the shadow, but would not see the keeled animal, which would blend into the lighted background. (Modified from Yu. G. Aleyev, Nekton, Dr. W. Junk BV., 1977. Reproduced by permission of Kluwer Academic Publishers.)

rounded edge, the shadow will be eliminated when viewed from below. This is because all surfaces on the body now are oriented so that none is illuminated exclusively by the diffuse scattered light; there is also at least some component of the intense surface light (Fig. 3.11). Without this conspicuous shadow, an animal viewed from below becomes virtually invisible in the downwelling light and hence is camouflaged from deeper-dwelling predators. The elimination of the shadow is enhanced if the keel has reflecting surfaces on it in the form of white pigment or scales.

Cryptic coloration is also characteristic of most nektonic animals. In the lighted upper waters of the ocean, the dominant spectral colors are blues and greens. When water is viewed from above at the surface, it appears greenish or bluish looking down into the depths. It should not be surprising then that many nektonic forms have dark blue or green on their dorsal surfaces so that potential predators looking down would be hard-pressed to make out their form in the general bluish green background. At the same time, when viewed from below, the water looks brighter or whiter toward the surface. Any dark organism swimming in this area would be conspicuous if viewed from below, even if keeled to prevent a shadow. However, if the ventral side is colored white or silver to maximize light reflectance or to blend with the downwelled light, the animal will tend to become invisible. Hence, we observe that many nekton are bicolored, or *countershaded,* dark green or blue above and white or silver below (Fig. 3.12). For certain of the highly surface-dwelling vertebrates such as porpoises, the color pattern is more complex, with irregular bands of light and dark that actually mimic the pattern of wave-roughened surface waters themselves (Fig. 3.13). Some por-

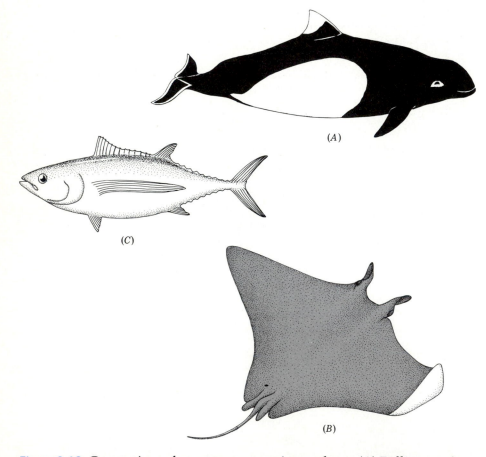

(A)

(C)

(B)

Figure 3.12 Contrasting color patterns on various nekton. *(A)* Dall's porpoise *(Phocoenoides dalli). (B)* Manta ray *(Manta hamiltoni). (C)* Albacore *(Thunnus alalunga).*

Figure 3.13 Cryptic coloring on the sides of a Pacific white-sided porpoise, *Lagenorhynchus obliquidens*, mimicking the wave-roughened surface of the water.

poises that live within tuna schools are countershaded gray above and white beneath, with white spots and speckles in the gray and black spots and speckles in the white. Colored in this manner, these animals are difficult to perceive from the side view, underwater, amid a school of tuna.

Among the abundant flying fishes there is an adaptation to produce large fins. These fishes escape predators by propelling themselves out of the water and gliding for long distances on these winglike fins.

Other than the above adaptations, there are few specialized morphological structures developed as defense against predators. Most likely this is because the development of various spines and armor would interfere with the ability to move quickly by increasing resistance. However, it may also be partly due to the fact that the larger nekton are the top predators in the system and have few or no other nekton consuming them.

Sensory Systems

Since nektonic animals are large, fast moving, and primarily predaceous, they might be expected to have well-developed sense organs. This is generally the case, and with few exceptions, such as lateral lines in fishes, the senses are no different from those possessed by other vertebrates in different habitats. Lateral lines are rows of small tubes, open to the water and containing sensory pits that are sensitive to pressure changes in the water.

Sharks and rays possess special sense organs called ampullae of Lorenzini, which are sensitive to minute electric currents in the water. Kalmijin (1971) has demonstrated that predaceous cartilaginous fishes can use this electroreception to find prey.

Another unusual sensory system is a geomagnetic sensory system in marine mammals that seems to play an important role in long-distance navigation and may also explain the various mass strandings of some toothed whales.

Most sensory information received by nekton comes via light (vision), sound (hearing), or chemicals (olfaction). Eyes tend to be well developed and complex in all forms, but the size of the eye relative to size of body varies greatly. The eyes are usually set laterally on the head in such a way that the fields of view for each do not overlap, yet encompass large areas on either side. The lack of overlap of fields of view means that binocular vision is small or absent among most nekton except for pinnipeds.

Echolocation

Among the mammals of the nekton, it is the sense of hearing that has generated the greatest number of specialized adaptations, which, in turn, suggests the great importance of hearing to these animals. The importance of sound to nektonic mammals resides in the fact that sound travels about five times faster in water than in air and has a much greater communication range than does sight. As a result, most nektonic animals show strong development of sound-receiving structures.

In the terrestrial environment, enhanced sound reception is usually indicated in external morphology through the enlarged external ears or pinna of mammals. However, such structures would create an excessive drag on aquatic vertebrates and hence are reduced or absent. To make up for their absence, there has been a tendency among aquatic mammals to develop other structures of the head to receive sound waves.

Sound reception and production are most highly developed in those cetaceans that use them for *echolocation* in much the same way we use sonar to determine depth. In echolocation or sonar, sound waves are sent out from a source in a particular direction. These sound waves pass uninterruptedly through the water until they impinge on a solid object. When they strike an object, they are reflected and return to the source. The time interval between the initial production of the sound and its movement to a target and subsequent return after reflection is a measure of the distance between the source and the object (Fig. 3.14). As the distance changes, so will the time necessary for the sound "echo" to return. Continual production of sound waves and sensory evaluation of the reflected waves during swimming give a nektonic animal a constant check on all objects in its path. Knowing the distance to objects makes it possible for the echolocating animal either to avoid them (predators) or to close in upon them (food source).

Low-frequency sound is used by echolocating animals to orient themselves in the water column with respect to objects around them. Low-frequency sound, however, does not produce information as to the fine structure of objects. In order to obtain that information, higher-frequency sound waves must be produced and reflected from the object, which will, in turn, discriminate considerable detail. Thus, most nekton animals that have highly developed echolocation also have the ability to vary the frequency of the sound produced. Discriminatory ability of high-frequency sound is really quite remarkable. Dolphins, for example, are reported by Kellogg (1958) to be able to distinguish between two fish species of similar size and shape; Norris et al. (1961) report porpoise able to distinguish even between similar objects differing only in thickness.

It is in the toothed whales that echolocation reaches its zenith. These animals possess elaborate morphological modifications of the head and respiratory systems that permit them to send and receive sound waves varying over a wide range of frequencies.

Toothed whales have a peculiarly bulging, rounded forehead and, associated with it, a dorsally placed external nasal opening or blowhole. Internally, a complex series of air sacs is associated with the nasal passages leading from the blowhole to the lungs. The rounded forehead is caused by the presence underneath of a large, fat-filled structure called the *melon.* This fatty organ reaches its greatest development in the sperm whales, where it is called the spermaceti organ; it constitutes as much as 40 percent of the entire animal's length. The relationship of these structures is diagrammed in Fig. 3.15.

Figure 3.14 Echolocation in a sperm whale. Sounds are sent out (solid lines) focused by the melon (spermaceti organ), and the returning echoes (broken lines) are received by the lower jaw.

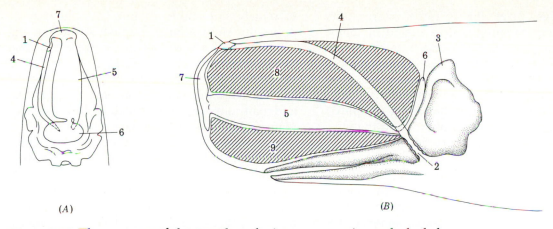

Figure 3.15 The anatomy of the sound-producing apparatus in toothed whales. *(A)* Dorsal view of the head of a sperm whale. *(B)* Side view of the head of a sperm whale *(Physeter macrocephalus).* (1) Nasal opening or nostril, (2) bony nasal duct, (3) sound-focusing surface of cranium, (4) left nasal passage, (5) right nasal passage, (6) frontal air sac, (7) distal air cavity, (8) and (9) upper and lower spermaceti sac (acoustic lens). (From Yu. G. Aleyev, Nekton, Dr. W. Junk BV., 1977. Reproduced by permission of Dr. W. Junk BV.)

Although we do not as yet completely understand how the above elaborate system operates in sound production and reception, enough is understood to permit a description of how this apparatus likely functions. Sounds are produced by the toothed whales by movement of air through the nasal passage and associated air sacs. Sound is produced by recirculating internal air during diving. Special muscles in and around the nasal passages and air sacs allow these channels to change shape and volume and thus change the frequency of the sound produced. The fatty melon is apparently used as an acoustical lens to focus and direct the train of outgoing sound waves, and this directionality increases with increasing frequency of the sound so that these animals can pinpoint objects with high-frequency sound. Directionality is further enhanced by the bones of the peculiarly shaped skull of these toothed whales. Reception of the reflected waves is centered in the bone and fat deposits of the lower jaw and in the middle ear. In contrast to most mammals, where the middle ear resides in bone attached to the skull, the toothed whale's middle ear is loosely attached to the skull by ligaments and provided with special air- and fat-filled spaces.

Thus, the elaborate morphological modifications of the head region of toothed whales are primarily directed to producing and receiving a wide range of sound frequencies, which in turn allow the animals both to navigate without bumping into objects and to pinpoint potential food organisms for subsequent capture. Toothed whales also possess very large brains relative to their body size, brains that are second only to humans' in development of the cerebral hemispheres. It is possible that these large brains are necessary to allow them to process rapidly the acoustical information received.

Reproduction and Life Cycle

Among epipelagic fishes, no special reproduction mechanisms are apparent that would sharply set them apart from their benthic or shallow-water relatives. Characteristically, however, holonektonic bony fishes such as tuna and marlin spawn eggs that float and undergo development in the open-ocean waters. Some even have threadlike structures associated with them that permit them to adhere to various kinds of floating debris. Because these floating eggs are planktonic, they are subject to tremendous losses due to predation. As a result, the fishes produce huge numbers to offset the equally large losses. Parin (1970) reports that skipjack and albacore tuna, for example, produce 2 and 2.6 million eggs, respectively, while striped marlin spawn over 13 million and ocean sunfish an incredible 300 million eggs. Spawning is often intermittent and is often extended over a period of many months.

Among the pelagic sharks, a different reproductive strategy is observed. These fishes produce only a few large eggs or embryos. Parin (1970) notes that the thresher shark (*Alopias*), for example, has but two embryos and the blue shark (*Prionace glauca*) up to 54. Obviously, if so few eggs were to start out as small as those of bony fishes and have to undergo development for a substantial time in the plankton, the chances that any would survive the tremendous predation are very poor. Thus, sharks enhance the chances for survival of their few offspring by producing very large eggs and by retaining the eggs in the female for a much longer period of time so that, at birth or hatching, the young are large in size and immune to most potential predators.

We know relatively little about the growth of pelagic fishes, but available knowledge suggests that the growth rate is very rapid. Tunas, for example, seem to increase their weights by 2–6 kg per year and length by 20–40 cm (Table 3.1). Correlated with this rapid growth, most nektonic fish seem to be short lived; even the large tunas appear to live only 5 to 10 years. By contrast, the pelagic sharks may live from 20 to 30 years.

Marine birds and turtles retain the reproductive characteristics of their terrestrial relatives. All produce shelled eggs, which are laid on land. Marine birds often congregate in large groups to nest on islands or cliffs, which are inaccessible to terrestrial predators. This ensures that the usually helpless young (*altricial*) will survive until old enough to fly. However, it also makes such birds extremely vulnerable to human predation or pollution, since a large fraction of the existing population of a species may be present in one small area and can be wiped out easily. For example, the Laysan albatross nests on Midway Island, where the Navy has airport installations, and many albatrosses are killed by collision with planes. Most marine birds have definite breeding seasons and may migrate thousands of miles from their feeding grounds to their breeding grounds.

Marine turtles all lay eggs in depressions that they excavate in the sand above the high-tide level on beaches in various parts of the tropics. This is the only time that these animals normally return to land. Immediately upon hatching, the young turtles instinctively head for the ocean, where their continued development occurs, but about which we know virtually nothing. As with birds, turtles tend to migrate thousands of miles and congregate off certain beaches for breeding. Females haul out to lay eggs only on certain beaches, and since both eggs and adults are considered excellent food by humans, most marine turtles have been drastically reduced in numbers in recent years in all areas of the world. Some sea snakes give birth to living young in the water; others lay eggs on beaches.

With respect to reproduction in marine mammals, there are two groups: those that give birth on land and those that give birth in the water. We know considerably more about reproduction in the group that breeds on land because they may be observed by humans at that time. Knowledge of the reproductive patterns of those giving birth in the water is generally limited to observations made on captive animals in aquaria.

Table 3.1   AGE AND LENGTH IN THREE SPECIES OF TUNA[a]

Species	Age, years							
	1	2	3	4	5	6	7	8
	Length, cm							
T. alalunga	18–66	32–84	54–89	73–94	70–95	82–100	98–105	—
T. obesus	45–50	62–70	74–94	85–116	97–138	115–155	—	—
T. albacares	54–103	70–136	85–155	99–154	109–134	110–130	127–145	135–160

[a]Albacore (*Thunnus alalunga*), big-eye tuna (*Thunnus obesus*), and yellow fin tuna (*Thunnus albacares*).

Source: After Parin, 1970.

Seals, sea lions, and walruses all give birth to live young on land or on floating ice. Their young are often unable to swim at birth and may require some time before they are capable of venturing into the water. Harbor seal (*Phoca vitulina*) pups, however, can swim immediately after birth (Plate 16). During this terrestrial period, the pups grow very rapidly and gain the strength and insulating layers of fat and fur they need in order to survive in the cold, open water. Many sea lions and seals such as fur seals, Steller's sea lions, and elephant seals are polygamous and territorial on the breeding grounds (see Plate 15). The largest and most aggressive males (harem bulls) tend to collect together varying numbers of females into a harem, occupying a small area of beach, which they then protect from other bulls (Fig. 3.16). Should another bull attempt to steal a female, take over a harem, or encroach upon the territory occupied by the bull and his harem, the resident "harem master" will fight. Although such fights may be mostly bluff and noise, serious fights occur in which one or the other bull may be seriously injured or killed. In the seals and sea lions where such territorial behavior occurs, the males are usually much larger in size than the females.

This territorial behavior and aggressiveness on the breeding ground do not extend to their life when away from the breeding grounds. This behavior suggests that breeding space is limiting, whereas pelagic food and space are not. This method of breeding also means that relatively few males, the harem masters, actually do the breeding, and most others are excluded (Fig. 3.16). As with the marine birds, many of these pinnipeds migrate long distances to their breeding grounds; for example, the northern fur seal (*Callorhinus ursinus*) is pelagic all over the

Figure 3.16 A breeding bull and females, northern fur seal (*Callorhinus ursinus*). (Photo by the author.)

northern Pacific Ocean, but most migrate back each summer to two small islands making up the Pribilofs in the Bering Sea, to breed.

In contrast to the pinnipeds, the cetaceans give birth to their offspring in the water. The young of whales must therefore be able to swim at birth and instinctively know how to surface for air. They also remain closely associated with the mother. Whereas young pups of sea lions may be left on the breeding grounds for days while the female forages for food in the open ocean, the juvenile whales always remain close to their mothers, where they are protected from potential predators. As with the pinnipeds, certain cetaceans may also undergo migration for breeding purposes. Often, this is a migration of thousands of miles from the feeding areas in cold water to the calving grounds in warmer waters (Figs. 3.17, 3.18). The reason for moving to warm waters for birth of the young is that the newly born young do not have the insulating blubber of the adults and hence will have a better chance of survival in warm water until such time as the additional insulating layers can be added.

The young of both pinnipeds and whales grow very rapidly, adding many kilograms per day. Blue whales, for example, can grow from 3 tons at birth to 23 tons at weaning, seven months later. The reason for such rapid growth is that the milk of pinnipeds and cetaceans is extraordinarily rich in fats (as much as ten times as much fat as in cow's milk) and is produced in large quantities.

Pinniped pups are nursed throughout the time they are on the rookeries. At the end of the season, in some species, they are usually abandoned by the females and must fend for themselves in the open ocean. Cetacean calves are nursed for varying times depending on the species. Nursing lasts 15 months in sperm whales but may last several years in pilot whales.

Because of the enormous amount of energy that must be put into milk production by these marine mammals to sustain a single offspring, only one young at a time is usually produced. The reproductive cycle is also such that young are produced annually (most pinnipeds) or at even longer intervals (walrus, certain whales;

Figure 3.17 Migratory routes of whales. *(A)* Migratory routes of humpback whales *(Megaptera novaeangliae)* in the Southern Hemisphere. *(B)* Migratory route of the gray whale *(Eschrichtius robustus)*. (A after E. J. Slijper, Whales and dolphins, University of Michigan Press, 1976; used by permission of Springer-Verlag, Heidelberg. B modified from V. B. Scheffer, A natural history of marine mammals, Scribner, 1976.)

Figure 3.18 The reproductive cycle of the fin whale *(Balenoptera physalis)* in the Southern Hemisphere. (Modified from N. A. Mackintosh, The stocks of whales, Fishing News Books.)

Fig. 3.18). This means that stocks of these animals can be easily reduced, and it may take a long time to recover their numbers.

Most marine mammals are long lived. Gray seals have been recorded to live 46 years and harbor seals 36 years, while small whales such as the bottlenose porpoise 32 years and large whales such as the sperm whale and fin whale up to 77 and 80 years, respectively. Correlated with the long life span is the delayed onset of sexual maturity and reproduction. Male fur seals, for example, do not become harem masters until 9 or 10 years of age, and sperm whales apparently do not breed until about age 20.

Migrations

As we noted in the previous section, a large number of marine mammals, birds, and reptiles undertake extensive migrations for breeding or other purposes. This migration for breeding is a common characteristic of air-breathing marine vertebrates.

Nektonic fishes also undertake extensive horizontal migrations that are equivalent in distance to those taken by the air breathers. These migrations are of great importance but, unfortunately,

have been little analyzed. Migrating holonektonic fishes include the various tunas and the saury.

Salmon, which spend most of their lives dispersed in the open ocean, migrate back to freshwater streams in which they were spawned to reproduce. In this respect, they are similar to the marine air-breathing vertebrates. Salmon have the ability to return to breed in the very same stream in which they were hatched. This requires considerable navigational abilities on the part of the fish. From recent studies, it appears that part of the key to this navigational ability resides in the fishes' sense of "smell"; that is, these fish follow various waterborne "scents" to locate their home streams. The suggested mechanism is that, when the young salmon migrate down the streams to enter the sea, they are imprinted with the "odors" of the various streams in succession. This sequential memory is then reversed on their return to allow them to find their way back to the same stream.

Equally remarkable migrations are made by the green turtles *(Chelonia mydas)*. As with the salmon, these animals migrate from distant feeding grounds to congregate on one or a few beaches to lay their eggs. How do they locate these beaches? In the case of beaches along the shores of continents, this is not as difficult to understand. Here, a turtle could simply follow along the shoreline until the beach with the right "smell" came along. However, many turtles along the Atlantic coast of South America regularly nest on tiny Ascension Island, some 1400 miles out in the middle of the Atlantic Ocean. How do they find such a small target? Since the turtles must navigate most of the distance to a mere pinpoint of an island out of sight of land and without any landmarks in the ocean, the most reasonable assumption is that they are navigating as we humans do, using information from celestial bodies! If this is true, we as yet do not know how it is done. One way would be to use the height of the sun to establish their latitude. Ascension lies due east of the bulge of Brazil. Thus, turtles could move along the coast to the tip of the bulge, then move off directly east, correcting for deviations through height of noonday sun. This should bring them within range of the island, where it either would be visible on the horizon or could be detected and homed in on through some chemical tasted in the water.

Other extensive and complex migrations are those undertaken by tunas and their relatives. Tunas are primarily tropical fishes that make extensive migrations across oceans within the tropical zone and also move into the temperate waters during the warmer seasons. Tunas tagged in Florida have been recovered across the Atlantic in the Bay of Biscay, and Pacific tunas tagged off California have been captured in Japanese waters. It is not entirely clear why these fishes should make such monumental journeys, but it appears that such migrations permit the fishes to exploit their food resources more completely and reduce the possibility of destroying or eating out their food in any one area. Such migrations enable these fast-moving fishes to take advantage of the rich food areas of the temperate zone. One of the most important factors that guides these migrations is water temperature. As primarily tropical fishes, tunas venture into the temperate waters when the water temperature rises above 20°C. Thus, off California, the cold upwelled water is very rich in food organisms, but the tunas usually enter these waters in the summer when surface temperatures reach 20–21°C. If the year is cold and temperatures of the surface waters do not rise to this point, the tunas are absent. However, the Atlantic bluefin tuna regularly summers off Newfoundland in water below 20°C, so not all tunas are restricted to warm water. Tunas apparently always return to tropical waters to spawn and to spend the early parts of their lives (Fig. 3.19).

The ability of migratory animals to make accurate journeys across what appear to us as featureless seas has generated in marine scientists a profound interest in how these animals manage to navigate accurately. Over the years a number of different hypotheses have been suggested to explain migration that have implicated a wide variety of sensory systems. Suggested mechanisms have included use of some sort of celestial compass as suggested above for turtles, odor or chemical imprinting as suggested for salmon, cueing on underwater topography, echolocation, detection of thermal structure, and most recently the ability to detect the flux density of the earth's magnetic field.

The ability of animals to use magnetic fields to find their way was not considered a possibility until recently because of the lack of this sense in humans and the apparent lack of any known biological mechanism capable of transmitting the weak geomagnetic field into an animal's nervous system. The discovery of magnotactic bacteria with linear chains of membrane-bound crystals of magnetite (Fe_3O_4) provided the biological mechanism for transducing the geomagnetic field to the nervous system. Subsequently, the finding of membrane-bound magnetite with abundant nervous connections in tuna by Walker et al. (1984) demonstrated the existence of an anatomical basis for the sense in a migrating nektonic animal.

Magnetic reception has now been suggested as the system used for long-distance navigation by turtles and marine mammals, but experimental proof has been difficult to come by due to the lack of good small-scale geographic maps of magnetic anomalies and the difficulty of setting up laboratory tests.

The most interesting data that seem to confirm the presence of this magnetic sense in whales come from the well-known phenomenon of *mass strandings.* If cetaceans are using the flux density of the earth's magnetic field to navigate by, they can use it in two ways. First,

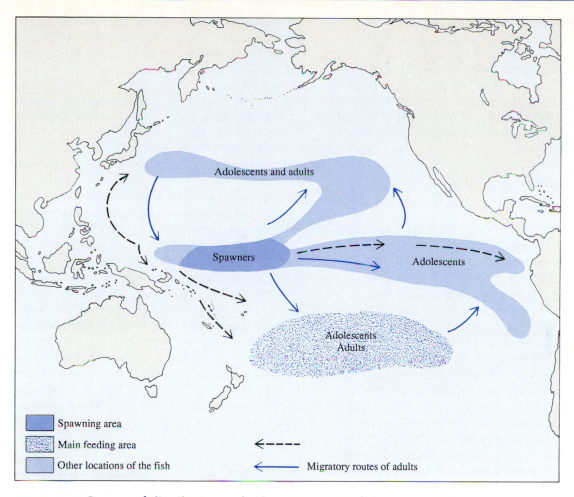

Figure 3.19 Centers of distribution and migratory routes of the big-eye tuna (*Thunnus obesus*) in the Pacific Ocean. (Modified from The Rand McNally atlas of the oceans, Rand McNally.)

the topography of the local field can be used like a road map, with the animals following certain magnetic contours. Second, a timer based on the regular changes or fluctuations of the magnetic field permits the animals to know their position and progress. Problems arise in this system whenever the geomagnetic contour lines cross land or when the pattern of the timer is disrupted by irregular fluctuations in the magnetic field. Thus, whenever these two conditions occur, magnetically navigating whales may well beach themselves. Analysis of mass stranding events by Kirschvink (1990) for the East Coast and Klinowska (1990) for England have demonstrated that all sites of mass stranding had only one physical feature in common, and that was that the geomagnetic contour lines crossed from the sea onto land. Thus, if the whales were following a strategy of traveling parallel to the geomagnetic lines, which would work well in the open ocean, but which gives no clue as to where land and water meet, they would strand themselves. Furthermore, the dates of such strandings were found to be consistent with times when there were irregular fluctuations in the normal patterns of geomagnetic fields.

There are also short-term movement patterns related to food availability and predator avoidance. Many dolphins, such as dusky dolphins, Hawaiian spinners, and Bay of Fundy harbor porpoises, for example, show offshore nighttime and onshore daytime movements.

Although many migrations are undertaken for breeding purposes, such as those of certain of the large baleen whales and the northern fur seal, there may be other reasons for migration as well. Predation is one possibility, and it may be that some baleen whales migrate to warmer locations to calve because killer whales are less abundant at low latitudes and thus such a ploy would protect the young.

Special Adaptations of Marine Birds and Mammals

The warm-blooded marine mammals and birds require some special adaptations to permit them to persist in ocean waters. These special adaptations are primarily concerned with maintaining temperature, diving, and osmotic regulations.

Water has a higher thermal conductivity than does air, which means that it is quick to extract heat from a warm body. Humans experience this when, as swimmers, they become chilled after a short time in water of even 80°F; in air of the same temperature, they would be comfortable indefinitely. Thus, marine mammals, which maintain elevated body temperatures with respect to the surrounding water, must somehow have adaptations to prevent their body heat from being drained away.

One means to slow the rate of heat loss is to have a large body. As you may recall from the plankton section (pp. 49–50), the ratio of surface area to volume of any body is lower for a large one than for a small one. The larger the body, then, the smaller the surface area and, correspondingly, the smaller the surface area in contact with the environment through which heat may escape. All nektonic marine mammals are of large size, and it may well be that the reason there are no mouse-sized marine mammals is that they would simply die from chilling. There are small marine birds (petrels, auklets), but these animals are never completely immersed in the water. Only a fraction of their body is in contact with the water at any one time (except during intermittent diving).

A second adaptation that prevents or reduces heat loss is a thick, insulating layer of blubber or fat just beneath the skin. This layer reaches its greatest thickness in whales, where it may be 2 ft thick. In pinnipeds such as the walrus and elephant seal, subcutaneous fat may constitute as much as 33 percent of the weight. Blubber and fat are poor conductors of heat and hence protect the animal from losing internal heat. The thicker the blubber or fat layer, the less the heat loss. Those marine mammals inhabiting polar waters therefore have thicker layers than do temperate and tropical species.

A final adaptation concerns the circulatory system. The areas of a marine mammal that offer the greatest surface area to the water and hence greatest heat loss, and also lack the protective layer of lipid, are the fins, flippers, and flukes. What adaptations prevent massive heat loss through these extremities? In cetaceans, the answer is that the arteries that bring the warm blood out to these extremities are surrounded by a number of smaller veins that bring blood back to the central core of the mammal. Because of this arrangement, the heat of the blood in the arteries can be absorbed by the cooler blood returning in the veins before it is lost to the external water through the thin flesh of the outer extremities (Fig. 3.20). This is a countercurrent system of circulation designed to save heat.

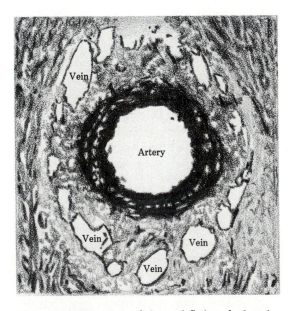

Figure 3.20 Section of the tail fluke of a bottlenosed dolphin, *Tursiops truncatus*, showing the arrangement of veins surrounding an artery (25×). (From P. F. Scholander and W. E. Schevill, J. Appl. Physiol. 8:281, 1955.)

Because most of their adaptations are designed to keep the body heat in, marine mammals (in particular, pinnipeds) may, on occasion, become too warm. Hot, still days may mean considerable stress from overheating. On these uncommon occasions, the animals must act to dissipate heat. They do this by waving their flippers in the air while increasing the blood flow out to the extremities and restricting the flow back to the core through the veins. The result is greater heat loss and subsequent cooling. Seals and sea lions may also open their mouths and pant like a dog.

Most nektonic marine mammals, particularly the pinnipeds and the whales, regularly dive, or are capable of diving, to depths far greater than that of humans. Whereas exceptional humans may be capable of free dives as deep as 60 m with a breath-hold duration of 6 minutes, various seals and porpoises are known to dive to depths of 160–900 m and hold their breaths for 6–48 minutes. The champion is the sperm whale, which can dive as deep as 2250 m and hold its breath for 80 minutes. How is it possible to do this? The answer rests with a number of anatomical and physiological adaptations.

One of the first questions to be asked regarding these prolonged deep dives is how do these animals avoid "the bends"? This is a serious affliction of human divers in which the nitrogen breathed under pressure in deep water as part of their air supply bubbles out of solution in the blood when a diver ascends too quickly. These bubbles then lodge in the joints, causing paralysis and even death. The key here is that humans get the bends due to breathing gas *under pressure,* as from a scuba tank. Since marine mammals do not inhale pressurized gas, but rather depend on gas inhaled at the surface at regular atmospheric pressure, this condition has less chance of occurring. The chance of its occurrence is further reduced by certain other adaptations, the most important of which is the collapse of the lungs. In a deep dive, the pressure of the outside water causes the gas exchange tissues of the lungs of the diving mammal to collapse. This collapse forces the residual gas in the absorptive area of the lung into the nonabsorptive cartilaginous air passageways, and thus diffusion of gas into the blood stops. With no gas entering the blood, there can be no nitrogen to

come back out when the animal surfaces. This collapse of the lungs is aided by the fact that diving marine mammals have few ribs that are attached to the sternum, and the sternum is shortened. Therefore, the rib cage is easily pushed in. Finally, it has also been suggested that the peculiar "foam" of emulsified fat droplets and mucus that many whales have in their respiratory passages serves as an absorptive material, absorbing the nitrogen gas so that it cannot enter the bloodstream where it could cause "bends."

A second problem with deep diving is to explain how the animals manage to survive so long without access to a supply of oxygen. A consideration of this problem will suggest that the only mechanisms for surviving are for the animals to store more oxygen than nondiving forms do and to conserve carefully the amount they have. Since the major organ system involved in oxygen storage and transport is the circulatory system, it is there we may look first for adaptations.

Many diving birds and mammals have a larger blood volume than their terrestrial relatives. The elephant seal, according to Elsner (1969), has a blood volume of 12 percent of its body weight, whereas humans have a blood volume of only 7 percent of body weight and domestic dogs only 9 percent. This larger volume thus allows more oxygen to be held in the body. This oxygen capacity is increased further, as Scholander (1940) notes, because marine mammals also possess a higher oxygen capacity per unit volume of blood than terrestrial mammals (40 ml O_2/100 ml blood in elephant seals versus 16–24 ml O_2/100 ml in humans), but it is not as great as might be expected. Neither the increased blood volume nor the increased O_2 capacity, however, is sufficient to account for the ability of pinnipeds and cetaceans to remain underwater for extended periods. For example, harbor seals have about the same weight as humans and possess about twice the amount of oxygen in the form of increased blood volume and oxygen capacity. However, rather than being able to submerge twice as long as humans, they can stay under five to ten times as long (see Plate 16). Other adaptations must also be working.

One of these additional adaptations is that of the marked slowing of the heartbeat during the period of submersion. This *bradycardia* is common to all diving, air-breathing vertebrates. The decrease in heartbeat is quite dramatic. For example, in Pacific bottle-nosed porpoises, *Tursiops truncatus,* the heartbeat under experimental conditions drops from about 90 beats/minute at the surface to about 20 beats/minute during a five-minute dive (Fig. 3.21).

Of more importance than slowing of the heartbeat, however, are two other adaptations. The first is that during a dive the circulatory system cuts off the blood supply to various organs and organ systems, including the muscles, digestive system, and kidneys. This cutoff has the effect of conserving the limited oxygen supply in the blood for those more sensitive and vital tissues such as the brain and central nervous system. Thus, a limited supply of oxygen is made to last longer by allowing it to be used only by selected organs. In this way the animal may remain submerged much longer than if it had to supply oxygen to all its tissues. The second adaptation is related to the first: It is that the muscular system and other organs are extremely tolerant of anaerobic conditions and continue to function when the blood flow is cut off. This results in the buildup of large amounts of lactic acid in the muscles during a dive. Lactic acid is the end product of anaerobic metabolism. The muscular system is able to continue functioning in the absence of a blood supply not only because it is very tolerant of lactic acid and anaerobic conditions, but also because of yet another adaptation. The muscular system of marine mammals is rich in an oxygen-containing compound called *myoglobin.* This compound has a structure very similar to hemoglobin, but it is better at storing oxygen. Thus, when the animal dives, the muscle blood supply is shut off. The muscles initially have a large supply of oxygen in the myoglobin, but as the dive continues, the oxygen of the myoglobin is depleted and the muscles continue to operate anaerobically, building up the product lactic acid. It now appears, however, that few dives are of a duration that exceeds the aerobic capacity of the animals.

When a dive has been completed and the animal returns to the surface and begins breathing again, the blood is recharged with oxygen, the blood supply is restored to the muscles and other organs, and oxygen replenishes the myoglobin. The blood picks up the accumulated lactic acid, where it is eliminated by oxidation. At this point, the animal is again ready to undertake another dive.

Marine mammals and birds generally have internal salt concentrations in their blood and body fluids that are lower than concentrations of surrounding seawater. This means that they face a potential osmoregulatory problem in that water has a tendency to pass from their bodies to the outside in order to equalize the salt concentrations between the inside and outside of the animal. The marine mammals and birds must compensate for this water loss either by drinking seawater or by obtaining water from their food. If they drink seawater, they gain an unwanted quantity of salt, which they then must get rid of in some way. Marine birds and turtles eliminate this excess salt through special salt-secreting glands in the region of the orbit of the eye. Marine mammals, however, do not have such

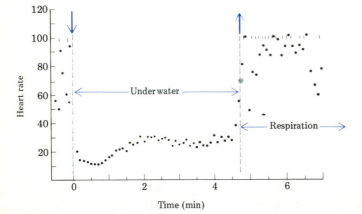

Figure 3.21 Change in heart rate in the bottle-nosed porpoise, *Tursiops truncatus,* during diving. Arrows indicate the beginning and end of the dive. (Reprinted by permission from Nature, Vol. 212, No. 407, p. 123. Copyright © 1966 Macmillan Journals Limited.)

glands, nor do they have sweat glands as do terrestrial mammals. The only organ left to them for elimination of salt is the kidney. Excess salt must be removed via the kidneys and washed out of the body with large amounts of water. The urine is very concentrated, and the large multi-lobular kidneys are very efficient at removing salts without losing much water.

ECOLOGY OF NEKTON

As we noted in the introduction, because of the difficulties of observation and experiment with natural populations of nekton, we know very little about the ecology of these forms. Most of our information concerns the feeding relation-ships and probable trophic links.

Feeding Ecology and Food Webs

Basically, all adult nekton are carnivores preying upon either the smaller plankton or other nek-ton. By far the largest number of nekton are predators on other nekton. The plankton feeders consume the larger zooplankton and include the fishes such as flying fish and sardines (Clupei-dae), but the best known are the large baleen whales. Baleen whales include all of the larger whales, such as the blue whale, the largest ani-mal that has existed on this planet. These tre-mendous animals have large mouths that are devoid of teeth; they have instead sheets of "baleen" or "whalebone," which hang down like a curtain from the roof of the mouth cavity. Baleen consists of a series of closely spaced parallel plates fringed on the free end. These plates form a sieve mechanism. In right and bowhead whales the mouth is very large with long, finely fringed baleen that permits them to sieve out small copepods and euphausiids (= krill) from large amounts of water while slow-ly moving forward. In the more streamlined ror-qual whales, which range in size from the 9-m minke to the 25-m blue, the feeding is somewhat different. These animals have relatively smaller mouths with short baleen. They can, however, open their mouths to a greater than 90° angle between the upper and lower jaw. These animals feed by lunging into masses of krill or schools of small fishes with widely open mouths and taking

in large volumes (up to 70 tons) of water and prey. The large volume distends the accordion-like furrows or pleats in the throat. The animals then contract the throat muscles and push the water out through the baleen while keeping the mouth partially closed. Prey are trapped against the baleen (Fig. 3.22). Baleen whales are not the only marine mammals to feed on krill; the crab-eater seal of the Antarctic does so also, but it filters out the krill on the cusps of its peculiarly shaped teeth (Fig. 3.23).

One exception to this plankton feeding among whales is the gray whale (*Eschrichtius robustus*), which is a bottom feeder. Gray whales dive down to the bottom and take in large quantities of sediment, which they subsequently filter through their baleen, retaining small inver-tebrates such as amphipods. Gray whale feeding leaves large, irregular depressions on the shallow seafloor, and the whales are important distur-bance forces of these areas.

Plankton-feeding fishes are varied and in-clude such common forms as flying fish, salmon, and the largest fishes, the whale sharks. As with the baleen whales, the dominant food organisms appear to be various crustaceans of the zoo-plankton, including euphausiids, copepods, and amphipods. The type of zooplankton taken var-ies among different areas of the ocean as well as with different seasons. In the case of the Pacific Ocean salmon, the diet varies considerably, not only with season, but also among different years. There is some evidence that this change in diet with years is related to the competition that exists for food among the several salmon species. Thus, when there are years of high abundance of certain of the salmon, the intensified competi-tion leads to an expanded range of food items taken and to changes in the composition of the diet. For example, Parin (1970) notes that in 1956, when pink salmon (*Oncorhynchus gor-buscha*) were low in numbers, 73.2 percent of the diet of chum salmon (*O. keta*) consisted of euphausiids, but in 1957, when pink salmon were abundant, chum consumed ctenophores and jellyfish (72.6 percent of the diet). It does not appear that planktivores are specialized on any one kind of plankton; rather, they are size selective.

Whereas few of the vertebrates of the epi-pelagic zone of the oceans are plankton feeders,

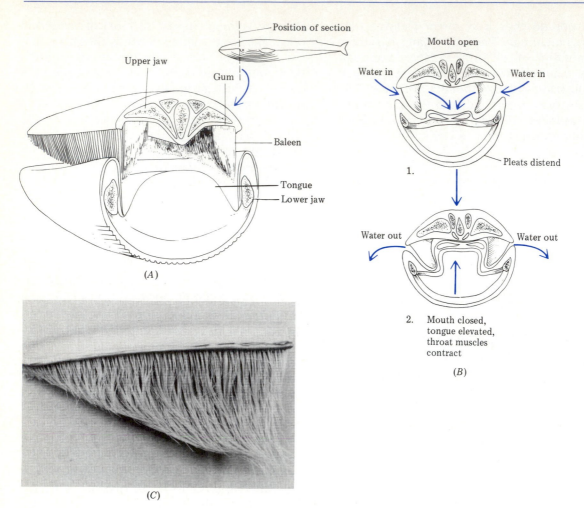

Figure 3.22 *(A)* Cross section of a baleen whale head to indicate the functional anatomy of the feeding apparatus. *(B)* The feeding process in a baleen whale: (1) mouth open and tongue depressed, water and plankton flow in; (2) mouth closed, tongue raised, water passes out through the baleen, which strains out the plankton. *(C)* Photograph of the baleen from a gray whale *(Eschrichtius robustus)*. (Modified from E. J. Slijper, Whales, Cornell University Press, Century Hutchinson. Photograph by the author.)

there are a large number of small fishes living in the mesopelagic zone (see Chapter 4) that are planktivores and that migrate into the epipelagic at night to feed. Thus, the competition for plankton may be greater than first anticipated, but our understanding of these relationships is poor at present. These same small fishes also form a food source for some of the carnivorous fishes of the epipelagic.

The nekton-feeding fishes, birds, and mammals dominate the open-ocean nekton. Their diet generally includes fishes, squids, or large crustaceans. The size of prey taken generally depends on the size of the predator, with the larger species taking progressively larger prey species. The largest carnivore in the oceans is the sperm whale (*Physeter macrocephalus*), which preys upon the largest squid, diving deeply to obtain them. Apparently, the sperm whales are the only animals preying upon the giant squid, *Architeuthis*. (The stomach of one sperm whale yielded a giant squid 10.5 m long and weighing 184 kg.) The second largest carnivore of the nekton is the killer whale (*Orcinus orca*), which

Figure 3.23 The teeth of the Antarctic crab-eater seal *(Lobodon carcinophagus)* used to filter euphausiids. (Photo courtesy of AIBS from September 1964 cover of *BioScience*. Used by permission of Dr. V. B. Scheffer.)

preys upon fishes, penguins, porpoises, seals, and sea lions (see Plate 14). These animals have also been known to attack the much larger baleen whales, which they do in packs of 3 to 40. In such attacks, the killer whales bite off pieces of the larger prey, as they do when feeding on smaller porpoises, seals, and birds. The capacity of these animals is truly enormous; one specimen, reported by Slijper (1976), was captured with parts of 13 porpoises and 14 seals in its stomach.

The smaller-toothed whales, seals, and sea lions tend to feed on various squids and fishes, which they capture and swallow whole. Both groups have been attacked by humans because of the belief that they feed heavily on commercially important fishes such as salmon and therefore compete with humans. Although they take commercially important fishes, it remains to be demonstrated whether such fishes constitute any significant part of the diet. The truly fish-eating whales include dolphins of the genera *Delphinus, Stenella, Tursiops,* and *Lagenorhynchus.* Examination of the stomachs of five species of delphinids revealed a diet composed 90 percent of mesopelagic lantern fish. These animals have a short, slender rostrum with many teeth. Those porpoises feeding on squid tend to have reduced numbers of teeth, but increased ridges in the palate to hold slippery squid. Included with squid eaters are the rare beaked whales (Ziphii-dae) and the pilot whales. The types of food and size ranges for most whales are summarized in Fig. 3.24.

Pelagic birds are generally more restricted in their feeding areas because they cannot dive as

deeply as the marine mammals. They appear to feed mainly in the surface waters. They consume various small fishes and squids. Since certain species often occur in tremendous numbers, it may be that they have a very significant effect on the fish and squid populations of the upper

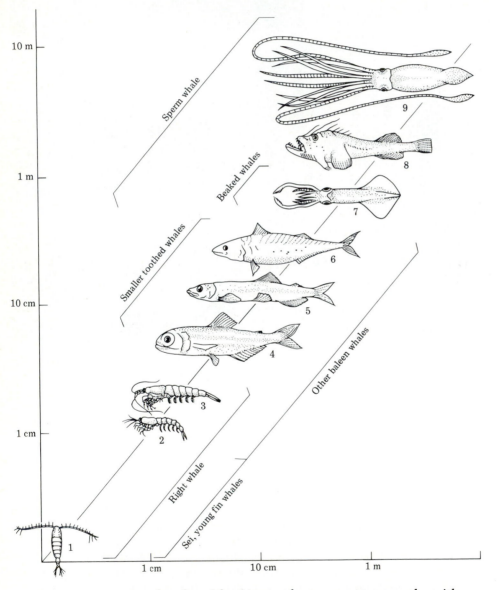

Figure 3.24 Size ranges of preferred food items of cetacean groups on a logarithmic scale: (1) calanoid copepods; (2, 3) various euphausiids; (4) myctophid (lantern) fish; (5) capelin; (6) mackerel; (7) onychoteuthid squid; (8) bathypelagic angler fish; (9) architeuthid squid, sporadically eaten by sperm whales. (From D. E. Gaskin, The ecology of whales and dolphins, Heinemann, 1982.)

layers, but we are ignorant of the details of this at present. They may also well be competing with various fishes and smaller marine mammals for the same food resource.

We know most about the food of various nektonic fishes; of these, the commercially important species such as tuna, albacore, and marlin are best known. They consume fishes, squids, and crustaceans.

Tunas have been intensively studied, and their food is extremely varied, with as many as 180 different food items reported for striped tuna (*Katsuwonus pelamis*). Part of this varied diet is due to the great areas over which they move, thus encountering different arrays of food items. Small tunas tend to feed on species in the surface layers of the epipelagic zone, whereas larger tunas obtain food from depths where they capture mesopelagic organisms. As with planktivores, the composition of the diet varies with season and location. Marlins and swordfish have diets similar to tunas'.

Nektonic sharks have a significant role in the feeding relationships of nekton, feeding on several different species of fishes, including tunas, marine birds, mammals, and squids. The largest carnivorous shark, the white shark, has the most varied diet, which includes not only fishes, birds, and mammals, but also other sharks. One was even found with remains of a basking shark in its stomach.

One of the most consistent features of feeding in nektonic fishes is the general lack of selectivity or specialization. Most fishes seem to feed on any food present of the appropriate size.

With the limited information now available, it is possible to outline the food webs that exist in the open epipelagic zones in the oceans of the world, but only in a very simplified and general way, as much more study is needed to define all of the links and their relative importance. These food webs are different among the tropical and polar oceans (see Fig. 3.25 on pp. 120–122). In the cold temperate oceans such as the North Pacific, the main net phytoplankton are diatoms, which are, in turn, consumed by various zooplankton. These serve as food for plankton-feeding fish, mainly salmon and migrating mesopelagic forms. Top carnivores include the sharks and marine birds and mammals.

In the Antarctic, primary production is also centered in the diatoms, and the foremost herbivores are copepods and euphausiids. These are consumed by baleen whales and certain fishes. The top-level carnivores are various marine mammals.

In the tropical waters, the primary producers are dinoflagellates and coccolithophores. The herbivores include a wide variety of zooplankton organisms. These, in turn, provide food for a number of planktivorous fishes such as flying fish at the surface and lantern fish from the mesopelagic. All serve as food for the larger, first-level predatory fishes and squids. This level of predators is preyed upon by larger predators such as marlin, swordfish, and sharks. Finally, at the top are the largest sharks, the white and mako, which attack tuna, swordfish, and marlin.

Although these are highly simplified food webs, which do not include all the newly discovered important nanoplankton producers and consumers (see pp. 79–82), they are included here to illustrate the differences between areas. It should be noted that there are marked differences between the tropics and temperate-polar areas. In the first place, the food webs are more complex in the tropics, with more links and more trophic levels. This is partly due to the greater number of species present in the tropics and the general absence in colder waters of the larger, swifter predaceous fishes (tuna, marlin). The second observation is that marine mammals and birds seem to play a larger role in the food webs of the polar regions than the tropics. Indeed, larger numbers of marine mammals occur in colder seas than in warmer seas.

Ecological Significance of Cetacea

Although there is considerable interest by people and the media in the plight of the world's whales and dolphins and their past and future decimation at the hands of humans, little scientific information has been brought to bear on the question of their ecological importance in either pelagic or inshore ecosystems. As indicated earlier in this chapter, the importance of many, if not most, nektonic organisms in the dynamics of

(A)

Figure 3.25 Food webs and trophic structure of the pelagic community in three different geographical areas. (A) Cold temperature waters. Algae: (1) diatoms. Herbivores: (2) euphausiids, (3) copepods, (4) pteropods. Filterers: (5) juvenile fish, (6) anchovy, (7) hyperiid amphipods, (8) lantern fish. Predators: (9) salmon, (10) squid, (11) lancetfish. Top predators: (12) mackerel shark, (13) porpoise, (14) seals and sea lions, (15) killer whale, (16) marine birds.

the open ocean is very hard to determine. Preliminary investigations, as reviewed by Katona and Whitehead (1988), suggest that although the productivity of cetaceans is low, the biomass is large enough to be significant on an ocean-wide basis. This significance can be appreciated by noting that it has been postulated by Kanwisher

and Ridgway (1983) that cetaceans probably consume more prey than the entire world's fisheries. If sustained, this suggests that the energy flow through the cetacean component of the open-ocean food webs may be extremely important. This is particularly true for areas such as the Antarctic seas, where Kanwisher and Ridg-

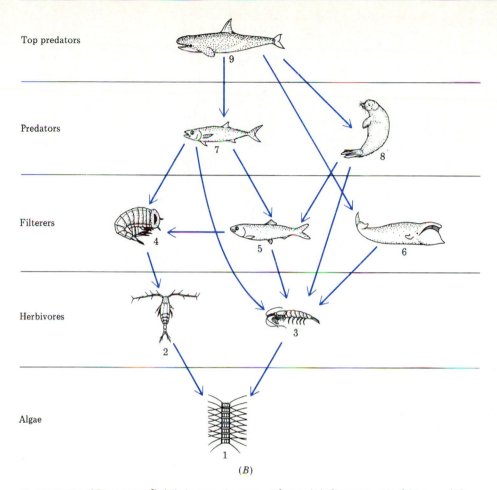

Figure 3.25 *(Continued) (B)* Antarctic seas. Algae: (1) diatoms. Herbivores: (2) copepods, (3) euphasiids. Filterers: (4) hyperiid amphipods, (5) planktivorous fish, (6) baleen whales. Predators: (7) predatory fish, (8) seals and sea lions. Top predators: (9) killer whale.

way (1983) estimated that the unexploited baleen whale population there consumed 190 million metric tons of krill (*Euphausia superba*) per year, more than two times the current world fisheries catch (see Chapter 11). Now that these Antarctic whales have been severely reduced in numbers (by perhaps 90 percent) by hunting, evidence to sustain their importance has come by looking at the results of this unparalleled overexploitation. Populations of Antarctic birds and pinnipeds increased by three times following the destruction of the whales. The reason for this population explosion is simply suggested to

be the availability of the excess krill not consumed by the whales.

Whales are important also in that their carcasses sink rapidly and provide a significant food resource for deep-sea benthic creatures (see following chapter). Gray whales disturb large areas of the subarctic Bering and Chukchi seas on a scale equivalent to large geological or meteorological forces (see Chapter 5). Finally, whales produce a considerable amount of the sound in the world's oceans and it is possible that such signals may be used as cues by other species. Clearly, while whale removal from the ocean

Figure 3.25 *(Continued) (C)* Tropical seas. Algae: (1) coccolithophores, (2) dinoflagellates. Herbivores: (3) euphausiids, (4) copepods, (5) shrimp. Filterers: (6) vertically migrating mesopelagic fishes, (7) flying fishes, (8) hyperiid amphipods, (9) lantern fish, (10) ocean sunfish. Predators: (11) mesopelagic fishes, (12) snake mackerel, (13) squid, (14) dolphin *(Coryphaena)*. Top predators: (15) tuna, (16) lancetfish, (17) marlin, (18) medium-sized sharks, (19) large sharks. (*A, B, C* modified from N. V. Parin, Ichthyofauna of the epipelagic zone, trans. from the Russian by the Israel Program for Scientific Translations, 1970.)

ecosystem would probably have little effect on the primary productivity or the total biomass of organisms, it would certainly alter some systems, perhaps more than we realize. Certainly, the richness of our human lives would be much decreased.

BIBLIOGRAPHY

Aleyev, Y. G. 1977. Nekton. Dr. W. Junk BV., The Hague. 435 pp.

Andersen, H. T. (ed.). 1969. The biology of marine mammals. Academic Press, N.Y. 511 pp.

Brooks, W. S. 1917. Notes on some Falkland Island birds. Bull. Mus. Comp. Zool. 61:135–160.

Carey, F. G. 1973. Fishes with warm bodies. Sci. Amer. 228(2):36–44.

Carr, A. 1965. The navigation of the green turtle. Sci. Amer. 212(5):79–86.

Clarke, M. R. 1979. The head of the sperm whale. Sci. Amer. 240(1):128–141.

Denton, E., and N. Marshall. 1958. The buoyancy of bathypelagic fishes without a gas-filled swim bladder. J. Mar. Biol. Assoc. 37(3):753–767.

Dunson, W. A. (ed.). 1975. The biology of sea snakes. University Park Press, Baltimore. 530 pp.

Elsner, R. 1969. Cardiovascular adjustments to diving, pp. 117–145. *In:* Andersen, H. T. (ed.). The biology of marine mammals. Academic Press, N.Y.

Gaskin, D. E. 1982. The ecology of whales and dolphins. Heinemann, London. 459 pp.

Hardy, A. 1965. The open sea: Its natural history. Vol. II, Fish and fisheries. Houghton Mifflin, Boston. 322 pp.

Kalmijin, A. J. 1971. The electric sense of sharks and rays. J. Exp. Biol. 55:371–383.

Kanwisher, J. W., and S. H. Ridgway. 1983. The physiological ecology of whales and porpoises. Sci. Amer. 248(6):110–121.

Katona, S., and H. Whitehead. 1988. Are cetaceans ecologically important? Oceanog. Mar. Biol. Ann. Rev. 26:553–568.

Kellogg, W. N. 1958. Echo ranging in the porpoise. Science 128:982–988.

Kirschvink, J. L. 1990. Geomagnetic sensitivity in cetaceans: An update with live stranding records in the United States, pp. 639–649. *In:* Thomas, J. A., and R. A. Kastelein (eds.). Sensory abilities of cetaceans, laboratory and field evidence. Plenum Press, N.Y.

Klinowska, M. 1990. Geomagnetic orientation in cetaceans: Behavioural evidence, pp. 651–663. *In:* Thomas, J. A., and R. A. Kastelein (eds.). Sensory abilities of cetaceans, laboratory and field evidence. Plenum Press, N.Y.

Kooyman, G. L., and R. J. Harrison. 1971. Diving in marine mammals. Oxford biology reader no. 6.

Norris, K. S., J. H. Prescott, P. V. Asa-Dorian, and P. Perkins. 1961. An experimental demonstration of echo locating behavior in the porpoise, *Tursiops truncatus* (Montagu). Biol. Bull. 120:163–176.

Norris, K. S. 1969. The echolocation of marine mammals, pp. 391–423. *In:* Andersen, H. T. (ed.). The biology of marine mammals. Academic Press, N.Y.

Parin, N. V. 1970. Ichthyofauna of the epipelagic zone. Jerusalem, Israel Program for Scientific Translations. 206 pp.

Scheffer, V. B. 1976. A natural history of marine mammals. Scribner, N.Y. 157 pp.

Scholander, P. F. 1940. Experimental investigations on the respiratory function in diving mammals and birds. Hvalradets Skrifter Norske Videnskaps-Akad., Oslo 22:1.

Slijper, E. J. 1976. Whales and dolphins. University of Michigan Press, Ann Arbor. 170 pp.

Walker, M. M., J. L. Kirschvink, S-B. R. Chang, and A. E. Dizon. 1984. A candidate magnetic sense organ in the yellowfin tuna, *Thunnus albacares*. Science 224:751–753.

Würsig, B. 1979. Dolphins. Sci. Amer. 240(3):136–148.

Würsig, B. 1989. Cetaceans. Science 244:1550–1557.

Deep-Sea Biology

*B*y far the most extensive habitat on the planet inhabited by living organisms consists of the permanently cold, dark waters of the deep oceans and the associated bottom. The shallow waters fringing the continents and islands of the world make up less than 10 percent of the total area of the world's oceans, and the lighted upper waters of all oceans are an even smaller fraction of the total volume of ocean living space. Thus, of the 70 percent of the planet's surface that is covered with water, perhaps 85 percent of the area and 90 percent of the volume constitute the dark, cold area we call the deep sea.

Although this area is the largest habitat on earth, its biology is the least known and explored. This is primarily due to the difficulty of sampling this remote area and to the fact that, until recently, it was virtually inaccessible to humans. The recent advent of deep manned and remote-operated submersible vehicles has now allowed much of this vast area to be observed by scientists.

Although this huge area is unlikely to be sampled, much less observed, by most users of this book, because of the very fact that it is such a large part of the ocean, which is still a frontier where new and exciting discoveries are being made, and because it may become increasingly important as a source of materials and as a dumping site for human wastes, it is well to consider it.

The object of this chapter is to outline our present understanding of the basic principles concerning adaptations and organization of life forms in this vast and intriguing area.

ZONATION

When we refer to the "deep sea" in this chapter, we mean that part of the marine environment that lies below the level of effective light penetration for phytoplankton photosynthesis in the open ocean and deeper than the depth of the continental shelves (> 200 m). The upper part of the deep sea retains some light intensity and has been called the *disphotic zone* (Fig. 4.1). The disphotic zone is a transition area. Below it, the entire area is permanently dark. One term for this area is the *aphotic* zone as opposed to the lighted, or *photic,* zone, where all primary production occurs. In tropical waters, the start of the aphotic zone is deeper (~600 m), whereas in temperate waters, it is shallower (~100 m).

Several schemes of zonation for the deep sea have been proposed over the years and reviewed by Menzies, George, and Rowe (1973), but none has been universally acceptable to all scientists working in the area. There are a number of reasons for this, but the primary one is the lack of sufficient ecological information. Hence, most schemes of zonation have been produced by simply dividing the water column into discrete zones based on depth, or temperature changes, or both. Another approach has been to divide the deep ocean on the basis of species abundance, distribution, or associations. A problem here is that certain species are distributed at different depths in different ocean basins, de-

pending on special hydrological or ecological characteristics of the region. Similarly, the changes in temperature occur at different depths in different areas. On the other hand, kinds of species may be better for indicating "zones" than the associated environmental parameters. We shall follow the zonation scheme outlined in Chapter 1 and illustrated in Fig. 1.20. The relevant deep-sea zones are listed in Table 4.1 and briefly reviewed here.

A primary division that can be made in the deep sea is into benthic (associated with the bottom) and pelagic (associated with the open water) zones. Since the physical environment differs between them, these two areas are inhabited by very different associations of organisms. We know perhaps more about the benthic deep-sea fauna than about the pelagic fauna. The benthic fauna, as noted in Chapter 1, can be subdivided into those that occupy a *bathyal* zone on the continental slope and those living in a large *abyssal* zone constituting most of the deep-sea floor. The animals of the trenches live in a separate *hadal* (ultra-abyssal) zone, but little is known of their biology.

In the pelagic division, there is an upper zone immediately below the photic zone, which contains animals, many of which migrate vertically into the photic zone at night. This is variously called the *twilight, disphotic,* or *mesopelagic* zone. This zone is inhabited by many species of animals with well-developed eyes and a great

Figure 4.1 Schematic diagram to show the penetration of sunlight into the clearest ocean water and into clear coastal water in relation to minimum intensity values for some biological light receptors. (From T. R. Parsons, M. Takahashi, and B. Hargrave, Biological oceanographic processes, 3rd ed., Pergamon, 1984.)

Table 4.1   DEEP-SEA FAUNAL ZONES

Light	Pelagic zones	Depth range	Benthic zones	Depth range
Present (photic)	Epipelagic or photic	0–200 m	Continental shelf or sublittoral	0–200 m
Present (disphotic)	Mesopelagic	200–1,000 m (?) ⎫	Bathyal	200–4,000 m (?)
Absent (aphotic)	Bathypelagic	(?) 1,000–4,000 m (?) ⎭		
	Abyssalpelagic	(?) 4,000–6,000 m (?)	Abyssal	4,000–6,000 m (?)
	Hadalpelagic	6,000–10,000 m	Hadal	6,000–10,000 m

Note: (?) = Boundary uncertain.

Source: After Hedgpeth, 1957.

variety of light organs. The dominant fishes here are black, and the crustaceans are red. Because this is the most accessible of the deep-sea pelagic zones, we know most about it. The numbers of organisms appear to be the highest of all the deep-sea pelagic zones. This zone extends down to about 700 to 1000 m, the depth varying with location, clarity of water, and other factors.

Below the mesopelagic, it is much more difficult to establish any pelagic zones that are universally acceptable. The region between the lower limit of the mesopelagic and the upper limit of the deep trenches (taken to be 6000 m) has been divided into two zones by Hedgpeth (1957): an upper *bathypelagic* and a lower *abyssalpelagic,* a division continued here. The boundaries of these zones are very uncertain and vary from one author to another. These areas are characterized by low numbers of species and individuals, considerably fewer than observed in the mesopelagic. The organisms tend to be white or colorless, with reduced eyes and bioluminescent organs.

The open water of the trenches is often called the *hadalpelagic* zone. It appears set off from the other areas, and we are so ignorant of it that we do not know whether a separate pelagic community exists in these areas.

SAMPLING THE DEEP SEA

In the preceding section we alluded to the problems of sampling and obtaining information about the deep sea. Before continuing this discussion, it is useful to indicate briefly how our information is obtained from the deep sea and why it is difficult to obtain and often biased. An analogy has been made that characterizing the deep sea and its communities using the gear that we have traditionally used is like characterizing the terrestrial communities using samples taken with a butterfly net towed behind an airplane! Exaggerated perhaps, but close enough to have a ring of truth.

The traditional types of gear used to sample the deep-sea benthos include various types of nets, grabs, and dredges of similar design to those used in shallow water. Representative types are illustrated in Fig. 4.2. Sampling of pelagic animals is done with several types of midwater trawls (Fig. 4.3). Because the density of organisms is so low in pelagic areas below the mesopelagic, these midwater nets must have a large gape and sieve a lot of water to obtain any number of animals and to reduce the problem of simple avoidance of nets by the animals.

The main problem in sampling the deep sea, pelagic or benthic, is that the deeper the sampling, the more cable necessary to get down to depth and the longer the time needed to make a single haul. Furthermore, as more and more cable is played out in a deep haul, the greater the weight and the greater the chance for snarling the cable or otherwise fouling the gear. In order that most nets towed on the bottom fish correctly, an amount of line equaling perhaps two to three times the depth must be played out. If one were trawling at only medium depths of, say, 4000 m, this would be about 8000–12,000 m of cable, or 8–12 km (= 5–7 miles). Not only do

(A)

(B)

Figure 4.2 Some types of deep-sea bottom sampling gear. *(A)* Epibenthic sled. *(B)* Anchor dredge. Both of these samplers are designed to work with either side down. *(A* from R. J. Menzies, R. Y. George, and G. T. Rowe, The abyssal environment and ecology of the world oceans, Wiley, 1973. *B* from J. Nybakken, Readings in marine ecology, Harper & Row, 1971.

few vessels have the capability of carrying such amounts of cable, but it takes a considerable amount of time to feed that amount out and then retrieve it. Twenty-four hours may be required to make one trawl in the trenches. One can thus appreciate why the number of samples from deep water is not great.

Still other problems plague the deep-sea biologist. In shallow waters of the continental shelf it is possible to ascertain whether the trawl or dredge is fishing on the bottom by "feeling" the vibrations on the cable. In the deep sea, so much cable is out that the weight of it far exceeds the drag of dredge or trawl, and vibration in the cable does not reflect action of the sampling device. Failure of the gear to contact the bottom or to fish correctly has been suggested by Menzies (1964) as responsible for up to 50 percent of

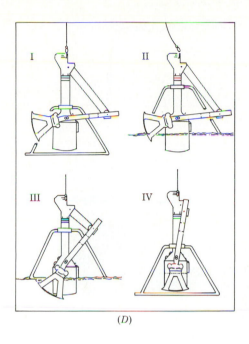

(D)

Figure 4.2 *(continued)* *(C)* Box core. *(D)* Sequence showing the box core taking a sample. C, D from J. D. Gage and P. A. Tyler, Deep sea biology, Cambridge University Press, 1991.)

deep-sea sample failures. Finally, with so much wire played out, the chance for snarling, knotting, or wrapping it up in the gear is considerably increased. Best results are obtained where depth-indicating devices, such as an acoustical "pinger," with shipboard readout are used with the sampling device so that the gear can be better controlled.

A major problem in sampling the deep-sea benthos is the basic qualitative nature of most dredging devices. They sample an unknown area of the bottom. Because densities of species and individuals are difficult or impossible to estimate from these samples, comparative studies of deep and shallow water have been difficult.

Fortunately, recent investigations in the deep sea have employed large box corers, which greatly overcome this problem by sampling a known bottom area (Fig. 4.2C).

Finally, in sampling the midwater animals, there is the problem of net avoidance. Trawling in deep water usually means that the ship must travel relatively slowly because of the great weight of the cable it drags. This means that many fast-swimming pelagic animals may simply avoid the net by swimming out of its path. If this happens to any great extent, then characterizing the communities from such net hauls gives a misleading idea of what is in fact there.

Figure 4.3 Rigging a large midwater trawl for fishing. (Courtesy of Dr. Greg Cailliet.)

In recent years, particularly since 1970, new instrumentation has permitted deep-sea biologists to consider aspects of the ecology of deep-sea animals that were not possible in earlier years. Deep-sea cameras of various designs have done much to increase our understanding of the real state of deep-sea communities, particularly bottom communities. The advent of deep manned and remotely operated submersibles with various devices for capturing organisms including pressure retrieval systems has further advanced our understanding of these areas by permitting actual observation and subsequent retrieval of living organisms for experimental studies (Figs. 4.4, 4.5).

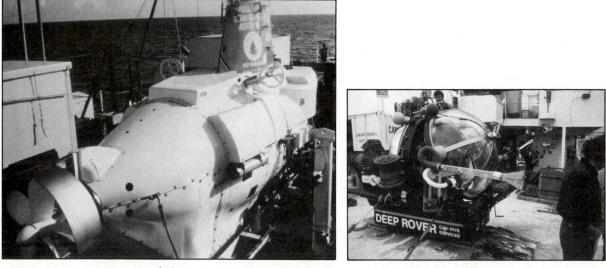

(*A*)   (*B*)

Figure 4.4 (*A*) The deep submersible *Alvin*. (*B*) The deep submersible *Deep Rover*. Note the suction sampler on the right side of the vehicle. (*A*, courtesy of Dr. H. Mullins. *B*, courtesy of T. Kieckhefer.)

Figure 4.5 Some characteristic deep-water benthic animals from the bathyal zone of the Gay Head–Bermuda transect in the Atlantic Ocean. (1) *Hygrosoma petersii* (echinoid). (2) *Phormosoma placenta* (echinoid). (3) *Geryon quinquedens* (decapod crustacean). (4) *Aldrovandia affinis* (fish). (5) *Ophiomusium lymani* (brittle star). (6) *Mesothuria lactea* (sea cucumber). (Photos courtesy of Dr. Fred Grassle; reprinted with permission from J. F. Grassle et al., Deep-Sea Res., Vol. 22, Pergamon Journals Ltd., 1975.)

ENVIRONMENTAL CHARACTERISTICS

Before considering the life in the deep ocean, it is necessary to establish the particular physical and chemical conditions under which the organisms must function. Perhaps the major point to be emphasized in considering these factors is that, at any given level or position in the deep ocean, these factors remain remarkably constant throughout long periods of time.

Light

In the deep ocean, the entire area is without light, except at the very upper parts of the mesopelagic, where there is some light that penetrates. Since this area is either dark at all times or has such extremely low light levels, photosynthesis is not possible, precluding any plant-based primary productivity. What light is present in this dark world is usually produced by the animals themselves (see pp. 144–146). This lack of light means that the organisms have to rely on senses other than vision to find food and mates and to maintain various interspecific and intraspecific associations.

Pressure

Of all the environmental factors acting on deep-sea organisms, pressure shows the greatest range. Pressure increases 1 atmosphere (14.7 lb/in.[2] or 1 kg/cm[2]) for each 10 m in depth. Since the deep sea varies from a few hundred meters down to the bottom of the trenches at more than 10,000 m, the range of pressure is from 20 to more than 1000 atm. Most of the deep sea is under pressure between 200 to 600 atm. In no other marine environment does pressure exhibit such great range or play such a potential role in the distribution of organisms.

Until recently, we lacked detailed information on the direct effects of pressure on most deep-sea organisms. This was primarily due to the fact that virtually all organisms trawled up from the deep sea arrived at the surface in a dead or dying state. Only in recent years, with the advent of traps that incorporate a special pressure-maintaining chamber, have undamaged larger metazoan animals been retrieved from the deep sea in good condition. Some of these have been maintained for experimental purposes, and we are obtaining more knowledge of the real biological effects of pressure.

That pressure must have a significant effect on deep-sea organisms can be extrapolated from experiments done on one group of organisms successfully retrieved alive from the deep sea. Bacteria have been retrieved alive and cultured from the deepest areas of the ocean. Figure 4.6 shows the results of culturing a deep-sea bacterium under different pressure regimes. The fact that bacteria from the deeper areas virtually cease growth and reproduction at lower pressures, but grow and reproduce actively at higher pressures corresponding to their natural environment, strongly suggests that there are special adaptations to pressure and that they can profoundly affect organisms. Since metazoan animals are anatomically and physiologically more complex than bacteria, it seems highly probable that multicellular animals will evince an even stricter pressure-dependent physiology.

Studies have been done by Siebenaller and Somero (1978) on the enzyme systems of two closely related fishes living at different depths. These studies have demonstrated that differ-

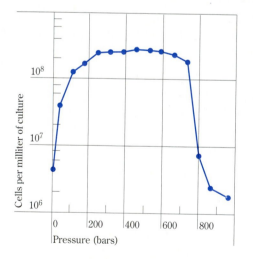

Figure 4.6 Graph of the amount of growth of a deep-sea bacterium at different pressures. (From A. A. Yayanos, A. S. Dietz, and R. Van Boxtel. Isolation of a deep sea barophilic bacterium and some of its growth characteristics. Science 205(4408): 808–809, 1979. Copyright © 1979 by the American Association for the Advancement of Science.)

ences in hydrostatic pressure as small as 100 atm or less are sufficient to alter markedly the functional properties of enzymes—namely, their ability to bind to the appropriate substrate and the rate at which the reaction proceeds. Such changes are comparable to the changes in enzyme activity caused by temperature differences. These results strongly suggest that hydrostatic pressure plays a major role in organisms' adaptation to the deep-sea environment.

More recently Somero (1982), in a review of physiological and biochemical adaptations of deep-sea fishes, has pointed out certain other differences in muscle enzymes between shallow- and deep-dwelling fishes. Deep-dwelling fishes tend to have muscle enzymes that are less efficient and are present in the cells in low concentrations. The combination of these two characteristics probably results in the low metabolic rates as well as the sluggish movements of deep-sea fishes. Most of these fishes are "float-and-wait" predators. Unlike the muscle enzymes, however, the enzymes of the nervous system seem not to be inefficient or reduced.

A great number of laboratory studies on various kinds of cells from protozoa to mammalian have demonstrated that pressure markedly affects the morphology of the cells, including ability to form mitotic spindles and to undergo mitosis. Zimmerman and Zimmerman (1972) report that amoebas tend to lose their pseudopods and ball up, and *Tetrahymena* loses its ability to locomote with cilia. Pressure also has a significant effect on certain physiological and biochemical processes such as typical muscle physiology. Perhaps the most dramatic effect of pressure is its effect on macromolecules such as proteins. Protein synthesis and function are markedly affected by pressure in laboratory situations, usually adversely. Since proteins are vital in living systems as enzymes and as major building blocks, it would seem that deep-water animals must have some special adaptation with respect to protein structure and function in order to survive.

The body of laboratory studies of the observed definite effect of pressure on shallow-water animals, coupled with the known effects of pressure on cells and macromolecules (= protein), and the recent experimental evidence from deep-sea animals themselves are strong indica-

tions that the physiology of deep-sea animals is profoundly affected by pressure and that pressure may be of considerable significance in zonal patterns of distribution.

There is another effect of pressure that affects organisms that secrete calcium carbonate shells. As pressure increases and temperature decreases, the solubility of calcium carbonate increases, as does its tendency to dissociate. This means that as one goes deeper into the ocean, it becomes more and more difficult for organisms to lay down shells. In the deep sea the number of shell-forming species decreases, and the remaining ones tend to have thin shells.

Salinity

Below the first few hundred meters in the world's oceans, the salinity is found to be remarkably constant throughout the depths (Fig. 4.7). There are some minor differences in salinity, but none that can be considered ecologically significant.

Temperature

The area of greatest and most rapid temperature change with depth in the oceans is the transition zone between the surface waters and the deep waters, the area known as the *thermocline.* Thermoclines vary in thickness from a few hundred meters to nearly a thousand (Fig. 4.8). Below the thermocline, the water mass of the deep ocean is cold and far more homogeneous. Thermoclines are strongest in the tropics, where the temperature of the epipelagic zone is usually above 20°C. From the base of the epipelagic the temperature drops over several hundred meters to 5 or 6°C at 1000 meters. It continues to decrease toward the bottom, but the rate of change is much slower (Fig. 4.8). Below 3000–4000 m, the water is essentially isothermal. What is ecologically significant is that, at any given depth, the temperature is practically unvarying over long periods of time. There are no seasonal temperature changes, nor are there any annual changes. Perhaps nowhere on earth is there another habitat with such a constant temperature. An isolated anomaly in this picture is the hydrothermal vents. Temperatures at these vents are often considerably above ambient (see pp. 153–158).

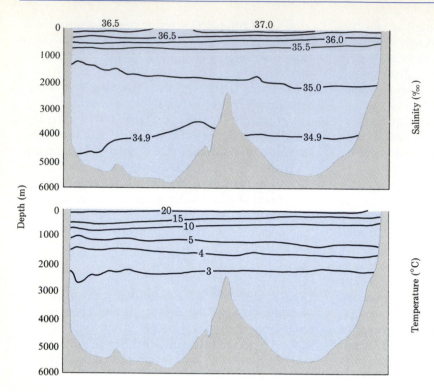

Figure 4.7 Salinity and temperature changes across the North Atlantic Ocean with respect to depth. (After D. A. Ross, Introduction to oceanography, copyright © 1982, p. 305. Adapted by permission of Prentice-Hall, Inc., Englewood Cliffs, N. J.; from F. C. Fuglister, Atlantic Ocean Atlas.)

Oxygen

Despite the fact that the deep ocean waters lie far removed from a source of oxygen replenishment, either by interaction with the atmosphere or through production by plants, there are essentially no abyssal or hadal areas that lack sufficient oxygen to support life. The few exceptions include the Carioca trench off Venezuela in the Caribbean and the Santa Barbara basin off California. They are anaerobic at the bottom. The oxygen present in the deep-water masses entered the water when this now deep-water mass was at the surface. Virtually all of the water of the deep sea has its origin at the surface in the Arctic or Antarctic areas (see Chapter 1, pp. 11–16). Here, the oxygen-rich cold water then sinks and flows north or south to make up the deep water of the world's oceans. Since it no longer gains oxygen after sinking, the reason that these deep masses are not depleted of oxygen from respiration of deep-water organisms is likely due to the fact that the density of deep-water organisms is so low and they have such low metabolic rates that together they are insufficient to deplete the resident oxygen supply. That this situation is true is supported by the fact that the oxygen concentration declines in the 20 m or so just

Figure 4.8 The thermocline layer in the ocean. (From D. E. Ingmanson and W. J. Wallace, Oceanology: An introduction, Wadsworth, 1973.)

above the bottom in the deep sea. It is also near the bottom that the most dense concentrations of organisms occur in the deep sea.

Another peculiar feature of oxygen concentration in the deep sea is the presence of the aforementioned *oxygen minimum* zone between 500 and 1000 m (Fig. 1.3). Oxygen values are higher both above and below this zone. In the oxygen minimum zone, oxygen values may fall to less than 0.5 ml/liter. The occurrence of this zone is mainly the result of respiration of organisms, coupled with the lack of interchange of the water mass with more oxygen-rich water. The reason that an oxygen minimum zone occurs at 500–1000 m and not deeper is that the numbers of animals in deeper waters are so low that they never deplete the available oxygen, whereas at 500–1000 m, organisms are abundant. Above 500 m, oxygen depletion does not occur, even though the animal biomass is high, because there is constant replenishment from air and by plants.

Food

The deep sea is removed from the photosynthetic zone and has no primary production except for certain areas where chemosynthetic bacteria are found (see pp. 153–158). All organisms living in the deep sea are thus dependent on food that ultimately is produced elsewhere where photosynthesis is possible and subsequently transported into the deep sea. The deep sea is therefore unique among the world ecosystems in that it has no indigenous primary productivity.

Potential food must come into the deep sea by sinking from the surface waters. Since the populations of organisms are greatest in the upper layers of the oceans, the chance that any food particle will be able to sink through all these voracious animals without being consumed or decaying is small. As a result, few food particles are available to the animals of the deep sea. This paucity of food is probably the reason that the density of deep-sea animals is very low and the size of most is small. Without sufficient energy in the form of food, large numbers of organisms cannot be sustained.

The probability that a particle of food will decay or be consumed during sinking increases with increasing depth, since there is more time and there are more organisms to consume it.

Thus, the deeper the organisms, the less the food available to them.

Perhaps a major portion of the particles sinking from the photic zone are particles of material that are not directly suitable as food. An example would be fecal pellets and shed chitinous exoskeletons of crustaceans. Most organisms cannot digest chitin. These materials, however, are acted upon by bacteria during descent into the deep sea and, after reaching the bottom, are converted into suitable food in the form of bacterial protoplasm. Because the residence time for a food particle is longer on the bottom than in the water column, there is an increase in bacteria in the bottom oozes of deep sea, which thus allows for increased numbers of larger organisms. Densities of the benthic infaunal organisms may exceed that of pelagic organisms at the same depth. This also explains the reduction in oxygen in the near-bottom water.

The amount of food available to the deep sea is correlated with the amount of primary productivity in the surface waters above or else with the proximity to a secondary source, such as organic debris from terrestrial habitats. Deep-sea areas under productive surface waters and near islands or continents have more food than deep waters under low surface productivity or far from land.

Several types of food sources are available in the deep sea. Directly available foods include those deep-sea organisms that spend their early or larval stages in the lighted upper waters and then migrate into the depths, where they furnish food to the predators. An example is the group of deep-sea angler fishes whose young can be found in the upper waters of the ocean but the adults of which are all restricted to the deep sea. Another directly available food source is the large dead bodies of marine mammals and fishes that may sink quickly before all readily available flesh is consumed (Fig. 4.9). This would seem to be a particularly unpredictable type of food, but Isaacs and Hessler of Scripps Institution of Oceanography have shown, by placing baited cans with cameras attached into the deep sea, that food is quickly detected and brings in consumers from great distances (Fig. 4.10). That this unlikely food source must be reliable is also supported by the fact that several groups of deep-sea gastropod mollusks have also evolved that live on whale and shark remains.

Figure 4.9 Food sources to the deep sea.

Indirect food sources include indigestible animal and plant remains, which first must be acted on by bacteria. Examples are chitin and wood and cellulose of land plants. Other potential sources are colloidal or dissolved organic mate-rial in the water and gelatinous plankton matter (= marine snow). At present, the relative importance of these sources cannot be evaluated.

We can summarize by saying that food is very scarce in the deep sea, when compared to other

(A)   (B)

Figure 4.10 Two photographs showing the rapid attraction of deep-sea animals to bait and its subsequent consumption. (A) At the time of reaching the bottom. (B) Six hours and 40 minutes after reaching the bottom, showing consumption of bait by the amphipod *Hirondella gigas* (at 9605 m in the Philippine trench). (Photos courtesy of Dr. Robert Hessler, Scripps Institution of Oceanography.)

areas of the oceans, and that it decreases with depth and distance from land. This food scarcity is responsible for the low densities and small sizes of deep-sea organisms.

ADAPTATIONS OF DEEP-SEA ORGANISMS

Correlated with the conditions of the environment in the deep sea, the organisms that live there possess a number of adaptations. Because our ability to experiment with living metazoan animals from the deep sea has been limited up to the present, we have little direct evidence of the function of many adaptations and must offer educated guesses based on the conditions under which we know they live. We also know little about physiological and biochemical adaptations at present.

One adaptation that can be observed, even in those organisms from the upper mesopelagic, is that of *color*. Mesopelagic fishes in particular tend to be either silvery gray or a deep black (Plates 17, 18). They are not countershaded, as are epipelagic fishes. The mesopelagic inverte-

brates, on the other hand, may be either purple or bright red in color. Mesopelagic jellyfish are often a deep, dark purple, whereas the crustaceans such as copepods, mysids, and shrimps are brilliant red (see Plates 20, 21, 23, 24). Since these organisms live in waters that are virtually unlighted, organisms that are black will be invisible. But why the red color? Young (1983) has suggested that since most bioluminescent light is blue in color, the red pigmentation would tend to protect these animals from the revealing rays of the bioluminescent flashes that are used by predators.

Organisms living even deeper, in the abyssal and bathyal zones, are often colorless or a dirty white. They seem to lack pigment. This is particularly true for animals dwelling on the bottom. Fishes, however, may be black at all depths.

Another adaptation seen particularly in mesopelagic and upper bathypelagic fishes is the presence of *large eyes*. The eyes in these fishes are much larger relative to body size than those in fishes of the lighted epipelagic zone (Fig. 4.11). Usually, the presence of large eyes is correlated with the presence of light organs (see pp. 144–146). These fishes dwell in the upper-

(A)

(B)

(C)

Figure 4.11 Large eyes in three mesopelagic fishes of the family Myctophidae. (A) *Myctophum punctatum.* (B) *Lampanyctus elongatus.* (C) *Diaphus metopoclampus.* (After N. B. Marshall, Aspects of deep-sea biology, Philosophical Library, 1954.)

Figure 4.12 Fish with reduced eyes, a gulper eel of the genus *Saccopharynx*. (Photo courtesy of Michael Kelly.)

Figure 4.13 A tubular-eyed fish of the genus *Argyropelecus*. (Photo courtesy of Michael Kelly.)

Figure 4.14 The dimorphic eyes of the squid *Histioteuthis*. (Photo courtesy of Dr. C. Recksiek.)

Figure 4.15 *Linophryne*, a genus of fish with a very large mouth. (Photo courtesy of Michael Kelly.)

most parts of the deep sea, where some small amounts of light may penetrate. Many also migrate vertically into the epipelagic zone at night. Thus, the large eyes give them maximum light-collecting abilities in an area in which light intensities are very low. The large eyes are also presumably needed to detect the low light intensities of the light organs. Increased surface area is only one visual adaptation. These fishes also have enhanced "twilight vision" derived from the pigment rhodopsin and an increase in the density of rods in the retinal area.

Fishes dwelling in the deepest parts of the ocean (abyssalpelagic, hadalpelagic) show a different trend. Many of these have very small eyes or lack eyes entirely (Fig. 4.12). Because they dwell in permanent darkness, eyes are not necessary. Generally, fishes in depths of 2000 m and above have eyes, often large; below 2000 m, eyes are small, degenerate, or lost. Often, bottom dwellers have no eyes.

Yet another adaptation in fishes is *tubular eyes* (Fig. 4.13; Plate 18). Fishes with peculiar tubular eyes are found in several families. They give the fishes an extremely bizarre appearance (Fig. 4.13). In these fishes the eye is a short black cylinder topped with a hemispherical, translucent lens. Each eye has two retinas, one at the base of the cylinder and one on the wall of the cylinder. Apparently, these eyes function such that the main retina at the base of the tube focuses on nearby objects and that on the wall on distance objects. However, just why they should be developed in the deep sea is not clear.

Among the invertebrates, certain squids of the family Histioteuthidae exhibit another peculiar adaptation in that they have one eye much larger than the other (Fig. 4.14). Such squids also display numerous light-producing organs called photophores. Work by Young (1975) has demonstrated that these squids, which live in the mesopelagic at depths between 500 and 700 m, usually orient with the arms down and with the large eye directed vertically upward, while the small one is oriented downward. In this position, presumably the large eye gathers the faint light downwelling from the surface, while the smaller eye responds to light from photophores. It has been further demonstrated with another midwater squid, *Abraliopsis,* that the squid responds to downwelling light by turning on its photo-phores just enough to match the downwelling light and hence becomes invisible when viewed from below (i.e., prevents silhouetting).

Food scarcity in the deep sea seems to be the reason for another series of adaptations. Most deep-sea fishes have large mouths, larger relative to their size than fishes of other areas (Fig. 4.15; Plates 17, 26). Furthermore, the mouth is often beset with long teeth recurved to the throat, an adaptation to ensure that whatever is caught does not escape. Even more bizarre is that, in some fishes, the mouth and skull are so hinged that the animals can open the mouths much wider than their own bodies (Fig. 4.16; Plate 26), enabling them to swallow prey larger than themselves (Fig. 4.17). Food is not passed up simply because of size! Still other fishes, such as the angler fish (Ceratoidea), have responded to the scarce food by establishing themselves as traps, using a highly modified part of the dorsal fin as a lure (Plate 27). In this case, the lure is a luminescent organ. Other lures are found in many other fishes, such as the stomiatoids, and occur on barbels attached to the chin (Fig. 4.18). One curious adaptation reported by McAllister (1961) is that many carnivorous fishes have black linings to their coeloms so that bioluminescence from ingested prey will not shine out and attract predators to them!

Because food is scarce and the resulting density of organisms is very low, there is a potential problem of finding a mate for reproductive purposes in this vast, dark area. One adaptation to this problem is found in angler fish, *Ceratias,* in which the large individuals are all female. The males are very tiny and are parasitic, attached to the body of the female (Fig. 4.19). Since the males are always present to provide the sperm, the search for a mate is negated. Of course, the male has first to find the female, and this is presumably accomplished via olfaction.

Body size of deep-sea organisms represents a peculiar paradox. Since food is scarce, it might be anticipated that most organisms would be small in size. Indeed, when considering fish, this appears to be the case. Most fishes captured in the deep sea are quite small, smaller than corresponding relatives in shallow water. As far as we now know, the deep sea contains few large fishes, though a few large ones have been photographed

Figure 4.16 The hinged jaw in *Chauloidus*. *(A)* Anatomical construction. *(B)* Operation during swallowing. (*A*, after A. Hardy, The open sea, Houghton Mifflin, 1956. Reprinted by permission. *B*, after N. B. Marshall, Aspects of deep-sea biology, Philosophical Library, 1954.)

by the aforementioned baited "monster camera." On the other hand, certain species of some invertebrate groups, particularly amphipods, isopods, ostracods, mysids, and copepods, attain a much greater size in deep waters than their relatives in shallow water. This phenomenon of larger size with increasing depth is called *abyssal*

gigantism (Plate 22). The size that certain of these "abyssal giants" attain is really quite remarkable. The isopod *Bathynomus giganteus* reaches 15 in. (42 cm) and the amphipod *Alicella gigantea* 6 in. (15 cm). *Gigantocypris,* the giant mesopelagic ostracod, has a diameter of several centimeters. The copepod *Gausia princeps*

Figure 4.17 Deep-water angler fish, *Melanocetus johnsoni*, before and after swallowing a larger fish. (From A. Hardy, The open sea, Houghton Mifflin, 1956. Reprinted by permission.)

delayed sexual maturity confer certain advantages to organisms in the deep sea. These include production of larger eggs and subsequently larger young, which can then feed on a wider range of food sizes, thus precluding special food in this food-short area. Large animals are also more mobile and can cover more area in search of both food and potential mates. Increased longevity means a longer period of sexual maturity and hence greater time to find mates.

Although these abyssal giants have captured much attention, they are rare. Most of the deep-sea benthic infauna (polychaete worms, crustaceans, and mollusks) are much smaller than their shallow-water counterparts. In fact, the small size of benthic invertebrates is actually the major characteristic of deep-sea fauna, not gigantism. Given the vast area of the deep sea, some giants should be expected, and similar examples of giant taxa can also be found in the less extensive shallow-water areas—for example, the giant clams found on tropical reefs.

The bottom substrate of most deep-ocean areas is a soft ooze. Benthic organisms inhabiting this soft material tend to have delicate bodies, to possess long legs or, in the case of sessile animals, to have long stalks to raise them above the ooze (Fig. 4.20). Some fishes have long, narrow fins that serve the same purpose.

In general, as one goes deeper in the oceans, the anatomical organization of the animals becomes more simplified. This is best seen in a comparison of the features of mesopelagic and bathypelagic fishes as outlined in Table 4.2. It is also reflected in the generally slow mode of life of most deep-sea creatures. Recent measurements of oxygen uptake on the deep-sea floor at several localities by Smith and Hinga (1983) gave results from 0.02 to 0.1 ml of oxygen per m^2 per hour, or more than 100 times less than measured in shallow water. Part of this is due to the much reduced density of organisms, but part also reflects the low metabolic activity of the organisms present.

Biochemical adaptations to the deep sea have been difficult to ascertain in the past because of the aforementioned lack of live animals. However, recent data on chemical composition suggest that profound changes may occur. In fishes, Childress and Nygaard (1973) report that the water content of the body tissues increases with

Figure 4.18 Luminescent barbel of the stomiatoid fish *Idiacanthus* sp. (Photo courtesy of Michael Kelly.)

reaches a size of 10 mm, nearly ten times the size of most calanoid copepods.

Scientists currently disagree as to the reasons for abyssal gigantism, but there are two main theories. The first attributes it to the peculiarities of metabolism under conditions of high pressure. The second suggests that the combination of low temperature and scarce food reduces the growth rates in these crustaceans and increases the longevity and the time it takes them to reach sexual maturity, such that larger size is obtained. The extreme expression of this combination would be the "abyssal giants."

The large size may also be the result of natural selection operating. Large size, long life, and

Figure 4.19 Large female angler fish, *Linophryne argyresca*, with attached small parasitic male. (From C. T. Regan and E. Trewavas, Deep sea angler-fishes *[Ceratoidea]*, Dana report no. 2, The Carlsberg Foundation and Oxford University Press, 1932.)

(A)

(B)

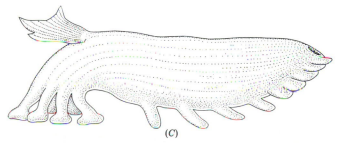

(C)

Figure 4.20 Adaptations to raise animals above the ooze. (A) Cirrate octopod. (B) A stalked crinoid. (C) A sea cucumber of the genus *Peniagone*. (A, courtesy of Dr. Clyde Roper. B, courtesy of Dr. H. Mullins. C, redrawn from Hansen, Galathea Report, Vol. 2.)

Table 4.2   ANATOMICAL DIFFERENCES IN MESOPELAGIC AND
BATHYPELAGIC FISHES

Features	Mesopelagic, plankton-consuming species	Bathypelagic species
Color	Many with silvery sides	Black
Photophores	Numerous and well developed in most species	Small or regressed in most; a single luminous lure on the female of most angler fish
Jaws	Relatively short	Relatively long
Eyes	Fairly large to very large with sensitive pure-rod retinas	Small or regressed, except in the males of some angler fishes
Olfactory organs	Moderately developed in both sexes of most species	Regressed in females but large in males of *Cyclothone* species and anger fish
Central nervous system	Well developed in all parts	Usually weakly developed
Myotomes	Well developed	Weakly developed
Skeleton	Well ossified, including scales	Weekly ossified; scales usually absent
Swim bladder	Usually present, highly developed	Absent or regressed
Gill system	Gill filaments numerous, bearing very many lamellae	Gill filaments relatively few, with a reduced lamellar surface
Kidneys	Relatively large, with numerous tubules	Relatively small, with few tubules
Heart	Large	Small

Source: From Marshall, 1980.

increasing depth, while the lipid and protein concentrations decrease. In other words, they become more like jellyfish. In crustaceans, Childress and Nygaard (1974) found that the protein content also decreased with depth. Presumably, this decrease in both groups is related to the scarcity of food from which to produce the protein. Caloric content also decreases.

A final set of adaptations in deep-sea organisms has to do with bioluminescent organs and will be taken up separately.

Bioluminescence in the Deep Sea

Marine bioluminescence is not confined to deep-sea organisms. Indeed, it is a phenomenon of widespread occurrence in the sea, and many people have observed it in the form of "phospho-rescent seas," which occur as a result of light production in the surface waters by untold millions of dinoflagellates. Whereas bioluminescence is of wide occurrence, it is in the deep-sea mesopelagic organisms that it reaches its highest and most complex development; there we find the greatest number of organisms that have the ability to produce light. The prevalence of bioluminescence in the deep sea and the large numbers of organisms that can produce light have been demonstrated abundantly in recent years by photos and videos taken by remote-operated vehicles (ROV) and manned submersibles. Hence it is appropriate to discuss the phenomenon at this time.

Bioluminescence is simply the production of light by living organisms. The mechanism of light production is well known from studies of terrestrial organisms like fireflies, and the same mechanism is used by aquatic organisms. In the reaction a substrate called luciferin is oxidized in the presence of the enzyme luciferase to produce a molecule that subsequently emits light. Luciferin and luciferase may differ from organism to organism, but the reaction is similar. The spectrum of color produced varies from species to species, but in general tends to be in the blue or green area.

Not only are large numbers of deep-sea organisms capable of producing light, they also have developed the most elaborate organs for producing this light. These light-producing organs are called *photophores* and are particularly abundant in fishes and squids (Fig. 4.21), but are present in other invertebrates as well (see Plates 18, 19, 26, 27).

The largest number of animals with photophores is found in the upper areas of the deep sea, the mesopelagic and upper bathypelagic. The incidence of organisms with bioluminescence decreases in the deepest parts of the sea.

Photophores in deep-sea organisms are of several types and range in structure from simple to complex. The simplest photophores consist of a series of glandularlike cells that produce the light or else a simple glandular cup holding a bacteria culture that produces the light (see also pp. 408–409). In either case, the cells or cup are surrounded with a screen of black pigment cells. The more elaborate photophores have in addition one or more of the following: lenses to focus the light, a color filter, and an adjustable diaphragm of pigment cells. In some fishes and crustaceans, the photophores may be covered by a flap of flesh, which can be adjusted to turn the light either on or off. Still others are able to move the photophores by muscular action. Some squids are known to combine their photophores with overlying layers of skin that contain chromatophores, thus allowing them to alter the color or intensity of the light produced.

That the production of light is of considerable adaptive significance to its possessor is demonstrated by (1) its great prevalence in the deep sea, and (2) the complex anatomical organs evolved to accommodate its production. It is also apparent that bioluminescence probably serves different functions in different animals. Several different hypotheses have been suggested for the use of bioluminescence in the deep sea. These have been reviewed by Young (1983), who suggests that most are used primarily as a means of food capture or defense against predators.

Many organisms, particularly fishes and squids, have their photophores placed ventrally.

Figure 4.21 Light organ patterns of some deep-sea squids. *(A) Nematolampus. (B) Pterygioteuthis. (C) Histioteuthis. (D) Ctenopteryx* male. (From M. Nixon and J. B. Messenger [eds.], The biology of cephalopods, Academic Press, 1977.)

For those animals living in the mesopelagic zone, where some light downwells from the surface, a predator looking up at them could probably see a silhouette. The photophores, when turned on to produce a similar intensity as the downwelled light, would cause the silhouette to disappear and make the animal less vulnerable to predation. This is, in effect, camouflage. Since some organisms have been shown to be able to adjust their downward photophores to the incident radiation level from above, this is a viable function.

Photophores might also be used to produce a "blinding" flash of light, which would momentarily startle a potential predator and thus allow disengagement and escape. However, since such light production may also have the opposite effect of attracting more distant predators, this use of bioluminescence contains an element of risk. Some deep-sea squids (*Histioteuthis dispar*) distract potential predators by shooting out a luminescent cloud rather than a black "ink" cloud as is done by squids in lighted waters. This cloud masks the escape of the squid.

As noted before, photophores might also be employed as lures to attract potential prey items within range of the predator. This is undoubtedly the function of the elaborate light organ (esca) on the modified dorsal fin (illicium) of the angler fish and the "lures" of other midwater fishes (Fig. 4.18; Plates 26, 27). Photophores may also be used for simple illumination of an area such that a predator may pick out prey items.

A final set of uses has to do with recognition. In a perpetually dark area such as the deep sea, the particular pattern of distribution of photophores over the body of the animal would be visible at a distance. Elaborate patterns of photophores occur on fishes in particular. Furthermore, the pattern is different for different species. The existence of these patterns would thus allow individuals to recognize their own species. This may be of importance in maintaining schools and also in finding potential mates for reproduction (Fig. 4.22). In the latter case, certain fishes also have photophore patterns that differ between males and females to further ensure mate selection. It is interesting to note here that one of the first scientists to observe living mesopelagic fishes *in situ*, William Beebe, who made his observations from a bathyscape, noted that he could tell the different myctophid (lantern fish) species in the dark from their photophore pattern. If he could, it seems likely that the fish can also.

Other uses of photophores may occur and more may come to be recognized with increased knowledge of this area, but the above represent the current most plausible hypotheses.

We can say in summary, paraphrasing Young (1983), that in this homogeneous twilight area without physical barriers or structures for concealment and with abundant predators, life becomes a peculiar battle in which the judicious use of bioluminescence is of major importance in concealing one from predators, finding and attracting prey, and ensuring survival. No other area on earth is quite like it.

(A)

(B)

(C)

(D)

(E)

Figure 4.22 Species-specific light organ patterns (open circles) in the lantern fish genus *Diaphus.* (A) *D. macrophus.* (B) *D. lucidus.* (C) *D. splendidus.* (D) *D. garmani.* (E) *D. effnlgeus.* (From N. B. Marshall, Aspects of deep-sea biology, Philosophical Library, 1954.)

Plate 17 *Anoplogaster cornuta,* a "daggertooth" fish from the mesopelagic zone. Notice the very large mouth and large, sharp teeth for seizing prey. *(Photo courtesy of Mr. Michael Kelly.)*

Plate 19 A mesopelagic amphipod, *Cystisoma fabricii,* showing the transparent body and huge white light organ. *(Photo courtesy of Michael Kelly.)*

Plate 18 *Argyropelecus lychnus,* a mesopelagic hatchetfish. Note the lines of black photophores along the ventral portion of the body, the tubular eyes, and silvery reflecting scales. *(Photo courtesy of Michael Kelly.)*

Plate 20 A mysid crustacean of the genus *Boreomysis.* *(Photo courtesy of Gary McDonald.)*

Plate 21 A deep-sea prawn of the genus *Notostomus.* Note the characteristic deep red-orange color of many deep-sea crustaceans. *(Photo courtesy of Dr. Roger Seapy.)*

Plate 22 The giant deep-sea mysid *Gnathophausia*, an example of abyssal gigantism. *(Photo courtesy of Dr. Roger Seapy.)*

Plate 24 The deep-water jellyfish *Arctopodema amplum. (Photo courtesy of Dr. George Matsumoto.)*

Plate 23 A deep-sea jellyfish, *Benthocodon hyalinus.* Note the characteristic deep purple color. *(Photo courtesy of Dr. George Matsumoto.)*

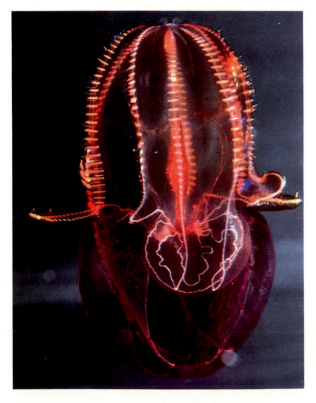

Plate 25 A deep-sea ctenophore, *Lampoctena sanguineventer. (Photo courtesy of Dr. George Matsumoto.)*

Plate 26 The deep-sea fish *Aristostomias scintillans.* Note the greatly expanded mouth cavity that allows the animal to swallow very large prey items. The long barbel from the jaw region has a luminescent organ at the end. *(Photo courtesy of Dr. Roger Seapy.)*

Plate 27 A deep-sea angler fish of the genus *Melanocetus.* Note the extremely large mouth and the luminescent "lure" above the head. *(Photo courtesy of Dr. Roger Seapy.)*

Plate 28 A group of deep-sea holothurians of the genus *Scotoplanes* on the bottom at 3000 m. Holothurians are one of the major groups of large "megafauna" organisms characteristic of sedimentary areas of the deep sea. *(Photo courtesy of Dr. Barbara Hecker.)*

Plate 31 The large vestimentiferan tube worm *Riftia pachyptila.* The red anterior portion of the body extending from the white tube is colored red because of the hemoglobin of the blood. *(Photo courtesy of Woods Hole Oceanographic Institution.)*

Plate 29 Sea pens of the genus *Kophoblemnon* in 3000 m of water. *(Photo courtesy of Dr. Barbara Hecker.)*

Plate 32 A bed of the large clam *Calyptogena magnifica* near a deep-sea hydrothermal vent. *(Photo courtesy of Woods Hole Oceanographic Institution.)*

Plate 30 A typical deep-sea brisingid starfish. *(Photo courtesy of Dr. Barbara Hecker.)*

COMMUNITY ECOLOGY OF THE BENTHOS

Although we have samples of organisms from the pelagic environment of the deep sea, the problems of capture and net avoidance make the available information on the community composition of the pelagic fauna incomplete at best and suspect at worst. The exception to this is the mesopelagic, where many samples and observations are available. Considerably more is known about the deep-sea benthic communities because the sedentary and sessile nature of the animals makes net avoidance less important. Thus, the samples taken and the observations of this area by deep submersibles and cameras have given us a reasonable idea of its composition. Ecologically all benthic organisms throughout the world's oceans can be divided into two groups on the basis of position. The *epifauna* are those benthic organisms that live on or are otherwise associated with the surface of the bottom. *Infauna* is the term used to describe those organisms that live within the substrate. We shall use these terms again in subsequent chapters.

Faunal Composition

Virtually all the major animal groups have representatives in the deep sea, but the relative abundance of groups varies. Crustaceans, particularly isopods, amphipods, tanaids, and cumaceans, are common in the deep sea. In the Atlantic abyssal area, they constitute 30 to 50 percent of the fauna. Polychaete worms are also abundant, making up from 40 to 80 percent of the fauna in the Atlantic. Particularly common in abyssal areas are sea cucumbers (Holothuroidea), usually of large size, and brittle stars (Ophiuroids). Holothurians are often the most common organisms in deep-sea photographs (Plate 28). This, plus the fact that they form 30 to 80 percent of the biomass of living organisms taken in some trawls, suggests they are a major component of the abyssal community. Since they are deposit feeders, the abyssal oozes are likely an excellent food source, hence their abundance.

Starfish (Asteroidea), sea lilies (Crinoidea), and sea urchins (Echinoidea) are also present, but not in abundance (Fig. 4.23; Plate 30). Among the sponges, the deep sea is populated by the glass sponges (Hexactinellida), a group rarely found in shallow water. Sea anemones (Anthozoa), sea pens (Pennatulacea), and sea fans (Gorgonacea) are the only common representatives of the phylum Cnidaria (Plate 29). A number of fishes are also present, but their general mobility makes it difficult to estimate relative abundance. Figure 4.24 diagrammatically summarizes zonation changes and typical abyssal animals.

(A)   *(B)*

Figure 4.23 Two views of the bottom at 1300 m on the Gay Head–Bermuda transect showing variation in abundance of organisms. *(A)* Area barren of large benthic animals. *(B)* Concentration of the echinoid *Phormosoma placenta.* (Photos courtesy of Dr. Fred Grassle; reprinted with permission from J. F. Grassle et al., Deep-Sea Res., Vol. 22, Pergamon Journals Ltd., 1975.)

Figure 4.24 Zonation of benthic animals with depth on the Atlantic continental margin of Europe. (From B. Heezen and C. Hollister, The face of the deep, Oxford University Press, 1971.)

In the trenches (hadal zone) the available data suggest that there is a higher percentage of pericaridean crustaceans, polychaetes, and holothurians and a lower percentage of sea stars, sea urchins, and brittle stars in comparison with the abyssal.

The major fishes of the bottom waters of the deep sea are rat tails (Macrouridae), brotulids, liparids, and certain eels (*Synaphobranchus, Cyema*).

One of the dominant features of the deep-sea benthos is the very small size and fragility of most individuals. For example, in exploration of the shallow waters of the continental shelf, a 1-mm sieve is often used to screen out infaunal organisms. Such a screen used in the deep sea

would retain practically no organisms. Furthermore, the deep-sea animals would fall apart unless carefully handled. The fragility of these animals is due to several circumstances, including the fact that the environment is virtually motionless and that calcium carbonate is difficult to deposit.

Community Structure

In contrast with the shallow-water benthos or the epipelagic, the deep-sea benthos is depauperate with respect to suspension feeders. According to Jumars and Gallagher (1982), no more than 7 percent of the numbers of species are suspension feeders. The reason for this lies in the fact that this is an area of very slow water movement and one in which there are few food particles in the water. As a result, there is little food to take out of the water and a filter feeder would have to expend more energy creating a water current than the animal would receive back in food—a losing situation.

Carnivores also appear to be less abundant among the invertebrates, although this is uncertain due to the difficulty of separating omnivores and scavengers from carnivores. Herbivores are present in low numbers, feeding mainly on plant remains such as wood. The majority of animals, however, some 80 percent by number, are deposit feeders ingesting the organic-rich sediment.

Most attempts at analyses of the age structure of these benthic organisms through size-frequency graphs have given the same results no matter what time of the year they were sampled. Such monotonous similarity in these graphs suggests that these animals reproduce throughout the year and have no seasonal breeding time. Life spans for most of the deep-sea animals are not known.

In an attempt to understand colonization and recruitment rates in the deep-sea benthos, Grassle and Morse-Porteous (1987) set down trays of defaunated sediments at two sites in the North Atlantic (1800 and 3600 m). Results indicated similar rates of colonization at the two depths, but at much slower rates than observed in shallow water. Furthermore, even after 5 years, the species composition and density of the experimental trays were different from that of the surrounding natural area. The most common colonizers were polychaete worms of the families Spionidae, Capitellidae, and Sigalionidae. Experiments with organically enriched but azoic sediments demonstrated that organisms would preferentially settle in large numbers in the organically enriched areas or patches. Screening out predators resulted in high densities of colonizers. It appears that normally heavy predation coupled with occasional escapes from predation may explain the patchiness of the fauna.

Diversity

Because of the low density of benthic populations, the small size of most invertebrates, and the paucity of adequate samples, the deep sea was considered a biological desert for many years. In the early 1960s, this notion was abandoned as sampling efforts intensified and methods of dredging and sample processing improved. As a result, the desert idea was replaced by the opposite belief, first postulated by Sanders (1968), that the deep sea is a highly diverse community. The concept of species diversity is founded on the intuition that large groups of co-occurring species are characterized by a greater complexity of biological interactions. As noted in Chapter 1, species diversity has two important components: the number of species and their patterns of relative abundance. For example, if two areas have the same number of species, one area has a greater diversity if it has a more even distribution of individuals among the resident species. The idea of a highly diverse deep-sea fauna is based entirely on their relative abundance patterns. While the total number of individuals is sparse, each deep-sea species is commonly represented by only a few individuals. In other words, most deep-sea species are rare. This pattern had led to the popular contention that deep-sea diversity is as high as that of coral reefs, or among the highest in the world. In contrast, shallow-water assemblages are often dominated by one or several very abundant animals, but there are also many rare species.

The suggestion that the deep sea is a highly diverse area had the salutory effect of generating even more interest in and sampling of the deep sea by other scientists, the more critical evaluation of the Sanders work, and the generation of alternative hypotheses.

If one accepts the idea of a highly diverse deep sea, there are several conflicting hypotheses attempting to explain this diversity. The first is termed the *stability-time hypothesis,* first suggested by Sanders (1968). This says that high diversity occurs because the highly stable environmental conditions have persisted over long periods of time and have allowed species to evolve that are highly specialized for a particular microhabitat or food source. Since most of the benthic animals are detritus feeders in the deep sea, this means that they have become very specialized for a particular narrow range of particle sizes. Such specialization is virtually unknown among deposit feeders elsewhere in the oceans and has not been demonstrated in deep-sea benthos. Moreover, the stability-time hypothesis has been falsified as an explanation of diversity patterns among the insects found in sugarcane fields and among the marine crustaceans inhabiting coral heads.

A second hypothesis is the *cropper* or *disturbance theory* suggested by Dayton and Hessler (1972). The basic premise here is that organisms generally increase in numbers until they reach the limit of some resource that is in least abundance. At that point, competition occurs, and in the ensuing competition, species are eliminated (principle of competitive exclusion). Since, in the deep sea, animals are very low in numbers compared to the total amount of space available to inhabit, competition for space does not occur. The only other resource likely to be competed for is food. As noted above, it is in very short supply. The cropper theory suggests that none of the deep-sea animals are food specialists, and that they are generalists feeding on anything they can engulf, not distinguishing whether it is living or dead; hence, the term "cropper," as one that feeds indiscriminately on anything smaller than itself. High diversity is maintained because the intense "cropping" by all levels and sizes of animals prevents any species from building up its population to the point where it would be competing with another for the food. High diversity is due, then, to intense predation at all levels, which allows a large number of species to persist, eating the same food, but never becoming abundant enough to enter into competition. A central argument against this theory is that heavy predation should result in the evolution of definite life history features. These include production of large numbers of young, a short life span, and fast growth. As far as is known (see Table 4.3), these are the exact opposite of the life history characteristics found in the deep-sea organisms, which grow slowly, live long, and produce few offspring. However, if cropping results in differential mortality of age groups, either set of life history traits can, in theory, evolve.

A final hypothesis, which does adequately explain the original reports that diversity increases with increasing depth, is the *area hypothesis*. This is a very simple idea. Since species number is positively correlated with area (i.e., more area equals more species), diversity is

Table 4.3 SUMMARY OF GENERAL BIOLOGICAL CHARACTERISTICS OF DEEP-SEA ORGANISMS

Reproduction and development

Few eggs, large, yolk rich
Slow gametogenesis
Late reproductive maturity
Reduced gonadal volume
Slow embryological development
High incidence of direct development
Breed usually once (semelparous)

Physiological

Low metabolic rate
Low activity level
Low enzyme concentration
High water content
Low protein content
Small size

Ecological

Slow, indeterminate growth
High longevity
Slow colonization rate
Low population densities
Low mortality due to low predation pressure

highest in the deep sea because it covers the greatest area. Both diversity and area of sea bottom decrease with decreasing depth. This correlation is highly significant. So, although the actual numbers of species per unit bottom area have not been established, trends in relative abundance patterns can be related to the size of large biogeographical areas. But this correlation applies only to samples taken in the Atlantic Ocean. Workers in the Pacific Ocean have not found a general increase in diversity with increasing depth. On the contrary, both components of diversity (species density and relative abundance) are highest at intermediate bathyl depths rather than in the abyss.

Additional data and analyses of various data sets have produced a picture that, though still not completely resolving the diversity issue, has helped to clarify some aspects of it and put the issue more in perspective. Rex (1981) has demonstrated for most benthic organisms that the diversity does not continue to increase as one goes deeper into the ocean. In fact, he has shown that diversity is higher at intermediate depths and decreases both downward and upward to the continental shelf, giving a curve of parabolic shape (Fig. 4.25).

The reasons for this parabolic diversity pattern and deep-sea diversity in general still remain unanswered. Data now available, however, suggest that there are various patterns of distribution of deep-sea benthos that change with depth, and that at different depths the patterns result from differing sets of interacting factors. For example, it is now apparent from photographs and observations that the deep ocean floor is not a homogeneous featureless plain but in fact has various mounds, holes, fecal pellets, and tracks. Furthermore, these features, because of the stability of the water, tend to persist for long periods of time and are therefore suitable "microhabitats" for various organisms. This can lead to increased diversity just as it does in the obviously topographically diverse rocky intertidal areas (Chapter 6) and coral reefs (Chapter 9). In other words, the deep sea is "patchy," and organisms are distributed patchily. If one samples several patches, one naturally gets higher diversity. Moreover, the patches exist on different scales of size and so do the organisms. This means that diversity also changes on different size scales. Thus Rex (1981) showed that for small animals maximum diversities occurred at greater depths. This perhaps is due to longer

persistence of the aforementioned microhabitat features or to the absence of competition with larger organisms. Predation and competition may also act differentially. In the deeper waters the density of animals is low, due to decreased energy input from above. This means that predators are rare because their prey is rare or absent. In such cases it may be that competition or microhabitat specialization is the dominant structuring force. In shallower depths, predators are more abundant due to increased prey density and may act strongly to structure the community, relegating competition or microhabitat specialization to a minor role.

Most recently, Grassle and Morse-Porteous (1987) have suggested that high diversity in the deep sea is maintained by a combination of three factors: (1) the patchy distribution of organically enriched areas in a background of low productivity; (2) the occurrence of discrete small-scale disturbance events (mainly from biogenic agents) in an area of otherwise great constancy; and (3) the lack of barriers to dispersal among species distributed over a very large area.

In summary, it appears that none of the early hypotheses are by themselves adequate to explain all deep-sea diversity patterns, and that the observed patterns are the result of many factors interacting simultaneously, but with certain factors having more influence in certain areas, at certain depths, or at certain times. These factors include microhabitat diversification due to a

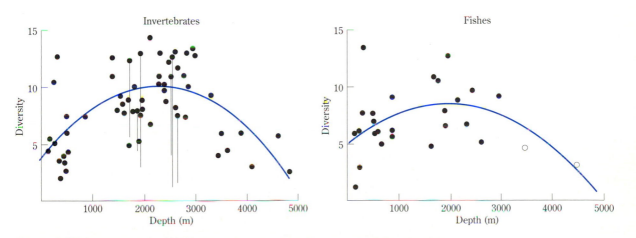

Figure 4.25 Patterns of species diversity in the deep-sea benthos based on samples from the northwest Atlantic Ocean. (From M. Rex, Community structure in the deep sea benthos. Ann. Rev. Ecol. Sys. 12:331–353, 1981.)

heterogeneous environment, energy input, competition, predation, and rate of disturbance. When we have more quantitative samples from more areas of the deep sea, perhaps this whole issue of diversity and the forces that structure the organization of communities will become more clear.

Life History Patterns

General life history patterns with respect to reproduction and development are more difficult to generalize, primarily because we still lack data about these aspects of the life history of most deep-sea animals. In concordance with the low metabolic rate and low activity rate, however, we find that perhaps most deep-sea animals tend to have slow or delayed sexual maturity, smaller gonads, and slow embryological development. Some, in fact, may reproduce only once during their life span (= semelparous). Meade et al. (1964) report hermaphroditism to be common among fishes.

In shallow-water areas and terrestrial areas, there are usually cyclic changes in some environmental parameter that serve to initiate and control reproduction. As a result, reproduction is usually confined to a particular time of year. In the deep sea, where light is absent and all environmental parameters are monotonously constant, it would appear that a cyclic pattern of reproductive activity should be absent. What little evidence we have thus far suggests that this indeed is the case. Reproductive data from benthic organisms in the deep sea studied by Rokop (1974) show that reproduction is constant and continuous throughout the year, with no seasonal peaks.

Since food is scarce, it might be suspected that most deep-sea organisms would produce few offspring that would not spend time as larvae in the open water (see Chapter 1, pp. 23–27, on larval strategies). However, our current state of knowledge has little information about eggs and larvae of deep-sea organisms except for fishes, and we can neither completely confirm nor deny that assumption. What we do know suggests that many deep-sea benthic animals do have few large eggs, suggesting they are hatched in an advanced state and have a short or no larval life. Most fishes, on the other hand, have pelagic eggs and larvae.

Two general reproductive patterns exist in the deep sea. In the first case, the early stages of the life history are spent in the lighted surface waters, and the juveniles migrate down to adult depths as they mature or metamorphose. Such is the pattern exhibited by certain deep-sea angler fish (Fig. 4.26), myctophids, stomiatids, and certain squids. In the other, no migration occurs and the young stages are spent in the same area as the adult. Such is the case with the unique

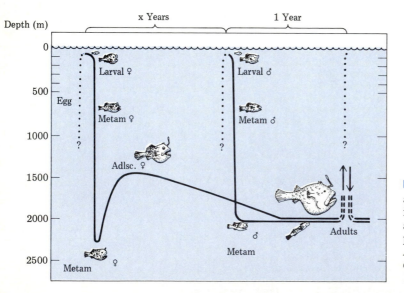

Figure 4.26 Life cycle of an angler fish. The fish spends its larval life at the surface and descends on metamorphosis. (From N. B. Marshall, Aspects of deep-sea biology, Philosophical Library, 1954.)

cephalopod *Vampyroteuthis infernalis* and certain prawns.

Table 4.3 summarizes the general biological features of deep-sea animals.

Hydrothermal Vent Communities

Perhaps one of the most significant and exciting discoveries in deep-sea biology, indeed in marine biology in general, occurred in 1977. At that time, a series of dives with the deep submersible *Alvin* was made in the Galápagos rift zone some 200 mi northeast of the islands. As reported by Ballard (1977), these dives, to 2700 m, revealed a spectacular abundance of hitherto unknown marine animals living in and around four hot-water geysers in an otherwise barren ocean floor. The water temperature in these vent areas was 8–16°C, considerably above the normal temperature prevalent at the bottom at these depths (about 2°C). This warm water was found to be very high in reduced sulfur compounds, namely, H_2S. The hydrogen sulfide was used as an energy source by bacteria (Fig. 4.27; see also Chapter 6 on sulfide bacteria in mud shores).

Since that time, further exploration with submersibles and with towed camera systems has revealed similar vent communities, at depths ranging from 1500 to 3200 m, at several additional locations in the eastern Pacific Ocean. In addition two other hydrothermal vent communities have been reported from the western Pacific, in the Marianas Back Arc Basin (3590–3660 m) and in the North Fiji Basin (1990 m). In the Gulf of Mexico at the base of the Florida Escarpment at 3200 m and on the continental slope off Louisiana at 600–700 m, additional vent communities have recently been discovered that are associated with "cold" hypersaline brines or other hydrocarbon seeps. The locations of the known hydrothermal vents, hydrocarbon seeps, and brine pools are shown in Fig. 4.28.

All these vent areas and cold seeps display an astonishing assemblage of large (megafauna) animals forming a unique association for the deep sea. Truly they are "oases" in an otherwise low-density, low-productivity area. Each hydrothermal vent system seems to have an ecologically similar set of dominant species, though not always a similar set taxonomically. Grassle

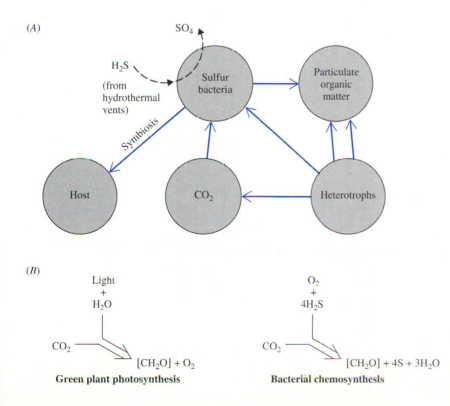

Figure 4.27   Chemosynthesis in hydrothermal vent communities. *(A)* Sulfur bacteria in the vents utilize the energy in hydrogen sulfide gas to build complex organic molecules and grow. The source of carbon is carbon dioxide gas, which is also dissolved in the water. The bacteria are either eaten directly by heterotrophs, live symbiotically with the larger animals (to which they furnish organic material), or die and form particulate organic matter (detritus) that, in turn, is used as a source of food by heterotrophs. *(B)* Comparison of photosynthesis and chemosynthesis. *(A,* after A. R. Longhurst, Analysis of marine ecosystems, Academic Press, 1981.)

Figure 4.28 Locations of hydrothermal vents, hydrocarbon seeps, and hydrothermal vent springs in the world oceans. (After H. H. Webber and H. V. Thurman, Marine biology, 2nd ed., HarperCollins, 1991.)

(1985) has reported a total of at least 16 previously unknown families of invertebrates at these vents, and many species remain undescribed. All vent organisms are endemic. The dominant animals, by size, at these vents usually include one or more of the following: large clams, mussels, and vestimentiferan worms (Figs. 4.29, 4.30, 4.31). The latter belong to a special deep-sea phylum known as the Pogonophora and are characterized by the complete lack of a digestive tract. Vestimentiferans appear to be absent from the West Pacific vent systems. Other associated large animals include crabs, a number of snails, a variety of polychaete worms, an odd benthic siphonophore, and enteropneust worms (called "spaghetti worms" from their appearance in photos; see Fig. 4.32). The cold seeps and hypersaline brine communities are less well known, but seem to be dominated by mussels rather than pogonophoran worms and large clams.

All of the vent communities are confined to an area within only a few meters of the vent system. The cold seeps and brine communities may be larger. They are separated from other such associations by hundreds or thousands of kilometers. The latter has aroused the interest of marine biologists curious as to how such communities not only persist, but how they disperse to find these vents; they are virtually looking for a needle in a haystack. Although we still have much to learn about these communities, we have made much progress since their discovery.

All vent systems and cold seeps are dependent for their existence on the primary productivity of chemosynthetic bacteria. Potential energy in the form of reduced sulfur compounds, such as hydrogen sulfide, is spewed forth from the hot vents. These sulfides are oxidized by the chemosynthetic bacteria to release energy, which in turn they use to form organic matter from carbon dioxide (Fig. 4.28). Productivity is very high. Karl et al. (1984) measured production rates at one vent of 19 μg C/g/h. The primary consumers are the various animals that filter out the bacteria from the water, graze the bacterial film from the rocks, or are locked in a symbiotic association with the bacteria (Fig. 4.30).

The animals that form a symbiotic association with bacteria are of particular interest. The large vestimentiferan worm, *Riftia pachyptila,* and the largest clam, *Calyptogena magnifica,* both contain symbiotic chemosynthetic bacteria in their tissues. *Riftia* has no digestive tract and there is evidence that it, in fact, obtains its nutrition from the bacteria. Although the clam retains its ability to filter feed, it too apparently derives most of its nutrition from the symbiotic bacteria. What is curious about this system is that for the bacteria to function they must have a reduced sulfide source. According to Arp and Childress (1983), this is brought to them by the blood of *Riftia* and *Calyptogena* (Plates 31, 32). Under normal circumstances this would poison or kill a typical aerobic animal, since sulfides bind to the cytochrome system, interfering with oxygen uptake and therefore respiration. However, the bloods of *Riftia* and *Calyptogena* seem to have in them a protein factor that binds to the sulfide to transport it, freeing it again in the presence of the bacteria that take it up. This is assumed since there is no inhibition of oxygen uptake in the presence of sulfide. How other vent animals may deal with sulfide is not as yet known. In the cold seeps and brine pools, by contrast, the dominant mussels (*Bathymodiolus*) have a methane-based symbiosis. The seeps and brine pools are rich in methane, which is taken up by the gills of the mussels, where intracellular bacteria oxidize the methane and provide energy for the mussels and the bacteria.

Hessler and Smithey (1983) have noted that vent animals are zoned in the vent systems with respect to water temperature (Fig. 4.33). *Riftia, Calyptogena,* and the mussels (*Bathymodiolus*) seem to be limited to the mouths of the vents at the highest temperatures, whereas others are spaced out away from the vents until those at the periphery are in water very near ambient deep-sea temperatures. All vent animals appear to be restricted to the vent areas, and we do not know what keeps them from invading the adjacent deep-sea area, nor do we know what prevents the normal deep-sea fauna from invading the vent systems.

According to Grassle (1985), the life span of individual vents is only in the tens of years, and scientists have observed inactive vents with the dead remains of vent animals. This suggests that these vent species have life spans also measured in decades. Recent estimates of the life span of *Calyptogena* by Grassle (1985) gave the ages of the oldest at 30 years. Given these circumstances, it would seem that the vent animals would experience strong selective pressure to grow rapidly to sexual maturity, produce many young, and have an efficient means of dispersal. What we know now suggests that the dominant animals do indeed grow much faster than surrounding deep-sea animals away from the vents, but we are uninformed yet about reproductive potential for most. In the case of dispersal, however, a strange situation seems to prevail. Most species seem to lack planktotrophic larvae and have nonplanktotrophic development and hence low dispersal. It is an enigma how such development could facilitate the finding of widely separated vent systems, but somehow it must because Grassle (1985) reports that close relatives of present-day vent animals have existed for 200 million years.

Certain areas of the hydrothermal vents in the Galápagos have been the object of follow-up studies, reported by Childress (1988), that have demonstrated how quickly these areas can change. The particular area called "the Rose Garden" when discovered in 1979 was dominated by large numbers of vestimentiferan worms and contained significant, but fewer, numbers of large clams and mussels. When revisited in 1985, a little over 5 years later, the worms had markedly declined, while the mussels had dramatically increased and dominated the area. The clams were also abundant. Among the associated fauna, filter feeders seemed to have generally declined, siphophores and enteropneusts were virtually absent, and mobile scavengers and carnivores such as galatheid crabs and gastropod mollusks were more abundant. Accord-

Figure 4.29 A vent clam field dominated by the giant white clam *Calyptogena magnifica*. (Photo courtesy of Woods Hole Oceanographic Institution. Photo by George Somero.)

Figure 4.30 Large mussels *(Bathymodiolus)* and galatheid crabs of the genus *Munidopsis* around a hydrothermal vent. (Photo courtesy of Woods Hole Oceanographic Institution. Photo by Robert Hessler.)

Figure 4.31 Giant vestimentiferan tube worms, *Riftia pachyptila*, near a Galápagos thermal vent. (Photo courtesy of Woods Hole Oceanographic Institution. Photo by John Donnelly.)

Figure 4.32 Mats of enteropneust worms, *Saxipendium coronatum*, called "spaghetti worms," draped over rocks around hydrothermal vents. (Photo courtesy of Woods Hole Oceanographic Institution. Photo by James Childress.)

Figure 4.33 A diagrammatic representation of a typical vent community showing the presence of the larger macrofaunal organisms. (Modified from R. M. Haymon and K. C. McDonald, The geology of deep sea hot springs, Amer. Sci. 73[5]:441–449, 1985.)

ing to Hessler et al. (1988), these striking community changes seem to be a result of two factors. The first was a change in the physical conditions that was manifested mainly by a decline in the vent water flux, which in turn led to a decline in the vestimentiferans, and probably also the filter feeders, but favored the mussels. The second was continuing recruitment into the area, which favored the crabs, mussels, and clams. It is also possible, however, that the vestimentiferans declined due to the fact that the mussels as they grew and increased effectively cut off the vestimentiferans from their sulfide source by absorbing the sulfide before it could get to the more distant absorbing crowns of the worms. This latter scenario does not require any change in the water circulation.

It would appear that mortality, at least among the vestimentiferan worms, is very high. Tunnicliffe et al. (1990) have shown that, for the vestimentiferan *Ridgeia pescesae* on the Juan de

Fuca vent system, 44 percent of the animals were removed over a 26-day period due to collapsing sulfide/sulfate chimneys on which they lived or to predation by fish.

As a result of observations at the Rose Garden, a general pattern of community change over time has been suggested by Hessler et al. (1988) for the Galápagos hydrothermal vents. Beginning with an established vent system dominated by vestimentiferans, clams, and mussels, one sees the vestimentiferans diminish with time, while the clam and mussel populations grow. This may be the result of a decrease in water flux or sulfide concentration or the result of competition with the more efficient mussels. As flow diminishes in the vents, suspension feeders will decline as will the clams, leaving the mussels as the most persistent. Since the same suite of dominant organisms is not present at all hydrothermal vents, the general applicability of the above successional scheme is simply unknown.

MIDWATER COMMUNITY ECOLOGY

Because of the increased difficulty in sampling pelagic deep-sea organisms due to net avoidance and migration, the lack of observations on these animals from deep submersibles, their avoidance of light, and their small size, we know less about the ecology of midwater animals than we do about the deeper but sedentary and sessile benthos.

Deep Scattering Layers

Early in World War II, when the use of echo sounders was being experimented with in the oceans as a means of detecting submarines and the bottom, the devices (now known as sonar) regularly began to pick up echoes. They came, not from the bottom, but from sources in the water column itself, hundreds or even thousands of meters above the known bottom. These areas of sound reflection occurring in midwater were termed *deep scattering layers* (DSLs). The depths of the DSLs varied. During the day, there were often two or three layers varying in depth from 200 to 700 m (Fig. 4.34). At night, these different layers migrated toward the surface, where they often merged into a single broad band. At dawn, the layers returned to depth. Although first interpreted as the result of some physical discontinuity in the water, the DSLs are now known to be concentrations of midwater animals, and the movement of the layers toward the surface at night is the result of a vertical

migration similar to that undertaken by surface zooplankton (see pp. 74–77). Presumably, these midwater animals are migrating into the surface waters at night in order to feed on the abundant plankton.

The major animals forming the DSLs include the lantern fish (Myctophidae) and other similar mesopelagic fishes (Fig. 4.11), shrimplike crustaceans such as euphausiids ("krill"), sergestid and pasaphaeid shrimp, various squids, and certain deep-floating hydrozoan cnidarians called siphonophores (Plates 20, 21, 22, 25). The composition of the scattering layer varies with geographical area.

Deep scattering layers have been found in all oceans except the Arctic. They are best developed in areas with high surface productivity and are faint in areas of low productivity.

Species Composition and Distribution

The pelagic zones of the deep sea are dominated primarily by small fishes of the families Myctophidae, Gonostomatidae, and Sternoptychidae, and euphausiid and decapod crustaceans (Fig. 4.35). Blackburn (1977) has noted that, together, these groups constitute more than 80 percent of the biomass, at least in the mesopelagic zone. One estimate for the Pacific Ocean has fishes making up 20 to 45 percent of the biomass, shrimps 15 to 25 percent, and euphausiids 35 to 50 percent in the upper 1000 m of water (Fig. 4.36). Minor components include other crustaceans such as mysids, amphipods, and copepods, chaetognaths, scyphozoan jellyfish,

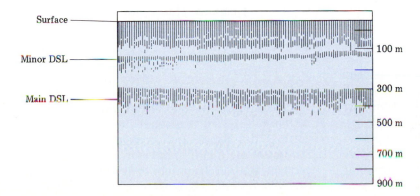

Figure 4.34 A sonogram recording showing the presence of the deep scattering layers in the Indian Ocean. (After G. B. Farquhar [ed.], Proceedings of an international symposium on biological sound scattering in the ocean. U.S. Government MC Report 005, 1970.)

Figure 4.35 Representatives of the major groups of organisms of the pelagic zones of the deep sea. *(A)* Lantern fish, *Tarletonbeania crenularis* (Myctophidae). *(B)* Gonostomatid fish, *Cyclothone pallida* (Gonostomatidae). *(C)* Hatchet fish, *Argyropelecus affinis* (Sternoptychidae). *(D)* Euphausiid, *Euphausia pacifica*. *(E)* Sergestid shrimp, *Sergestes similis*.

and cephalopods. This dominance of the pelagic deep-sea fauna by small fishes, euphausiids, and shrimps seems to be widespread in all oceans, but as Omori (1974) has noted, the relative abundance of each group changes with latitude and with depth (Fig. 4.36). Seasonal changes in the composition of the assemblage appear to be minor but have received little study.

There are two components of the pelagic species assemblages in the mesopelagic zone. One includes those animals that do not migrate vertically—certain fishes (*Cyclothone, Sternoptyx*), mysids, some shrimps (sergestids, penaeids, carideans), and a few euphausiids, amphipods, cephalopods, and cnidarians. The other component is called the *interzonal fauna* and undergoes a diurnal vertical migration. These

animals live below 450 m by day and migrate into surface waters by night. This component includes most of the mesopelagic fishes (Myctophidae, Gonostomatidae) and most euphausiids and decapods. Hence, the composition of the community changes depending on whether it is sampled at night or during the day (Fig. 4.36).

Food and Feeding

Information on the food of midwater organisms is sparse, but is most adequate for two dominant groups, the fishes and decapod crustaceans. Studies of the diet of midwater fishes have been concentrated on the numerically abundant families Myctophidae, Gonostomatidae, and Sternoptychidae. Whereas there is often considerable

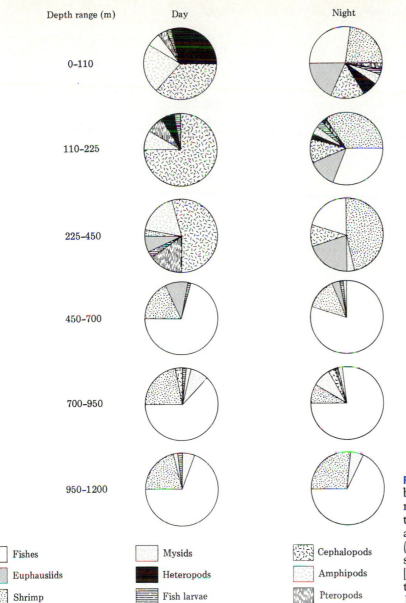

Depth range (m)   Day   Night

0–110

110–225

225–450

450–700

700–950

950–1200

	Fishes		Mysids		Cephalopods
	Euphausiids		Heteropods		Amphipods
	Shrimp		Fish larvae		Pteropods

Figure 4.36 Differences in biomass composition of the midwater community in the tropical Pacific with depth and between day and night. (Adapted from N. R. Anderson and B. J. Zahuranec [eds.], Oceanic sound scattering prediction, Plenum, 1977.)

variation in species composition in the diets of these fishes, the principal components in all cases, as noted by Hopkins and Baird (1977), are smaller crustaceans such as copepods, euphausiids, ostracods, amphipods, and small decapod shrimps (Table 4.4).

Changes in the diet of these fishes occur with age. In general, the larger and older individuals take a greater proportion of the larger crustaceans than do the smaller fishes (Table 4.4).

Seasonal changes in diet have also been observed for some fishes and they may reflect availability of prey. Geographical variation in diet composition also occurs in widely distributed species. Midwater fishes are thus best characterized as opportunistic predators, in which the diet composition strongly reflects the prey availability.

Omori (1974) reports most decapod crustaceans are also predators or omnivores utilizing various available food items. Analyses of stom-

Table 4.4   CHANGES IN DIET WITH SIZE AND AGE IN SIX MIDWATER FISHES

	Size range (mm)	Percent of food < 5 mm	Percent of food > 5 mm	Trend in diet composition (increasing percent)
Argyropelecus aculeatus	10–20	99	1	Ostracods, copepods,
(WN central Atlantic)	50–60	12	88	→ fish
Argyropelecus affinis	21–34	94	6	Copepods, ostracods,
(off southern California)	52–68	68	33	→ euphausiids, salps, chaetognaths
Argyropelecus sladeni	10–20	66	34	Ostracods, copepods,
(off southern California)	30–50	24	76	chaetognaths → euphausiids
Sternoptyx diaphana	10–24	100	0	Copepods, ostracods,
(Pacific subantarctic)	40–80	60	40	→ euphausiids
Sternoptyx obscura	10–20	86	14	Copepods → ostracods,
(off southern California)	30–50	63	37	amphipods, chaetognaths
Valenciennellus tripunctulatus	15–20	85	15	Small copepods →
(E. Gulf of Mexico)	30–35	58	42	larger copepods

Source: Modified from Hopkins and Baird, 1977.

achs of midwater shrimps suggest that copepods, euphausiids, and other small crustaceans constitute the majority of the food. There is also evidence for a certain amount of cannibalism. As with fishes, the shrimps are not specialists, but opportunists, taking what is available within a certain size range.

In spite of the paucity of food in the pelagic areas of the deep sea—a condition that, as we have seen, causes the organisms to have the varied diets observed among fishes and crustaceans—there is a certain amount of selectivity and food resource partitioning among animals. The most common form of food resource partitioning occurs when species feed in different depth zones in the water column. Different prey items or different prey abundances occur in different depth zones; thus, predators feeding in these different zones encounter and consume prey populations of differing composition. Even within a given depth zone, the prey composition of the resident predators varies, indicating a certain degree of resource partitioning. This is probably due to anatomical differences in the prey capture mechanisms of the predator. Such structural modifications allow them to prey only on certain prey sizes or shapes. Since many organisms migrate through depth zones on a diurnal basis, the available prey composition varies over a 24-hour period at any given depth. Therefore, a predator that feeds during the day at a given depth may encounter different prey than one feeding at the same depth at night.

Life History Patterns of Midwater Organisms

Most fishes and decapod crustaceans of the mesopelagic zone are short-lived creatures. According to Mauchline (1977), myctopid and gonostomatid fishes tend to become sexually mature at from one to three years and probably live no more than two to four years.

The reproductive cycle among midwater fishes is not as yet well known. Data from Clarke (1973, 1974) on species of myctopids and stomiatoid fishes from the central North Pacific Ocean suggest that the fishes have a seasonal reproductive cycle and spawn in the spring and summer. Juveniles do not appear to undergo vertical migration as do adults, and they are distributed in shallower water than are the adults.

Decapod shrimps of the dominant mesopelagic genera *Sergestes, Pasiphaea,* and *Gennadas* mature in their first or second year and have life spans of one to three years. Euphausiids of midwater zones generally mature in their first year and live perhaps two years. On the other hand, the bathypelagic and abyssalpelagic crustaceans are estimated to live considerably longer, as much as two to seven times longer.

The lower temperatures and the greater paucity of food in the bathypelagic and abyssalpelagic zones probably decrease the rate of growth, delay the onset of sexual maturity, and, therefore, increase longevity of these animals. The extreme expression of this trend is the aforementioned "abyssal gigantism" (pp. 140–142).

Mesopelagic decapod shrimps also show seasonal reproductive activity, usually spawning in late winter or spring in temperate-zone seas. In bathypelagic shrimps, it appears that spawning is prolonged over more of the year. Little is known about the movements and distribution of larval shrimps.

BIBLIOGRAPHY

Andersen, N. R., and B. J. Zahuranec (eds.). 1977. Oceanic sound scattering prediction. Plenum, N.Y. 859 pp.

Angel, M. V., and M. Sibuet (eds.). 1990. Deep sea biology. Prog. Oceanog. 24(1–4):1–343.

Anonymous. 1984. Deep-sea hot springs and cold seeps. Oceanus 27:3–78.

Arp, A. J., and J. J. Childress. 1983. Sulfide binding by the blood of the hydrothermal vent tube worm *Riftia pachyptila.* Science 219:295–297.

Ballard, R. D. 1977. Notes on a major oceanographic find. Oceanus 20(3):35–44.

Blackburn, M. 1977. Studies on pelagic animal biomasses, pp. 283–299. *In:* Andersen, N. R., and B. J. Zahuranec (eds.). Oceanic sound scattering prediction. Plenum, N.Y.

Brauer, R. (ed.). 1972. Barobiology and the experimental biology of the deep sea. North Carolina Sea Grant Program, University of North Carolina, Chapel Hill. 428 pp.

Bruun, A. 1957. Deep-sea and abyssal depths, chpt. 22, pp. 641–672. *In:* Hedgpeth, J. E. (ed.). The treatise on marine ecology and paleoecology. Vol. I, Ecology. Memoir 67, Geol. Soc. Amer.

Childress, J. (ed.). 1988. Hydrothermal vents: A case study of the biology and chemistry of a deep sea hydrothermal vent of the Galápagos Rift. The Rose Garden in 1985. Deep-Sea Res. 35(10–11A):1677–1849.

Childress, J., and M. Nygaard. 1973. The chemical composition of midwater fishes as a function of depth of occurrence off southern California. Deep-Sea Res. 20:1093–1111.

Childress, J., and M. Nygaard. 1974. Chemical composition and buoyancy of midwater crustaceans as a function of depth of occurrence off southern California. Mar. Biol. 27:225–238.

Clarke, T. S. 1973. Some aspects of the ecology of lantern fishes (Myctophidae) in the Pacific Ocean near Hawaii. Fish. Bull. U.S. 71:401–434.

Clarke, T. S. 1974. Some aspects of the ecology of stomiatoid fishes in the Pacific Ocean near Hawaii. Fish. Bull. U.S. 72:337–351.

Dayton, P., and R. H. Hessler. 1972. The role of biological disturbance in maintaining diversity in the deep sea. Deep-Sea Res. 19:199–208.

Dietz, R. S. 1962. The sea's deep scattering layers. Sci. Amer. 207(2):44–50.

Ernst, W. G., and J. G. Morin (eds.). 1982. The environment of the deep sea. Prentice-Hall, Englewood Cliffs, N.J. 371 pp.

Gage, J. D., and P. A. Tyler. 1991. Deep sea biology. Cambridge University Press, N.Y. 504 pp.

Grassle, J. F. 1985. Hydrothermal vent animals: Distribution and biology. Science 229:713–717.

Grassle, J. F., and L. Morse-Porteous. 1987. Macrofaunal colonization of disturbed deep sea environments and the structure of deep sea benthic communities. Deep-Sea Res. 34:1911–1950.

Hardy, A. 1956. The open sea. Vol. I, The world of plankton; chpt. 12, Life in the depths. Houghton Mifflin, N.Y. 385 pp.

Hedgpeth, J. 1957. Classification of marine environments, chpt. 2, pp. 18–27. *In:* The treatise on marine ecology and paleoecology. Vol. I, Ecology. Memoir 67, Geol. Soc. of Amer.

Heezen, B. C., and C. D. Hollister. 1971. The face of the deep. Oxford University Press, N.Y. 659 pp.

Hessler, R. R., and W. M. Smithey, Jr. 1983. The community structure of megafauna, pp. 735–770. *In:* Rona, P. A., et al. (eds.). Hydrothermal processes at seafloor spreading centers. Plenum, N.Y.

Hessler, R. R., W. M. Smithey, M. A. Boudrias, C. H. Keller, R. A. Lutz, and J. Childress. 1988. Temporal changes in the megafauna at the Rose Garden hydrothermal vent (Galápagos rift; eastern tropical Pacific). Deep-Sea Res. 35(10–11):1681–1709.

Hopkins, T. L., and R. C. Baird. 1977. Aspects of the feeding ecology of oceanic midwater fishes, pp.

325–360. *In:* Andersen, N. R., and B. J. Zahuranec (eds.). Oceanic sound scattering prediction. Plenum, N.Y.

Idyll, C. P. 1976. Abyss. Crowell, N.Y. 428 pp.

Jones, M. E. (ed.). 1985. Hydrothermal vents of the eastern Pacific, An overview. Bull. Biol. Soc. Washington No. 6. 547 pp.

Jumars, P. A., and E. D. Gallagher. 1982. Deep-sea community structure: Three plays on the benthic proscenium, pp. 218–255. *In:* Ernst, W. G., and J. G. Morin (eds.). The environment of the deep sea. Prentice-Hall, N.Y.

Karl, D. M., D. J. Burns, K. Orrett, and H. W. Jannasch. 1984. Thermophilic microbial activity in samples from deep sea hydrothermal vents. Mar. Biol. Lett. 5:227–231.

MacDonald, A. G. 1975. Physiological aspects of deep sea biology. Cambridge University Press, N.Y. 450 pp.

Marshall, N. B. 1954. Aspects of deep-sea biology. Philosophical Library, N.Y. 380 pp.

Marshall, N. B. 1980. Deep sea biology: Developments and perspectives. Garland STPM Press, N.Y. 566 pp.

Mauchline, J. 1972. The biology of bathypelagic organisms, especially crustacea. Deep-Sea Res. 19:753–780.

Mauchline, J. 1977. Estimating production of midwater organisms, pp. 177–215. *In:* Anderson, N. R., and B. J. Zahuranec (eds.). Oceanic sound scattering prediction. Plenum, N.Y.

McAllister, D. E. 1961. A collection of oceanic fishes from off British Columbia with a discussion of the evolution of the black peritoneum. Bull. Nat. Mus. Canada 172:39–43.

Meade, G. W., E. Bertelson, and D. M. Cohen. 1964. Reproduction among deep sea fishes. Deep-Sea Res. 11:569–596.

Menzies, R. J. 1964. Improved techniques for benthic trawling at depths greater than 2000 meters. Biol. Antarctic Seas. Antarctic Res. Ser., Amer. Geophys. Union 1:93–109.

Menzies, R. J., R. Y. George, and G. T. Rowe. 1973. The abyssal environment and ecology of the world oceans. Wiley, N.Y. 488 pp.

Nixon, M., and J. B. Messenger (eds.). 1977. The biology of cephalopods. Academic Press, N.Y. 615 pp.

Omori, M. 1974. The biology of pelagic shrimp in the ocean. Adv. Mar. Biol. 12:233–324.

Rex, M. A. 1981. Community structure in the deep sea benthos. Ann. Rev. Ecol. Sys. 12:331–353.

Rokop, F. 1974. Reproductive patterns in the deep-sea benthos. Science 186:743–745.

Rona, R. P., K. Bostrum, L. Laubier, and K. L. Smith (eds.). 1983. Hydrothermal processes at seafloor spreading centers. Plenum, N.Y. 796 pp.

Rowe, G. T. (ed.). 1983. The sea. Vol. 8, Deep sea biology. Wiley, New York. 560 pp.

Sanders, H. L. 1968. Marine benthic diversity, a comparative study. Amer. Nat. 102:243–282.

Sanders, H. L., and R. R. Hessler. 1969. Ecology of the deep-sea benthos. Science 163:1419–1424.

Siebenaller, J., and G. N. Somero. 1978. Pressure adaptive differences in lactate dehydrogenases of congeneric fishes living at different depths. Science 201(4352):255–257.

Smith, K. L., and K. R. Hinga. 1983. Sediment community respiration in the deep sea, pp. 331–370. *In:* Rowe, G. T. (ed.). The sea. Vol. 8, Deep sea biology.

Somero, G. N. 1982. Physiological and biochemical adaptations of deep sea fishes: Adaptive responses to the physical and biological characteristics of the abyss, pp. 256–278. *In:* Ernst, W. G., and J. G. Morin (eds.). The environment of the deep sea. Prentice-Hall, Englewood Cliffs, N.J.

Tunnicliffe, V., J. F. Garrett, and H. P. Johnson. 1990. Physical and biological factors affecting the behaviour and mortality of hydrothermal vent tubeworms (vestimentiferans). Deep-Sea Res. 37 (1):103–125.

Yayanos, A. A., A. S. Dietz, and R. van Boxtel. 1979. Isolation of a deep sea barophilic bacterium and some of its growth characteristics. Science 205(4408):808–809.

Young, R. E. 1975. Function of the dimorphic eyes in the midwater squid, *Histioteuthis dofleini.* Pac. Sci. 29(2):211–218.

Young, R. E. 1983. Oceanic bioluminescence: An overview of general functions. Bull. Mar. Sci. 33:829–845.

Zimmerman, A. M., and S. B. Zimmerman. 1972. Commentary on high pressure effects on cellular systems, pp. 140–147. *In:* Brauer, R. W. (ed.). Barobiology and the experimental biology of the deep sea. North Carolina Sea Grant Program, University of North Carolina. 428 pp.

Shallow-Water Subtidal Benthic Associations

*I*n this chapter we deal with the bottom communities and conditions for life in the shallow waters fringing the landmasses of the world— namely, the continental shelf region of the world's oceans. This is a region about which we have considerably more knowledge than we have about the deep sea. It is more accessible and has been the object of much study because the continental shelf areas are the major fishing grounds of the world. This chapter covers major subtidal associations except for coral reefs, which, because of their many unique properties, are covered separately in Chapter 9. The shallow subtidal area is contiguous at its upper end with the intertidal regions of the world's oceans, and

at least some of the principles that we discuss in this chapter will be applicable to the intertidal zone also.

Coverage and Definitions

This chapter covers that area of the oceans that lies between the area of lowest low water on the shore to the edge of the continental shelf at a depth of about 200 m. In our scheme of classification, this area is known as the *sublittoral*. Overlying it are the waters of the *neritic* zone (see Chapter 1, Fig. 1.20, on the scheme of zonation). Most of this zone is composed of soft sediments, sand and mud, and a much lesser area of hard substrate. Therefore on an areal and numerical basis, the bottom is dominated by infaunal organisms.

Infaunal organisms are usually divided into categories based on size. *Macrofauna* consists of those organisms that are greater than 0.5 mm in size. The term *meiofauna* is applied to those organisms that lie within the size range of 0.5 to 0.062 mm. A final size class is called *microfauna,* which includes organisms below 0.062 mm in size. This group is primarily composed of protozoa and bacteria. This chapter will deal primarily with macrofauna organisms. Meiofauna and microfauna are covered in Chapter 7.

ENVIRONMENTAL CONDITIONS

The continental shelf waters show more variability in environmental conditions than do either the epipelagic of the open ocean or the deep sea. Perhaps the most important physical factor that acts on the bottom communities is turbulence or wave action. In these shallower waters, the interaction of waves, currents, and upwelling all act to create turbulence. This turbulence generally keeps inshore waters from becoming thermally stratified except for brief periods, at least in the temperate zone. Thus nutrients are rarely limiting or locked up in a bottom reservoir. Productivity is generally higher than in similar waters offshore because of nutrient abundance, both from runoff from land and recycling (see Table 2.2). This high productivity sustains high populations of both zooplankton and benthic organisms.

Wave action is an important factor in this area. Long-period ocean swell and storm waves have an effect that can extend to the bottom in these shallow waters. In soft bottoms, the passage of such waves may cause large surging motions in the bottom water, which greatly affects the stability of the substrate. The substrate particles may be moved around and resuspended. This has a profound effect on the infaunal animals in the substrate. Wave action also determines the type of particles that are present. Heavy wave action removes fine particles by keeping them in suspension, leaving mainly sand; thus, fine silt sediments can occur only in areas that have low wave action or are too deep to be affected by wave action.

Salinity in this region is more variable than in the open ocean or deep ocean, but except for areas where large rivers discharge massive amounts of fresh water, the salinity does not change enough to be of ecological significance.

Temperature is also more variable in inshore waters and shows a definite seasonal change in the temperate zone. These temperature changes may be used by organisms as cues to begin or end various activities such as reproduction.

Light penetration in these turbulent waters is reduced when contrasted with open ocean areas. The combination of large amounts of debris, both from land and from breakup of kelp and seagrasses, plus the high plankton densities due to abundant nutrients, act to reduce light penetration from 10 to 20 m to a few meters.

The food supply is abundant in this area, partly because of the increased productivity by the plankton, but also because of the production of attached plants such as kelp and seagrasses. This is one of the few areas in the sea where macroscopic plants have any significance in production. A final food source is runoff from land. Although there are large plants in the sublittoral shelf, there are relatively few large grazing animals, and the major use of kelps and seagrasses as food occurs only after they have been broken up into detrital particles.

The soft bottoms in the sublittoral are essentially without large topographic feature diversity, and the vast expanses extend monotonously for long distances. Small feature diversity exists in the form of ripple marks, worm tubes, fecal mounds, and so on (Fig. 5.1). Lacking large topographic relief, the only apparent difference from one place to another is that of substrate grain size and composition (silica vs. calcium carbonate). Subtidal hard substrates, on the other hand, may have considerable relief with many potential habitats. The lack of major topographic relief in the infaunal areas generally means fewer different habitats for animals to occupy and fewer potential ways for making a living. The number of infaunal species is thus generally less than the number of epifauna species. There are, if you wish, fewer niches available. The most abundant infaunal animals are deposit feeders,

Figure 5.1 A shallow-water soft bottom. Note the ripple marks. (Photo courtesy of Dr. Glen VanBlaricom.)

ingesting the abundant detritus that rains down, or else are suspension feeders, filtering out of the water column the abundant plankton or floating detritus. A lesser number, mainly worms, crustaceans, mollusks, and echinoderms, are predators. Bottom fishes, on the other hand, are predominantly carnivores.

It is convenient to divide the subtidal continental shelf region into four major habitat types, which are readily recognizable and topographically or physiognomically different, and discuss each of these separately. The four are: open unvegetated sedimentary environments (the most common areally), hard substrates dominated by low-encrusting plants and animals, kelp beds and forests, and seagrass beds.

UNVEGETATED SEDIMENTARY ENVIRONMENTS

Sampling

From the earliest times of Petersen (see subsequent section) until about the 1960s, the common method for sampling the organisms of the soft sediments of the continental shelf was to drop a *grab* (Fig. 5.2). Various types of benthic grabs have been developed over the years, and each has certain advantages and disadvantages. Most have the advantage of being quantitative; that is, they take a sample of known areal extent from the bottom. Standard treatment of these grab samples has been to screen out the organisms on one or a series of screens. Mesh sizes of 1.0 mm and 0.5 mm are the most commonly employed.

Although such sampling was quantitative and gave a good feeling for the kinds of organisms present and their relative abundance, it was a remote method in which the benthic biologist never actually saw the community. It was also not possible to do any manipulative experiments with the community. This was changed, in the 1960s, with the advent of scuba (self-contained underwater breathing apparatus), which allowed the biologist to investigate a community firsthand and to undertake manipulative experiments such as using cages to keep various organisms in or out of the community. Thus changes in time could be followed. Experimental ecology arrived with scuba, but because it was not used until later in sedimentary environments as opposed to the hard substrates, we have as yet fewer data for the soft sediments.

Grabs continue to be employed today, particularly in areas below safe diving depths (about 30 m), but within safe diving depths, actual

(A) *(B)* *(C)*

Figure 5.2 Various types of grab samplers. *(A)* Peterson grab. The jaws close as the cable is pulled upward, thus securing the sample. *(B)* Smith-McIntyre grab. This is a spring-loaded grab, in open position here, which will close when it hits the bottom. *(C)* Reineck box corer. The "box" is closed by a knife edge actuated by pulling up on the cable. (All after N. A. Holme and A. D. McIntyre [eds.], Methods for the study of marine benthos, 2nd ed., Blackwell Scientific Publications, 1984.)

firsthand work by divers seems to be more popular and more productive.

Other sources of data include remote, towed cameras or video systems, and side-scan sonar. The latter is capable of giving very accurate small-scale detail of the bottom topography.

Infaunal Associations

The dominant groups of organisms that constitute the macrofauna present in sublittoral soft bottoms belong to four main taxonomic groups: class Polychaeta, subphylum Crustacea, phylum Echinodermata, and phylum Mollusca. Polychaete worms are the most abundant and are represented by numerous tube-building and burrowing species. The dominant crustaceans are the larger ostracods, amphipods, isopods, tanaids, mysids, and a few of the smaller decapods. They mainly inhabit the surface of the sand and mud. Mollusks are mainly represented by various burrowing bivalve species with a few gastropods at the surface. Echinoderms particularly common in the subtidal benthos include brittle stars, heart urchins, sand dollars, sea cucumbers, and a few predatory sea stars.

It is of interest that the first quantitative work done in marine ecology was done on the sublittoral benthic infauna. Back in the first years of

this century, C. G. Joh. Petersen, a Danish biologist, began his investigations on the benthos of the Danish seas. He was interested in evaluating the role of bottom organisms in supporting the fish populations on which the Danish fishing industry depended. In order to do this on a quantitative basis, he invented a grab, the now famous "Petersen grab," which picked up a definite area of the bottom, usually 0.5 or 0.1 m^2 (Fig. 5.2). Since he had a sample that covered a known area of bottom, he could then count the numbers of organisms in those samples and from them extrapolate the numbers in the whole bottom area. Once this was done, he would have some estimate of the amount of food available to the fishes and could guess at the biomass of fish that could be supported.

Analyzing many of these grab samples over the years, Petersen (1918, 1924) observed that extensive bottom areas were occupied by recurrent groups of species and that other areas were inhabited by different associations of species. In all cases, he observed that a relatively few species made up most of the individuals and biomass. This was in contrast to nonquantitative dredge samples taken in the same waters. In the latter case, the faunal lists might be nearly identical in the two areas. The value of the quantitative study was that for the first time it gave a way to evaluate differences between communities based on relative abundance, not just presence or absence of species.

After several years of evaluating these quantitative samples, Petersen observed that the different areas and their dominant organisms remained relatively constant and uniform over time. He then proposed these associations as communities and named them on the basis of the dominant animals. For example, he had a *Macoma balthica* community from inner Danish waters in 8–10 m of water, which was composed of *M. balthica* as the dominant but

was also characterized by *Arenicola marina, Mya arenaria,* and *Cardium edule* (Fig. 5.3). In this same manner, he named and characterized a whole series of communities from the sublittoral in Scandinavian waters.

Following Petersen's work, various benthic biologists began to investigate other shallow-water benthic areas around the world. The leader of this school was another Dane, G. Thorson. What Thorson (1955) discovered, particularly for the temperate-zone waters, was that communities of organisms similar to those found by Petersen were found in similar habitats around

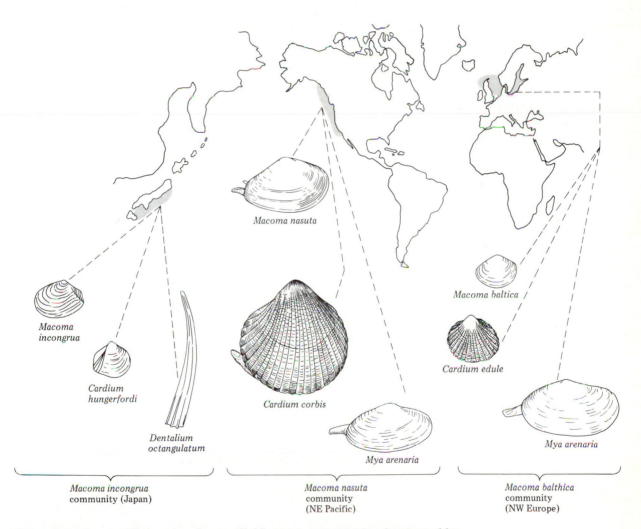

Figure 5.3 Diagram illustrating the parallel bottom communities dominated by the bivalve *Macoma* in three different areas in the North Temperate Zone. (Modified from G. Thorson, Bottom communities. *In:* The treatise on marine ecology and paleoecology, Vol. 1, Ecology, edited by J. Hedgpeth, The Geological Society of America, 1957.)

the world. Thus arose the concept of *parallel bottom communities.* In this ecological concept, similar sediment types at the same depths around the world contain similar communities. Granted, the species are not the same, but they are closely similar ecologically and taxonomically (Fig. 5.3). Additional work in tropical waters by Thorson (1966) revealed that this concept did not extend to such areas, at least not in the same way.

That certain recurrent groups occupy the same substrate types over large areas of the world's ocean seems established. This suggests that such associations are not random, but represent real interacting systems in which some combination of factors acts to ensure continued persistence of the community.

Petersen described the communities, and others later validated their continued existence over time and in various parts of the world. What was lacking in all this descriptive material was an explanation of how such communities persisted and what kinds of interactions ensured continued success. Since Petersen's time, answers to these questions have come to light and will be discussed in the following sections.

Community Pattern and Structure

Most marine organisms are distributed in a patchy manner. We noted this with respect to the plankton in Chapter 2 (pp. 82–85). It is also the situation that one encounters when analyzing benthic organisms. Even in the early days of Petersen not all quantitative samples had either the same species or the same relative abundances. They also varied with season. Petersen was able to characterize his communities primarily on a statistical basis dependent on the presence of a large number of samples. In other words, even in these open sedimentary environments that are removed from the more physically variable surface or intertidal areas, and which may appear monotonous to our eyes, patchiness of organisms exists—both in time and space. It is the result of the various physical factors and interactions of organisms and gives rise to the diverse patterns that we observe in the distribution of benthic associations. Certain cycles that may occur on a daily, seasonal, yearly, or longer

basis also have their origins in the interactions producing patchiness. A further complicating factor is that variability in benthic associations is not only on a horizontal scale, but on a vertical scale as well, since many of these infaunal organisms burrow to different depths in the substrate.

Within any sedimentary benthic community there is a remarkable number of macrofaunal (greater than 0.5 mm in size) organisms, usually on the order of hundreds, such that it is difficult to work with individual species in any attempt to understand the various ecological relationships. Fortunately, as Woodin (1983) has pointed out, the various organisms in these environments can be classified into a limited number of ecological categories, and one can then use the functional groups to compare assemblages of species and to attempt to arrive at generalizations regarding how the communities operate. A functional group includes all the various species of different animal taxa that use and affect the environment in similar ways. For example, in Chapter 1 we noted the division of animals into opportunistic or equilibrium species. That is one functional classification.

Another approach would be to classify various species as *sediment destabilizers* and *sediment stabilizers.* Destabilizers or *bioturbators* include both motile and sedentary organisms whose activities, either physical movement in the case of the motile ones or biological activities in the case of the sedentary ones, cause the sediment to move, become resuspended, erode, or otherwise change. Examples of this kind of organism are mobile deposit-feeding clams such as *Nucula, Yoldia,* and *Macoma* studied by Rhoads and Young (1970) and the sedentary deposit-feeding sea cucumber *Molpadia oolitica,* which Rhoads and Young (1971) found to destabilize the sediment by its feeding and defecation (Fig. 5.4). Sediment stabilizers include plants such as various seagrasses, whose roots bind sediments and whose upper parts change the local hydrographic regime (see pp. 196–197). They also include various invertebrate tube builders, which, when in high density, form mats or areas in which sediment stability is high. Examples of tube dwellers that stabilize sediments include amphipod crustaceans, phoronid worms, and polychaetes (Fig. 5.4; Plate 34). This

Phoronopsis   *Owenia*   *Zostera*   *Yoldia*   *Macoma*   *Molpadia*

Sediment stabilizers   Sediment destabilizers

Figure 5.4 Representatives of the two functional groups of sedimentary organisms: sediment stabilizers and sediment destabilizers or bioturbators.

classification of sediment stabilizers and destabilizers is an improvement on Woodin's (1976) earlier classification of burrowing deposit feeders, tube builders, and suspension feeders.

Still another functional division could be the more traditional trophic division of herbivore, omnivore, and carnivore. Whereas it is usually easy to classify most infaunal organisms into such functional divisions, there are those that are intermediate. For example, maldanid polychaetes are tube builders, but they are also deposit feeders. To place such an organism into a stabilizing or destabilizing category will require more information on their effect on the substrate. It is also important to note here that any functional group includes a number of different taxa of invertebrates and may well include different nutritional types. Thus, for example, we may have mobile burrowers that are predators, deposit feeders, or suspension feeders, and they may also be divided into those that are opportunistic and those that are equilibrium in their life history features. In other words, categorization by functional groups can be done several ways and does not necessarily exclude certain other features of the life history of the organism. The only trophic mode that is absent from open sedimentary communities is macroautotrophy. There are no large plants, and any herbivores in this system must feed upon plant material that is either in the plankton (suspension feeders) or

benthic microalgae (diatoms) or that is brought in from outside as debris (deposit feeders or scavengers).

Community Structure and Change

Basically, changes in sedimentary community structure can come about through either the action of physical factors or biological factors, or some combination of the two. The simplest cases involve the dominance of physical factors. For example, along the Pacific coast of California, Oliver et al. (1979) have demonstrated that subtidal benthic communities are zoned along a gradient of increasing wave disturbance to the bottom. In shallow water (less than 14 m) the bottom sediments are regularly disrupted by waves, and the dominant inhabitants are various small, mobile crustaceans, whereas in the deeper zones (greater than 14 m) the fauna is dominated by tube-building polychaete worms. Evidence showed that wave action prevented the establishment of worm tubes in the shallow water. Wherever wave action was diminished, the polychaete zone became shallower, and wherever wave action was stronger, the worm zone was driven deeper.

A more complicated example in which both biological and physical factors act is seen in an East Coast example. Mills (1969) studied a sandy area in Barnstable Harbor, Massachusetts, that had for some time been dominated by the snail *Ilyanassa obsoletus.* This community had low biomass and low numbers of individuals of associated organisms due probably to the mechanical disturbance of the sediments by the snails. Because of the low density of infaunal animals, this sandy area was colonized in spring by the tube-building amphipod *Ampelisca abdita.* The *Ampelisca* quickly built a dense mat of tubes that excluded *Ilyanassa.* Selective deposit feeding by the *Ampelisca* allowed fine sediment particles to accumulate, and the increased topographic diversity created by the tubes also changed the physical environment of the area. This led to a change in the biological environment and the colonization of the mats and sediments by an abundance of polychaete worms favoring fine sediments. The *Ampelisca* tube mats are, however, not stable, and stormy weather, particularly in winter, causes large areas to wash out. This permits the return of *Ilyanassa,* and the cycle can presumably start all over again.

The role of biological factors in deciding community structure is more complex. The dominant biological factors that act to alter community structure are competition, predation, and recruitment. We shall consider each separately and then consider some test cases where two or more interact.

Competitive interactions among organisms can be direct or indirect. Direct competition involves either contact between organisms or the direct interference in some aspect of the other organism's life cycle. Indirect competition usually involves one organism interrupting or preventing resource use by another.

Direct interactions, including those often seen between organisms on hard substrates, such as overgrowth, pushing off other organisms, and so forth (see Chapter 6), are not as common in sedimentary environments. This is probably because such activity involves having a solid substrate to push against. It also may be inherent in colonial organisms, which are conspicuous by their absence in sedimentary substrates. Direct competition, which involves aggressive activities leading to exclusion or spacing out, has been reported by several authors for various polychaetes and small crustaceans, but most studies

were done in the laboratory. Still, the finding of a regular distribution pattern such as found by Levin (1981), Reise (1979), Wilson (1981), and others suggests that these kinds of actions occur. Some direct interactions that displace other species have been reported by Levinton (1977), who has shown that movement of the mobile bivalve *Yoldia limatula* within the substrate disrupts the burrows of the more sedentary bivalve *Solemya velum,* leading to its replacement. Myers (1977) reports a similar situation in which a burrowing sea cucumber does the same thing to small bivalves.

Indirect interactions leading to changes in community structure and organization appear to be more common in sedimentary communities. These indirect actions usually involve some sort of change in the sediment condition.

Burrowing deposit feeders tend to be most abundant in soft, muddy sediments, areas with a high concentration of organic matter. It has been observed in several areas that, where burrowing deposit feeders are prevalent, suspension feeders are rare or absent. What happens is that the deposit-feeding organisms burrow through the top few centimeters of bottom and, in so doing, create a very loose, unstable layer of fine particles including fecal pellets. This layer is then easily resuspended by the slightest water motion. The resuspended sediment has the effect of clogging up the fine filtering structures of the suspension feeders, rendering them unusable. Furthermore, the constant reworking and settlement of resuspended particles tend to lead to burial of the newly settled larvae of suspension feeders, thus killing them. It does not kill the larvae of the deposit feeders because they burrow into the more compact substrate underneath. The clogging of filtering surfaces makes feeding impossible and leads to the suffocation of settling larvae that excludes the suspension feeders. This exclusion of one trophic group by modification of the environment by another is called by Rhoads and Young (1970) *trophic group amensalism.* In this way, then, deposit feeders maintain their own communities and exclude suspension feeders. An example of this is seen in Rhoads and Young's (1970) work in Buzzard's Bay, Massachusetts, where they showed that reworking of the sediments by three deposit-feeding clams, *Nucula proxima, Macoma tenta,* and *Yoldia limatula,* effectively excluded the suspension feeders.

Suspension-feeding organisms seem to be more abundant where substrate is more sandy, where the organic matter is less, and thus where deposit feeders would find less food and more difficult burrowing. In the absence of deposit feeders, the suspension feeders can establish themselves. Once established, they may also exclude any potential deposit feeders by filtering larvae out of the water.

A more complicated situation occurs over large areas of Cape Cod Bay, Massachusetts, where a large sea cucumber, *Molpadia oolitica,* occurs in high densities (2 to 6 per square meter) in silt-clay muds. This animal is a sedentary deposit feeder that lives head down vertically in the substrate. It ingests sediment at depth and deposits loose fecal matter at the surface, forming a mound around the anal opening. The reworking of the sediment produces a loose, high-water-content, easily suspended sediment layer on the surface of the substrate except in the immediate vicinity of the fecal cone. The areas between the cones are unstable and exclude suspension feeders (= amensalism). On the other hand, the stable cone areas attract the tube-building, suspension-feeding polychaete *Euchone incolor* and other tube-dwelling polychaetes. These tubes, in turn, help to further stabilize the fecal cone area and attract other suspension feeders (Fig. 5.5). It is important to note in this example that *Molpadia* is a relatively large animal. In order to effect changes that will influence other animals, an animal must be large enough or abundant enough to change large areas of the sediment.

Tube-building organisms have the ability to stabilize the substrate with their tubes (Plate 34). They also extend down into the substrate. By stabilizing the substrate, they prevent resuspension of fine particles. At the same time, the presence of the tubes in the substrate in sufficiently high densities serves to restrict the available space for burrowing animals. The burrowers cannot burrow through because they collide with the rigid tubes. Thus, the tube builders act indirectly to restrict burrowing animals. Such exclusion by interference with normal activities

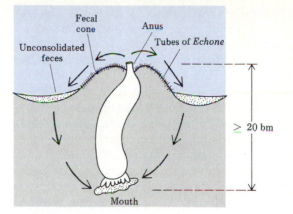

Figure 5.5 *Molpadia oolitica.* Diagrammatic representation of feeding position, showing surface cone and depression and subsurface excavation produced by intensive feeding. (From D. C. Rhoads and D. K. Young, Animal-sediment relations in Cape Cod Bay, Massachusetts, II. Reworking by *Molpadia oolitica* [Holothuroidea], Mar. Biol. 11[3]:255–261, 1971.)

is termed *competitive interference.* Although burrowing deposit feeders rarely co-occur with suspension feeders or with tube dwellers, and vice versa, it is possible to find these different functional groups in the same geographical area, usually with rather sharp boundaries between them. These boundaries usually do not coincide with any physical discontinuity and are presumed to be the result of either predation or some other disturbance, physical or biological, that has removed the resident organisms. Since most sedimentary animals are solitary and do not reproduce asexually, any patch created by disturbance must be reoccupied either by immigration of adults or settlement of larvae (see pp. 23–27 in Chapter 1).

The second biological factor that is important in determining the species structure of infaunal communities is predation. Both invertebrate and vertebrate predators take a toll on the infaunal organisms. Predator activity may be responsible for clearing certain small areas of macrofauna and thus creating a disturbance that will be followed by a recolonization sequence. This is one way that patchy distributions occur on the bottom. Such clearing of areas is known to result from feeding activities of certain flatfish, rays, whales, and walruses (see Chapter 3).

Sediments, as we have noted, are profoundly three-dimensional substrates, and it may be possible for some prey organisms to use this three-dimensionality as a refuge from predators. If, for example, the predator is unable to dig into the substrate after its prey and the prey is able to burrow deeply, then the depth becomes a refuge for the prey. By the same token, however, if a predator can dig as far down into the substrate as the prey can, then there is no refuge. Moreover, the predator in its digging activities may have a much more widespread effect on the other members of the infauna community by destroying or disrupting the surrounding sediments. For example, VanBlaricom (1982) showed, in a subtidal area off southern California, that large rays created pits in the sand in the process of removing resident clams. These large pits had most other infauna destroyed and were places where resettlement could take place.

Predators, depending on their position and effect on the sediment, have been classified into three different types by Woodin (1983). *Surface predators* are those that are exposed at the surface and take organisms at or near the surface without disrupting the sediment structure (Fig. 5.6). Included here are most fishes and various crabs, as well as certain invertebrates that may

be infaunal themselves but that capture and ingest other infaunal adults, juveniles, or larvae. A peculiar group that needs to be included here consists of "browsers" that take off only parts of the prey. An example would be the nipping off of siphons from clams by various fish species, as reported by deVlas (1979) and Peterson and Quammen (1982). An interesting "cannibalistic" relationship exists between various suspension feeders and their own and other larvae that may be captured and ingested as they settle to the substrate and are intercepted by the feeding nets of the suspension feeders. A special case includes predators such as worms of the genus *Glycera,* which form burrows in the substrate but in fact feed mainly at the surface, emerging from their burrows to seize prey items (Ockelmann and Vahl, 1970).

Burrowing predators include those that Woodin (1983) terms "weasel" predators, meaning that they move down various tubes or channels provided by the deeper-dwelling prey and attack them. An example would be various nemertean worms and perhaps some gastropods that have a long proboscis that can be extended down into the tubes. An interesting case is that of the sea star *Pisaster brevispinus,* which feeds on deep-dwelling clams on the Pacific coast of North

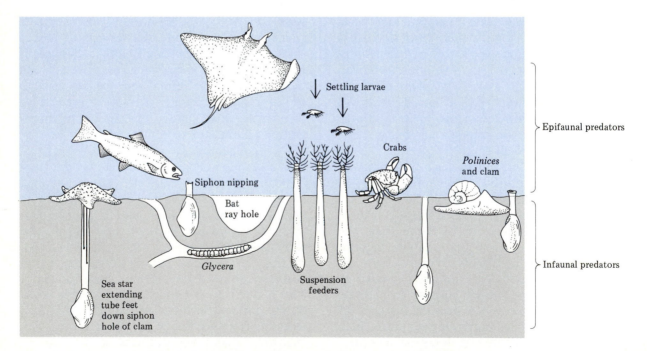

Figure 5.6 Various predator types and their effects on shallow-water sedimentary communities.

America. According to Van Veldhuizen and Phillips (1978), *Pisaster* somehow senses the buried clams as it moves across the surface. It then extends the tube feet from around the mouth area deep into the sediment (it can extend them to a length of 20 cm), attaches them to the clam, and pulls the clam to the surface without greatly disturbing the sediment. It can also simply extend its stomach down to the clam to digest it (Fig. 5.6).

Digging predators excavate holes to get to their food and include such different animals as the blue crab *Callinectes sapidus,* the gastropod *Buccinum undatum,* and the aforementioned rays. Digging predators have a greater effect on the general infaunal organisms than do "weasel" predators because they also disturb the sediment.

A final category includes those predators that burrow through the substrate and live within it at all times. Moon snails of the family Naticidae are a good example because they move just beneath the surface to capture bivalves.

Different predatory types have different effects on their environment and create diverse patterns in sedimentary communities. In general, the digging predators probably have the most widespread and recognizable effects because they create a sediment change. Predation by surface predators seems most likely to have an effect on the size classes of the prey, particularly if there is a deep refuge for the adults, which leaves the juveniles most vulnerable. The siphon nippers would seem to have most effect on the growth of the prey. The effects of other predators are difficult to predict a priori without knowledge of prey preferences, relative predator densities, and other factors.

It is difficult to determine the relative roles of predator-induced and wave-induced disturbances on the structure of soft-bottom communities. This is mainly because it has been difficult in the past to devise good experiments that will exclude predators from areas of the bottom while allowing wave action to proceed. Virnstein (1977) and other ecologists have devised various "cages" that, when placed over areas of the bottom, exclude or include predaceous fishes and invertebrates. When carefully done, such studies have produced increased numbers of infauna organisms within the "cage" as compared to outside (Fig. 5.7). This suggests that predation is important in determining at least the numbers of infaunal organisms present and perhaps also the species composition.

We can summarize our present understanding regarding the relative roles of competition and predation on the structure of sedimentary communities by paraphrasing from a recent review by Wilson (1990), who concludes there is no evidence for both predation and competition affecting community structure but curiously no evidence of any competitive dominant in any soft sediment system. Whereas we have made

Figure 5.7 Effects of a cage that excludes predaceous starfish and fish from the infauna, thus increasing the infauna density inside the cage as compared to outside.

much progress in understanding the organization of infaunal communities in the last 20 years, we still lack the ability to generate a priori predictions of the effects of either competition or predation. We still clearly lack knowledge of growth rates, life spans, and the nature of the food and population dynamics of the dominant soft-bodied invertebrates, particularly deposit feeders. Without these data, it is still too early to come up with any unified theory.

Most benthic infauna and epifaunal invertebrates are recruited from larvae settling out from the plankton. Thus a primary process that could produce changes in infaunal adult abundances and community structure is recruitment. Despite this fundamental relationship, there are few studies analyzing the role of recruitment in determining adult abundances and community structure. This is probably because it is difficult to separate or isolate the effect of recruitment from various postrecruitment mortality sources. In a review of the role of recruitment, Menge and Farrell (1989) suggest that recruitment may be an influence in structuring adult communities when one or more of three conditions prevail. The first is that the recruiting organisms are generally free of other sources of mortality, such as predation and competition. The second is that the recruiting organisms settle in such numbers that they either swamp the predators with sheer numbers or keep pace with the predation rate. Finally, recruitment may be significant if the recruitment rates of all species in the community are low and none of the organisms settling are competitive dominants. It is probably safe to conclude here that recruitment certainly has the potential to influence community structure in these sedimentary communities, or in any benthic community, but that the lack of definitive studies does not permit us to do more than speculate.

Vertical Distribution and Competition

Competition among benthic marine invertebrates is generally restricted to food and space. In the subtidal soft-bottom areas of the oceans, where the dominant species are suspension and deposit feeders, competition is usually for space. Food, in the form of either detritus or plankton, is abundant. Organisms that are suspension feeders must, of course, have their feeding structures in the water column. However, it is possible to get more suspension feeders in a given area of bottom if, instead of placing them all on the surface, they are distributed vertically in the substrate, with only the feeding organ at the surface (Fig. 5.8). Depending on the orientation of the animal in the substrate, the feeding structure may take up less room than the whole animal. Deposit feeders may feed not only at the surface but at any level in the substrate that contains organic material. Such animals do, however, require a connection to the surface to obtain oxygen from the overlying water. In a similar manner then, more deposit feeders could be accommodated and direct space competition reduced if the different species were distributed vertically in the substrate.

Such vertical distribution of species is a well-known feature of soft bottoms. It occurs in both mud and sand (Fig. 5.9) and in both suspension-

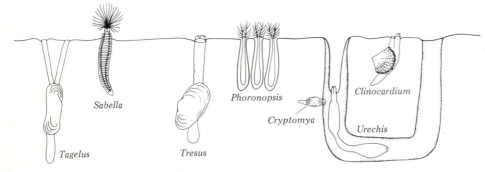

Figure 5.8 Various infaunal suspension-feeding types distributed at different depths in the substrate.

(A) East Coast

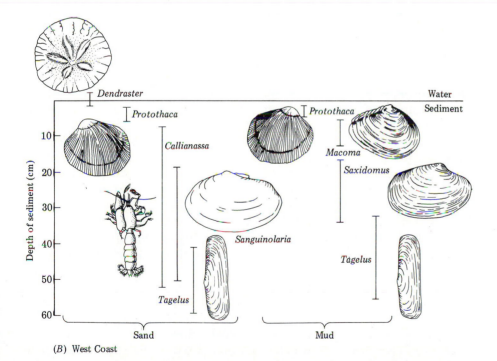

(B) West Coast

Figure 5.9 Vertical distribution of benthic infauna in (A) an Atlantic Coast bay, Quisset Harbor, Massachusetts, and (B) a Pacific Coast lagoon, Mugu Lagoon, California. (Modified from J. Levinton, Ecology of shallow water deposit feeding communities, Quisset Harbor, Mass. *In:* Ecology of marine benthos, edited by B. Coull, University of South Carolina Press, 1977; and from C. H. Peterson, Competitive organization of the soft bottom macrobenthic communities of southern California lagoons, Mar. Biol. 43:343–359, 1977.)

and deposit-feeding organisms. Experiments by Peterson (1977), in which certain species were removed from given depth strata, resulted in an increase in abundance of other species occupying the same depth level. Still other experiments, in which the density of given species at given levels was increased above the natural levels, have resulted in rapid emigration out of the high-density areas by the species. These experiments together suggest that there is definite competition for space among species, and it results not only in vertical spacing of species but also in maintenance of certain densities.

Competition for space may also be the reason for the occurrence of *commensalism* among certain species (see Chapter 10 for discussion). That is to say, competition is so severe that it selects for organisms that share space. An example is the clam *Cryptomya californica,* which lives with its siphons in burrows of the shrimp *Callianassa californiensis* or the worm *Urechis caupo* (Fig. 5.8).

Community Patterns, Stability, and Variability

Community structure in open sedimentary environments changes over differing time periods and produces certain patterns or patches that recur on differing areal and time scales. We shall now take what we know of physical and biological factors and their roles in structuring communities and see how these act together over time to give varying community patterns. It is our purpose here to document the types of patterns, explain how they likely come about, and why they change, using some examples from the scientific literature. We note, as Woodin (1983) does, that there is a definite geographical bias to our fund of knowledge. Most well-studied cases come either from the temperate Pacific coast of North America or from the temperate Atlantic coast of North America or Europe. These two areas seem to have definite differences in infaunal and predator abundances, the Pacific coast being most dense and, in general, relatively greater in abundance of large predators than the Atlantic coast. As a result, the relative importance of competition and predation in structuring these communities appears to differ between the two areas.

In some instances communities are very stable over periods of several years. This stability seems to result from the dominance of one or a few physical factors, or a dominant animal that controls its own environment and is long lived. A good example is the previously discussed sea cucumber *Molpadia oolitica.* As we noted earlier, *Molpadia* by its feeding activities creates a bottom topography of stable cones or mounds interspersed with unstable depressions. By creating these unstable areas, *Molpadia* prevents other organisms, mainly suspension feeders, from settling in the area. The exception is the more stable area of the fecal cones, where the suspension feeders can and do settle and add to the stability of the cones. Such areas, once established, remain for years, perhaps because of the life span of *Molpadia.* On the southern California coast, Fager (1968) found that a shallow subtidal sand community dominated by nine large epifaunal invertebrates (two sea anemones, one sea pansy, one brittle star, three snails, one starfish, and one hermit crab) remained constant for a six-year period. He also found that this community seemed to lack substantial interspecific interaction, suggesting that the community was structured by the physical factors alone.

There are also situations in which the communities show cyclical oscillations on a regular basis between alternate states. For example, in the aforementioned sedimentary system in Barnstable Harbor, Massachusetts, Mills (1969) demonstrated that the area was dominated at certain times by the gastropod *Nassarius obsoletus* but at other times was replaced by the tube-dwelling amphipod *Ampelsica abdita.* In this case both physical and biological factors act, but alternately. *Ampelisca* is apparently a dominant space competitor. It also interferes with *Nassarius* feeding, so it outcompetes *Nassarius* in seasons with low storm activity, leading to the mat development. However, in stormy weather, usually in winter, the tube mats are rolled up by the heavy wave action, freeing the substrate for *Nassarius.* Thus this community alternates between two states. Such regular patterns may also result from pulses of reproductive activity and subsequent settlement of young. Typically the pattern is to have high numbers of juveniles in spring and summer. This is followed by more or less massive mortality produced by competitive interactions or predation, and hence there are smaller numbers in winter.

The importance of predation in sedimentary environments is demonstrated by a study by Virnstein (1977) in subtidal sand communities in Chesapeake Bay. Virnstein excluded by cages several large predators, including the blue crab (*Callinectes sapidus*) and two bottom-feeding fish, the spot (*Leiostomus xanthurus*) and the hogchoker (*Trinectes maculatus*). He found that inside the cages, where predators were absent, the infaunal invertebrates increased in density and diversity. The species that had the largest population increases were the shallow-living op-portunistic ones, those most likely to be preyed upon. The deep-dwelling species showed little change, primarily because they probably avoid predation by their habits.

Although we lack much data for multiyear changes in sedimentary communities, we do know that in some cases there are changes that occur at rather long intervals in time. One way of obtaining such a pattern would be through the variable recruitment success of the dominants.

If, for example, a dominant, long-lived species produces many juveniles each year, but in most years all or nearly all die, and then in one year all or most live to grow up, the resulting adult population and the community will show oscil-lations spaced out in years. Such a situation seems to prevail for the amphipod *Pontoporeia affinis* in the Baltic Sea (Fig. 5.10). Such a long-term cycle may affect other dominants also. For example, in the Baltic Sea, *Macoma balthica* oscillates in dominance with *Pontoporeia* because at high densities of *Pontoporeia* the juveniles of *Macoma* are outcompeted. The oscillatory period here is about six or seven years.

The above discussion has centered on the role of a single factor—or of two or more factors acting alternately—to impose a pattern of sta-bility. In most benthic sedimentary environ-ments, however, there is a mosaic of patches, each with a somewhat different structure of species. These are usually the result of several factors acting simultaneously or in some se-quence. They are the most complicated situa-

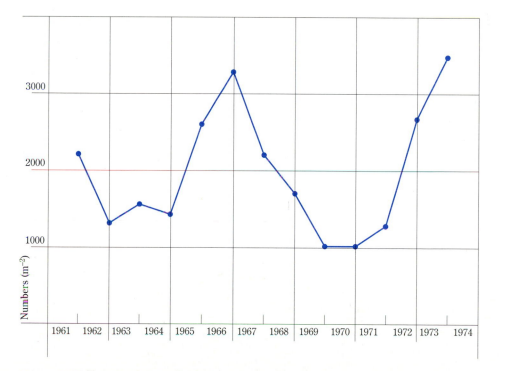

Figure 5.10 Graph of the population changes in the density of the amphipod *Pontoporeia affinis* in the Baltic Sea. (From J. S. Gray, The ecology of marine sediments, Cambridge University Press, 1981.)

tions and can be understood best by again resorting to some well-studied examples.

One of the simplest situations was that studied by Kastendiek (1976) at depths of 2–13 m on sand at Zuma Beach in southern California. Here the dominant species are the sea pansy *Renilla kollikeri;* the sand dollar *Dendraster excentricus;* the nudibranch *Armina californica;* and the starfish *Astropecten armatus.* Off Zuma Beach, *D. excentricus* forms a dense band (up to 1500 individuals per square meter) that lies parallel to the shore and runs the entire length of the beach (Plate 39). The band moves into more shallow water during calm periods of the year and offshore during stormy times. *Renilla kollikeri* occupies an area inshore of the sand dollar bed. *Armina californica* is a specialist predator of *R. kollikeri,* and *A. armatus* feeds on both *R. kollikeri* and *D. excentricus* (Fig. 5.11).

In this system, *Renilla kollikeri* is prevented from moving further shoreward because the increased wave motion in shallower water prevents it from anchoring itself. It cannot extend further seaward because of the sand dollar bed. When *R. kollikeri* enter the sand dollar bed, the *Dendraster excentricus* uproot and eliminate them. Hence, during that time of year when the sand dollars move shoreward, that portion of the *R. kollikeri* bed it occupies is destroyed.

Armina californica and *Astropecten armatus* have a synergistic effect. *Astropecten armatus* is repelled by the expanded autozooids of the *Renilla kollikeri.* However, *R. kollikeri* has a characteristic behavior when attacked by *A. californica.* This is to close up all polyps and raise itself up off the substrate. The colony thus acts like a sail. The turbulent water then uproots the *R. kollikeri* and tumbles it away from the *A. californica.* Thus, *A. californica* gets only one or two bites before the escape response. However, a closed-up *R. kollikeri* is readily eaten by *A. armatus;* so escape from one predator is not escape from the other. *Astropecten armatus,* however, is capable of ingesting only *R. kollikeri* up to 40 mm in rachis diameter. *Armina californica,* in turn, is able to consume *R. kollikeri* colonies if they become lodged against objects such as sand dollars.

The sand dollar bed forms a barrier to *Astropecten armatus,* which normally is distributed seaward of the bed and is not able to penetrate shoreward of it. *Astropecten armatus* also consumes sand dollars. What controls the edges of the sand dollar beds is not known.

Hence, in this system, there are both physical and biological controls on the community. *Renilla kollikeri* is controlled seaward by *Dendraster excentricus* and shoreward by waves, and the population is further regulated by predators, of which one is unable to attack it due to the sand dollar bed. Should the sand dollar bed disappear, Kastendiek predicts that *Astropecten armatus* and *Armina californica* would destroy the *R. kollikeri* population.

A final example is that of VanBlaricom (1982), which demonstrates how a temporal

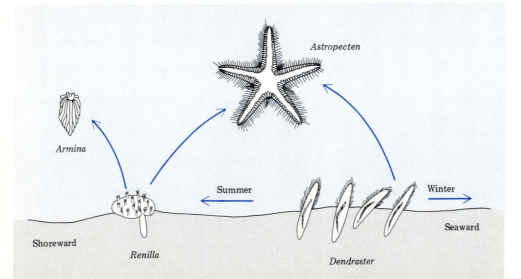

Figure 5.11   The food web of a southern California subtidal sand flat.

mosaic of patches of different groups of infauna may come to exist in sedimentary environments. VanBlaricom investigated the community structure of an open sedimentary environment at 17 m off La Jolla in southern California. In this area the infaunal community is disturbed by two large rays—the round sting ray *Urolophus halleri* and the bat ray *Myliobatis californica*—that dig pits to expose their prey. In so doing they also destroy most of the other infauna in the area of the pit or expose them to other predators. As a result, the pits represent open areas that are available for settlement. These pits also accumulate food material in the form of organic detritus. VanBlaricom showed that the pit areas attract first a set of opportunistic invertebrates that are nocturnal swimmers and feed on detritus. Later animals tend to be subsurface deposit feeders that displace the early colonists. Depending on the time since its creation, each pit may have a different set of organisms, leading to a mosaic on the bottom of differing infaunal associations representing various stages in recolonization. The picture is further complicated by the activities of still other predators. The starfish *Astropecten verrilli* consumes crab larvae soon after they settle. This reduces the abundances of two crabs, *Cancer gracilis* and *Portunus xantusii*. The two crabs, in turn, feed on infauna organisms, so that sea star predation is important to maintenance of the infaunal community. Finally, the speckled sand dab (*Citharichthys stigmaea*), a fish that is a visual predator, ingests invertebrates stirred up by the ray digging, thereby further depleting the infauna in the pits.

In summary, marine subtidal infaunal communities of unvegetated sedimentary environments are characterized by patchy distributions of organisms and certain variability in species abundances and composition over time. The communities are maintained by settlement of larvae from the plankton (see Chapter 1). The ability of the larvae to choose areas for settlement, coupled with ability to delay metamorphosis, ensures that settlement is not random. The patchiness and variability are the result of continual random disturbances created by water motion or biological activity such as predation, which eliminates the population in a small area (= patch). This is followed by recolonization and change with time due to a complex series of biological interactions involving different life history strategies, opportunist versus equilibrium; competitive interference; and exclusion and predation. The result is that different areas have different species abundance and composition depending on the time from the last disturbance and the relative stability of the substrate. Competition for space is ameliorated through vertical stratification of the organisms.

ROCKY SUBTIDAL COMMUNITIES

Subtidal rocky surfaces are probably not as abundant throughout the world as are sedimentary substrates, but they are common in some areas. Included in the category of hard substrate communities are those that tend to have a set of low-growing encrusting plants and animals, as well as kelp forests and coral reefs. We discuss only the first here. Kelp forests are covered in the next section, and coral reefs are dealt with in detail in Chapter 9.

Usually the subtidal hard substrates are densely covered with low-growing plants and sessile or sedentary animals (Plate 33). The prominent motile animals are usually sea urchins, crabs, and gastropod mollusks. There is little, if any, open primary space (Fig. 5.12). Most of these surfaces are without significant topographic relief, giving a homogeneous two-dimensional surface. Nevertheless, they have a high species diversity or species richness, leading biologists to ask how so many species can coexist in such a relatively unstructured flat area. In contrast to the sedimentary environment, there is no third dimension (= depth) to be occupied by additional species. Furthermore, there does not appear to be a fine partitioning of the ecological niches as in other environments. Most of the animals of this area are either colonial encrusting species able to spread over surfaces by asexual means or single individuals that settle gregariously, forming large space-occupying clumps. Both groups feed by suspension feeding. Included in the first group, as dominants, are animals such as sponges, tunicates, bryozoans, and various cnidarians such as anemones and hydroids. The second group comprises bivalve mollusks, algae, sea urchins, and gastropod mollusks. These animals do not appear to be food limited.

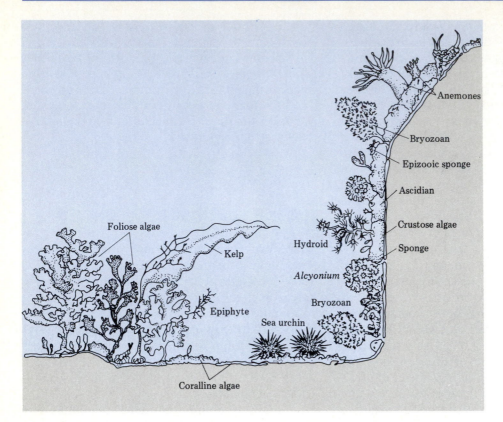

Figure 5.12 A diagrammatic representation of the characteristic flora and fauna inhabiting the vertical and horizontal surfaces of the Massachusetts subtidal rocks. Horizontal surfaces are dominated by a canopy of foliose algae and other, epiphytic species, unless the surfaces are being actively grazed by sea urchins, in which case coralline algae predominate. Vertical surfaces support diverse assemblages of invertebrates and algae that compete for space; some of these species are seasonal, such as the erect bryozoans and hydroids that form a canopy layer during the warm months. (Modified from K. Sebens, The ecology of the rocky subtidal zone, Amer. Sci. 73:548–557, 1985.)

It now appears from studies done primarily in Massachusetts by Sebens (1985, 1986), in Maine by Witman (1987), and in Australia by Fletcher (1987) that the organization and persistence of these areas, and the co-occurrence of many species, are explainable by a combination of biological and physical factors that include storm disturbance, competition for space, grazing, recruitment, and mutualism. The disturbances generate patches of free space that are then subsequently recolonized by larvae or juveniles of nearby adults, or by spread of the adjacent colonial organisms. Subsequent competition among these organisms for the limited space, coupled with seasonality of certain organisms and mutualistic interactions among certain species, completes the cycle. This means that in

any one area at any one time there is either a mosaic of patches of different ages and with different sets of or different abundances of species, resulting in the high species diversity, or a different zonation pattern. Now let us examine how this comes about using a few examples.

In the Massachusetts subtidal, Sebens (1985, 1986) found the most significant biological disturbance was the sea urchin *Strongylocentrotus droebachiensis.* Where these urchins have been abundant for a period of years on flat or sloping surfaces, they will have grazed off most sessile and sedentary organisms, leaving only a thin layer of encrusting coralline red algae (*Lithothamnion, Phymatolithon, Clathromorphum*). Although other animals also graze the bottom areas, none appear to have the effect of the urchin. Within this generally depauperate grazed area, there exists one refuge from the urchin grazing and from other predation: the beds of the mussel *Modiolus modiolus.* Experiments by Witman (1985) demonstrated that consumption of several prey items was much greater outside than inside the mussel beds. Two groups of predators patrol these areas, a diurnal set consisting of several fishes, and a nocturnal set of crabs (*Cancer irroratus* and *Cancer borealis*) and the lobster (*Homarus americanus*).

Where the urchins are rare or absent, the hard bottom consists of a 10- to 20-cm thick canopy of fleshy algae and, below that, encrusting coralline algae and patches of various colonial animals such as ascidians, plus various small, motile crustaceans and the tubes of polychaete worms and amphipod crustaceans.

Vertical rock walls in this area usually are inaccessible to urchins and have quite a different association of organisms. On these walls are a variety of encrusting, mostly colonial invertebrates dominated by a colonial compound ascidian *Aplidium pallidum,* the sea anemone *Metridium senile,* the octocoral *Alcyonium siderium,* and several sponges. This community appears to be structured by competition among the organisms for space. Since most of these organisms are colonial, competition tends to involve direct overgrowth of one by another whenever their edges meet. The species on the vertical faces have been shown by Sebens (1985, 1986) to exist in a competitive hierarchy that ranges from those able to overgrow all others to those that can be overgrown by all others. In general, the colonial species with the larger or more massive growth form are competitively dominants. Thin, flat colonies are competitively inferior.

Experimental addition of urchins to vertical faces results in the removal of all the encrusting forms except the coralline algae. Caging urchins out of horizontal surfaces results in a dense growth of foliose algae.

The above situation would seem to lead to rather monotonous landscapes of sea urchins and encrusting coralline algae on horizontal surfaces and to massive aggregations of the dominant space competitor, *Aplidium,* on vertical walls. This, however, does not happen, and there are horizontal areas of foliose algae and vertical walls with various numbers of other inferior space competitors. How does this come about? It has to do with a number of life history and growth features of some of these invertebrates. The sea urchins, for example, are subject to predation by fishes, several crabs, and (uncommonly) by lobsters. They are also destroyed by an undetermined disease and by winter storms. Whenever massive urchin mortality occurs in an area, the community changes and will remain in that alternate state until urchin grazing again reduces the area to crustose corallines. This may, however, take some time, so we get patches. Also, urchins are not able to gain access to vertical walls because they cannot effectively attach themselves to the colonial animals. If some disturbance does clear the vertical surface of *Aplidium,* the urchins may gain access and may keep that patch of vertical surface grazed down to corallines for a long period of time. Such an area may exist next to a vertical face, with a well-developed association of colonial animals dominated by *Aplidium.*

Inferior colonial competitors may exist with *Aplidium* for several reasons. In the case of the erect octocoral *Alcyonium,* it tends to grow upward as a juvenile. At the same time, *Aplidium* refrains from reproducing for its first two years. On an open surface, therefore, if *Alcyonium* and *Aplidium* settle at the same time, the time lag in *Aplidium* growth will allow *Alcyonium* to grow up high enough to avoid overgrowth. Although *Aplidium* will eventually grow over the surface, it will not displace the *Alcyonium,* which expands above it, maintaining only a narrow attachment to the rock under the *Aplidium. Alcyonium* also inhibits urchin grazing either physically, by presenting a vertical obstruction, or through action of its nematocysts on the urchins. This in turn may aid the other members of the wall community. *Alcyonium* has crawling larvae, which will not grow up unless they settle more than 2 cm away from the parent colony. They further need a certain type of substrate—either rock or coralline algae. Since neither is usually present where the adults are, there is usually no recruitment. Since the dispersal is so space limited, there may be walls that lack any of the species only meters away from adult *Alcyonium.* The anemone *Metridium,* also an inferior space competitor, is found in patches, but because of a different set of factors. *Metridium,* once established, holds its space by using its special "sweeper tentacles," heavily armed with nematocysts, for actively deterring any other potential competitor. Meanwhile it reproduces asexually by fragmenting, and so forms a dense clone that can persist for years.

Somewhat further north, in the Gulf of Maine, Witman (1987) investigated subtidal rocky surfaces on extremely exposed sites at depths of 4–20 m. Here the deeper surfaces (11–18 m) were dominated by the mussel *Modiolus modiolus* while the shallower areas from 4 to 8 m were dominated by the kelps *Laminaria digitata* and *L. saccharina*. Experimental manipulations in these two communities demonstrated the causes of this dramatic zonation pattern. The key species was again the urchin *Strongylocentrotus droebachiensis*. The shallower-dwelling kelps were prevented from moving deeper by grazing by the sea urchins, thus maintaining the integrity of the mussel bed. Storm-generated dislodgment was found to be the most significant factor in the mortality of the mussels. In the absence of urchins, the kelps would attach to the mussels, making them 30 times more subject to dislodgment and mortality by storms. Therefore, the urchins increase the *Modiolus* survivorship by removing the kelp and thus decrease the probability of dislodgment. This *Modiolus–Strongylocentrotus* interaction was found to be mutualistic in that it took significantly more force to dislodge the urchins from the *Modiolus* surfaces than from surrounding bare rock surfaces or kelp beds. The *Modiolus* beds also provided a refuge from predation for the recently recruited small urchins.

Finally, the ability of both kelp and mussels to recover from dislodgment due to storm disturbance indicated that kelp recovery was much faster, occupying all open patches, suggesting the kelps were competitively superior and in the absence of removal by urchins would probably dominate the entire depth range. Similar subtidal zonation patterns, with a shallow zone dominated by kelps and a deeper zone dominated by urchins, have been reported for other temperate regions in Nova Scotia and New Zealand and suggest a more universal pattern, perhaps maintained by a similar mechanism.

A somewhat similar mechanism was found by Fletcher (1987) to maintain subtidal rocky areas on the east coast of Australia. In this area the subtidal rock is covered with encrusting coralline algae on which are found high densities of a suite of grazers that includes a sea urchin, *Centrostephanus rodgersii*; three limpets, *Patelloida alticostata, P. mufria*, and *Cellana tramoserica*; and two turbinid gastropods, *Australium tentiforme* and *Turbo torquata*. Manipulative experiments here, however, gave different results depending on which group of grazers was removed. Removal of all grazers resulted in the rapid colonization of the crustose algae by fleshy, foliose algae. Yet if only the turbinid snails were removed, there was no significant change in the community. If limpets and urchins were removed, the fleshy algae covered 80 to 100 percent of the area within a year. But if only the urchins were removed, the fleshy algae increase was slowed and only approached 100 percent after 2 years. Finally, where only limpets were removed, there was an initial algae increase that was followed by a decline to near conditions found where no grazers were removed.

The differential responses of the community, which depended on which set of grazers was removed, resulted from the more complex interactions among the suite of limpets, turbinids, and urchins. Whereas it seemed that the presence of the sea urchins was necessary to the maintenance of the encrusting coralline community, it also appeared that the continued maintenance of at least the populations of the three species of limpets was also somehow dependent on the presence of the urchins. However, the mechanism or mechanisms responsible for this are unknown at present.

Although these studies emphasize the importance of urchins as the dominant grazers and the "keystone" species in the determination of the type of community formed on subtidal hard substrates, one cannot exclude the possibility that other grazing animals may be equally important. This has been demonstrated by Ayling (1981), who investigated three communities: one dominated by encrusting coralline algae, one dominated by sponges, and one by ascidians in the subtidal of New Zealand. There were three groups of grazers: an urchin, *Evechinus chloroticus*, a suite of herbivorous gastropods, and a grazing fish, *Parika scaber*. At high densities the urchin grazing was responsible for the perpetuation of a low-diversity community consisting primarily of crustose coralline algae. However, removal of the suite of grazing gastropods resulted in a change in the community to one with more fleshy ephemeral algae, even in the presence of the urchins. Removal of the grazing fish, which primarily fed on sponges and ascidians, caused a major shift in the community structure from one dominated by crustose coralline algae to one dominated by ascidians and sponges. Finally, there also appeared to be a role played by disease and parasites, because Ayling also reported episodic outbreaks of a bacterial or fungal disease among the sponges that destroyed large numbers of the otherwise long-lived sponges.

In summary, we have on hard substrates a two-dimensional surface. In those areas so far studied, all in cold or warm temperate seas, the communities are dominated either by an encrusting community consisting of coralline algae, ascidians, or sponges that potentially compete for primary space among themselves by overgrowth, or by beds of mussels forming large clumps, or by kelps. In all cases sea urchins appear to be the "keystone" species, for without them the community changes. Where urchin densities are high, the communities are dominated by crustose coralline algae or by mussels, both very resistent to urchin grazing. Where urchins are absent or in fewer numbers, communities of fleshy algae or kelp, or both, dominate. Physical disturbance, primarily in the form of storms but also in the form of disease and predation, causes mortality among the urchins, thereby freeing the more ephemeral algae or kelps to occupy more space. This "typical" picture may be modified by suites of smaller gastropods and fish grazers, both of which may alter the community structure to favor certain of the encrusting organisms. Finally, the differences in the life histories of various competitively inferior component organisms may lead them to persist even in the face of their superior competitors.

The existence in the same area of different groups of organisms or patches of organisms is what Sutherland (1974) has called *alternate stable states.* These alternate stable states result when some factor or set of factors disturb the dominant species, causing a community to switch to a new composition, which in turn remains stable until another significant disturbance occurs.

Although there has been some argument among ecologists as to whether or not alternative stable states is a real phenomenon, the fact remains that very different community structures can and do exist near one another under virtually identical environmental conditions. This suggests that some perturbing factor has caused a community to change from one configuration to another, and that there is some mechanism which allows the two different configurations to persist. One of the most intriguing examples of this phenomenon is that reported by Barkai and McQuaid (1988) for two subtidal rocky communities in South Africa. Off the west coast of South Africa there are two islands, Malgas and Marcus. They are separated by 4 km, so both experience similar environmental conditions. However, the rocky subtidal communities are quite different in each. At Malgas Island, the rock lobster *Jasus lalandii* and algae dominate the community and the lobsters comprise 70 percent of the biomass. Few other organisms, save for a few gastropods of the species *Burnupena papyracea,* are found. At Marcus Island, by contrast, rock lobsters are rare and the community is dominated by extensive beds of the mussel *Choromytilus meridionalis,* large populations of the carnivorous gastropod *B. papyracea,* and numerous holothurians and sea urchins. Why such extreme differences, and how are they maintained? Barkai and McQuaid demonstrated through experimental manipulations that on Malgas Island the rock lobsters prey on the settling mussels, preventing their establishment. They also prey on the gastropod *Burnupena,* but are deterred from preying on those gastropods that have a protective bryozoan (*Alcyonidium nodosum*) on their shell. Hence some gastropods continue to persist in the face of heavy lobster predation. At Marcus Island, when rock lobsters were transferred into the subtidal, all were attacked and killed immediately by the carnivorous gastropod *Burnupena,* which thus reversed the normal predator-prey relationship established at Malgas Island. Since there are normally no lobsters at Marcus Island, what do the *Burnupena* prey on? The answer seems to be that they scavenge on dead or dying mussels but are unable to successfully attack healthy mussels. Therefore, it appears that on both islands predation maintains the current communities, but instead of differences in predators at each, the change is maintained by a reversal of predator-prey roles (Fig. 5.13). These two different com-

Figure 5.13 Schematic summary of rock lobster and whelk predator-prey role reversal at Marcus and Malgas islands. (From A. Barkai and C. McQuaid, Predator-prey role reversal in a marine benthic ecosystem, Science 242:62–64, 1988.)

munities persisted through the 4 years of the study as alternative stable states maintained by the very unusual mechanism of predator-prey reversal. Since the local fishermen have reported that the two islands had similar lobster populations some 20 years ago, the unanswered question is how the change came about.

The concept of alternate stable states is not limited to subtidal communities. It is applicable to many marine communities, and we shall see it again in our considerations of intertidal and coral reef systems.

KELP BEDS AND FORESTS

Throughout a large part of the cold temperate regions of the world, hard subtidal substrates are inhabited by a community dominated by very large brown algae known collectively as *kelps* (Fig. 5.14). These associations are technically known as *kelp beds* if the algae do not form a surface canopy and *kelp forests* where there is a floating surface canopy (Plates 37, 38). These terms are, however, often used interchangeably.

Structure and Distribution

Kelps are attached to the substrate by a structure called a *holdfast* rather than by true roots (Fig. 5.15). From the holdfast arises a stemlike or trunklike *stipe,* which ends in one or more broad, flat *blades.* At the base of the blade is a *pneumatocyst* or *float,* which keeps the blade at the surface. Kelps obtain their nutrients directly from the seawater as do the organisms of the phytoplankton. They depend on the constant movement of water past them to avoid nutrient exhaustion. Since these shallow waters are constantly moved by wave action and currents and the nutrients are replenished by turbulence, upwelling, and runoff from land, nutrient depletion rarely occurs and luxuriant growth occurs, forming what may justifiably be called "kelp forests."

In contrast to the monotonous, level landscape of the soft bottom or the short-growing colonial organisms of other subtidal rock surfaces, the kelps form an extensive three-dimensional habitat composed of several vertical layers or strata (Fig. 5.16). As a result, a large number of potential habitats are available and the variety of life is greater.

The major kelps that dominate the structure of the "kelp forests" are the genera *Macrocystis*

Figure 5.14 Distribution of kelp beds in the world, indicating the dominant genera. (L) *Laminaria;* (M) *Macrocystis.* (Modified from K. H. Mann, Seaweeds: Their productivity and strategy for growth, Science 182[4116]:975–983, 1973. Copyright © 1973 by the American Association for the Advancement of Science.)

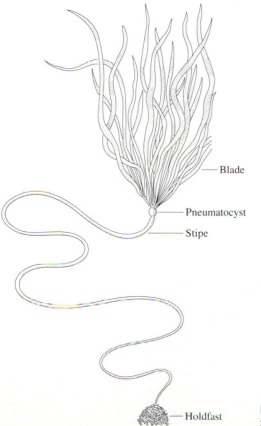

Blade

Pneumatocyst

Stipe

Holdfast

Figure 5.15 A kelp plant *(Nereocystis)* and its structure.

and *Nereocystis.* "Kelp beds" are dominated by the genera *Laminaria, Pterygophora,* and *Ecklonia.* The Pacific coast of both North and South America is dominated by *Macrocystis,* whereas *Laminaria* is dominant in Atlantic waters (Fig. 5.14). In contrast to most algae, which are small, generally never exceeding a few centimeters or a decimeter in length, the major kelps are giants, with lengths that are the equivalent of trees on land. *Macrocystis* and *Nereocystis* on the Pacific coast of North America may reach 20–30 m in length. Such massive plants grow upward from the bottom and spread their blades at the surface of the water, where they obtain the maximum amount of light. These blades form a "canopy" similar to that of terrestrial forests, cutting off light to the substrate below (Fig. 5.17). Below the canopy, extending up from the bottom, is a set of "understory" algae forming another layer (Figs. 5.16, 5.18).

Figure 5.16 The distribution of some common seaweeds within giant kelp forests on the Pacific coast of North America. Plants in the four zones of vegetation layering include: (1) small filamentous species and encrusting coralline algae; (2) bottom-canopy plants such as *Gelidium, Calliarthron,* and *Plocamium;* (3) understory-canopy kelps such as *Pterygophora, Eisenia,* and *Laminaria;* and (4) midwater and surface-canopy plants such as *Egregia, Macrocystis,* and *Nereocystis.* This is a generalized diagram and some species do not co-occur at the same site. (Modified from M. S. Foster and D. R. Schiel, The ecology of giant kelp forests in California: A community profile, U.S. Fish & Wildlife Serv. Biol. Rept. 85[7.2], 1985.)

The extent of the kelp beds and forests on various coasts depends on several factors. In the first place, a hard substrate must be available for attachment. Second, the kelps must have light and can establish themselves only in water depths where the young, small plants receive enough light to grow. The amount of light received at any depth is a function of water clarity, and where clear water prevails, kelp beds and forests extend out from shore to depths of 20–30

m. Where such shallow water is extensive, the beds may extend out as much as several kilometers from shore. In more turbulent, murky water, the depth of the beds may be more restricted. Finally, kelps seem limited by high temperature and associated low nutrients, as well as by grazing. Kelp beds and forests occur throughout the world wherever the water is cool, but are absent from warm temperate and tropical areas. Their considerable extent along the Pacific coast

(A)

(B)

Figure 5.17 *(A)* The surface of a kelp forest showing the thick canopy of blades. *(B)* The interior of a California kelp bed. (*A*, photo courtesy of Ron McPeak; *B*, photo courtesy of Dan Reed.)

of North and South America is due to the cold upwelling on the North American coast and the cold water of the Humboldt Current along the South American coast.

Ecology and Life Cycle

Kelps have a phenomenal growth rate. *Nereocystis luetkeana* has been recorded by Scagel (1947) as growing in length by 6 cm/day, while, according to North (1971), *Macrocystis pyrifera* commonly grows 50 cm/day on the California coast. Such growth rates mean that plants grow from the bottom to the surface in a short time. Kelp forests and beds are also extremely productive. Mann (1973) reports that the net annual produc-

tivity ranges from 800 g C/m^2 in California kelp beds to perhaps 2000 g C/m^2 in Indian Ocean kelp beds. These figures are several times the production of the phytoplankton on the same per-area basis.

Kelps tend to be perennial plants, often losing stipes and blades but regrowing new ones from the holdfast. The life span for *Macrocystis pyrifera* plants in California ranges from three to seven years maximum. The life cycle of kelp alternates between an asexual macroscopic stage termed the *sporophyte* and a sexual microscopic stage termed the *gametophyte* (Fig. 5.19). Only the sporophyte stage is considered here because little is known about the gametophyte in the natural environment.

Figure 5.18 Cross section showing the inhabitants of a generalized giant kelp forest. The numbers to the right indicate vegetation layers (see legend for Figure 5.16). Three broad zonal associations along the depth gradient are shown: (Z1) inshore of the giant kelp community; (Z2) within the giant kelp community; and (Z3) offshore from the giant kelp community. Various subcommunities or associations are indicated by the circular diagrams: *(A)* animals associated with the surface of *Macrocystis* and other seaweeds (polychaetes, isopods, bryozoans); *(B)* animals found in giant kelp holdfasts (small sea urchins, brittle stars, crustaceans, polychaetes (although shown on the outside, these organisms occupy the spaces within the holdfast); *(C)* plants and animals characteristic of horizontal surfaces (various sea stars, urchins, benthic fishes, understory algae); *(D)* organisms most common on vertical surfaces (primarily sessile animals such as sponges, ascidians, bryozoans, sea anemones). Some of the organisms shown do not necessarily co-occur at any one site. (From M. S. Foster and D. R. Schiel, The ecology of giant kelp forests in California: A community profile, U.S. Fish & Wildlife Serv. Biol. Rept. 85[7.2], 1985.)

The kelps provide the framework for the kelp forest or kelp bed community. Associated with these dominants are many other species of algae, invertebrates, and fishes.

Despite the enormous productivity of these kelps, relatively few herbivores graze directly on the plants. It has been estimated that only 10 percent of the net production enters the food

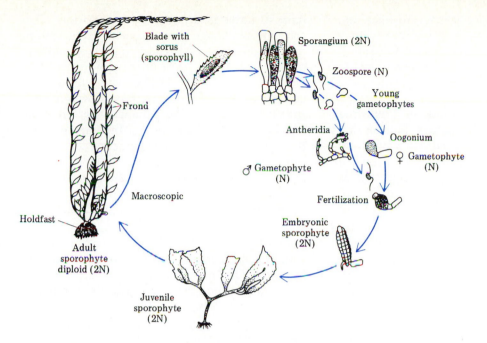

Figure 5.19 Life cycle of a typical kelp plant *(Macrocystis).* (From M. S. Foster and D. R. Schiel, The ecology of giant kelp forests in California: A community profile, U.S. Fish & Wildlife Serv. Biol. Rept. 85[7.2], 1985.)

webs of the kelp bed through grazing. The remaining 90 percent enters the food chains in the form of detritus or dissolved organic matter.

Since the general economy of kelp beds and forests depends on the continued existence of the giant kelps, it is of interest to know the causes of kelp mortality. Adult plants are only occasionally destroyed by grazing herbivores, but they are vulnerable to destruction by mechanical forces, mainly wave action, and to nutrient depletion, mainly nitrogen. Since they occur in quite shallow water, often on open coasts, storm waves can have a devastating effect. The waves rip the plants up from the substrate and cast them up on the beach. Furthermore, once a plant has been uprooted, its movement through a bed, as Rosenthal and associates (1974) have demonstrated, often results in entangling among the stipes and blades of other plants. The result is that even more plants are pulled away from the substrate and destroyed. Very severe storms can virtually denude kelp beds of the stipes and blades, but if the holdfast and some meristem tissue from the kelps remain, new stipes and blades can be produced.

The very large El Niño event of 1982–83 that produced devastating effects on the coral reef communities in the Pacific Ocean (see p. 370) also had catastrophic effects on the extratropical kelp forests, of which those of the California coast have been best documented. The storm season of 1982–83 was the most severe in decades, perhaps the worst of this century, according to Tegner and Dayton (1987). These storms caused massive mechanical damage to the kelp forests. In San Diego the Point Loma kelp forest was reduced from 600 hectares in 1982 to 40 hectares, and off the Palos Verdes Peninsula in Los Angeles the cover was reduced from 196 hectares to 18. In addition to the loss of canopy, the severity of the waves also ripped up holdfasts, leading to further loss. Finally, recovery in many places, particularly in southern California, was delayed because plants not killed by the storms were retarded in growth by the high water temperatures and nutrient-poor water that accompanied the storms. Exceptions to this were the sandy areas around Santa Barbara, where old holdfast material served as recruitment and growth centers, and most of these forests recovered in a few years.

Whereas storms, especially severe ones, are usually devastating to existing kelp forests, they may have vastly different effects on the subsequent community that develops, which is dependent on the state of the community before the advent of the storms. Ebeling et al. (1985) made regular observations on a southern California kelp forest over a five-year period. During that time the forest was subjected to two severe storms. The first storm removed the canopy of *Macrocystis* but left the understory kelps, such as *Pterogophera californica,* intact. This resulted in a drastic reduction of the drift kelp used as food by the urchins *Strongylocentrotus purpuratus* and *S. franciscanus.* The urchins then emerged from their sheltered cracks and crevices and, in the absence of effective predators, consumed most of the remaining living kelp. This, in turn, weakened the other parts of the food web and led to further declines in kelp forest associates, such as fish. The second storm in the area caught the urchins in the open and destroyed them, as well as clearing surfaces. The newly opened surfaces were then subsequently colonized by kelp plants that produced a fully developed kelp forest a year later. Thus two storms led to dramatically different communities, which in turn were dependent on the type of community existing immediately prior to the storm.

Another potential cause of mortality is grazing, particularly by various sea urchins and fishes. Here the evidence is more equivocal. There seems to be little doubt that sea urchins of many species can have significant effects on kelp beds and forests. These effects may range from wholesale removal of plants to selective feeding and even to provision of suitably cleared space in which the settlement of new kelp plants may take place. The creation by urchins of so-called "barren grounds" has been reported for many regions in which kelp occur. In recent years, greatly increased numbers of urchins in certain areas, such as New England and southern California, has coincided with diminished kelp abundance—undoubtedly because of urchin-grazing pressure. However, the causes of the increased urchin abundance—a kind of population explosion—are not always clear. In southern California, for example, the urchins do not

have to undergo a population explosion. They can simply change their behavior from feeding on drift algae to active foraging on live plants, or vice versa. This has been demonstrated by Harrold and Reed (1985), who studied the grazing activity of red urchins (*Stronglylocentrotus franciscanus*) in adjacent barren areas and kelp-dominated areas. In barren areas they found drift algae sparse; hence the urchins were poorly nourished and were actively out and about grazing all surfaces. In the kelp areas the urchins were well nourished, lived mainly in cracks moving little, and fed on abundant drift kelp.

During the study a substantial recruitment of kelp occurred on both the barren site and the kelp site. On the barren site the large recruitment transformed the site into a kelp-dominated area. This increased the amount of drift algae, and the urchin population underwent a behavioral transformation, moving into cracks and crevices, feeding on drift algae, and reducing active surface grazing to levels found in kelp-dominated areas. Thus there was a change in the community structure that was facilitated by a behavioral change in the dominant urchin, which in turn was triggered by a switch in the mode of feeding. The switch was, in turn, dependent on the availability of drift algae, which in turn was dependent on the abundance of new kelp. In this case the grazing intensity was independent of urchin density, and the ability of the kelp to colonize a barren area successfully seemed to be regulated by a favorable set of hydrographic conditions (Fig. 5.20).

The above studies demonstrate that it is not necessary to alter the number of urchins to effect a change in the kelp forest community structure, and that storms can have a dramatic effect on both urchins and kelp. What about predators? Until recently, it had been assumed that urchin populations are kept in check by various predators and that many, if not all, urchin population explosions are the result of dramatic decreases in the major predator. In other words, urchins are capable of destroying kelp beds, but under normal conditions predation checks prevent this destruction from occurring. In most cases, however, the evidence to support this contention is

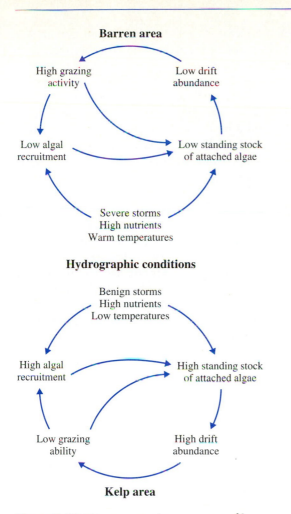

Figure 5.20 Diagrammatic summary of how barren areas and kelp areas are maintained at San Nicolas Island, California. (After C. Harrold and D. Reed, Food availability, sea urchin grazing and kelp forest community structure, Ecology 66[4]:1160–1169, 1985.)

very poor or cannot be separated from other effects. For instance, it was originally thought that increases in urchin densities in eastern Canada and New England were due to the reduced numbers of lobsters, which were believed to prey on the urchins. It now appears that lobsters feed mainly on crabs, and only rarely on urchins. Similarly, in southern California, the sheephead fish (*Semicossyphus pulcher*) was

thought to be the major predator on urchins, but more recent studies have suggested that it is not. It is likely that storms or changes in the inshore waters in the area, due to sewage outfalls and perhaps to hydrographic changes caused by El Niño, have led to changes in numbers of both kelp and urchins or to a change in behavior on the part of the urchins that has manifested itself in changes in community structure.

The only well-documented case where urchins have been shown to be controlled by a predator is in Alaska. Here work by Estes and Palmisano (1974) demonstrated that the sea otter (*Enhydra lutris*) is the dominant predator on urchins. In the Aleutian Islands these investigators found that islands with large otter populations had low urchin densities and a lush growth of kelps. Islands where otters were absent were virtually without kelps and had enormous urchin densities. The almost complete destruction of the sea otter during the nineteenth century along most of the Pacific coast probably allowed the increase in sea urchin numbers and could have permitted the gradual decrease in the kelp beds. The recovery of the sea otter in California and Alaska has been followed by decreased sea urchin numbers and increased extent of kelp forests in a few areas studied.

The importance of urchins to the integrity of kelp beds in California can be suggested by making a comparison of the recovery of the beds from the two most devastating El Niño events in this century, those of 1957–59 and 1982–83. In the former case, according to Tegner and Dayton (1987), the kelp forests had dense populations of the sea urchins *Strongylocentrotus pupuratus* and *S. franciscanus*, which normally feed on drift kelp. After the 1957–59 event these urchins survived and, because of lack of drift, moved out to feed on the adult plants, reducing them even more. Persistence of these urchins precluded kelp forest recovery until well into the 1960s. By contrast, the 1982–83 El Niño was a stronger event with a more devastating effect on the kelp forests (see preceding section). Yet most of these forests recovered in a shorter time, probably because, by 1982–83, the fishery for sea urchins had reduced the populations so much that urchin grazing could not prevent the recruitment of new kelp or the regrowth of surviving plants. It is also possible that the anomalous environmental conditions associated with the 1982–83 El Niño may have had deleterious effects on the urchin recruitment.

Other predators of sea urchins exist in kelp beds, notably several starfish species. In some areas they have been suggested as important in controlling urchin populations (Fig. 5.21). The role of other organisms in structuring the kelp bed community is not as well known.

A generalized food web for invertebrates is given in Fig. 5.21 for a California kelp forest. In this qualitative food web, it can be noted that the primary producers have only the two species of urchins as major grazers. The other grazer, *Astraea undosa,* is a herbivorous gastropod that is unlikely to exert any considerable effect on the kelp. Sea otters are not present in this area, and urchin control is presumably through the activities of the four predaceous starfish, octopus, and one fish.

Kelp forests also show various patterns of zonation, but the zonation is not consistent from one area to another (Fig. 5.22). Zonation is determined by changes in physical factors with depth, such as light and wave action, by differences in topographic relief, and partly by competition among species. A general zonation pattern would have a shallow zone nearest the shore that lacks kelp and is dominated by surf grass and small coralline algae. Kelp are absent because the wave surge removes them. At about 5 m, depth conditions are right for kelp, and the forest develops and extends to 20–30 m. As greater depths are approached, the light intensity becomes limiting, and the edge of the kelp forest seems to be set at that point where there is insufficient light for kelp growth.

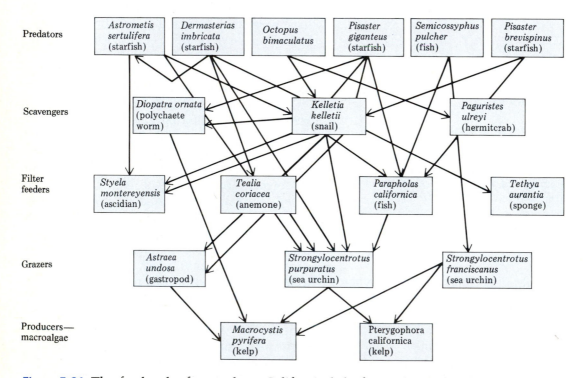

Figure 5.21 The food web of a southern California kelp forest. (From R. J. Rosenthal, W. D. Clarke, and P. K. Dayton, U.S. National Marine Fisheries Service Fisheries Bulletin, Vol. 72, 1974.)

Figure 5.22 Different zonation patterns of the dominant organisms of five southern California kelp forests. (From M. S. Foster and D. R. Schiel, The ecology of giant kelp forests in California: A community profile, U.S. Fish & Wildlife Serv. Biol. Rept. 85[7.2], 1985.)

SEAGRASS COMMUNITIES

Many areas of the shallow sea bottom are covered with a lush growth of aquatic "grasses," which are collectively called *seagrasses*. Seagrasses are flowering plants that are adapted to live submerged in seawater.

Composition and Distribution

Seagrasses worldwide encompass only about 50 species, according to den Hartog (1977), a small number in comparison to their ecological importance. Seagrass beds form dense carpets of as many as 4000 blades per square meter over extensive areas of the bottom, making them one of the most conspicuous communities of the shallow waters of temperate and tropical seas. They may have a standing biomass of 2 kg/m². Seagrass beds or meadows are usually well defined, by a visible boundary, from surrounding unvegetated areas. They can vary in size from

isolated patches to a continuous carpet that covers many square kilometers. They also vary in the density of the plants.

Most seagrass species are very similar in gross external morphology. They have long, thin, straplike leaves that have air channels and a monopodial growth form. The plants arise from a creeping rhizome (Fig. 5.23). When compared to freshwater aquatic plants, the seagrasses are considerably fewer in number of species and also very much less diverse in morphology.

Seagrasses occur from the midintertidal region down to depths of 50 or 60 m. They seem to be most abundant, however, in the immediate sublittoral area. The number of species is greater in the tropics than in the temperate zones. All types of substrates are inhabited by these grasses, from soupy mud to granitic rock, but the most extensive beds occur on soft substrates. The dense intertwined rhizomes and roots of the grasses form a strong mat that penetrates the substrate and secures the plant against water

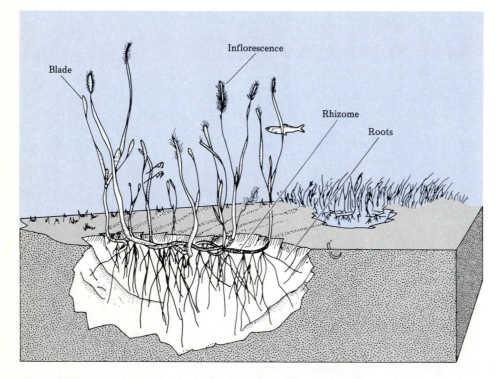

Figure 5.23 Diagrammatic three-dimensional illustration of an eelgrass meadow. (Redrawn from G. W. Thayer, W. J. Kenworthy, and M. S. Fonseca, The ecology of eelgrass meadows of the Atlantic coast: A community profile. U.S. Fish & Wildlife Serv., FWS/OBS–84/02, 1984.)

motion, thereby stabilizing the bottom sediments.

Environmental Conditions

Seagrass beds are strongly influenced by several physical factors. Perhaps the most significant is water motion—currents and waves. Since seagrass systems exist in sheltered as well as relatively open areas, they are subject to differing amounts of water motion. But for any given seagrass system the water motion is relatively constant over time. Seagrass meadows under high water motion tend to form a mosaic of individual mounds, whereas meadows in areas of low water movement tend to form flat, extensive carpets. The seagrass beds, in turn, act to dampen out wave action, particularly if the blades reach the water surface. This damping effect can be significant such that 1 m into a seagrass bed the wave motion can be reduced to zero. Currents are also slowed as they move into the bed.

The slowing of wave action and currents means that seagrass beds tend to accumulate sediments within them. However, this is not universal and depends greatly on the current regime under which the bed exists. Seagrass beds under the influence of strong currents tend to have much of the lighter particles, including seagrass debris, moved out, whereas beds in weak current areas accumulate lighter detrital material. It is interesting that temperate seagrass beds (*Zostera*) seem to accumulate sediments that are from sources outside the beds, whereas tropical seagrass beds derive most of their sediments from within.

Since most seagrass systems are depositional environments, they eventually accumulate organic material that leads to the creation of fine-grained sediments with a much higher organic content than that of the surrounding unvegetated areas. This, in turn, reduces the interstitial water movement and the oxygen supply. Moreover, the high rate of metabolism of the microorganisms in the sediments causes the sediments to be anaerobic below the first few millimeters. According to Kenworthy et al. (1982), anaerobic processes of the sediment mi-

croorganisms seem to be an important mechanism for regenerating and recycling nutrients and carbon, thus assuring the high rates of productivity measured in these beds (Fig. 5.24).

Other physical factors that seem to have an effect on seagrass beds include light, temperature, and desiccation. Water depth and turbidity, for example, together or separately control the amount of light available to the plants and hence the depth to which the seagrasses may extend. Although it was early suggested by Setchell (1929) that temperature was critical to growth and reproduction of eelgrass, *Zostera marina,* it has since been shown that this particularly widespread seagrass grows and reproduces at temperatures between 2 and 4°C in the Arctic (Hudson's Bay) and at temperatures up to 28°C in Chesapeake Bay. Still, extreme temperatures in combination with other factors may have dramatic deleterious effects. For example, in areas of the subarctic and cold temperate North Atlantic, ice may form in winter. Robertson and Mann (1984) note that, when the ice begins to break up, the wind and tides may move the ice around, thus scouring the bottom and uprooting the eelgrass. In contrast, at the southern end of the eelgrass range, in North Carolina, temperatures over 30°C in summer cause excessive mortality. Seagrass beds also decline if subjected to too much exposure to the air. The effect of desiccation is often difficult to separate from the effect of temperature. Most seagrass beds seem to be tolerant of considerable changes in salinity and can be found in brackish waters as well as full-strength seawater.

Productivity

Because seagrass beds are densely covered with plants, the result of extensive nutrient cycling and regeneration, and because they cover such extensive areas in continental shelf waters, they have a very high productivity rate and contribute significantly to the total production of inshore waters. McRoy and McMillan (1977) have estimated the production of temperate beds to be from 500 to 1000 g $C/m^2/year$, and McRoy and Helfferich (1977) report values of over 4000 g $C/m^2/day$ for *Thalassia* beds in the tropics.

Figure 5.24 Diagrammatic representation of the phosphorus and nitrogen cycles in a seagrass bed. (Modified from G. W. Thayer, W. J. Kenworthy, and M. S. Fonseca, The ecology of eelgrass meadows of the Atlantic coast: A community profile, U.S. Fish & Wildlife Serv., FWS/OBS–84/02, 1984.)

They are thus among the most productive areas of the oceans (compare with plankton productivity, Table 2.2; and that of kelp, p. 189). Most of these data on productivity, however, are for the two most studied species: *Zostera marina* (eelgrass) of the North Temperate Zone and *Thalassia testudinum* (turtle grass) of the tropics. Little is known about the other species.

In contrast to other productivity in the ocean, which is confined to various species of algae dependent on nutrient concentrations in the water column, seagrasses are rooted plants that absorb nutrients from the sediment or substrate. They are, therefore, capable of recycling into the ecosystem nutrients that would otherwise be unavailable because of being trapped in the bottom.

Structure and Biological Interactions

Most seagrass beds generally have only one or at most a few dominant species of seagrasses. Communities in which two or more grasses are present have been little studied and may poten-

tially be different in biological interactions. In general, the number of seagrass species increases from temperate zone to the tropics. On the Atlantic coast of the United States the only seagrass present in temperate waters is eelgrass, *Zostera marina.* In the Caribbean Sea three main species are found: *Thalassia testudinum, Halodule wrighti,* and *Syringodinium filiforme.* They co-occur in various mixed stands.

Seagrass beds expand at different rates depending on the environmental conditions, but it can apparently be quite fast. Thayer et al. (1975) found in North Carolina that a bed covering 30 percent of a bay in 1969 had increased to cover 55 percent by 1973. Unfortunately, we do not have data on how long these beds remain in any given environment.

Seagrass beds are complex systems whose physical structure is dominated by the leaves, roots, and detritus of the seagrasses themselves. Furthermore, they contain large numbers of epiphytic and epizoic organisms, burrowers, and motile animals, all interacting in ways that are only partly known at present. The roots and leaves provide horizontal and vertical complexity, in addition to abundant food, such that the diversity of organisms within seagrass beds is much higher than in the unvegetated surrounding areas.

The various trophic levels represented in the seagrass beds undergo shifts in abundance in response to various changes in the seagrasses as well as to seasonal changes in environmental factors. Therefore, there may be considerable temporal variation in the composition and abundance of the associated organisms.

On the seagrass blades there is usually a community of epiphytic and epizoic organisms that are so abundant that the biomass of these organisms exceeds that of the seagrass blade on which they reside. This community may also reduce productivity of the seagrass by preventing or reducing light intensity at the surface of the grass blade. As yet little is known of the importance of this component of the community to the general economy of the beds.

Despite the obvious position of seagrass beds as key primary production units in inshore waters (and in direct contrast to the situation in the terrestrial environment, where grasses are heavily grazed by a variety of vertebrate and invertebrate herbivores), seagrasses are consumed directly by surprisingly. few animals. It appears that the major grazers are certain birds, a few fishes, sea urchins, and the green sea turtle (*Chelonia mydas*). Despite the small numbers of grazers, their effect may be significant in certain areas. Among the birds, the black brant (*Branta nigricans*) and American brant (*Branta bernicla*) geese have been established as major grazers on seagrasses. McRoy (1966) reported that the black brant consumes 4 percent of the standing crop of eelgrass in Izembeck Lagoon in Alaska. Moreover, Cottam (1934) indicated that on the East Coast 80 percent of the diet of American brant consisted of eelgrass. Feeding by these and other herbivorous birds can have a dramatic effect on seagrass beds. For example, Wilkins (1982) estimated that in one year Canada geese (*Branta canadensis*) consumed 20 percent of the eelgrass standing crop in lower Chesapeake Bay, while in Australia black swans (*Cygnus atratus*) uprooted 94 percent and consumed 82 percent of net production of *Zostera muelleri*. In many

of these cases the beds are almost cleared of parts of the seagrasses that are above the sediments. Similarly in the tropics, urchins such as *Lytechinus variegatus* have been reported by Camp et al. (1973) to denude large areas of seagrass. The widespread tropical urchin *Diadema* has also been implicated in the destruction of tropical seagrass beds. Green turtles are known to feed on seagrasses, but because of their current low numbers and absence of historical data, we know little of what effect this grazing may have had. Ogden (1980) reports, however, that in the few areas where these turtles are still abundant they create very visible grazing scars in seagrass beds. This suggests that at high population densities in the past they may have had significant effects, especially since in the Caribbean Sea the diet of the turtle seems to be almost exclusively seagrasses. Among fishes, the pinfish (*Lagodon rhomboides*) of the American East Coast appears to feed heavily on eelgrass and can reduce small beds to a short "stubble."

A wide variety of predators, both resident and migratory, is associated with seagrass beds and they have diverse feeding strategies. Despite their abundance, there is little experimental evidence that they regulate either the structure of the seagrass beds or the abundance of prey. Various experiments where predators have been caged out of areas have usually not demonstrated significant increases of prey inside the cages. There is, however, evidence that seagrass beds may serve as refuges from predation when contrasted with unvegetated areas around them. Peterson (1982), for example, found that densities of two clams, *Mercenaria mercenaria* and *Chione cancellata,* were reduced in number in areas where seagrasses were removed relative to their densities in control areas where there was seagrass cover. He attributed this to greater predation by three species of whelks of the genus *Busycon.* In this case the seagrass bed provided a refuge, probably by binding the sediment with its roots, thereby preventing the whelks from gaining access to the clams.

Fishes are abundant in seagrass beds, and food habit studies on the American East Coast by Thayer et al. (1975) indicate that many of them feed within the bed, removing considerable biomass (Fig. 5.25). In North Carolina,

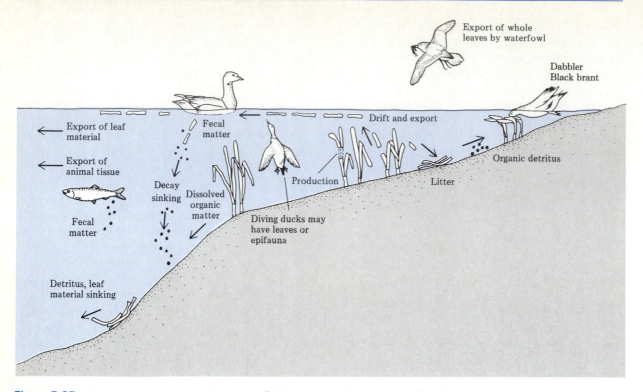

Figure 5.25 Diagrammatic representation of the principal pathways of movement of organic material out of a seagrass bed. (Modified from J. C. Zieman, The ecology of the seagrasses of South Florida: A community profile, U.S. Fish & Wildlife Serv. FWS/OBS–82/25, 124, 1982.)

Thayer et al. (1975) found that the total mass of food in the guts of fishes leaving a meadow was three times that of the fish entering. In the tropics, on the other hand, Ogden (1980) reports that most herbivorous fishes are not resident in the turtle grass beds, but migrate in during the night from surrounding reefs. Grazing by such fishes, as well as by urchins, is responsible for the formation of "halos" of grazed areas around reefs.

Certain of the larger fishes, such as sharks and rays, are actually or potentially of considerable importance in structuring seagrass communities. For example, Orth (1975) reports that the feeding of the cownose ray (*Rhinoptera bonasus*) destroys large areas of seagrass beds on the East Coast. These fish dig into the bottom to pull out clams and in so doing destroy seagrasses and alter the sediments. Not only do these rays drastically reduce clam densities (from 60 to 100

Mya arenaria per square meter to zero) but that of the associated fauna as well.

Relationships to Other Systems

With the exceptions noted above, relatively little of the living seagrass tissue is consumed directly by grazers. Most of the seagrass organic production either is transformed into detritus and is consumed or further broken down in situ or else is transported out of the system to be used by other shallow-water communities. How is this large food source channeled out to other shallow-water systems? There are at least four ways. Organic material can be transported out as whole plant parts with associated organisms, as detritus, as dissolved organic matter, or as living tissue or fecal material of various motile vertebrates (Figs. 5.25, 5.26). The importance of seagrass beds to adjacent communities varies,

Figure 5.26 The pathway for the channeling of eelgrass into the food web. (Modified from C. P. McRoy and C. Helfferich [eds.], Seagrass ecosystems, Dekker, 1977; by courtesy of Marcel Dekker, Inc.)

depending on geography, hydrographic conditions, and depth. Shallow-water beds tend to export more material because they are more often disturbed by water movement, and by ice scour, and because they can be reached more easily by various birds. Deep-water beds, on the other hand, are usually inaccessible to birds and also less disturbed by water motion. In these beds, however, movement by fishes may be more significant. The detrital material produced in seagrass beds may be transported to enrich areas at considerable distance from the seagrass beds themselves. It has been suggested as a source of energy for organisms as far removed as the abyssal benthos.

Ecology

Ecologically, seagrass beds serve a number of important functions in inshore areas. They are a major source of primary productivity in shallow waters around the world and thus an important source of food for many organisms (in the form of detritus). In addition, they function to stabilize the soft bottoms, on which most species grow, primarily through the dense, matted root system. This stabilization of the bottom by roots is sufficiently durable to withstand storms as severe as hurricanes. This system, in turn, serves to shelter many organisms. Thus, there are many animals that are commonly found in seagrass beds but that do not have a direct trophic relationship to the grasses themselves. The beds serve as nursery grounds for many species that spend their adult lives in other areas. Included among these latter species are several of commercial importance, such as the shrimp *Penaeus duorarum* of southern Florida. These beds may also harbor adults of commercially important species such as the bay scallop, *Aquipecten irrad-*

ians, off the East Coast. Destruction of seagrass beds, therefore, may have a serious effect on certain commercial fisheries.

Seagrass beds may also act as traps for sediment and, therefore, act to build up the bottom. Such a situation has been recorded for the *Posidonia* beds in the Mediterranean Sea, which build extensive terraces up from bottom. When such a situation prevails and the buildup approaches the surface, the floating leaves act to break the force of the waves, forming a calm water habitat on the bed.

The leaves of the seagrasses also serve as a protective canopy, shielding the inhabitants of the bed from the effects of strong sunlight. Where the beds become intertidal, the leaves may cover the bottom substrate at low tide, protecting the inhabitants from desiccation. In Florida, seagrasses are the primary substrate for most attached macroalgae. Very few algae colonize soft sediments, so with limited hard substrate in the sublittoral zone, this is an important function.

The importance of seagrass beds and what

might happen when they are destroyed was given dramatic emphasis by a natural destructive event that occurred to the dominant seagrass *Zostera marina* in the North Atlantic Ocean. In the early 1930s, the *Zostera* beds on both sides of the Atlantic Ocean were massively destroyed by a mysterious "wasting disease" of epidemic character. Although there had been no universal agreement as to the causative agent for this destruction up until recently, Short et al. (1987) have now confirmed that a pathogenic strain of the protistan *Labyrinthula* causes the symptoms. During the epidemic of the 1930s, the *Zostera* beds in the Pacific Ocean were not affected.

The effects of the *Zostera* disappearance were different in different areas. In Europe, the most immediate results were that the level of the beach fell; with the loss of sediment, many bared rocks appeared that sprouted a new growth of algae in place of *Zostera;* and there was the formation of many new sandbars (Fig. 5.27). The organisms also changed, but less than was anticipated. Whereas the eelgrass beds had been dominated by detritus-feeding animals, after the

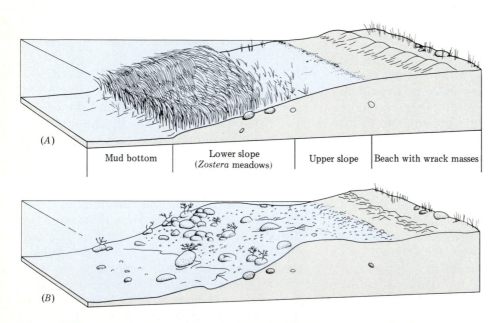

| Mud bottom | Lower slope (*Zostera* meadows) | Upper slope | Beach with wrack masses |

Figure 5.27 Representation of: *(A)* the conditions of the eelgrass beds of Scandinavia before the onset of the wasting disease, and *(B)* the extreme conditions following the complete destruction of the beds. (From C. P. McRoy and C. Helfferich [eds.], Seagrass ecosystems, Dekker, 1977; by courtesy of Marcel Dekker, Inc.)

loss the same bottom area had more suspension feeders and in general more species. In Europe, very few annual species associated with *Zostera* actually disappeared completely. In the United States, on the other hand, destruction of the *Zostera* beds meant the almost complete disappearance of the animals associated with the beds. The reason for this difference probably has to do with the fact that, in European waters, the alga *Fucus* quickly replaced *Zostera,* thus offering the same protected environment, whereas this was not the case in the northeastern United States. Since the 1930s, the *Zostera* beds have slowly returned to reoccupy their former habitat. Recently, however, Robblee et al. (1991) have reported the widespread demise of the

eelgrass beds of *Thalassia testudinum* in Florida Bay. This catastrophic loss appears to have the same characteristics as the "wasting disease" of *Zostera* and may have a similar causal agent. The extent and ecological results of this new outbreak are not as yet known.

For some seagrass communities, den Hartog (1977) reports a successional pattern such as has been found for terrestrial communities (Fig. 5.28). An example is the seagrass beds of the tropical Caribbean. Here, the climax community is dominated by *Thalassia testudinum,* but there is a series of stages leading up to *Thalassia* that are different, depending primarily on the substrate. On the other hand, in the temperate Atlantic where the dominant seagrass is *Zostera*

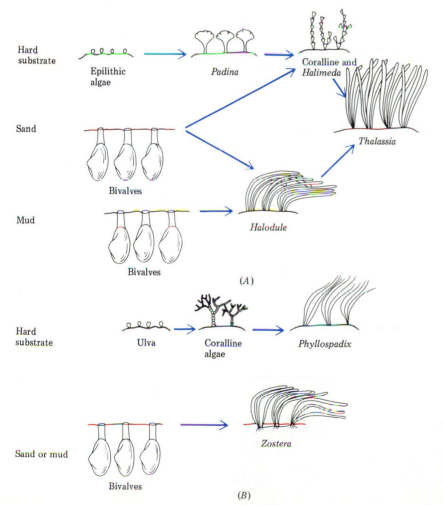

Figure 5.28 Successional sequences in seagrass beds. *(A)* Tropical areas. *(B)* Temperate areas.

marina, no series of successional stages has been uncovered and *Z. marina* colonizes directly; that is, it is both pioneer and terminal stage. In the temperate Pacific Ocean, the seagrass *Phyllospadix scouleri* occurs on hard substrates on wave-swept shores (*Zostera* occurs on soft bottoms in protected areas). Turner (1983) has shown that there is an obligate succession leading to the development of *P. scouleri* beds. Open areas in surf grass beds are settled first by *Ulva,* a green ephemeral alga. This is replaced by several red algae. Finally, surf grass colonizes either by the rhizomes growing into the space or by seed recruitment. Rhizome invasion is not dependent on any preexisting species and may occur directly, but for seeds to establish the surf grass there must be present macroalgae for the seeds to attach to. Furthermore, the seeds attach preferentially, or at least have a better survival, if the algae are turf forming with bushy branches and a narrow central axis. Thus a true facilitated succession occurs.

SOME SPECIAL COMMUNITIES

In addition to the widespread communities described above, there are a few of more restricted occurrence that have received considerable study; hence, they deserve mention here as examples of how communities are organized, how they may vary temporally, and how they are maintained.

A Sea Pen Community

In Puget Sound, Washington, at depths of 10–50 m on soft bottoms, there exists a community dominated by dense stands of the long-lived cnidarian sea pen *Ptilosarcus gurneyi* (Figs. 5.29, 5.30). In an extensive study, Birkeland (1974) found that *Ptilosarcus gurneyi* provides the major food source for seven predators: four starfish (*Hippasteria spinosa, Dermasterias imbricata, Crossaster papposus,* and *Mediaster aequalis*) and three opisthobranch mollusks (*Armina californica, Tritonia festiva,* and *Hermissenda crassicornis*). *Ptilosarcus* itself has a life span of about 15 years; takes several years to reach sexual maturity; has irregular, unpredictable recruitment, resulting in stands of differing age structure; and never reaches sizes at which it is immune to predation.

Adult *Ptilosarcus* are readily consumed by three starfish—*Hippasteria, Dermasterias,* and *Mediaster*—plus the opisthobranch *Armina,* which seems to prefer the largest sea pens. Of these four predators, *Hippasteria* has the most restricted diet, feeding only upon *Ptilosarcus.* The others are capable of consuming other prey,

Figure 5.29 A North Pacific sea pen *(Ptilosarcus)* bed. (Photo courtesy of Dr. Charles Birkeland.)

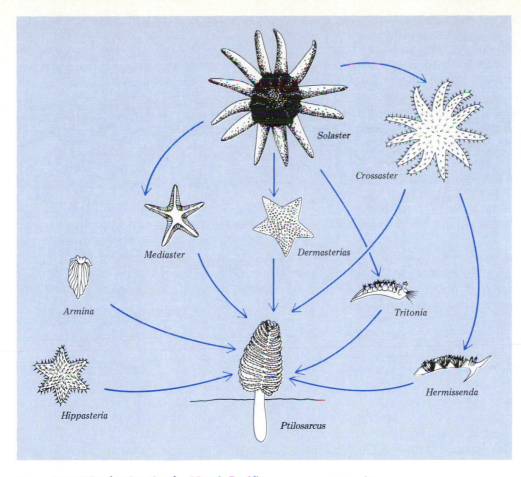

Figure 5.30 The food web of a North Pacific sea pen community.

although, in Puget Sound at least, *Armina* seems to feed exclusively on *Ptilosarcus*. Juvenile *Ptilosarcus* are preyed upon by *Hermissenda*, *Tritonia*, and *Crossaster*, but this is uncommon.

In this community one might expect that the generalists (*Dermasterias*, *Mediaster*, and *Crossaster*), because they are not dependent completely on *Ptilosarcus*, would maintain high population numbers and thus reduce the *Ptilosarcus* population so low that the specialists (*Armina* and *Hippasteria*) would have nothing left. In other words, they could outcompete them. What is to prevent this? The answer lies with yet another starfish, the top predator *Solaster dawsoni*, which preys upon *Mediaster*, *Dermasterias*, and *Crossaster*, reducing their numbers so sufficiently that competitive exclusion does not occur (Fig. 5.30), allowing *Hippasteria* to persist.

This community is thus maintained by a complex food web relationship that has as its key industry species the sea pen *Ptilosarcus*, which is preyed upon by several specialist and generalist predators. The generalists (*Mediaster*, *Dermasterias*, and *Crossaster*) are unable to obtain numerical dominance and thus outcompete the specialists (*Hippasteria* and *Armina*) because a top predator, *Solaster*, keeps their numbers in check.

A Sea Pansy Community

Off San Diego in southern California in 5–10 m on a sand bottom exists quite a different community. As noted earlier on p. 178, this community was first studied by Fager (1968) between 1957 and 1963, when it was dominated by cnidarians, primarily the sea pansy *Renilla kolli-*

keri and the anemones *Harenactis attenuata* and *Zaolutus actius.* Associated with these were three gastropods (*Nassarius fossatus, Nassarius perpinguis,* and *Polinices recluzianus*) and two echinoderms (*Amphiodia occidentalis* and *Astropecten armatus*). In this community, Fager found the pattern of distribution of the major organisms to be primarily aggregated, and the populations remained constant over the six-year period. Furthermore, there was little evidence of any interaction among the species, and no evidence that these large macroinvertebrates were either preyed upon by mobile fishes or themselves preyed upon the smaller infauna organisms.

Eleven years later, in 1974, Davis and Van-Blaricom (1978) repeated the study at the same site. Whereas Fager had observed essentially nonvariant densities of the major species over six years, Davis and VanBlaricom found both long- and short-term changes in the abundances and densities of seven of the dominant species. These studies suggest that shallow-water benthic communities are not completely stable over long periods, even though they may appear so for several years. Furthermore, it appears that a much longer time interval is necessary to evaluate what is happening in such areas.

BENTHIC BIOLOGY OF POLAR SEAS

The Arctic and Antarctic regions of the earth appear physiognomically similar in that both are characterized by the dominating presence of ice and snow, year-round cold temperatures, and drastic changes in photoperiod that prevent photosynthesis during a significant part of the year. Most of these outward similarities, however, mask more fundamental differences that make the two polar regions very different both physically and biologically.

In this section we discuss, first, the differences between the two polar regions that give rise to the different biological associations. We then discuss the biological differences and conclude with a brief look at some examples.

Perhaps because of the profound interest and support of the U.S. government through its Antarctic research programs over the last 25 years, as well as a similar commitment from other world powers to Antarctic research, we have a better understanding of shallow-water Antarctic communities than we have of those of the Arctic. This is reflected here in the coverage allotted to each.

Physical Conditions of the Arctic and Antarctic Regions

The Arctic Ocean system is an isolated sea surrounded by landmasses that leave only two outlets to other oceans, the Bering Strait to the Pacific and Fram Strait to the Atlantic. Both of these outlets have relatively shallow sill depths, 70 m in the Bering Strait and 400 m in the Fram Strait (Fig. 5.31). The Arctic Ocean Basin is surrounded by extensive shallow shelves such that the deep water of the central basin is cut off from other oceans.

The circulation pattern consists of two major anticyclonic currents, the Beaufort gyre over the Canadian Basin and a transpolar current across the Eurasian part of the basin that exits through the Fram Strait.

The landmasses surrounding the Arctic Ocean have several large rivers that discharge significant amounts of sediment into the basin such that the basic substrate overlying the extensive continental shelves is particulate matter. The large freshwater discharges also lead to the development of a low-saline, stratified surface layer.

The Arctic Ocean pack ice is a persistent multiyear system that is harder and thicker than that of the Antarctic. Antarctic pack ice is usually seasonal and only a year old or younger. The whole central Arctic Ocean is permanently covered by ice (Fig. 5.31).

By contrast, the Antarctic marine ecosystem is open to all oceans and exists as a circumpolar ring of water surrounding a central landmass. The Antarctic continental shelf is narrow and falls off rapidly to deep water. The northernmost marine boundary is defined by the circular Antarctic Convergence, which is situated over deep water in a zone of persistent and strong westerly winds (Fig. 5.32). Immediately south of the

Occurrence of Ice

☐ Permanent during entire year	▨ Occasional in winter and spring
▨ Always or frequently during year	■ Occasional icebergs

Figure 5.31 Geography of the Arctic and Antarctic regions, showing the seasonality of ice cover. (From G. Hempel, On the biology of polar seas, particularly the southern ocean, pp. 3–33. *In:* J. S. Gray and M. E. Christiansen [eds.], Marine biology of polar regions and effects of stress on marine organisms, Wiley, 1985.)

Antarctic Convergence, nutrient-rich circumpolar water rises toward the surface and fertilizes the Antarctic surface waters (Fig. 5.32). Near the continent itself, and beneath the ice shelves, seawater is cooled to freezing or below and flows downslope to form the bottom water of the earth's deep ocean basins. Since the Antarctic continent is permanently covered by ice and lacks any rivers, there is little or no sediment input and no formation of stratified surface layers.

Seasonal variations in temperature compared to the Arctic are small in the Antarctic, both horizontally and vertically. In winter pack ice covers more than half of the Antarctic marine area but is reduced to only 10 percent in late summer (Fig. 5.31). Antarctic pack ice is only 1–2 m thick in contrast to the permanent 3–4 m thick Arctic pack ice. Both the Arctic and Ant-

arctic have pronounced year to year variations in ice cover.

These physical differences between the two polar regions are summarized in Table 5.1.

Because the polar inshore waters are covered with ice most of the year, benthic organisms and communities are rarely disturbed by wave action. However, polar shallow-water areas are subject to a stress factor unknown in other areas. That factor is ice. During the winter months, ice forms a thick layer over the continental shelf areas, extending down several meters. Wherever this ice layer makes contact with the bottom, the fauna is destroyed, either by mechanical action of the ice grinding against the bottom or by freezing the animals. This means that in polar areas, the shallowest subtidal areas are devoid of permanent communities down to the level to which the ice extends. This area may, however,

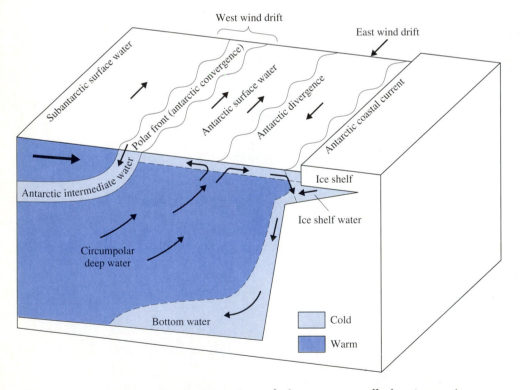

Figure 5.32 Diagrammatic representation of the currents off the Antarctic continent. (After G. Hempel, On the biology of polar seas, particularly the southern ocean, pp. 3–33. *In:* J. S. Gray and M. E. Christiansen [eds.], Marine biology of polar regions and effects of stress on marine organisms, Wiley, 1985.)

Table 5.1 OCEANOGRAPHIC FEATURES OF ARCTIC AND ANTARCTIC OCEANS

	Antarctic	Arctic
Area ($\times 10^6$ km)	35–38	14.6
Pack ice		
Mean maximum	22	12–13
Mean minimum	2.6	6–7
General features		
Form	Circumpolar ring between 50–60°S and 65–70°S	Mediterranean Sea around North Pole, almost enclosed by land at 70–80°N
Shelf	Narrow, open to all oceans, with large exchange in deep layers	Broad, two narrow openings (Bering Strait, Fram Strait)
Current systems	Circular currents (west wind drift, east wind drift), large eddies (Weddell Sea, Ross Sea)	Transpolar current, Beaufort gyre
Pack ice	High seasonality	Little seasonality
Cover in winter	c.50%	90%
Cover in summer	c.10%	80%
Age, thickness	Mainly one year, 1.5 m	Mainly multiyear, 3.5 m
Icebergs	Large, tabular icebergs, abundant and carrying coarse material	Small, irregular icebergs, only in Greenland and Bering Sea, not in Arctic Basin
River input	None	Much
Stratification	Low vertical stability	Surface layer (20–50 m) stable all year
Nutrients in euphotic zone	High all year	Seasonal depletion

Source: Modified from Hempel, 1985.

be inhabited during ice-free times of the year by mobile or transient organisms.

In Antarctic seas, in addition to the action of the thick layer of annual sea ice affecting the bottom, there is an additional area below that subject to a different kind of ice action. This area, which extends down to 30 m or so, is one in which ice platelets begin to form on the bottom around any convenient nucleus. As these platelets grow and increase in size, they may surround various sessile and sedentary invertebrates living on the bottom. Since ice is generally less dense than the surrounding water, as these grow they also tend to lift off the bottom and rise toward the thick sea ice above, carrying with them the entrapped invertebrates. Such organisms become trapped in the sea ice and are permanently lost from the community. Such ice formation is termed *anchor ice* (Fig. 5.33). The result of anchor ice formation, and its tendency to remove organisms, is the creation of a definite zonation of the benthic fauna with depth.

Biological Comparisons of Arctic and Antarctic Seas

As might be anticipated, the above differences in physical characteristics are reflected in differences in biology as well. Perhaps the most significant difference in the two polar regions is that the Antarctic region is far richer in terms of species of benthic organisms (Table 5.2). It also has an extremely high degree of endemism (Table 5.3). The high diversity of Antarctic species is matched by a high biomass. By contrast, the Arctic fauna is impoverished and consists mainly of organisms derived from the Atlantic Ocean. In general, the biomass figures for any given depth in the Antarctic are one to two orders of magnitude higher than for the Arctic. The high biomass of the Antarctic benthos is probably related to the nutrient-rich water and high offshore productivity, while the high species richness and endemism, according to Hempel (1985), are related to a long evolutionary history and isolation as well as to the presence of both hard and soft substrates.

Arctic benthic communities are often dominated by one or only a few species, but Antarctic communities usually have several species that are common or dominant. Because of the extensive areas of sediments, the Arctic benthic fauna is mainly an infauna, whereas the Antarctic benthos is characterized by large numbers of epifaunal species. According to Hempel (1985), fish are less important in Antarctic benthic com-

Figure 5.33 Anchor ice area in the Antarctic. (From P. K. Dayton, G. Robilliard, and A. L. DeVries, Anchor ice formation in McMurdo Sound, Antarctica, and its biological effects, Science, Vol. 163, 1969. Copyright © 1969 by the American Association for the Advancement of Science.)

Table 5.2 COMPARISON OF THE NUMBER OF SPECIES OF MACROBENTHIC ORGANISMS FROM THE ARCTIC AND THE ANTARCTIC

	Arctic	Antarctic
Sponges	200	300
Polychaetes	c.300	650
Amphipods	262 (Barents Sea)	470
Pycnogonids	29 (Barents Sea)	100
Ascidians	47	129
Bryozoans	200 (Barents Sea)	310
Mollusks	224 (Barents Sea)	875
Isopods	49 (Kara Sea)	299

Source: After Hempel, 1985.

Table 5.3 DEGREE OF ENDEMISM IN SELECTED GROUPS OF ANTARCTIC MARINE ORGANISMS

	Genera (%)	Species (%)
Isopoda and Tanaidacea	10%	66%
Fishes	70	95
Pycnogonida	14	90
Echinodermata	27	73
Echinoidea	25	77
Holothuroidea	5	58
Bryozoa	—	58
Polychaeta	5	57
Amphipoda	39	90

Source: After Hempel, 1985.

munities and are mainly restricted to the near bottom.

In both polar regions birds and mammals are the top carnivores (see pp. 115–119). Most polar benthic invertebrates appear to be long lived, slow-reproducing animals, but it is unclear as to whether this is due to some inherent response to low temperatures or to some other feature of the polar environment, such as low food resources.

Among the benthic invertebrates in the Antarctic, there is little evidence for any temperature adaptation with respect to reproduction, development, or growth. Pearse et al. (1991) also note that, contrary to "Thorson's rule," that cold-water benthic organisms should show more nonpelagic development, pelagic planktotrophic development is common. The only probable difference from temperate and tropical planktotrophic organisms is that the Antarctic pelagic larvae take a very long time to develop to metamorphosis. How these larvae survive the poor food conditions prevalent during the Antarctic winter is still not known.

Ecological Communities of Polar Seas

Sea Ice Communities A marine environment unique to polar seas is that of the organisms that inhabit the sea ice. Sea ice provides a very extensive habitat for a set of organisms ranging in size from bacteria to marine birds and mammals. We are here, however, mainly interested in the organisms that inhabit the ice on a long-term basis and that interact with each other.

Over 200 species of marine organisms have been reported living in association with sea ice according to Garrison (1991). The major groups include bacteria, certain algae, heterotroph protistans, and small metazoans. There is still disagreement as to whether or not these ice associations are true communities, since many of the species are also found in adjacent open waters. It has been suggested that the ice may serve as a temporary habitat for planktonic species and may be important in providing a "seed" population for plankton blooms at the sea edge.

There does appear to be a well-developed microbial food web in the sea ice. This microbial community or association may be significant to polar benthic communities because there is evidence that benthic organisms feed on the sea ice biota.

A suggested conceptual food web for the ice community is given in Fig. 5.34.

An Antarctic Soft-Sediment Community Oliver and Slattery (1985) studied a rare soft-bottom community in McMurdo Sound, Antarctica. This was a community composed pri-

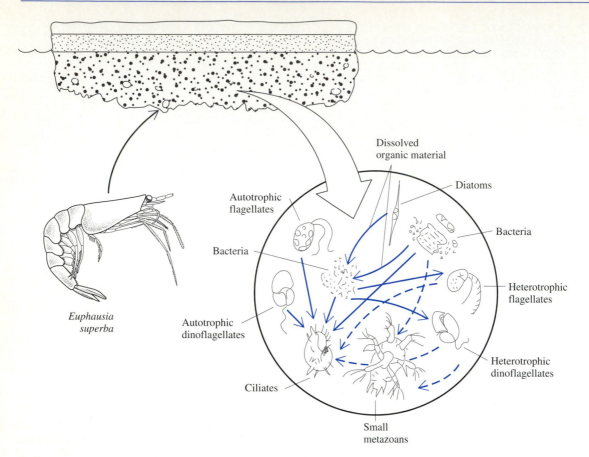

Figure 5.34 A conceptual food web for the pack-ice microbial community. Solid arrows show feeding relationships that have been documented. Dashed arrows show predator-prey relationships that are known in other systems but have not been confirmed in sea ice. None of the feeding rates is known. (After D. L. Garrison, Antarctic sea ice biota, Amer. Zool. 31:17–33.)

marily of burrowing and tube-building polychaete worms and small crustaceans that occurred in densities of over 100,000 individuals per square meter. This dense assemblage was divided into three faunal groups of species that had similar characteristics. The three groups were canopy, understory, and subsurface (Fig. 5.35). The canopy was dominated by the anemone *Edwardsia* and consisted primarily of suspension feeders. The understory included mainly mobile crustaceans that were deposit feeders and predators. The subsurface species group was composed primarily of tube-dwel-

ling polychaetes that were mainly deposit feeders.

Experimental manipulations of this community suggested that the community structure was regulated by the understory predatory small crustaceans. These animals consumed the juveniles of the dominant polychaete species. The adults of the dominant polychaetes had a refuge in size from the predators. Therefore, the dense assemblage was dominated by large and long-lived species.

This community probably persists in time because of several co-occurring factors. First,

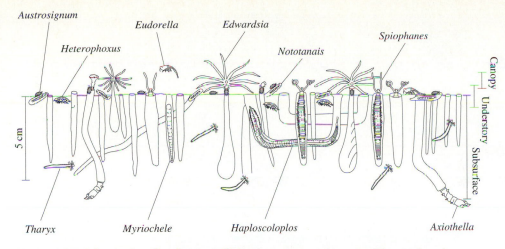

Figure 5.35 Schematic of a dense soft-bottom Antarctic assemblage, showing the three faunal guilds: canopy, understory, and subsurface species. (After J. S. Oliver and P. N. Slattery, Effects of crustacean predators on species composition and population structure of soft bodied infauna from McMurdo Sound, Antarctica, Ophelia 24[3]:155–175.)

there are no large organisms or predators, such as walruses, gray whales, skates, or rays, that could introduce biological disturbances. These can, as we have seen earlier (p. 115), remove large areas of infauna in other shallow-water areas. In this community, such a biological disturbance would destroy the assemblage, because the adults would be removed while the motile crustacean predators would likely remain and remove the settling juveniles. Second, the adults are long lived and are immune to predation from the small crustaceans when they reach a certain size. This means that there is a long time over which they reproduce, enough time to allow a few juveniles to escape the predation of the crustaceans and replace the very few adults that may die.

In contrast to other soft-bottom infaunal communities that seem to be structured by competition (see pp. 171–176), there is no evidence of a competitive dominant.

An Antarctic Hard-Bottom Community

Antarctic hard-bottom communities appear to be more common than soft-bottom communities. They have also had a longer period of investigation, beginning in the late 1960s. Indeed, the more than 20 years of marine ecological work in the McMurdo Sound area of Antarctica has given us a better understanding of community interaction in this "remote" area than we have for many temperate or tropical areas close to home.

In the shallow inshore waters of McMurdo Sound, Dayton, Robilliard, and Paine (1970) recognize three zones (Fig. 5.36). Zone I extends from 0 to 15 m and is essentially devoid of life, due to scouring by sea ice and the almost universal coverage by anchor ice. Only in the ice-free periods are mobile organisms able to enter.

Zone II extends from about 15 m down to the limit of anchor ice formation at 30 m or so. This zone experiences less anchor ice and has a fauna of numerous sessile animals, mainly anemones and other cnidarians. The anemones here often feed upon rafted-in jellyfish or sea urchins, while the other cnidarians are suspension feeders. The anemones appear to be territorial.

Zone III begins abruptly where anchor ice ceases and continues down to an undetermined depth. The substrate here changes from a cobbly-rocky bottom to one that is a thick mat of

Zone I: 0–15 m

Zone II: 15–30 m

Zone III: below 33 m

Figure 5.36 Vertical zonation of invertebrates in Antarctic shelf areas. A few motile animals forage into Zone I, which is otherwise barren of sessile animals. Cnidarians dominate the sessile animals of Zone II and sponges, Zone III. (With permission from M. W. Holdgate, Antarctic biology, Vol. I, Academic Press, 1970. Copyright © by Academic Press [London] Ltd.)

sponge spicules. Sponges dominate this area, but the area has a rich and diverse fauna of other invertebrates as well, of which starfish are the most conspicuous motile animals. This zone is noteworthy because of its extreme physical stability, which has resulted in a complex, highly diverse community that, according to Dayton and associates (1974), seems to be regulated by biological factors (predation) rather than by physical factors as in Zones I and II.

The dominant organisms in this area are very slow growing, so slow, in fact, that in many cases the growth rates are too low to measure in one year. One exception is the sponge *Mycale acerata*, which has a very rapid growth rate and seems capable of outcompeting all other sessile invertebrates and taking over the substrate space. *Mycale* is prevented from dominating the substrate because it is preyed upon by two sea stars, *Perknaster fuscus* and *Acodonaster conspicuus*, which keep it in check. *Acodonaster conspicuus* also preys on several other dominant sponges but is prevented from reaching population densities that would allow it to destroy the sponge community by yet another asteroid, *Odontaster validus*, which consumes the larvae, young, and even adults of *A. conspicuus* (Fig. 5.37; Plate 35). Thus, we have a community that is regulated through finely tuned biological interactions, primarily predation.

(A)

(B)

(C)

(D)

Figure 5.37 Sequence showing the consumption of the large Antarctic starfish *Acodonaster conspicuus* by the small starfish *Odontaster validus* and the nemertean *Lineus corrugatus*. (From P. K. Dayton, G. A. Robilliard, R. T. Paine, and L. B. Dayton, Biological accommodation in the benthic community at McMurdo Sound, Antarctica, Ecol. Monogr. 44[1]:105–128.)

BIBLIOGRAPHY

Ayling, A. M. 1981. The role of biological disturbance in temperate subtidal encrusting communities. Ecology 62(3):830–847.

Barkai, A., and C. McQuaid. 1988. Predator-prey role reversal in a marine benthic ecosystem. Science 242:62–64.

Birkeland, C. 1974. Interactions between a sea pen and seven of its predators. Ecol. Monogr. 44:211–232.

Camp, D. K., S. P. Cobb, and J. F. Van Breedveld. 1973. Overgrazing of seagrasses by a regular urchin, *Lytechinus variegatus.* Bioscience 23:37–38.

Cottam, C. 1934. Past periods of eelgrass scarcity. Rhodora 36:261–264.

Coull, B. C. (ed.). 1977. Ecology of marine benthos. University of South Carolina Press, Columbia. 467 pp.

Crisp, D. J., and P. S. Meadows. 1962. The chemical bases of gregariousness in cirripedes. Proc. Roy. Soc. London (Biol.) 150:500–520.

Davis, N., and G. VanBlaricom. 1978. Spatial and temporal heterogeneity in a sand bottom epifaunal community of invertebrates in shallow water. Limnol. Oceanogr. 23(3):417–427.

DeVlas, J. 1979. Annual food intake by plaice and flounder in a tidal flat area in the Dutch Wadden Sea with special reference to consumption of regenerating parts of macrobenthic prey. Neth. J. Sea Res. 13:117–153.

Dayton, P., G. Robilliard, and R. T. Paine. 1970. Benthic faunal zonation as a result of anchor ice at McMurdo Sound, Antarctica, pp. 244–258. *In:* Antarctic ecology, Vol. I. Academic Press, N.Y.

Dayton, P. K., G. A. Robilliard, R. T. Paine, and L. B. Dayton. 1974. Biological accommodation in the benthic community at McMurdo Sound, Antarctica. Ecol. Monogr. 44(1):105–128.

den Hartog, C. 1977. Structure, function and classification in sea grass communities, pp. 89–121. *In:* McRoy, C. P., and C. Helfferich (eds.). Seagrass ecosystems. Dekker, N.Y.

Ebeling, A. W., D. R. Laur, and R. J. Rowley. 1985. Severe storm disturbance and reversal of community structure in a southern California kelp forest. Mar. Biol. 84:287–294.

Estes, J., and J. Palmisano. 1974. Sea otters, their role in structuring near-shore communities. Science 185:1058–1060.

Fager, E. W. 1968. A sand bottom epifaunal community of invertebrates in shallow water. Limnol. Oceanogr. 13(3):448–464.

Fenchel, T. 1977. Aspects of the decomposition of seagrasses, pp. 123–146. *In:* McRoy, C. P., and C. Helfferich (eds.). Seagrass ecosystems. Dekker, N.Y.

Fletcher, W. J. 1987. Interactions among subtidal Australian sea urchins, gastropods and algae: Effects of experimental removals. Ecol. Monogr. 57(1):89–109.

Foster, M. S., and D. R. Schiel. 1985. The ecology of giant kelp forests in California: A community profile. U.S. Fish & Wildlife Serv. Biol. Rept. 85(7.2). 152 pp.

Garrison, D. L. 1991. Antarctic sea ice biota. Amer. Zool. 31:17–33.

Gray, J. S. 1981. The ecology of marine sediments. Cambridge University Press, N.Y. 185 pp.

Harrold, C., and D. Reed. 1985. Food availability, sea urchin grazing and kelp forest community structure. Ecology 66(4):1160–1169.

Hempel, G. 1985. On the biology of polar seas, particularly the southern ocean, pp. 3–33. *In:* Gray, J. S., and M. E. Christiansen (eds.). Marine biology of polar regions and effects of stress on marine organisms. Wiley, N.Y.

Holme, N. A., and A. D. McIntyre (eds.). 1984. Methods for the study of marine benthos. 2nd ed. Blackwell Scientific Publications, Oxford. 387 pp.

Jones, N. S. 1950. Marine bottom communities. Biol. Rev. 25:283–313.

Kastendiek, J. 1976. Behavior of the sea pansy, *Renilla kollikeri* Pfeffer (Coelenterata, Pennatulacea), and its influence on the distribution and biological interaction of the species. Biol. Bull. 151:518–537.

Keegan, B. F., P. O. Ceidigh, and P. J. S. Boaden (eds.). 1977. Biology of benthic organisms. Pergamon, N.Y. 630 pp.

Kenworthy, W. J., J. C. Zieman, and G. W. Thayer. 1982. Evidence for the influence of seagrass on the benthic nitrogen cycle in a coastal plain estuary near Beaufort, North Carolina. Oecologia 54:152–158.

Levin, L. A. 1981. Dispersion, feeding behavior and competition in two spionid polychaetes. J. Mar. Res. 39:99–117.

Levinton, J. S. 1977. Ecology of shallow water deposit feeding communities, Quisset Harbor, Massachusetts, pp. 191–227. *In:* Coull, B. (ed.). Ecology of marine benthos. University of South Carolina Press, Columbia.

Mann, K. H. 1973. Seaweeds: Their productivity and strategy for growth. Science 182:975–983.

McRoy, C. P. 1966. The standing stock and ecology of eelgrass in Izembeck Lagoon, Alaska. M.S. Thesis, University of Washington. 137 pp.

McRoy, C. P., and C. Helfferich (eds.). 1977. Seagrass ecosystems. Dekker, N.Y. 314 pp.

McRoy, C. P., and C. McMillan. 1977. Production ecology and physiology of sea grasses, pp. 53–88. *In:* McRoy, C. P., and C. Helfferich (eds.). Seagrass ecosystems. Dekker, N.Y.

Menge, B. A., and T. M. Farrell. 1989. Community structure and interaction webs in shallow marine hard bottom communities: Tests of an environmental model. Adv. Ecol. Res. 19:189–262.

Mills, E. L. 1969. The community concept in marine zoology, with comments on continua and instability in some marine communities: A review. J. Fish. Res. Bd. Can. 26:1415–1428.

Myers, A. C. 1977. Sediment processing in a marine subtidal sandy bottom community. II. Biological consequences. J. Mar. Res. 35:633–647.

North, W. 1971. Growth of individual fronds of the mature giant kelp, *Macrocystis,* pp. 123–168. *In:* North, W. (ed.). The biology of giant kelp beds (*Macrocystis*) in California. Beihefte zur Nova Hedwigia Heft 32, Verlag von J. Cramer. 600 pp.

Ockelmann, K. W., and O. Vahl. 1970. On the biology of the polychaete *Glycera alba,* especially its burrowing and feeding. Ophelia 8:275–294.

Ogden, J. C. 1980. Faunal relationships in Caribbean seagrass beds, pp. 173–198. *In:* Phillips, R. C., and C. P. McRoy (eds.). Handbook of seagrass biology: An ecosystem perspective. Garland STPM Press, N.Y. 353 pp.

Oliver, J. S., and P. N. Slattery. 1985. Effects of crustacean predators on species composition and population structure of soft bodied infauna from McMurdo Sound, Antarctica. Ophelia 24(3):155–175.

Oliver, J. S., P. N. Slattery, L. W. Hulberg, and J. W. Nybakken. 1979. Relationships between wave disturbance and zonation of benthic invertebrate communities along a subtidal high energy beach in Monterey Bay, California. Fish. Bull. 78:437–454.

Orth, R. J. 1975. Destruction of eelgrass *Zostera marina* by the cownose ray *Rhinoptera bonasus* in the Chesapeake Bay. Chesapeake Sci. 16:205–208.

Pearse, J. S., J. B. McClintock, and I. Bosch. 1991. Reproduction of Antarctic benthic marine invertebrates: Tempos, modes and timing. Amer. Zool. 31:65–80.

Petersen, C. G. Joh. 1918. The sea bottom and its production of fish food. A survey of the work done in connection with the valuation of the Danish waters from 1883–1917. Rap. Danish Biol. Stat., Vol. 25. 62 pp.

Petersen, C. G. Joh. 1924. A brief survey of the animal communities in Danish waters. Amer. J. Sci., Ser. 5. 7(41):343–354.

Peterson, C. H. 1977. Competitive organization of the soft bottom macrobenthic communities of southern California lagoons. Mar. Biol. 43:343–359.

Peterson, C. H. 1982. Clam predation by whelks (*Busycon* spp.): Experimental tests of the importance of prey size, prey density and seagrass cover. Mar. Biol. 66:159–170.

Peterson, C. H., and M. L. Quammen. 1982. Siphon nipping: Its importance to small fishes and its impact on growth of the bivalve *Protothaca staminea* (Conrad). J. Exp. Mar. Biol. Ecol. 63:249–268.

Phillips, R. C., and C. P. McRoy (eds.). 1980. Handbook of seagrass biology. Garland STPM Press, N.Y. 353 pp.

Rasmussen, E. 1973. Systematics and ecology of the Isefjord marine fauna (Denmark). Ophelia 11(1–2):1–495.

Reise, K. 1979. Spatial configurations generated by motile benthic polychaetes. Helgol. Wiss. Meeresunters. 32:55–72.

Rhoads, D. C., and D. K. Young. 1970. The influence of deposit feeding organisms on sediment stability and community structure. J. Mar. Res. 28(2):150–178.

Rhoads, D. C., and D. K. Young. 1971. Animal-sediment relations in Cape Cod Bay, Massachusetts. II. Reworking by *Molpadia oolitica.* Mar. Biol. 11(3):255–261.

Robblee, M. B., T. R. Barber, P. R. Carlson, Jr., M. J. Durako, J. W. Fourqurean, L. K. Muehlstein, D. Porter, L. A. Yarbro, R. T. Zieman, and J. C. Zieman. 1991. Mass mortality of the tropical seagrass *Thalassia testudinum* in Florida Bay (U.S.A.). Mar. Ecol. Progr. Ser. 71:297–299.

Robertson, A. I., and K. H. Mann. 1984. Disturbance by ice and life history adaptations of the seagrass *Zostera marina.* Mar. Biol. 80:131–142.

Rosenthal, R. J., W. D. Clarke, and P. K. Dayton. 1974. Ecology and natural history of a stand of giant kelp, *Macrocystis pyrifera,* off Del Mar, California. Fish. Bull. 72(3):670–684.

Scagel, R. F. 1947. An investigation on marine plants near Hardy Bay, B.C. Provincial Dept. Fisheries, Victoria, B.C., Canada. pp. 1–70.

Sebens, K. 1985. The ecology of the rocky subtidal zone. Amer. Sci. 73:548–557.

Sebens, K. 1986. Spatial relationships among encrusting marine organisms in the New England subtidal zone. Ecol. Monogr. 56(1):73–96.

Setchell, W. A. 1929. Morphological and phenological notes on *Zostera marina.* Univ. Calif. Pub. Botany 14:389–452.

Short, F. T., L. K. Muehlstein, and D. Porter. 1987. Eelgrass wasting disease: Cause and recurrence of a marine epidemic. Biol. Bull. 173(3):557–562.

Sutherland, J. P. 1974. Multiple stable points in natural communities. Amer. Nat. 108:859–873.

Tegner, M. J., and P. K. Dayton. 1987. El Niño effects on southern California kelp forest communities. Adv. Ecol. Res. 17:243–279.

Tevesz, M. J. S., and P. L. McCall (eds.). 1983. Biotic interactions in recent and fossil benthic communities. Plenum, N.Y. 837 pp.

Thayer, G. W., S. M. Adams, and M. W. LaCroix. 1975. Structural and functional aspects of a recently established *Zostera marina* community, pp. 517–540. *In:* Cronin, L. E. (ed.). Estuarine research, Vol. 1. Academic Press, N.Y.

Thayer, G. W., W. J. Kenworthy, and M. S. Fonseca. 1984. The ecology of eelgrass meadows of the Atlantic coast: A community profile. U.S. Fish and Wildlife Service. FWS/OBS–84/02. 147 pp.

Thorson, G. 1950. Reproductive and larval ecology of marine bottom invertebrates. Biol. Rev. 25:1–45.

Thorson, G. 1955. Modern aspects of marine level bottom animal communities. J. Mar. Res. 14:387–397.

Thorson, G. 1957. Bottom communities (sublittoral or shallow shelf), pp. 461–534. *In:* The treatise on marine ecology and paleoecology, Vol. I. Memoir 67, Geol. Soc. Amer.

Thorson, G. 1966. Some factors influencing the recruitment and establishment of marine benthic communities. Neth. J. Sea. Res. 3(2):267–293.

Turner, T. 1983. Facilitation as a successional mechanism in a rocky intertidal community. Amer. Nat. 121:729–738.

VanBlaricom, G. R. 1982. Experimental analyses of structural regulation in a marine sand community exposed to ocean swell. Ecol. Monogr. 52:283–305.

Van Veldhuizen, H., and D. Phillips. 1978. Prey capture by *Pisaster brevispinus* (Asteroidea, Echinodermata) on soft substrate. Mar. Biol. 48:89–97.

Virnstein, R. W. 1977. The importance of predation by crabs and fishes on benthic infauna in Chesapeake Bay. Ecology 58:1199–1217.

Warner, G. F. 1984. Diving and marine biology, The ecology of the sublittoral. Cambridge University Press, New York. 210 pp.

Wilkins, E. W. 1982. Waterfowl utilization of a submerged vegetation (*Zostera marina* and *Ruppia maritima*) bed in the lower Chesapeake Bay. M. A. thesis, College of William and Mary, 83 pp.

Williams, S. 1987. Competition between seagrasses *Thalassia testudinum* and *Syringodium filiforme* in a Caribbean lagoon. Mar. Ecol. Progr. Ser. 35:91–98.

Wilson, W. H., Jr. 1981. Sediment mediated interactions in a densely populated infaunal assemblage: The effects of the polychaete *Abarenicola pacifica.* J. Mar. Res. 39:735–748.

Wilson, W. H. 1990. Competition and predation in marine soft sediment communities. Ann. Rev. Ecol. Sys. 21:221–241.

Witman, J. D. 1985. Refuges, biological disturbance, and rocky subtidal community structure in New England. Ecol. Monogr. 55:421–445.

Witman, J. D. 1987. Subtidal coexistence: Storms, grazing, mutualism and the zonation of kelps and mussels. Ecol. Monogr. 57(2):167–187.

Woodin, S. A. 1976. Adult-larval interactions in dense infaunal assemblages: Patterns of abundance. J. Mar. Res. 34:25–41.

Woodin, S. A. 1983. Biotic interactions in recent marine sedimentary environments, pp. 3–38. *In:* Tevesz, M. J. S., and P. L. McCall (eds.). Biotic interactions in recent and fossil benthic communities. Plenum, N.Y.

Zieman, J. C. 1982. The ecology of the seagrasses of South Florida: A community profile. U.S. Fish & Wildlife Service Program FWS/OBS–82/25. 124, 26 pp.

Intertidal Ecology

Although it constitutes by far the smallest area of all in the world's oceans, existing as an extremely narrow fringe a few meters in extent between high and low water, the intertidal zone is perhaps the best known. Because it is the most accessible to humans, it has been studied more intensively than any other marine habitat. Here, and only here, can direct observations be made of aquatic organisms during the period of low tide, without special equipment. The intertidal zone has been of interest and use to humans since prehistoric time.

Despite being restricted, the intertidal has within it the greatest variations in environmental factors of any marine area, and these can occur within centimeters of each other. Coupled with this is a tremendous diversity of life, which may be as great as or greater than that found in the more extensive subtidal habitats.

It should be emphasized that this area is truly an extension of the marine environment and is inhabited almost exclusively by marine organisms. Although it may be exposed to air as much as half the time, the terrestrial fauna and flora have not invaded it to any extent, except at the uppermost fringes.

The richness, the diversity of environmental factors, and the ease of access act together to attract to this area a disproportionate amount of scientific attention. More knowledge is thus available for this small area and its organisms, and their interactions are better known than in any other area. Thanks to such interest and knowledge, the intertidal has produced more unifying concepts with respect to the organization of marine communities than any other. As a result, a disproportionate amount of space in this book is devoted to a discussion of it.

ENVIRONMENTAL CONDITIONS

In part, the tremendous array of environmental factors and ranges found in the intertidal results from the fact that this zone is exposed to the air for a certain amount of time during a day and also because most physical factors show a wider range in air than in water (see Chapter 1, pp. 1–3).

Tides

The periodic, predictable rise and fall of the level of the sea over a given time interval is called a *tide*. Certainly, this is the most important environmental factor influencing life in the intertidal zone. Without the presence of the tide or some other means of inducing a periodic rise and fall in the water level, this zone would not exist as it is, and many of the other factors would cease to be of influence.

With very few exceptions, most shore areas of the world experience tides. The only major seas that virtually lack tidal action are the Mediterranean and the Baltic. In these areas, the water level on the shoreline fluctuates primarily due to the effects of wind action pushing water. This does not mean, however, that all seashores experience the same tidal range, or even the same type of tide. The reasons for the occurrence of different tides and different ranges are complex, having to do with the interaction of the tide-generating forces—the sun and moon, the rotation of the earth, the geomorphology of the ocean basins, and the natural oscillations of the various ocean basins. Most of the explanation is beyond the scope of this text, but a simplified explanation will be given here.

Tides occur due to the interaction between the gravitational attraction of the sun and the moon on the earth and the centrifugal force generated by the rotating earth and moon system. As a result of these forces, the water in the ocean basins is pulled into bulges. Gravitational attraction of one body for another is a function of the mass of each body and their distance apart. In the case of the sun and the moon, the gravitational generating force of the moon on the earth is about twice that of the sun, even though the sun is many times more massive than the moon. This is due to the greater distance of the sun from the earth.

The earth and moon form an orbiting system that revolves around their common center of mass. Because of the large size of the earth relative to the moon, this point is located inside the earth. The revolution of the earth-moon system creates a centrifugal force (acting outward) that is balanced overall by the gravitational force acting between the two bodies. However, gravitational force is much stronger than centrifugal force on the side of the earth facing the moon than on the side opposite. As a result, the side facing the moon has the water pulled into a bulge (= high tide). On the opposite side of the earth, the gravitational force of the moon will be the least, and the stronger centrifugal force will pull water into a bulge away from earth (another high tide; Fig. 6.1). Thus, we have the two high tides. These then circle the earth following the position of the moon as the earth revolves on its axis once every 24 hours. Low tides are about halfway between the high tides.

The system as described above would give two high tides and two low tides of equal magnitude each day. As we know, however, tides are not equal, nor are there always four tides per day. How do these discrepancies arise? In part, this is due to the fact that the earth is not vertical with respect to its plane of orbit around the sun. It actually is inclined 23½° from the vertical. As a result, during the earth's rotation on its axis, a given point on the earth's surface is subjected to different tidal heights, as Fig. 6.1 shows. It should also be noted from Fig. 6.1 that, at the highest latitudes (polar seas), only a single high tide would occur. Further changes in the height of the tides result from changes in the moon relative to the earth as the moon moves in its orbit around the earth. Since the orbit of the moon is not circular but elliptical, there are times when the moon is closer to the earth (*perigee*) and others when it is farther away (*apogee*). Tides are greater at perigee and diminished at apogee.

(A)

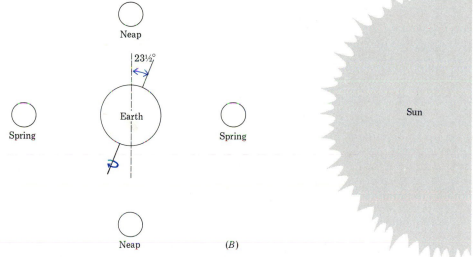

(B)

Figure 6.1 Origin of tides. *(A)* Moon acts to raise a bulge on the side of earth nearest it due to greater gravitational force than counteracting centrifugal force. On the opposite side the centrifugal force is stronger and throws another bulge outward. Because of the inclination of the earth on its rotational axis, point A, as it rotates, will experience two high tides of different height. *(B)* Position of the moon and sun at neap and spring tides.

The effect of the sun is seen in spring and neap tides. Spring tides are tides that show the greatest range (both high and low) and result when the moon and sun are directly aligned and thus these two forces are combined. Neap tides, on the other hand, are tides showing minimum range and result when sun and moon are at right angles to each other and thus counteract each other.

Differences both in number of high and low tides per day and in height in various parts of the world are due to peculiarities of the various ocean basins in which the tides occur. The extremely high tides of the Bay of Fundy and of Cook Inlet in Alaska are the result of the basic tide forces acting in the geometry of the basin.

Tides that are expressed as a single low and high per day are called *diurnal tides* (Fig. 6.2). Tides that have two highs and lows per lunar day are termed *semidiurnal tides*. Where there is a mixture of diurnal and semidiurnal, we have what are called *mixed tides*. The heights of the highs and lows vary from day to day as the positions of the sun and moon relative to each other change, each being alternately fully aligned and 90° out of alignment every 14 days.

Tides in combination with time can have two significant direct effects on the presence and organization of intertidal communities. The first effect results from the time that a given area of the intertidal is exposed to the air relative to the

Figure 6.2 Mixed, semidiurnal, and diurnal tide curves. (Modified from D. A. Ross, Introduction to oceanography, 4th ed., copyright © 1988, p. 287. Adapted by permission of Prentice-Hall, Inc., Englewood Cliffs, New Jersey.)

time that it is submerged in water. It is the *duration of exposure* to air that is most important, because that is when the marine organisms will be subjected to the greatest temperature ranges and the possibility of desiccation (= water loss). The longer the duration of exposure, the greater the chance for encountering a lethal temperature or desiccating beyond tolerable limits. Since most of these animals also must wait until covered with water to feed, the longer the duration, the less opportunity to feed and thus obtain sufficient energy. Animals and plants of the tidal zone vary in their ability to tolerate immersion in air, and this differential is part of the reason for the different pattern of distribution of organisms observed on many rocky shores (see pp. 229–232).

In certain areas of the world there are two high and two low tides per day, and where the two highs or two lows are not equal (areas with mixed tides), it can be shown that the duration of immersion is not smooth going from low to high tide levels and that there are break points where the duration changes greatly in a short distance. What this means is that an organism living just a few centimeters above one of these points would be subjected to a greatly increased exposure to air relative to one just a few centimeters below. The existence of these *critical tide levels* was early suggested as the reason for the sharp changes in fauna and flora distribution observed on rocky shores (see Fig. 6.10), since they were thought to delineate an important break point.

The second effect is the result of the time of day during which the intertidal is exposed to the air. As before, the potentially lethal ranges for physical parameters are encountered by marine organisms only when exposed to air, and in any daily cycle these vary greatly at different times of the day. If, for example, the low tides in the tropics occurred only during the hours of darkness, the intertidal fauna and flora would be exposed to much lower temperatures and less desiccation than if exposed during midday. Hence, we might expect a greater diversity of organisms in the intertidal of a tropical area where low tides regularly occurred during the night or early morning or evening than in an area where they occurred at midday. In the cold temperate zones, the reverse would be true be-cause in winter the lowest temperatures would occur in early morning or at night and might act to freeze the animals, whereas at midday the temperatures might be high enough to prevent freezing.

A final effect of the tide is that, because of its great predictability, it tends to induce certain rhythms in the activities of shore organisms. These rhythms may be with respect to time of spawning such as that seen in the grunion, a fish of the Pacific coast of the United States that spawns on the beach only on certain nights of the highest spring tides (Fig. 6.3), or the horseshoe crab (*Limulus polyphemus*) of the Atlantic and Gulf coasts, or they may be with respect to feeding or other activity. Many intertidal organisms are quiescent when the tide is out and resume normal activity, such as feeding, only when the tide is in.

Temperature

Because of its inherent physical characteristics (see Chapter 1), water—particularly in a large body such as an ocean—shows a minimum range of change in temperature. Furthermore, this range rarely exceeds the lethal limits for the organisms. Intertidal areas are, however, regularly subjected to aerial temperatures for varying periods where the range may often exceed the lethal limits of the marine organisms. Even if death does not occur immediately, the organisms may be so weakened by the extreme temperatures that they cannot resume normal activities and will suffer mortality from secondary causes.

Temperature also has an indirect effect on marine organisms. They are subject to death by desiccation, which may be hastened by high temperatures.

Wave Action

It is in the intertidal zone that wave action exerts the most influence on organisms and communities of any area in the sea. This influence is manifested directly and indirectly. Wave action affects shore life directly in two major ways. In the first place, it has a mechanical effect, which acts to smash and tear away objects with which it

Figure 6.3 Grunion spawning at high tide. (Photo courtesy of Western Marine Laboratory, Santa Barbara, California.)

makes contact. One can appreciate this action by noting the many pictures and stories of destruction wrought upon man-made structures due to storm waves of varying types. The action is similar in the intertidal zone, and hence any creatures that inhabit this zone must be adapted to this persistent force in some way (see p. 228). On shores that are made up of loose sand or gravel, the wave action is even more profound in that it moves the entire substrate around (see pp. 255–258), influencing the actual shape of the zone. Wave exposure may thus prove to be limiting for certain organisms that are unable to sustain the force, and necessary to others that cannot exist other than in heavy wave areas.

Second, wave action acts to extend the limits of the intertidal zone. It does this by throwing water higher on the shore than would normally occur as a result of the tide alone. This continual splashing allows the marine organisms to live higher in exposed wave-swept areas than in sheltered areas that are within the same tidal range (Fig. 6.4).

Wave action has some other, perhaps minor, effects also. It acts to mix atmospheric gases into the water, thus increasing the oxygen content so wave-washed areas never lack for oxygen. Because of the constant interaction with the atmos-phere, the formation of bubbles, and the stirring up of the substrate, wave-swept areas may have decreased penetration of light. This is, however, of little ecological significance.

Salinity

Salinity changes may occur in the intertidal in two situations that may affect the organisms. In the first place, the intertidal may be exposed at low tide and subsequently flooded by heavy rains or runoff from heavy rains. This means organisms will be subjected to greatly reduced salinities. In certain cases this reduction in salinity may exceed the limits and the organisms may die, since most intertidal organisms show a limited tolerance to decreased salinity. The second situation has to do with tidepools, those areas that retain seawater at low tide. These may either be flooded with freshwater runoff from heavy rains, thus reducing salinity, or they may show increased salinity when exposed to heavy evaporation during the day (see pp. 252–254).

Other Factors

Different substrates—rock, sand, mud—have very different faunas and community structure in the intertidal, but these are taken up separate-

Figure 6.4 Changes in the extent of vertical zonation with change in exposure to wave action.

ly in this chapter. Factors of pH and nutrients are also unimportant to organisms and community structure.

ADAPTATIONS OF INTERTIDAL ORGANISMS

Since the intertidal organisms are primarily marine in origin, the adaptations that we observe have to do primarily with avoiding or minimizing the stresses imposed by the daily exposure to air. The major stress from the marine environment is wave action.

Resistance to Water Loss

As soon as marine organisms are removed from the water into air, they begin to lose water. If they are to survive in the intertidal, this loss must somehow be minimized or the organisms must have body systems that will tolerate considerable losses during the hours of exposure to the air.

The simplest mechanism for avoiding water loss is seen in mobile animals such as crabs. These animals simply move from the exposed surface areas of the intertidal into very moist cracks, crevices, or burrows, where water loss is reduced or absent. Alternatively, they may seek refuge under the moist covering of algae. In other words, these animals avoid the adverse environmental condition of the shore by actively selecting suitable microhabitats (Fig. 6.5). A similar situation occurs in some species of anemones such as *Anthopleura xanthogrammica* of the Pacific coast of North America (Plate 42). It is soft bodied, with no way of preventing water loss. However, it normally is found only among barnacles or in crevices where water loss is

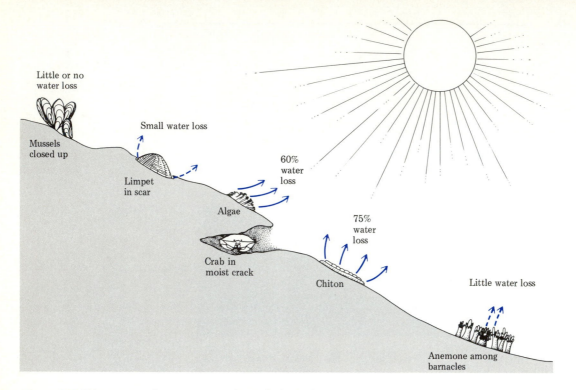

Figure 6.5 Diagrammatic representation of the adaptations to water loss in intertidal organisms.

reduced. Physiological adaptation is not needed. On the Atlantic coast the anemone *Anthopleura varioarmatus* covers itself with shell fragments to shield itself from drying when exposed to air.

Another simple mechanism is that seen in several of the high intertidal algae genera such as *Porphyra, Fucus,* and *Enteromorpha.* These plants cannot move and have no mechanism to avoid water loss. They simply are adapted to withstand a severe loss of water from their tissues. Specimens of these genera can be found that are dry and brittle after a long exposure at low tide, yet will quickly take up water and resume normal body processes when the tide returns. Kanwisher (1957) found that they can tolerate as much as a 60 to 90 percent loss of water and still recover. A similar tolerance to great water loss is seen in some intertidal sedentary animals. Boyle (1969), for example, reports chitons can tolerate a 75 percent water loss and limpets, according to Davis (1969), from 30 to 70 percent, depending on the species (Fig. 6.5).

Many species of intertidal animals, in contrast to the above, have mechanisms for preventing water loss. These mechanisms may be structural, behavioral, or both. Barnacles of many species are predominant members of the intertidal zone all over the world. These animals are sessile and avoid water loss by closing up within their shells at low tide. Here, the presence of an impermeable shell cuts down the evaporative water loss. Limpets of the genera *Patella, Acmaea, Lottia,* and *Collisella* are also dominant animals of the rocky intertidal. Certain species of these limpets have a "home scar" into which their shells exactly fit. At low tide, they return to these "homes" and by fitting into these grooves, greatly reduce water loss. Other limpets that have no scar simply clamp down tightly against the rock so no tissue is exposed, only shell. Other gastropods such as the periwinkles (*Littorina*) have opercula, which completely seal off the aperture to their shells. At low tide, they may pull into the shell, closing the aperture with the operculum, and thus reduce water loss. Certain bivalves, such as *Mytilus edulis* and oysters such as *Crassostrea virginica,* are able to survive in the intertidal because of their ability to close their valves tightly to prevent water loss. Still other organisms, such as the anemone *Actinia* and the hydroid *Clava squamata,* produce mucus that reduces water loss. Many inhabitants of sand or mud simply burrow into the substrate to prevent desiccation.

Maintenance of Heat Balance

Intertidal organisms that are exposed to extremes of heat and cold show behavioral and structural adaptations to maintain their internal heat balance. Though deaths by freezing have been recorded for intertidal organisms, extreme low temperatures seem to be less of a problem to shore organisms than high temperatures. This may be due to the fact that organisms are often living much closer to their upper lethal temperature than to their lower lethal one. Hence, most mechanisms for heat balance concern the avoidance of too high temperatures. This may be accomplished by (1) reducing heat gain from the environment and (2) increasing the heat loss from the body of the animal. Heat gain from the environment is reduced in several ways. One way is to have a relatively large body size when compared to similar species either lower in the intertidal or in the subtidal. Large body size means less surface area relative to volume and hence less area to gain heat. At the same time, a larger body takes longer to heat up than a smaller one. It has been recorded for gastropod mollusks such as *Littorina littorea* and *Olivella biplicata* that the larger individuals occur higher in the tidal zone and the smaller ones lower. Yet another mechanism to reduce heat gain is to reduce the area of body tissue in contact with the substrate. This is, however, difficult to achieve for most intertidal animals, since they need the extra purchase provided by a large area of tissue in contact with the substrate to avoid being swept off by waves. Yet there are some shore animals that do have such a small area of attachment or none at all. These occur where wave action is not severe, and, in those animals that lose attachment, the ability to reattach as the tide comes in is well developed.

Heat loss can be attributed to a number of causes. One mechanism, found in hard-shelled organisms such as mollusks, is greater elaboration of ridges and other sculpturing on the shell.

These act as radiator fins and facilitate heat loss. Examples of such sculptured mollusks are best found in the tropics and include *Tectarius muricata* and *Nodolittorina tuberculata.* Enhancement of heat loss occurs when the organism is light in color. (Dark organisms gain heat by absorption.) Many tropical and subtropical snails of the high intertidal such as *Nerita peleronta* (Caribbean) and *Littorina unifasciata* (New Zealand) are much lighter in color than their lower level relatives, presumably, then, facilitating heat loss (Fig. 6.6).

Finally, and most important, heat loss can occur through evaporation of water. Unfortunately, again, intertidal organisms must face the problem of desiccation and can ill afford to cool themselves through evaporation or else chance a lethal desiccation. Many intertidal organisms show a strategy of adaptation that allows them some cooling by evaporating water and, at the same time, avoiding excessive desiccation (Segal and Dehnel, 1962). To facilitate this balance, many intertidal animals have an extra water supply from which cooling may take place. This extra water is held in the mantle cavity of barnacles and limpets, and it is above the amount the animal needs to survive desiccation. Moreover, most high intertidal animals can withstand a remarkable amount of desiccation (see previous section). In addition, many ani-

Figure 6.6 Differences in heat absorption between smooth, dark shells and sculptured, light shells.

mals have the ability to reduce the rate of water loss from the tissues when necessary. Those animals living in intertidal sedimentary environments escape most of the problems of water loss simply because they reside within the substrate that contains water even at low tide.

Mechanical Stress

As noted previously, wave action reaches a zenith in the intertidal zone. As a result, it is necessary for any organism that lives intertidally to adapt in order to resist the smashing and tearing effects of waves. Wave action has different effects on rocky shores as opposed to sandy shores and requires different adaptations, which are discussed later. We discuss here some adaptations for dwellers of rocky substrates.

In order to maintain position in the face of wave action, several adaptations have been developed by intertidal organisms. One such adaptation, observed in barnacles, oysters, and serpulid polychaete worms, is simply to become fixed to the substrate. Similarly, the majority of algae in the intertidal are fused via a holdfast to the bottom.

Still other organisms develop attachments that are strong but not permanent, thus giving limited movement. Such attachments are the byssal threads of *Mytilus,* which anchor the animals securely but still can be broken and remade to allow limited slow movement.

Most intertidal mollusks are adapted to wave shock by possessing thicker shells than their subtidal relatives and lack delicate sculpturing, which is easily broken in the waves. The dominant intertidal mollusks, such as the various limpets and chitons, resist wave action through a strong, enlarged foot that clamps them to the substrate.

Motile organisms such as crabs have no structural mechanism to resist being swept away and survive only because they seek shelter from the waves in crevices or under rocks. Many, however, such as grapsids, are adapted for living in wave-swept areas by being flattened, thereby providing less resistance to wave action.

Respiration

Since most animals inhabiting the intertidal are marine, they have respiratory surfaces, *gills,*

adapted to extract O_2 from water. Usually, these are thin-walled extensions from the body surface. These respiratory organs are thus very susceptible to desiccation in air and do not function unless immersed in water. Such organs would be a distinct disadvantage during aerial exposure in the intertidal. Among intertidal animals, there is a tendency to enclose the respiratory surfaces in a protective cavity to prevent them from drying. This is seen particularly in various mollusks where the gills are in a mantle cavity, which itself is protected by the shell. Another adaptation, especially of high intertidal gastropods, is the reduction of the gill and formation of a vascularized mantle cavity, which then serves as a lung for aerial respiration. A similar situation prevails in barnacles where the mantle tissue acts as a respiratory organ.

Because animals with protected respiratory organs must also conserve water at low tide, they often close up (operculum) or clamp down (chitons, limpets) such that gaseous exchange is reduced. Thus, to conserve O_2 and water, most animals are quiescent during low tide.

Intertidal fishes often are specialized for cutaneous respiration through the reduction of gills and the proliferation of blood vessels in the skin. Bridges (1988) suggests that many fish may satisfy over half their oxygen needs by gas exchange through the skin.

Feeding

All intertidal animals must expose the fleshy parts of their bodies to feed. For those on hard substrates this means exposing those parts most susceptible to desiccation. As a result, virtually all diurnal rocky intertidal animals are active only during the time the tide is in and they are covered with water. This is true whether the animals are grazers, filter feeders, detritus feeders, or predators. By contrast, organisms living in soft substrates are often protected by the water-filled substrate and may be active feeders at low tide. Nocturnal animals may also be active at low tides occurring at night.

Salinity Stress

As noted above, the intertidal zone may be subjected to flooding by fresh water, which would create an osmotic stress on the intertidal organisms adapted to be submerged only in seawater. This is particularly true since most intertidal organisms do not show adaptations to tolerate salinity changes, certainly not like estuarine organisms do (see Chapter 8). Most have no mechanism to control the salt content of their body fluids and hence are *osmoconformers*. The only adaptations observed are the same as those for preventing desiccation, such as closing up valves or shells in barnacles and mollusks. Perhaps this is why there are records of catastrophic mortality of intertidal organisms following heavy rain or runoff. Apparently, however, such events are so rare or unpredictable that special mechanisms have not evolved.

Reproduction

Because so many of the intertidal organisms are sedentary or sessile, they must rely upon reproductive products or larvae that are free-floating in the plankton for dispersal. Hence, we see a very high incidence of planktonic larvae among intertidal organisms. These larvae are of many types (see Fig. 2.10 on meroplankton).

A second reproductive adaptation attributable to the intertidal position is that most organisms have a breeding cycle that is closely synchronized with the occurrence of certain tides, such as spring tides, in order to ensure fertilization. An example is *Mytilus edulis,* in which gonads mature during periods of spring tides and spawning occurs on subsequent neap tides. In *Littorina neritoides,* the eggs are spawned at the time of spring tides.

ROCKY SHORES

Of all the intertidal shores, those composed of hard material, the rocky shores—particularly those of the temperate zones—are the most densely inhabited by macroorganisms and have the greatest diversity of animal and plant species. They stand in striking contrast to the almost barren appearance of the surface of sand and mud shores. It is these densely populated, topographically diverse, and species-rich rocky areas that have fascinated marine biologists and ecologists for many years. In recent years these areas have been the subject of several classic studies that have enhanced our understanding of

how these associations of species interact with each other to maintain or change the community. It should be noted at the outset that these intertidal rocky areas differ in a fundamental way from the subtidal rocky areas discussed in Chapter 5. Most of the dominant organisms of intertidal rocks are solitary or clonal animals, whereas subtidal rocks seem dominated by colonial encrusting animals.

Zonation

One of the most striking features of any rocky shore anywhere in the world, when seen at low tide, is the prominent horizontal banding or *zonation* of the organisms. Each zone or band is set off from those adjacent by differences in color, morphology of the major organism, or some combination of color and morphology. These horizontal bands or zones succeed each other vertically as one progresses up from the level of the lowest low tides to true terrestrial conditions (Figs. 6.7, 6.8). This zonation observed on intertidal rocky shores is similar to the zonation pattern that one observes with increasing elevation on a mountain, where the different horizontal zones of trees and shrubs succeed each other vertically until, if one progresses far enough, permanent snow cover is reached. The major difference between these two areas is the scale, mountain zones being perhaps kilometers in height as opposed to intertidal zones extending a few meters vertically.

Rocky intertidal zones vary in vertical extent, depending on the slope of the rocky surface, the tidal range, and the exposure to wave action. Where there is a gradual slope to the rock, individual zones may be broad, whereas under similar tidal and exposure conditions on a vertical face, the same zone would be narrow. In the same manner, exposed areas tend to have broader zones than protected shores, and shores with greater tidal ranges have broader vertical zones (Figs. 6.4, 6.8).

Of course, these striking bands may be interrupted or altered in various places wherever the rock substrate itself shows changes in slope, composition, or irregularities that change its exposure or position relative to the prevailing water movement.

The fact that these prominent zones can be observed on nearly all rocky shores throughout

Figure 6.7 An intertidal area at low tide illustrating the conspicuous banding or zonation of organisms. (Photo courtesy Dr. Michael Foster.)

the world under many different tidal regimes led Stephenson and Stephenson (1949) to propose, after some 30 years of study, a "universal" scheme of zonation for rocky shores (Fig. 6.9). This "universal" scheme was really a framework using common terms that would allow comparison of diverse areas. It is characterized by establishing the zones on the basis of the distributional limits of certain common groups of organisms and not on the basis of tides. It thus reflects the knowledge of the Stephensons, and other intertidal ecologists, that distribution patterns of the organisms and zones vary not only with tides, but with slopes and exposure. Therefore, under similar tidal conditions there could be different bandwidths due to different exposures or slopes of rocks. It was this universal scheme that established a standard format for describing shore zonation, replacing a bewildering host of schemes and names that had been established by earlier biologists.

The Stephensons' scheme has three main divisions of the intertidal area. The uppermost is termed the *supralittoral fringe.* Its lower limit is the upper limit of barnacles, and it extends to the upper limit of snails of the genus *Littorina* (periwinkles). The dominant organisms are the littorine snails and black encrusting lichens (*Verrucaria* type). Part of this zone is reached by the extreme high water of spring tides, but most of its water comes from wave splash. Above this zone is the terrestrial or supralittoral zone.

The middle part of the intertidal is termed the *midlittoral* zone and is the broadest in extent. Its upper limit is coincident with the upper limit of barnacles, while its lower limit is that point where large kelps (*Laminaria,* etc.) reach their uppermost distribution. This zone is often subdivided and contains a host of different organisms. Perhaps the only universally present dominant group are the barnacles.

The lowermost zone of this scheme is the *infralittoral fringe,* which extends from the lowest low tide up to the upper limit of the large kelps. This is an extremely rich zone composed of organisms that can tolerate only limited exposure to air. It is really an intertidal extension of the *infralittoral zone* (Stephensons' term) or what we know as the sublittoral area.

(A)

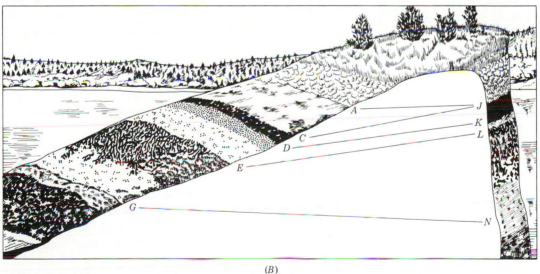

(B)

Figure 6.8 Zonation in the rocky intertidal. *(A)* An intertidal area in Scotland. *(B)* An intertidal area in British Columbia. Note in *(B)* the difference in the extent of zones. In *(B)*, line *A–J* indicates the lower limit of lichens; line *C–J*, the upper limit of the "black" zone; line *D–K*, the upper limit of barnacles; line *E–L*, the upper limit of *Fucus*; line *G–N*, the lower limit of barnacles. (*A*, from J. R. Lewis, The ecology of rocky shores, Hodder & Stoughton Educational, 1964. *B*, from T. A. Stephenson and A. Stephenson, Life between the tidemarks on rocky shores, Freeman, 1972.)

Upper-limit barnacles

Upper-limit laminarians

Equinoctial high-water spring tides

Upper-limit littorina, snails, etc.

Equinoctial low-water spring tides

A = Supralittoral (spray) zone
A′ = Supralittoral fringe
B = Littoral zone
B′ = Midlittoral
C = Infralittoral zone
C′ = Infralittoral fringe

Figure 6.9 Stephenson "universal" scheme of zonation for rocky shores. (From R. L. Smith, Ecology and field biology, 3rd ed., Figure 8-4. Copyright © 1980 by Robert Leo Smith. Reprinted by permission of Harper & Row, Publishers, Inc.)

Although the above scheme did set forth a means for describing zonation on rocky shores, it does not offer an explanation as to why the zonation occurs. It is this explanation of zonal patterns that intrigues many marine biologists.

Causes of Zonation

Whereas it is fairly easy to recognize and measure the extent of the zones on a rocky shore, it is more difficult to come up with suitable explanations as to why organisms are distributed into these zones. Two sets of factors can be considered to explain the phenomenon: physical and biological. We shall take each up in turn.

Physical Factors   The most obvious explanation for the occurrence of the zones is that they are a result of the tidal action on the shore and therefore reflect the different tolerances of the organisms to increasing exposure to the air and the resultant desiccation and temperature extremes. One difficulty with this explanation is that the rise and fall of a tide tend to follow a smooth curve with no obvious sharp breaks, which would correspond to the often sharp boundaries observed in the intertidal zones. If, however, one observes a whole series of such

tidal curves, such a series of breaks does become apparent. For example, Fig. 6.10*A* gives a typical tidal curve for a mixed tide area on the California coast and Fig. 6.10*B* gives the maximum time of continuous submergence for various tide levels. As can be seen from the graph, there are certain points on this curve that reflect sharp increases in exposure to air. This can also be deduced from Fig. 6.10*A,* which represents a single typical lunar tide cycle. Consider an organism at point X on the graph. At this point, the tide will cover the organism at least once every six hours. If, however, the organism moves up only a few inches, the tide will not cover it for 12 hours. Thus, there is a great change in exposure time with a very short vertical movement. Points exhibiting sharp increases in exposure time over short distances have been termed *critical tide levels* by Doty (1946) and were offered as one of the early explanations for the zonation patterns described above. In this explanation, it is important to remember that it is not the tide per se that causes the limit but rather the fact that, at these critical points, the organisms are subjected to greatly increased time in the air and hence to greater temperature fluctuations and desiccation.

The critical-tide hypothesis has been tested in various places by several scientists since its original promulgation by Doty in 1946. In general, it is difficult to find good correlations, particularly at the lower tide levels. This lack of good correlation can be partly attributed to the diverse topography of the different shores and to the variation in exposure. Thus, a species may well be able to exist above the critical tide level if the rocky shore is exposed to persistent violent wave action, which would throw water up higher and thus decrease desiccation. Similarly, caves, overhangs, and crevices remain moist when exposed areas are dried out, thus allowing persistence of organisms above the level predicted by the critical tides. The effects of these topography and exposure differences suggest, however, that the ultimate upper limits are set by certain physical factors—namely, desiccation and temperature.

We have previously noted that desiccation is a serious problem. Several natural observations as well as field experiments have suggested that desiccation can set the upper limits to organisms

Figure 6.10 *(A)* Typical tidal curve for a mixed tide area on the Pacific coast of North America. *(B)* Maximum time of continuous submergence for various tide levels on the Pacific coast of North America. (*B*, from J. Nybakken, Readings in marine ecology, Harper & Row, 1971, by permission of Ecological Society of America.)

and zones. In the North Temperate Zone, for example, rocks that have a north-facing slope often have the same organism occurring higher than do adjacent south-facing slopes. Since such rocks experience no difference in wave exposure or tides, the only reason for the height difference is that north faces dry out more slowly than south faces. Similarly, field experiments by Frank (1965) and others in which water was slowly dripped down slopes have resulted in the organisms extending higher up (Fig. 6.11). In other experiments, organisms were transplanted above their normal positions. Notable in this respect are barnacles, which of course cannot move. When transplanted by Foster (1971) above their normal tidal height, they died, with younger ones dying more quickly than older ones. Again, desiccation can be suspected.

Along with desiccation, and often acting in concert with it, is temperature. As noted previously, aerial temperatures have ranges that may exceed the lethal limits, and hence intertidal organisms may die from either freezing or "cooking." Upper limits to zones may therefore be partly attributable to the tolerable temperature limits of the intertidal organisms. In addition, high temperature promotes desiccation, and the synergistic effect of these factors may be even more devastating than each acting alone. The lack of abundant organisms on rocky inter-

tidal shores of the tropics may well be at least partly the result of desiccation combined with heat and ultraviolet light stress.

Finally, sunlight itself may act adversely to limit organisms on the shore. Sunlight includes wavelengths in the ultraviolet region that are

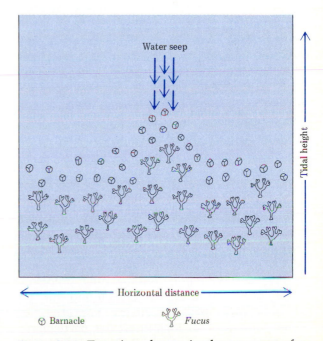

Figure 6.11 Zonation change in the presence of a water seep.

deleterious to living tissue. Water absorbs these wavelengths and so serves to protect most marine animals. However, intertidal animals have direct exposure to such rays at low tide. The higher an organism is in the intertidal, the greater the exposure to these rays. At the present time, there is little information concerning whether such UV radiation is active in controlling distribution of organisms, but the recent finding of UV-absorbing compounds in shallow-water organisms suggests its importance. Light also has been suggested as a regulator of the distribution of intertidal algae. This has to do primarily with the spectral quality of the light. As noted in Chapter 1, different wavelengths of light are absorbed differentially by water. In the case of intertidal algae, those that need light of longer wavelength (reds) that are absorbed most quickly by water would tend to be found higher in the intertidal, so when submerged, they would not be too deep for the penetration of red light (about 2 m). Since the main intertidal algae belong to three different groups—reds, browns, and greens—each of which has a slightly different absorption spectrum, it might be thought that they would be arranged along a depth gradient. In such a gradient, the green algae would be expected to be the highest, since they absorb mainly red light; the brown next; and the red algae, which absorb mainly the deep, penetrating green light, deepest. This, however, has not been found to be the case, and the intertidal algae are a mixture of all types at most levels. This is likely due to the interaction of other factors and the physiology of the algae, and it points out again the fallacy of attempting to explain patterns of species distribution with only single factors when a multitude are acting at all times.

Although there exists a considerable body of evidence that physical factors seem to be the strongest determinants of the upper limits of organisms, one should not lose sight of the fact that there may be alternative explanations, many of which have not been investigated. Underwood and Denley (1984) point out that other processes are at work in the high intertidal and may be confused with physical effects. For example, some animals in the high intertidal, particularly sessile filter feeders, may be killed by starvation during long periods of calm weather because they are not covered with water for long enough to enable them to feed. Similarly, grazing animals may be absent, not because they cannot tolerate the physical factors, but because their algal food is absent, perhaps killed by severe physical factors. Finally, in the case of larvae settling out of the plankton, they may choose not to settle in the area, in which case the upper limit is set by processes affecting larval settlement. In one instance reported by Underwood (1980), the upper limit of several algal species in Australia was shown to be directly due to the grazing by gastropods. In each of these cases, the upper limits of the organisms are the result of biological and physical factors acting either independently or together.

We can summarize the above by stating that, although much current evidence suggests that physical factors are strong determinants of the upper limits of the distribution of intertidal organisms, other explanations are viable and still others remain to be investigated.

Biological Factors   Although the early work in the intertidal focused on the importance of physical factors in setting the zonal patterns, more recent work has begun to clearly establish the great importance of a number of biological factors in setting various observed distribution patterns. In general, these biological factors are more complex, often subtle, and closely linked to other factors, which is probably why we have only recently begun to understand how they act. The major biological factors are *competition, predation, grazing* (= herbivory), and *larval settlement.* We shall take them up in order.

Competition for a certain resource does not occur if the resource is so plentiful that adequate supplies of it are available for all species or individuals. In the rocky intertidal zone, only one resource is commonly in limited supply and that is space. This is perhaps the most restricted area in the marine environment in terms of physical dimensions; at the same time, it is densely populated, at least in the temperate zone. As a result, there is an intense competition for space. This has resulted in observed zonal patterns. On the intertidal shores of Scotland, there is a distinct zonation of barnacles with the small *Chthamalus stellatus* living in the highest zone and the larger *Balanus balanoides* occupying the major portion of the midintertidal. Studies done by Connell (1961) showed that *Chthamalus* larvae settled out throughout the zones occupied by both barnacles, but survived to adulthood only in upper zones. The reason for the disappearance in the midlittoral region was due to competition from *Balanus balanoides,* which either overgrew, uplifted, or crushed the young *Chthamalus. Balanus* was prevented from completely eliminating *Chthamalus* because *Balanus* did not have the tolerance to drying and high temperatures that prevailed at the higher tidal levels and hence did not survive there, whereas *Chthamalus* could. Here, then, is a case where the zonation is at least partially a function of biological competition (Fig. 6.12).

Intertidal algae, particularly those on temperate shores, also often show abrupt limits to their upper and lower distribution. Earlier, these limits were ascribed to critical-tide levels, but it has also been suggested that this could be due to competition for space or access to light. In two experimental studies done on the Pacific coast of North America, evidence was found for competition. Dayton (1975) found that the dominant kelps—*Hedophyllum sessile, Laminaria setchelli,* and *Lessionopsis littoralis*—all outgrew and outcompeted certain smaller species in the lower intertidal. These smaller species were generally fast-growing species, which quickly colonized open areas. As noted in Chapter 1, they are *opportunistic* or *fugitive* species. In turn, among the three dominants, *Hedophyllum* was outcom-

Figure 6.12 Intertidal zonation as a result of the interaction of physical and biological factors. The larvae of two barnacles, *Chthamalus stellatus* and *Balanus balanoides,* settle out over a broad area. Physical factors, mainly desiccation, then act to limit survival of *B. balanoides* above mean high water of neap tides. Competition between *B. balanoides* and *C. stellatus* in the zone between mean tide and mean high water of neap tides then eliminates *C. stellatus.*

peted by the other two such that they came to dominate the areas. The second study was actually done in the subtidal, where Vadas (1968) found that the giant kelp *Nereocystis* outcompeted and overgrew the brown alga *Agarum*. On the New England coast, studies by Lubchenco (1978) directed at tidepool algae indicated that *Enteromorpha intestinalis* was a dominant space competitor as opposed to *Chondrus crispus;* in the absence of grazers, *E. intestinalis* would quickly outcompete *Chondrus* and take over the space. These few studies suggest that competition among algal species may be more widespread than originally assumed and may be a fertile ground for future ecological work.

A more complex example concerning competition is that among the mussel *Mytilus californianus* and several species of barnacles on the Pacific coast of Washington (Plate 40). In this case, studies by Dayton (1971) and Paine (1966, 1974) have shown that *M. californianus* is the dominant space competitor on open coast shores. Given enough time and freedom from predators, the *M. californianus* eventually overgrow and outcompete all other macroorganisms and take over the substrates throughout most of the midintertidal. *Mytilus californianus* takeover is, however, slow. Wherever open space occurs, it may be rapidly colonized by other organisms, including three species of barnacles, *Balanus glandula, B. cariosus,* and *Pollicipes polymerus.* These, in turn, displace any rapidly growing algal species. The barnacles persist only until the mussels enter. The mussels outcompete and destroy the barnacles by settling on top of them and smothering them. Since nothing appears to be large enough to settle and smother the *M. californianus,* they remain in control of the intertidal space. Given this competitive edge, it would appear that eventually the rocky Pacific coast of Washington could be a monotonous band of *M. californianus.* It is also curious that *M. californianus* forms dense clumps or bands only in the intertidal, though it is perfectly capable of living subtidally. Since it is a premier space competitor, why this abrupt lower limit? The reasons for this have to do with other biological factors that act to prevent such resource monopolization and may be grouped under the term *predation.*

The role of predators in determining the distribution of organisms in the intertidal, and hence, the zonal patterns, is best documented for the Pacific coast of Washington and is discussed here as an example of how complex biological interactions serve to give the prevailing observed distributions (Plates 41 and 47).

The dominant abundant intertidal animals on the Pacific coast other than the space-dominating *Mytilus californianus* are the barnacles *Balanus cariosus* and *Balanus glandula.* These latter two species occur abundantly in the intertidal region, despite the fact that they are competitively inferior to *M. californianus,* because a predatory starfish, *Pisaster ochraceus,* preferentially preys upon *M. californianus,* thus preventing it from completely overgrowing the barnacles (see Plate 40). *Pisaster ochraceus* is a voracious predator of mussels, consuming them at a rate that effectively prevents them from occupying all the space.

At the same time, *Balanus glandula* is found primarily as a band of adults in the high intertidal, while *Balanus cariosus* occurs as scattered, large individuals or clumps in the midintertidal. This pattern, as Connell (1970) has shown, is also due to predation. *Balanus glandula,* like *Chthamalus* in England, is capable of living throughout the intertidal zone and indeed settles throughout. The same is true for *B. cariosus.* That both show a restricted distribution is due to predation by three species of predatory gastropods of the genus *Nucella: N. lamellosa, N. emarginata,* and *N. canaliculata.* The abundance and motility of these predators is such that they are capable of completely consuming all the young *B. glandula* settling out in the midintertidal in 12 to 15 months. *Balanus glandula* survives only in a narrow band at the top of the intertidal, where the *Nucella* species are prevented from entering because of excessive desiccation. In the case of *B. cariosus,* however, the situation is somewhat different. There is no high level refuge for this barnacle, and as with *B. glandula,* the young are consumed by the *Nucella.* The refuge for *B. cariosus* is that, once it reaches two years of age, it is too large for *Nucella* to attack (Fig. 6.13). The only problem, as yet unresolved, though it may have to do with periodic events like big freezes at low tide, is

Time ⟶

| 0 | 12-15 Months | 2 Years |

High intertidal zone

Limit of effective *Thais* + *Piaster* predation

Midintertidal zone

Low–tide level

Refuge zone

Zone of complete consumption

1. Barnacles settle out throughout intertidal. *Thais* and *Pisaster* present.

2. *Nucella* consume most young barnacles in midintertidal. *Mytilus* begins to outcompete barnacles

3. Few *B. cariosus* become too large for *Nucella* and persist in midintertidal. *B. glandula* persists only in high intertidal. *Pisaster* prevents *Mytilus* monopolization. Typical situation.

- *B. glandula*
- Juvenile
- *B. cariosus*
- Juvenile
- *Mytilus*
- *Pisaster*
- *Nucella* spp.

Figure 6.13 Interaction of predation and physical factors in establishing the zonation of the dominant intertidal organisms on the rocky shores of the Northwest Pacific coast of North America.

how *B. cariosus* manages to survive the *Nucella* for two years. As a result, the pattern of distribution for *B. cariosus* is random clumps of large, older barnacles.

It is not known for certain what regulates *Nucella* populations, but they are preyed upon by large *Pisaster* and are also vulnerable to periodic events such as freezing during winter low tides. Such periodic events may reduce the population by such a significant amount that the *Balanus cariosus* could successfully establish themselves in an area. This may explain the differences among various shores with respect to the abundance of this barnacle.

Although it prefers mussels, *Pisaster ochraceus* can consume *Nucella* and barnacles of any size and is the primary predator of small- and medium-sized *Mytilus*. It is the top predator in

the system. Because of its ability to influence the structure of the entire community in the intertidal—through its ability to consume *Mytilus* and therefore prevent monopolization of the space—it has been called by Paine (1966) a *keystone species* as defined in Chapter 1.

The reason that refuges from predators occur in the upper intertidal is that both *Pisaster* and *Nucella* can feed only when the tide is in, and they require a long period to attack their prey successfully. The short period of immersion of the upper intertidal does not allow sufficient time for them to make successful forays into that area. In subtidal areas, unlimited time is available; hence, the starfish have sufficient time to attack and consume any prey item. It is probably for that reason that *Mytilus*, for example, do not extend into the subtidal areas, even though they can certainly live there—hence the reason for the sharp lower boundaries to the zonation of this species (Fig. 6.13).

The situation regarding the importance of the interaction of mussels and starfish in structuring intertidal rocky communities of the Pacific coast of Washington has a confounding factor that suggests just how difficult it is to understand how communities are structured and to extrapolate results from one area to another. Paine et al. (1985) have followed up on the earlier studies in Washington with similar studies of starfish removal from intertidal shores in Chile and New Zealand as a means to assess the universality of the results obtained earlier in Washington. In Chile the dominant space occupier was the mussel *Perumytilus purpuratus,* and its predator the sun star *Heliaster helianthus.* In New Zealand the mussel was *Perna canaliculus* and the predatory sea star *Stichaster australis.* When predator removal studies were done in these areas and compared to Washington, a notable difference occurred. In Washington and New Zealand, when the starfish were permitted to return to the areas from which they were formerly excluded, the mussel community persisted in monopolizing the space they had occupied when the starfish were excluded for as long as 14 to 17 years. In other words, there was no quick return to the conditions seen in control areas where predators were always present. By contrast, in Chile, when starfish were returned to the excluded areas, the community very rapidly returned to the condition of the undisturbed control. What are the reasons for this difference? Paine et al. (1985) suggest that the main reason is that in Washington and New Zealand the mussels were able to attain a size during the absence of their main predator such that, when the predators were reintroduced, the mussels had grown to such a size that they were immune to predation. In the case of Chile, *Perumytilus* never reaches a size at which it is immune to predation and hence the quick return to a "normal" pattern. These results, in turn, suggest that in Washington and New Zealand the community will not return to the pre-exclusion state until the current large mussels either die or are removed by some other destructive event, such as wave action or crush-

ing by wave-borne objects (see p. 248). This may take years. Hence, whereas we may expect all three of these communities to eventually converge to their "normal" condition, the time involved may be markedly different, thus leading, in the case of Washington and New Zealand, to an intertidal with a mosaic of patches—some completely dominated by mussels and others with scattered clumps, depending on the conditions and length of time the dominant predator was excluded.

Menge (1976) and Lubchenco and Menge (1978) have also demonstrated that predation is important in setting zonal patterns in the intertidal of the North Atlantic Ocean. In the low intertidal of New England, the competitive dominant species is *Mytilus edulis. Mytilus edulis* is able to outcompete and eliminate the barnacle *Balanus* (*Semibalanus*) *balanoides* and the alga *Chondrus crispus.* It is prevented from doing this by the starfish *Asterias forbesi* and *A. vulgaris* and the snail *Nucella lapillus,* all of which prey upon *M. edulis.* Where these predators are absent—namely, in the most exposed, wave-beaten areas—*M. edulis* eliminates *Balanus* and *Chondrus.* Thus, the exposed wave-beaten areas exhibit a zonal pattern in which the low intertidal is an *M. edulis* band, while in protected areas a diverse lower zone exists with *Mytilus, Balanus,* and *Chondrus* present (Fig. 6.14).

An unexplained anomaly of the above pattern is found in the very sheltered bays of New England. In these areas Petraitis (1987) has demonstrated that the barnacle *Balanus balanoides* and the mussel *Mytilus edulis* are the most common organisms in the lower intertidal, where perennial algae are rare. This is a condition similar to the very exposed outer shores, where predators are absent. However, here predators are abundant. The herbivorous gastropod *Littorina littorea* is thought to be responsible for the lack of algae, but the presence of the barnacles and mussels cannot be accounted for by a lack of asteroid predators nor by a refuge in size, since it appears that the starfish are capable of consuming all size classes.

A central theme of marine intertidal ecology, as argued by Connell (1975), is that wherever predation is reduced, competition will be in-

Without predators (exposed)

With predators (protected)

Mytilus

Balanus balanoides

Chondrus crispus

Asterias

Nucella

Figure 6.14 Composition of the low intertidal in the New England area with and without mussel predators.

creased. Indeed, the aforementioned examples demonstrate this. It is well, however, to remember that this is not the inevitable result of predator removal. For example, Keough and Butler (1979) removed predatory starfish from pier pilings in Australia and found no changes in the invertebrate populations or any increase in competition. This points up an important principle: that the relative importance of biological factors and physical factors on rocky shores varies over relatively short distances. The variation is both vertical and horizontal. Therefore, one cannot extrapolate from single experimental studies, done on one small area, to form models for the organization of large geographic areas of the intertidal.

The role of *grazers* or herbivores in regulating upper and lower limits of algal species is well studied and documented, and evidence suggests this process may also be of considerable importance. Whereas a number of animals graze upon intertidal algae, relatively few grazers seem to be abundant enough to be significant in altering or

determining community structure. The dominant groups are various gastropod mollusks, certain crustaceans, sea urchins, and fishes. The relative importance of these groups varies latitudinally and vertically in the intertidal zone (Fig. 6.15). Their relative abilities to keep algae in check may also vary.

Although data on the importance of grazing were late in being reported (not until well after World War II), in the past 20 years considerable evidence has been accumulated by various methods—experimental removals of grazers, caging experiments, natural disasters—such that our understanding of the role of grazing is now fairly well advanced. Grazing affects a number of parameters, including algal zonation, species diversity, patchiness, and succession.

Prior to the implementation of various experimental methods, it was assumed that in the intertidal the upper limit of the dominant algal species on rocky shores were a function of the species' ability to tolerate various physical factors. On the other hand, lower limits of algal

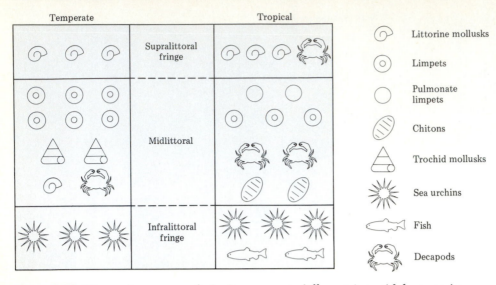

Figure 6.15 The main groups of algal grazers at different intertidal zones in temperate and tropical systems. Relative numbers of symbols indicate relative importance.

species distribution seem not to be set by physical factors, because most algae seem to grow faster when experimentally transplanted to lower levels or when they are grown constantly submerged.

Since most algae do grow better when set lower in the intertidal or submerged, why don't we find them there naturally? The answer that seems to be coming from many workers is that they are grazed out. For example, in various experiments that have removed grazing animals the result has been the appearance lower on the shore of a number of algae otherwise restricted to higher levels where grazing is less intense. This has been shown by Lubchenco (1978, 1980) to be true for *Chondrus* in New England, where the main grazer is the urchin *Strongylocentrotus droebachiensis*. Himmelman and Steele (1971) found the same thing for the alga *Alaria esculenta* in Great Britain, where *S. droebachiensis* is also responsible (Fig. 6.16).

Grazing may in some cases also set the upper limits of algal distribution. One of the more interesting pieces of evidence supporting this comes from the Torrey Canyon oil spill disaster. That 1968 spill killed off the main intertidal molluscan grazers on the shores of southern England, and Southward and Southward (1978) documented a subsequent rise in the upper limits of a number of intertidal algae. Then, following the return of the limpets and other grazing gastropods, the algae were once again grazed down.

The mechanisms by which the removal of grazers facilitates the upward or downward extension of algae vary. In the simplest case, the algae are directly limited by the grazer. In other cases, however, the absence of the grazers may permit a rapid establishment of ephemeral algae, in turn trapping moisture and thereby permitting the survival of the newly settled sporelings of the dominant algae. In still another scenario, the upward extension of algae may be seasonal due to changes in grazing pressure by certain grazers. For example, Castenholz (1961) showed that a summer decrease in high intertidal diatom cover was due to a summer increase in *Littorina* grazing activity in contrast to the less active winter condition. A more complicated situation is that in which both grazing and physical factors interact. Thus, Hay (1979) found that in New Zealand the alga *Durvillea* extended its range upshore in the winter, when grazers were absent, but was cut back in the summer by a combination of increased grazing and desiccation at the higher levels.

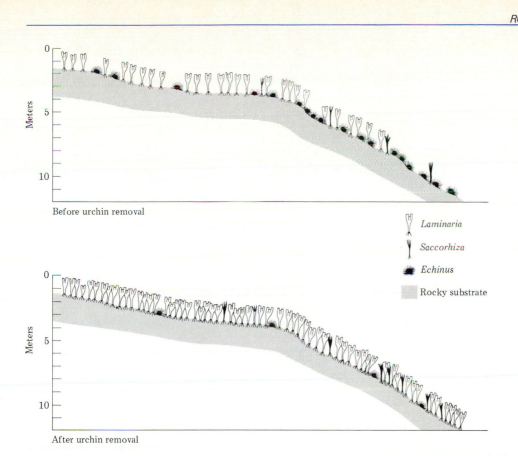

Before urchin removal

⚘ *Laminaria*

⚘ *Saccorhiza*

🦔 *Echinus*

▨ Rocky substrate

After urchin removal

Figure 6.16 Effect of sea urchin removal on kelp growth on the Isle of Man, Great Britain. (Modified from T. Carefoot, Pacific seashores, University of Washington Press, 1977.)

Surprisingly enough, some of the limits of algae are set by the effect of the algae themselves on potential grazers. In New Zealand, for instance, Underwood and Jernakoff (1981) showed that the downshore extensions of the main grazing limpets, *Cellana* and *Patelloida,* are limited by certain algae. In this case the algae settle and grow faster than the limpet grazers can remove them. As a result, the algae form a substrate that prevents the limpets from attaching, so they cannot enter the area and graze. Obviously, in this kind of situation there is a continuum. At some level there is a balance point defining one area where grazers are more efficient at grazing than the algae are at growing, thus eliminating the algae. At the other extreme, the algae grow faster and exclude the limpets due to the creation of an unsuitable surface for attachment.

Another consideration in the zonation of algae species is the ability of some algae to grow fast enough to escape grazing by becoming too large for the grazers to deal with. In these cases the algae will develop to adulthood only in areas where the grazers are low in density or somehow restricted from grazing during some seasons, thus allowing the algae to grow up and reach a size at which they are immune from the effects of grazing.

So far we have considered grazing and its effects on vertical zonation patterns. Grazing may, however, also affect the horizontal patterns of zonation—patterns that have as the dominant physical factor the variation in wave action. The best studied examples come from the Atlantic coast of Europe. Here the more exposed coastlines tend to have the intertidal dominated by mussels, barnacles, and limpets, whereas the more sheltered shores are covered with dense stands of fucoid brown algae. From various removal experiments and natural experiments, such as the Torrey Canyon oil spill, a conceptual model to explain this has been developed by Southward and Southward (1978). Very simply, this model says that on sheltered shores limpets and other grazers are rare (but see preceding section on New England sheltered shores). This permits the algae to grow and flourish. Furthermore, the algal canopy reduces the ability of limpets to settle on the shore. On exposed shores the limpets are abundant and graze out the algae, leaving the shore dominated by mussels and barnacles. On the opposite side of the Atlantic, in New England, limpets are rare and the dominant grazer, the snail *Littorina littorea,* decreases in abundance in wave-swept areas. As a result, the community changes here, going from exposed to sheltered conditions, are attributable not to grazing but to physical factors and predation, as noted previously.

We should not leave this issue of grazing without discussing a final subject—namely, the defenses of the algae to grazers. We cannot assume that algae are completely without protection from grazing. Indeed, they are not. The most obvious adaptations to grazing are morphological, and one of the most common is the laying down of calcium carbonate in the tissues. Calcium carbonate reduces the palatability of algae to grazers and makes the grazer exert more energy in feeding. Furthermore, Steneck (1982) has even shown for one coralline alga, *Clathromorphum,* that the meristem tissue and reproductive sites are all on the underside and thus protected from grazing. Growth form may also discourage grazing. Hay (1981) has shown, for example, that formation of short turfs in algae reduces grazing mortality. Still other algae, such as the Pacific coast *Egregia,* have considerable structural or "woody" tissue when mature; this may deter grazers.

Chemical defenses are also common. Many algae contain noxious or toxic compounds. For example, on the Pacific coast of North America the genus *Desmarestia* contains sulfuric acid in concentrations sufficient to erode the calcium carbonate teeth of the grazing urchin *Strongylocentrotus franciscanus.* Other algae contain various alkaloids, phenolic compounds, and halogenated metabolites.

Finally, there are evolved defenses in the life histories of certain algae. Some algal species appear to be able to exist in at least two growth forms, one a low "crust" and the other an upright frondose form. In areas with heavy grazing pressure, the slow-growing but grazer-resistant crusts predominate, but in areas of low grazing pressure or in areas from which grazers have been excluded, the frondose forms increase. Littler and Littler (1980) have suggested that having these two alternative growth forms in a single species is a strategy of *bet hedging.* The frondose form has fast growth and high competitive ability but low grazing resistance. The crust form, on the other hand, has slow growth rates and competitive ability but high grazing resistance. Another life history feature that may have evolved in algae to cope with grazing is reproduction. Many algae, particularly those that are favored by grazers because of the high fraction of digestible material and low fraction of structural tissue, have evolved the opportunistic habit of high reproductive output. This makes them also successful in colonizing newly available space. Quick maturity and high reproductive output mean that these species should be able to colonize an open area and reach maturity quickly before the grazers find them and decimate them. Other species have evolved the opposite, or K-selected, features. They are relatively slower growing, perennial, and long lived. To survive pressure from grazers, these algae either have evolved chemicals that make them unpalatable or grow large enough to be immune from grazers. Large size usually also means an enhanced competitive ability, at least for light.

We can conclude this section on grazing by looking briefly at the geographical variation in grazing and its effects on various kinds of communities. Beginning with the temperate zone, grazing is one of the main structuring agents of the rocky intertidal on the Atlantic shores of Europe and the Pacific shores of North America. On the Pacific coast of North America, a large number of species of limpets plus a few species of crabs and sea urchins are responsible for controlling community structure. Fewer grazers,

Plate 33 Anemones of the genus *Tealia* and the hydrocoral *Stylantheca (Allopora) porphyra* with surrounding tunicates, sponges, and coralline algae on the subtidal rocks of central California. *(Photo courtesy of Drs. Lovell and Libby Langstroth.)*

Plate 34 Subtidal tube-dwelling anemone-like cnidarian *Pachycerianthus fimbriatus* from the Pacific coast of North America. *(Photo courtesy of Drs. Lovell and Libby Langstroth.)*

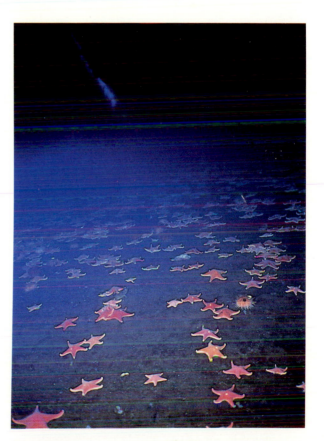

Plate 35 Abundant starfish of the genus *Odontaster* on the bottom in McMurdo Sound, Antarctica. *(Photo courtesy of Hunter Lenihan.)*

Plate 36 A giant sea spider of the genus *Colossendeis* on the bottom in McMurdo Sound, Antarctica. *(Photo courtesy of Dr. John Oliver.)*

Plate 38 A kelp bed composed of an overstory of *Laminaria dentigera* and an understory of the surfgrass *Phyllospadix*. *(Photo courtesy of Drs. Lovell and Libby Langstroth.)*

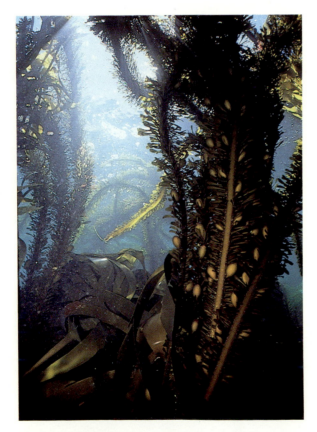

Plate 37 A view inside a giant kelp forest, dominated by the kelp genus *Macrocystis*, on the California coast. *(Photo courtesy of Drs. Lovell and Libby Langstroth.)*

Plate 39 Sand dollars, *Dendraster excentricus,* in a characteristic upright feeding position in a subtidal bed off California. *(Photo courtesy of Drs. Lovell and Libby Langstroth.)*

Plate 40 A large mass of the mussel *Mytilus californianus* with a large aggregation of its major predator, the starfish *Pisaster ochraceus,* at Neptune Beach, Oregon. Notice the large areas of clean rock at the lower right where the starfish have already consumed the mussels. *(Photo courtesy of Dr. Richard Mariscal.)*

Plate 42 The large green Pacific Coast anemone *Anthopleura xanthogrammica.* This anemone contains symbiotic zooxanthellae and zoochlorellae in its tissues, which provide part of its nutritional requirements. *(Photo courtesy of Drs. Lovell and Libby Langstroth.)*

Plate 41 The sea star *Pisaster giganteus* with the small aggregating anemones *Corynactis californicus* on the rocks of the Monterey Peninsula in California. *(Photo courtesy of Drs. Lovell and Libby Langstroth.)*

Plate 43 The sea palm *Postelsia palmaeformis,* characteristic only of areas of direct wave exposure and high wave shock, on the Pacific coast of North America. *(Photo courtesy of Dr. Richard Mariscal.)*

Plate 44 A brilliantly colored intertidal nudibranch mollusk, *Chromodoris macfarlandi.* *(Photo by the author.)*

Plate 45 The corallimorpharian anemone *Corynactis californicus* extruding out its mesenterial filaments to deter the predaceous starfish *Dermasterias imbricata. (Photo courtesy of Drs. Lovell and Libby Langstroth.)*

Plate 46 The aggregating anemone *Anthopleura elegantissima* showing the "no man's land" (bare brown area) between two genetically different clones. *(Photo courtesy of Drs. Lovell and Libby Langstroth.)*

Plate 47 An anemone of the genus *Tealia* preying on the starfish *Asterina (Patiria) miniata. (Photo courtesy of Drs. Lovell and Libby Langstroth.)*

Plate 48 The eolid nudibranch mollusk *Phidiana (Hermissenda) crassicornis. (Photo courtesy of Drs. Lovell and Libby Langstroth.)*

primarily limpets and other snails, exist on the European shore, but they are efficient at structuring the communities. Grazing is less important on the Atlantic coast of North America where limpets are rare and the major grazer, the snail *Littorina littorea,* seems less efficient than *Patella* in Europe. Grazing declines in importance as one moves poleward in the Atlantic. In the Southern Hemisphere, the intertidal zones of Australia and New Zealand are strongly structured by grazing and also have a high diversity of grazing species. Generally, as one moves toward the equator, the importance of grazing increases and so does the variety of grazing organisms. Crabs and fishes become more important as grazers in the tropics. The rocky intertidal in the tropics has been considered relatively barren due to the extremes of the physical factors acting there, but the large number of grazers suggests that grazing may be more important (see following section on tropical shores). Additional work is needed to clarify this situation.

We can summarize the effects of grazing in the three intertidal zones defined by the Stephensons to give a general framework for the North Atlantic Ocean as a model or example of how grazing acts. In the supralittoral fringe, grazers are either uncommon or have little time to graze. The rigorous and stressful physical environment is thus the most important agency causing changes in algal composition. The midlittoral zone seems dominated by barnacles and mussels but is strongly influenced by gastropod grazers, primarily limpets, and may vary in algal cover depending on exposure. This, in turn, may limit the numbers of gastropods. However, barnacles may also be present due to the ability of limpets and other grazers to keep the competing algae in check. In the sublittoral fringe, the algal growth exceeds the capabilities of the grazers to control it. Therefore, the competition for space and light among the various algae is the dominant interaction that structures communities at this level. These various factors and their relative importance are summarized in Fig. 6.17.

The ability of larvae of various benthic invertebrates to select areas on which to settle has been known for more than half a century (see Chapter 1). However, the importance of this choice to the eventual structure of intertidal communities has remained obscure due to the difficulty of working with larvae under natural conditions. Several studies recently have, however, demonstrated the importance of this selection ability. In Australia, the larvae of the barnacle *Tesseropora rosea* settle only where adults occur. Thus the distribution of adults is not the result of competition with other barnacles outside their range. Moreover, the upper and lower limits on the shore are not set by physical factors, but by settlement of the larvae. Furthermore, the barnacle larvae were found by Denley and Underwood (1979) to avoid settling on substrates already occupied by other organisms. A more comprehensive example, also from Australia, includes a second barnacle, *Chamaesipho columna.* In New South Wales, Underwood and Denley (1984) report that shores protected from wave shock have, as dominants, grazing gastropods, but where there is moderate wave action, the midshore is dominated by *Chamaesipho.* On the most exposed areas *Tesseropora* occupies midtide levels and *Chamaesipho* the upper reaches. Neither barnacle settles into the areas dominated by grazing gastropods, even if the grazers are removed. On exposed shores neither settles into areas occupied by the adults of the other species. In other words, this whole intertidal zonation pattern is structured by the patterns of larval choice and not by predation, competition, or grazing.

Larval recruitment can be very variable in time and in space. This variability has been documented, and its importance to the intertidal community structure can be significant. In some years any given intertidal area may experience poor recruitment due to particularly bad combinations of weather, reproduction by adults, water conditions such as wave action, and so forth. The result will be few survivors and a decrease in biological interactions such as predation, competition, and grazing. In other years favorable physical conditions may swamp an area with recruits, leading to increased competition or predation and probable changes in the adult population and the community. Thus the same area may have a community that not only differs in composition over time, but also may change in relative numbers of individuals and, therefore, in the importance of competi-

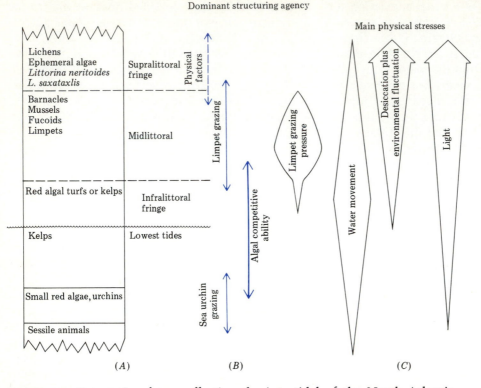

Figure 6.17 Structuring forces affecting the intertidal of the North Atlantic Ocean. *(A)* The main intertidal zones and their dominant organisms (after the Stephensons' universal scheme). *(B)* Dominant structuring agencies; overlap suggests a balance between factors. *(C)* Main physical stresses for marine algae and animals. (Modified from S. J. Hawkins and R. G. Hartnoll, Grazing of intertidal algae by marine invertebrates, Oceanogr. Mar. Biol. Ann. Rev., Vol. 21, 1983.)

tion, predation, and grazing in determining the structure of the community.

This great spatial and temporal fluctuation in the settlement of larvae of the dominant intertidal organisms means that the importance of competition, predation, and grazing are in turn likely to be variable. It is difficult to generalize about the effects of competition, predation, and grazing if the numbers of prey or potential competitors are also unpredictable. Although it is considered by some to be a "nuisance" in the setting up of experiments, this irregularity of appearance of various dominant intertidal invertebrates is a fact of nature and must be considered in any attempt to understand the structuring process in the intertidal zone. In other words, the irregular occurrence in a community of a numerically important, competi-

tively dominant, or "keystone" predator should lead to a variety of different outcomes from experimental studies done in the area because of variations in the intensity of biological interactions resulting from the larval recruitment patterns.

In an attempt to sort out the relative importance of competition, predation, and larval recruitment to the structure of rocky intertidal communities, Menge (1991) undertook a multiple regression analysis of these factors for New England and Panama shores. The results indicated that in New England recruitment explained 11 percent and predation and competition 50 to 78 percent of the structure, whereas in Panama recruitment explained 39 to 87 percent and predation and competition only 8 to 10 percent of the structure. If these figures hold true for other areas as well, this may indicate a fundamental difference in the structuring forces in temperate and tropical intertidal areas.

Tropical Intertidal Shores

Thus far, we have discussed the biological factors that determine community structure employing examples that have been exclusively from temperate regions of the world. What about the tropical intertidal? Are the communities and organisms similar? Unfortunately, there are few studies of the tropical intertidal from which to draw generalities. The most extensive are those of Menge et al. (1986) and Lubchenco et al. (1984) on the rocky intertidal of the Pacific coast of Panama. The general composition of the communities is similar to that of temperate rocky shores. The major differences are the lack of large algae and an abundance of herbivorous fishes and crabs as well as predaceous fishes in Panama. Other differences are evident. For example, studies indicate that the high intertidal of Panama has 90 to 98 percent bare substrate that is dominated by small barnacles. The mid- to low intertidal is dominated by encrusting algae (26 to 90 percent cover), while sessile animals are scarce (1 to 10 percent cover). Only in the very lowest parts of the intertidal is a diversity of organisms encountered. As might be expected from these figures, both species number and diversity increase with decreasing tide level, as does the maximum cover by organisms.

Working in the mid- and low intertidal, Menge et al. (1986) found that space is dominated by crustose algae. Community structure is determined by complex interactions among a diverse assemblage of herbivores and predators that Menge et al. (1986) divided into four "consumer" groups: large fishes, small fishes and crabs, herbivorous mollusks, and predaceous gastropods (Fig. 6.18). Experimental treatments involving deletions of various combinations of the four consumer groups and comparing with controls suggested that the effects of each consumer group on its prey species or group were often indirect and included both positive and negative effects. Both the small fishes and crab group and the large fish group had the effect of reducing the abundance and cover of sessile

colonial and solitary invertebrates as well as fleshy algae, but were incapable of totally removing them. The molluscan herbivores, in turn, grazed down the fleshy algae such that only the very resistant algal crusts survived, a group otherwise competitively inferior to the fleshy algae. The molluscan herbivores also inhibited recruitment of any sessile invertebrates. Finally, the predaceous gastropods also preyed upon the sessile invertebrates and reduced their numbers. These group interactions and results were further compounded by a combination of both predator-prey interactions among the consumer groups and various competitive or symbiotic interactions among the prey.

The results of these studies indicate that there was no "keystone" species in this area that prevented space monopolization by a superior space competitor as we saw in the case of the mussel-dominated communities in the temperate intertidal. Instead, the continued dominance of the area by crustose algae appears to be maintained by persistent intense predation and herbivory by a diverse assemblage of fishes and invertebrates on potentially dominant sessile invertebrates and fleshy algae. The dominance of this area by crustose algae and the failure of any organism or group of organisms to occupy even a majority of the available space are also likely maintained by a combination of extreme heat, desiccation, and ultraviolet ray stress that simply prevents occupation of much of the area by marine organisms. It seems obvious that more studies are needed on tropical rocky shores before we can assess the relative importance of the various biological and physical factors to community structure and the relative amount of variation over time and space.

Patchiness

Although the major large-scale patterns of organism distribution in the rocky intertidal—the zonation—can be explained through some combination of exposure, desiccation, temperature, competition, predation, larval settlement, or grazing, acting separately or together, these vertical bands are never perfect. Often small-scale local variations occur that act to alter the bands and affect the immediate local distribution of

Figure 6.18 Diagrammatic illustration of the organisms and physical heterogeneity at Taboguilla Island, Panama. The distant views are shown for the high *(A)*, mid *(D)*, and low *(G)* zones; each of these is ≈ *2 m* across the foreground. Closeup views are illustrated for open surfaces *(B)* and crevices or holes *(C)* for the high, mid *(E, F)*, and low *(H, I)* zones. Species identifications are: (1) Chthamaloid barnacles (live and dead); (2) *Littorina* spp.; (3) *Nerita scabricosta*; (4) *Tetraclita*; (5) *Siphonaria gigas*; (6) *Ostrea palmula*; (7) mixed algal crusts: *Schizothrix, Ralfsia,* and *Hildenbrandia*; (8) *Fissurella* spp.; (9) *Thais melones*; (10) *Acanthina*; (11) *Ceratozona*; (12) anemones; (13) coralline crust; (14) *Balanus inexpectatus* (live and dead); (15) *Pachygrapsus*; (16) *Siphonaria maura*; (17) encrusting bryozoan; (18) colonial tunicate; (19) mixture of *Abietinaria* and filamentous algae; (20) *Chama*; (21) *Echinometra*; (22) sponge; (23) *Ascidia*; (24) rasp marks of grazing fishes. (After J. Lubchenco et al., Structure, persistence, and role of consumers in a tropical intertidal community [Taboguilla Island, Bay of Panama], J. Exp. Mar. Biol. Ecol. 78:23–73.)

certain organisms. Such small-scale *patchiness* or *local distribution* patterns can result from either biological or physical factors.

Patchiness is a basic feature of the structure of rocky intertidal areas just as it is of plankton (Chapter 2, pp. 82–85) and of subtidal communities (Chapter 5, pp. 170–171). Patchiness is usually temporally and spatially variable. It is generated by any physical or biological activity that creates either an open space for recolonization, produces a different substrate, or alters the prevailing physical conditions acting on an area.

Perhaps the easiest of these local distribution patterns to explain are those that result from physical factors, primarily the variation in the nature, slope, and exposure of the rock surface. Existence of a different rock type in a local area as compared to the surrounding shore almost always means the presence of different organisms, or at least different abundances. An outcrop of sandstone among an otherwise granite rock shore usually means a different complement of organisms.

Similarly, slopes facing south and those facing north often have different organisms, or the densities and levels to which the organisms are found differ. This reflects the different amount of desiccation and harmful effects of direct exposure to the sun's rays (= insolation). On the Pacific coast of North America, for example, two limpet species, *Collisella digitalis* and *C. scabra,* inhabit the high intertidal zone, but, as Haven (1971) has shown, their local distribution within the zone reflects their reaction to the small-scale changes in the substrate. *Collisella scabra* is found primarily on horizontal surfaces where it has a home "scar" or depression in which it remains during periods of low tide. *Collisella digitalis* has no scar and is found on the cooler, moist, shaded vertical slopes. *Collisella scabra* is able to survive on the drier horizontal rocks because of its depression or scar, in which the fit is tight enough to prevent desiccation at low tide. *Collisella digitalis,* on the other hand, without the increased desiccation protection of the scar, must live where desiccation is not as intense (i.e., north-facing slopes and crevices). Similarly, certain species do not prefer to settle on vertical faces or overhangs. These areas are then settled by other species unable to compete on the flat surfaces. Also, the presence of cracks or crevices in the rock may aid the survival of algal spores, helping them to "escape" from grazers and create a patch of algae where others are absent due to grazing (Fig. 6.19). Cracks and crevices may shelter predators, which in turn affect the community. For example, Moran (1980) filled the cracks in an intertidal area with concrete. This reduced the density of the predatory gastropod *Morula marginalba,* and that in

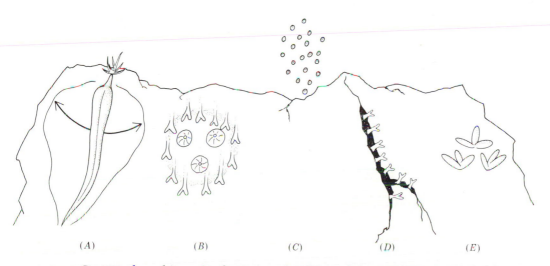

Figure 6.19 Causes of patchiness in algae on rocky shores. *(A)* Sweeping action of algal fronds. *(B)* Irregular spatial and temporal distribution of grazers. *(C)* Fluctuations in recruitment. *(D)* Refuge from grazing provided by pits and cracks in rock. *(E)* Escape of sporelings from grazers.

turn led to increases in barnacle numbers and altered community structure.

Much small-scale patchiness probably results from the destructive effects of wave action and wave-borne objects, such as logs, that may be hurled against the shore, smashing into small areas and clearing them of organisms. Such battering was studied by Dayton (1971) in the state of Washington, and it was found that any point on an open shore had a 5 to 30 percent chance of being hit by a log within a three-year period. An additional effect of wave action is that it produces aggregations of mobile species, such as snails and crabs, in crevices or other protected areas. Heavy wave action from storms, even if unaccompanied by battering of objects, also is responsible for removal of organisms from the intertidal. Thus, Menge (1976) found in the exposed intertidal in New England that winter storms could clear as much as 90 percent of the intertidal, whereas during summer and fall less than 10 percent of the same space was free of organisms.

The premier space competitors, *Mytilus edulis* in New England, or *M. californianus* on the Pacific coast of Washington, both tend to form large clumps that grow out from the substrate. As a result, as the clumps get larger, fewer mussels are attached directly to the substrate through their byssal threads and more are attached to each other. This creates an unstable situation in which the battering waves are more likely to tear off a clump, and hence older, larger clumps become more vulnerable (Fig. 6.20).

Small-scale distribution anomalies may also be the result of biological interactions of various types. Some of these are *passive,* such as the fact that many species of animals must inhabit moist areas in order to survive the exposure to air and hence shelter under the fronds of various intertidal algae. The patchiness of these forms follows the distribution of the algae, but it is not due to any direct action of the algae. Removal of the algal canopy also results in the disappearance of the sheltering organisms. Similarly, in the case of the larger algae with substantial fronds, the movement of the frond by the water causes a "sweeping" effect, which may be significant in removing larvae and spores from the rock near the alga, thus creating a "patch" (Fig. 6.19).

Other biological effects are more direct and lead to definite distribution patterns, which can be observed. Thus, for example, the intertidal anemone *Anthopleura elegantissima* of the Pacific coast of North America tends to occur in large clumps, uniformly separated from each other. Each clump has resulted from the continued asexual division of the original anemone, such that each clump consists of genetically identical individuals, a *clone.* Francis (1973) discovered that the sharp boundaries that exist between adjacent clones are the result of active fighting between the outer individuals of each of the clones (Fig. 6.21). In these fights, special tentacles called acrorhagi are employed and are able to inflict massive tissue damage to other anemones. The result of such clonal "border wars" is the separation of clones by anemone-free strips, which are just the width that prevent

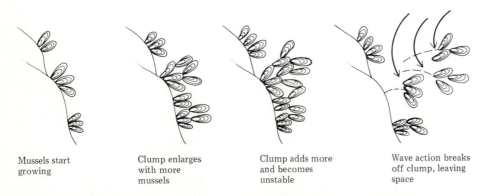

Mussels start growing

Clump enlarges with more mussels

Clump adds more and becomes unstable

Wave action breaks off clump, leaving space

Figure 6.20 Buildup and removal of mussel clumps by wave action.

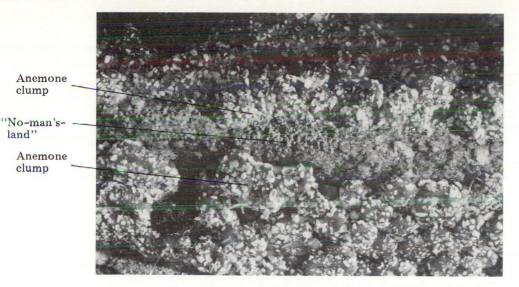

Anemone
clump

"No-man's-
land"

Anemone
clump

Figure 6.21 Clones of the anemone *Anthopleura elegantissima,* showing the separation caused by fighting between adjacent clones. (Photo by the author.)

two anemones from reaching each other (Plate 46). Thus, clumps of anemones are always separated from other such clumps.

Yet another example of biological interaction that produces spacing is the phenomenon of *territoriality.* Territoriality is defined as a pattern of behavior of an individual or group of individuals in actively keeping other members of the same or different species out of an area. This phenomenon is well known among various birds and mammals, but it also occurs among certain invertebrate organisms. On the north Pacific coast of America, the largest limpet in the intertidal is *Lottia gigantea.* Stimson (1970) has shown that individuals of this species create a territory in the intertidal, which they actively defend against other *Lottia* and other limpets. Should other limpets invade this territory, the resident *Lottia* pushes the interloper out. In this way, each *Lottia* keeps a small patch of rock free of potential competition and ensures that all algae growing on the site will be consumed by itself.

Another more complicated combination of factors producing small patches is that observed by Dayton (1973) and Paine (1988) in the alga *Postelsia palmaeformis,* which occurs in isolated clumps in the most wave-swept portions of the

Pacific coast (see Plate 43). *Postelsia* is an annual alga that inhabits areas that are dominated by the space-monopolizing mussel *Mytilus californianus.* The adult plants are immune to overgrowth by the mussels. The alga has a very limited dispersal of its propagules such that most settle within a few meters of the adult. Whereas they may settle and begin to grow on almost any hard surface, they are much more likely to survive to reproductive size if they settle on bare rock surfaces. *Postelsia,* although capable of settling, seems not to persist to adulthood in the presence of the common turflike algae *Halosaccion* and *Corallina.* Therefore, the continued persistence and successful reproduction of *Postelsia* seem to be dependent on a cyclical set of relationships. The heavy wave action is necessary to clear away patches of *Mytilus* back to the bare rock and to permit the settlement of the *Postelsia* propagules. The *Mytilus* are, in turn, necessary to eliminate the algal turf (*Halosaccion* and *Corallina*) that are too low to be removed by wave action and that outcompete *Postelsia.* The mussels do this by simply invading and overgrowing the lower-growing algae. If the mussels are not removed, then the mussels by their competitive ability will completely cover the intertidal. Removal of small patches of mussels by the violent wave action opens an area for *Postelsia* to occupy, providing its propagules colonize before the low-growing *Halosaccion* and *Corallina* are able to take over the patches. Therefore, the wave-swept area becomes a mosaic of patches of *Halosaccion* or *Corallina* and *Postelsia,* where patches of *Postelsia* result from its settling earlier than the two turflike algae and *Halosaccion* or *Corallina* result from its first settlement of a patch. *Postelsia* may then persist in the patch it occupies only if the adult plants are ripped out after reproducing, which leaves bare rock, or if another patch opens up within the dispersal range of the *Postelsia* propagules.

Predation and grazing can also create small patches or patchy distribution. Predation is the reason for the occurrence of *Balanus cariosus* (see p. 236) in small clumps in the intertidal; grazing by sea urchins may result in tidepools free of algae, while others without urchins have an abundant growth. On the New England coast, as noted above, tidepools without *Littorina littorea* are dominated by *Enteromorpha intestinalis,* whereas those with the snail have *Chondrus crispus.*

Another biological factor contributing to both large- and small-scale patchiness is the afore-mentioned preference for settlement shown by various intertidal invertebrate larvae. A final biological factor to consider is the phenomenon of succession, to which we have alluded and now take up in detail.

Succession

Creation of open spaces in any zone usually results in quick colonization by a series of *opportunistic* or *fugitive* species. These species quickly settle an open area, mature, and reproduce before the slower-growing dominant species retake the open space and force them out through competition. The most common of these opportunists are the smaller, filamentous green and red algae and smaller animals such as hydroids. These initial species are then replaced successively by competitively better species until the competitive dominant in the area takes over (Fig. 6.22). In any one zone, then, there may be many patches, which, depending on the time since their initial clearance, will be in differing stages of *succession* toward the dominant community characterizing the zone. This accounts for the differences one observes in different patches in the same zone in the same area.

Time ⟶

First stage bare rock with diatom/bacterial film

Small algal species

Barnacles replace algae

Mussels smother barnacles and eventually dominate

Barnacles   Mussels   Algae   Diatoms   Bacteria

Figure 6.22 Succession in a northwest Pacific coast intertidal mussel bed in the absence of *Pisaster*.

The ecological concept of succession, as we noted in Chapter 1, has several different models or hypotheses to explain the sequence of events following a disturbance in a community. In the rocky intertidal, succession seems to be a combination of the inhibition and facilitation models (see Chapter 1, pp. 19–22).

All successional sequences in the rocky intertidal are dependent on the time of the initial disturbance that frees up the space. That is because the first colonizers of a space are usually those that have propagules ready to settle. Generally, the first organisms to settle on freed space are those with the life history characteristics of "opportunistic species" (see Table 1.2).

The subsequent sequence of events in succession is variable and dependent on various combinations of events in the rocky intertidal. On the Pacific coast of North America, Sousa (1980) found that succession in boulder fields seemed to follow the inhibition model. He found that, once the early colonizer *Ulva* settled, it inhibited colonization by the perennial red algae, such as *Gigartina.* So long as *Ulva* remained healthy, it would prevent colonization by any other alga. However, this inhibition is broken by grazing of the crab *Pachygrapsus crassipes,* which then permits the succession to proceed to a community of perennial red algae. These middle successional red algae, in turn, slow the invasion by the late successional dominant, *Gigartina canaliculata,* but eventually *Gigartina* colonizes because it is more resistant to desiccation and overgrowth.

In other instances there does appear to be a successional sequence in which "pioneer" species are necessary before the later species can colonize (= the facilitation model of succession). For example, Hawkins (1981), working on English rocky shores, showed that successional sequences were quite different on barnacle-covered rocks kept free of limpets than on areas scraped free of barnacles and also kept free of limpets. In the case of the barnacle-covered rocks, the colonization sequence was wholly determined by the availability of propagules, but in the scraped areas there was evidence that the larger green algae and *Fucus* could not settle until after surfaces had been "prepared" by growth of diatoms and the green alga *Ulothrix.* The reasons for these differences in sequence are not altogether clear, but it is suggested that the barnacle shells may have provided spatial refuges free from grazers, thereby facilitating "escapes" of the later successional algae, which could then recruit directly. Another suggestion is that although a mat of green algae enhances *Fucus* settlement, early *Fucus* growth is inhibited by the mat. As a result, *Fucus* does not begin to take over until either the mat dies or the slow-growing sporelings of *Fucus* finally reach a height or size where they are superior to the green algae mat and take over. In this example, succession is a mixture of both the inhibition and facilitation models.

We can conclude this section by noting that continued physical disturbance of differing degrees of severity and physical extent coupled with seasonal variations in the types of propagules available, along with the fluctuations in the spatial distribution of dominant animals, would seem to ensure that some kind of succession is always occurring in the intertidal. This whole sequence in turn—and on a larger scale—is certainly one of the major sources of patchiness in the intertidal.

Horizontal Distribution Patterns

As we noted in the preceding sections, there are various changes in community structure as one moves from areas of high wave exposure to protected and quiet water. This is true for many geographical areas. Partly, these changes are caused by the action of water restricting certain species in or out of wave-swept areas. In other words, some species of algae or invertebrates simply cannot tolerate the wave action and are absent from wave-swept areas. Usually, these are species with delicate structure or with poor means of holding onto the substrate. In a similar fashion, certain species cannot tolerate quiet water areas. This may be due to their inability to compete, to lack of enough or proper food, or to the presence of predators that are not present in wave-swept areas. Thus, we have seen that *Mytilus edulis* outcompetes *Balanus balanoides* and *Chondrus crispus* in exposed areas of New England because no predators can tolerate the wave action; in protected areas, predators reduce competition, allowing the barnacle and algae to exist and thus altering the appearance of the zones. In Australia, protected shores have larger numbers of grazers, whereas exposed shores have two barnacle zones and few grazers. European coasts present the best examples of horizonal changes. On these coasts the limpets dominate the exposed areas and large algae are uncommon, whereas in sheltered areas limpets are rare and large algae flourish.

Age Structure

As we noted in Chapter 1, terrestrial communities or associations consist of a matrix of long-lived plant species, which allow us to characterize terrestrial communities on the basis of the dominant plant species. In the marine environment, especially the intertidal, the long-lived species tend to be various invertebrate animals, whereas the plants and the various algal species tend to be more short lived. We, in turn, tend to name zones in the intertidal on the basis of animals—*Mytilus* zone, barnacle zone, and so forth. Even though the potential longevity is great for various marine invertebrates, it is well to consider under what conditions such longevity is achieved and what the age structure of the animals actually is.

As noted previously, on the Pacific coast of North America the high intertidal has a band of *Balanus glandula;* below that, any of these species found are usually less than two years old. If the age structure of the large barnacles in the band is analyzed, it is generally found to consist primarily of a single age class. Why? Because the physical conditions are so harsh in this area that, most years, the juvenile barnacles simply do not survive. As a consequence, the barnacles there result from a single good year in which conditions allowed survival through the vulnerable juvenile stage. Once adult, the barnacles can survive succeeding harsh years. The reason that all of the lower level *B. glandula* are less than two years old is that, in two years, *Nucella* completely consumes all midtide barnacles. Similarly, *B. cariosus* populations in the midintertidal consist primarily of animals less than two years old. However, there are a few that may be large and many years old. Again, the majority are less than two years old, because that is the time during which *Nucella* consume them. The few large, old individuals are those that somehow avoided

predation until they were two years old and thus invulnerable to *Nucella.* Unless consumed by *Pisaster,* these individuals may live for many years. Thus, under very stringent environmental conditions or under optimal conditions low in the intertidal, the age structure of the dominant organisms is mainly of single-year classes.

Predators also seem to consist mainly of older individuals. Examples are *Pisaster ochraceus* and *Nucella lamellosa.* Herbivores such as *Lottia gigantea* and the turban snail *Tegula funebralis* also have populations dominated by older individuals. Why this should be the case is not known at present.

Interactions Among Factors: A Summary

We have seen how various physical and biological factors each act to cause the large-scale vertical distribution patterns of algae and invertebrates that we call vertical zonation on rocky shores. We have also seen how some of these same factors, plus others, act to introduce small-scale distribution anomalies into this pattern. In general, different factors tend to operate in different areas and at different levels on the shore. The high intertidal tends to have the distribution pattern of its characteristic species set by physical factors, whereas biological factors become more important in setting limits in the low intertidal. It is important to realize that these factors do not act in isolation and the final distribution pattern observed, either on a small or large scale, is often the result of the interaction among two or more factors (Fig. 6.23).

Tidepools

A characteristic feature of many rocky shores is the presence of tidepools of various sizes, depths, and locations. Certain conditions affecting life of tidepools differ markedly from the surrounding intertidal and hence necessitate a separate discussion here. Tidepools can, of course, occur at any level in the intertidal zone; however, low-level pools often are little different from the surrounding sea and are essentially extensions of the sublittoral. They will not be considered here. Only in midlittoral and infralit-

Creates spaces

Spaces settled by
small barnacles

Pisaster eats mussels

Nucella eats
barnacles

Pisaster eats *Nucella*

Some barnacles
persist to reproduce

Figure 6.23 Interactions among mussels *(Mytilus)*, barnacles, and their predators on the northwest Pacific coast of North America, which allow barnacles to
persist in the intertidal zone.

toral pools do we observe the conditions that
characterize these areas. Our remarks in this
section will be restricted to those pools that
undergo a complete interchange with the ocean
water during the tidal cycle (Fig. 6.24).

At first glance, tidepools would appear to be
more or less ideal places for aquatic organisms
seeking to escape the harshness of the intertidal
during its exposure to air. In reality, however,
escape from certain physical factors such as
desiccation may mean exposure to others that
operate more severely in tidepools.

Tidepools vary a great deal in size and hence
in the volume of water that they contain. Since
water is a great moderator of harsh physical
conditions, the larger the pool and the greater
the water volume, the less the fluctuations in
physical factors.

Three major physical factors are subject to
variation in tidepools. The first is temperature.
Whereas the ocean itself is a vast reservoir that
heats and cools very slowly and usually within
very narrow limits, the same is not true of
tidepools. These relatively small bodies of water
are subject to considerably more rapid changes.
Shallow tidepools exposed to the sun on warm
days may quickly reach lethal or near lethal
temperatures. Similarly, tidepools in cold temperate or subpolar regions may have temperatures in the freezing range in winter. An additional problem is this: The pool may either heat
up or cool down over a several-hour period while
exposed to the air, but when the tide returns, it
will be flooded at some point with ocean water at
quite a different temperature; this will suddenly
change the temperature of the whole pool. Thus,
any organism inhabiting such pools must still be
adapted to considerable sudden fluctuation.

The second factor to vary in pools is salinity.
During exposure at low tide, tidepools may heat
up. When that happens, evaporation occurs and
the salinity increases. Under hot tropical conditions, the salinity increase can be dramatic
enough in some instances to reach the point of

Figure 6.24 A high intertidal rock pool at low tide. (Photo by the author.)

precipitating out salt. The opposite situation is the case when heavy rains occur at low tide and flood pools with fresh water, dramatically lowering the salinity. Again, tidepool animals and plants may have to be adapted to wider ranges in salinity than typical marine or intertidal organisms. As before, when the tide returns, the pool will be flooded with seawater at some point, and again there will be a sudden, abrupt return to normal conditions.

The final physical factor undergoing change in pools is oxygen concentration. Since the amount of oxygen that can be held in seawater is a function of temperature, it follows that those pools that heat up during exposure to the air will lose oxygen. Under normal conditions, this may not be serious enough to produce oxygen stress, but if the pool is crowded with organisms, it may well produce a stress situation. For example, a pool filled with algae that was exposed at night would produce a situation in which the lack of photosynthesis coupled with high respiration could reduce the oxygen level significantly. Oxygen level has been recorded as falling to but 18 percent of saturation in tropical tidepools.

Tidepools are thus areas of refuge from desiccation for intertidal organisms, but in turn these organisms suffer from rapid changes in tempera-ture, salinity, and occasionally oxygen, thus restricting the fauna and flora to those organisms that are able to tolerate such ranges.

SANDY SHORES

Intertidal exposed sand beaches and protected sand flats are common throughout the world and are certainly better known than rocky shores to most humans, because they are the substrate of choice for various recreational activities. At the same time, they present a very marked change from the previously described temperate rocky shores because, in contrast to the crowded life on the latter, exposed sand beaches appear devoid of macroscopic life because the environmental factors acting on these shores create conditions where virtually all organisms bury themselves in the substrate. Intertidal protected sand flats by contrast are more often populated, with large numbers of macroorganisms that are visible on the surface.

In this section, we explore the action of physical factors on sand beaches and intertidal protected sand flats and see how these factors act to impose certain adaptations on the larger organisms found in this area. The very specialized conditions and organisms existing in the tiny spaces between the sand grains are the subject of the next chapter.

Environmental Conditions

As with rocky intertidal shores, sand beaches and flats are subject to a similar array of physical factors, but the relative importance of these factors in structuring the community and their effect on the substrate differ.

Perhaps the most important physical factor governing life on exposed sand beaches is *wave action* and its attendant correlation with *particle size* and *slope*. Anyone who has visited such marine beaches is aware of the fact that the particle size and slope of the sand can differ among beaches and also among seasons on the same beach. Particularly in the temperate zone,

the profile of the beach also changes between winter and summer. Beaches are defined by three factors: particle size, wave action, and slope. All three are interrelated such that defining two fixes the third. The importance of the particle size to organism distribution and abundance rests with its effect on *water retention* and its suitability for burrowing. Fine sand, through its capillary action, tends to hold much water above the tide level in its interstices after the tide has retreated. Coarse sand and gravel, on the other hand, allow water to drain away quickly as the tide retreats. Since the organisms inhabiting the intertidal are aquatic, they are well protected against desiccation in a fine sand beach but subject to desiccation in a coarse gravel beach. This makes the latter less hospitable. Fine sand is also more amenable to burrowing than coarse gravel (Fig. 6.25).

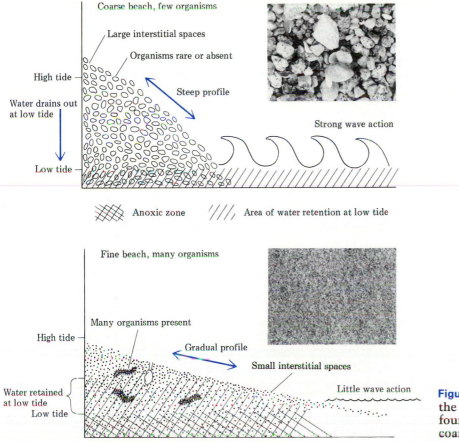

Figure 6.25 Comparison of the physical conditions found in fine-grained and coarse-grained beaches.

The *slope* of a beach is the result of the interaction between the particle size, wave action, and the relative importance of swash and backwash water. *Swash* is water running up a beach after a wave breaks; this action tends to carry particles with it, which may cause accretion of the beach if they remain there. *Backwash* is the water flowing back down the beach; this action tends to remove particles from the beach depending on the particle size.

Protected sand flats are much less seasonally variable, and usually consist of finer-grained sand at all times. In protected sand flats, both waves and water currents affect grain size. The magnitude of the wind is largely a function of the size of the body of water over which the wind blows, and the large embayments on the East Coast of the United States, such as Chesapeake Bay and Pamlico Sound, are large enough to generate waves that can have an effect on grain size. In New England, because the embayments are all small, wind and wave action is less important.

A second important physical factor in sand beaches and tidal flats, which was not of concern in rocky shores, is also the product of wave action. That is wave-induced *substrate movement.* The particles on sand or gravel beaches are not large enough to be stable when waves strike the shore. As a result, with each passing wave, the substrate particles are picked up, churned in the water, and redeposited (Fig. 6.26). Particles are thus being continually moved and sorted. The reason that fine sand occurs only where wave action is light and coarse sand where it is heavy is that, in heavy wave action, the smaller particles remain in suspension so long that they are carried away from the beach, leaving only those heavy enough (= coarse) to settle immediately. Light wave action also means that a smaller depth of particles is affected by the passage of a wave, whereas heavy wave action disturbs the substrate to a greater depth.

Within this continuum of grain size, slope, and wave action, a number of different beach types have been described. The extremes of these beach types are the dissipative and reflective (Fig. 6.27). A *dissipative beach* occurs where wave action is strong and sediments are fine. The beach is flat and maximally eroded, with much

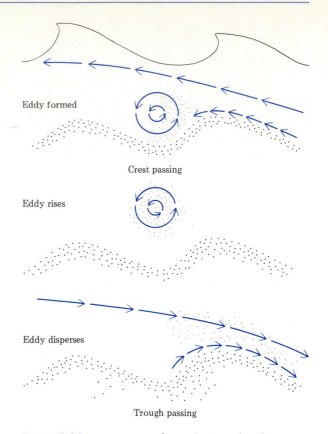

Eddy formed

Crest passing

Eddy rises

Eddy disperses

Trough passing

Figure 6.26 Transport of sand particles by wave action. (After J. Hedgpeth [ed.], Treatise on marine ecology and paleoecology, Geol. Soc. Amer. Memoir 67, Vol. I, Ecology, 1957).

of the sediment occurring in a broad surf zone and in berms or bars parallel to the beach. Most wave energy is consumed in the surf zone and swash is gentle. A *reflective beach* is most fully developed where wave action is minimal and the sediment is coarse. In these beaches there is no surf zone and waves produce large swashes up the beach. The beach slope is steep, and the backwash and swash collide to deposit sediment. Wave energy is directed against the face of the shore and reflected off the face.

Studies by McLachlan (1983) and others have demonstrated that the abundance and diversity of macrofauna organisms are more strongly correlated with particle size and slope of open beaches than with wave action. Very exposed beaches often have more abundant faunas than less exposed beaches if the latter have coarse grains and steep slopes. Fine sand beaches usual-

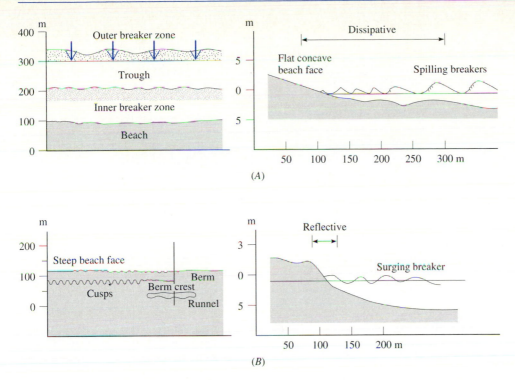

Figure 6.27 *(A)* Dissipative beach. The left side of the figure shows the beach from above; the right side is a cross-sectional view. *(B)* Reflective beach. The left side of the figure shows the beach from above; the right side is a cross-sectional view. (After A. C. Brown and A. McLachlan, Ecology of sandy shores, Elsevier, 1990.)

ly have flatter slopes, where wave action is dissipated in the surf zone rather than on the beach itself. This means that a very exposed beach can support a rich macrofauna if it is dissipative and not reflective. Indeed, a high biomass of filter feeders seems strongly correlated with dissipative conditions. Intertidal protected sand flats show much less local variability in grain size.

Since the substrate itself, at least the surface layers, is in constant motion and may be picked up and held for greater or lesser amounts of time in suspension, depending on the strength of the waves, it follows that any change in wave intensity on a given beach will mean a change not only in grain size but also in the profile or shape of the beach. This comes about when gentle wave action is replaced by heavier waves, which dig deeper into the beach, pick up more and heavier grains, and hold them in suspension longer, so they are deposited some distance away, usually

offshore. This changed profile is a common seasonal occurrence on many temperate zone beaches, where a gentle slope of fine sand occurs during the summer months and is replaced by a steep, coarse beach during winter storms (Fig. 6.28). Significant for organisms is the fact that the depth of substrate that may be physically moved can be a meter or more!

Because of this condition of ceaseless movement of the surface layers of the sediment, few large organisms have the capability of permanently occupying the surface of open sand or gravel beaches. This is the reason for the barren appearance of such beaches.

In contrast to rocky shores, sand beaches and sand flats usually have a rather smooth, uniform profile and thus lack the great topographic diversity of rocky shores. There are, for example, no crevices, no slopes facing different directions and offering different moisture conditions, nor are there overhangs or permanent tidepools. As

Figure 6.28 Beach profiles showing the difference between winter and summer.

a result, the environmental factors such as temperature, desiccation, wave action, and isolation act uniformly at each tidal level on the beach or flat.

Wave action is the dominant environmental factor acting on the exposed sand beaches, creating the special conditions that make it difficult, or impossible, for many organisms to inhabit this area. The much reduced wave action in protected sand flats, on the other hand, makes them suitable to a host of epifaunal and infaunal invertebrates. Both sand flats and sand beaches offer some advantages to marine organisms with respect to certain physical factors. Sand is an excellent buffer against large temperature and salinity changes. Measurements of temperature with depth on sand beaches at low tide have demonstrated that, below the first few centimeters, the temperature is very nearly that of the surrounding seawater. This amelioration of temperature is partly due to the insulating properties of sand itself and partly due to the water held in the interstices of the deeper layers. In a similar manner, salinity changes are also minimal below 10 or 15 cm on beaches, even if the upper layers have fresh water flowing over or falling on them. Again, this is due to the fact that the water held in the interstices is salt water, which has a higher density than fresh water. As a result, the fresh water remains perched on the surface. Furthermore, sand is a barrier to any harmful effects of exposure to direct sunlight (insolation) for any organism living in the sand,

simply because the sand is opaque to light and reflects or absorbs it in the surface layers. Desiccation is not a problem as long as the beach sand is fine enough to hold water by capillary action during periods of low tide. Thus, any organism burrowing into the sand will be constantly moist.

A final physical factor acting on beaches is *oxygen content.* Oxygen is never limiting in the water bathing the beach because the turbulence of wave action ensures constant saturation. Where it might become limiting is in the substrate itself. The water held in the beach or sand flat and responsible for ameliorating the changes in temperature and salinity also contains oxygen, which is available for use by the organisms. This supply, however, is used up by respiration of the organisms and must be replenished. This replenishment comes through interchange of the water with the sea above, and is dependent on the fineness of the sediments. Fine sediments have a slow rate of exchange and coarse sediments a rapid exchange. Thus, in fine-grained beaches, and particularly protected sand flats, water interchange is slow and may result in reduced oxygen supplies.

Adaptations of Organisms

The dominant environmental factor acting on open sand beaches is wave action, which creates the unstable, constantly moving substrate. In order to inhabit this area, organisms must first be adapted to tolerate these features. There are two routes that may be taken by organisms in adapting. The first is to burrow deeply enough into the substrate such that the organism is deeper than the depth of sediment affected by the passing wave. This strategy is employed by many large clams such as *Tivela stultorum,* the Pismo clam (Fig. 6.29). Such animals are usually also aided by developing a heavy shell, which helps in keeping the animal situated in the substrate, and very often have long siphons that enable them to burrow deeply. The only difficulty with this type of adaptation is that a severe storm may generate waves large enough to pull these animals out and throw them on the beach. That is the reason for the catastrophic destruction that occasionally occurs in these forms.

The second route of adaptation is the ability to burrow in very quickly as soon as the passing wave has removed the animal from the substrate. This appears to be the more common mechanism and is employed by many annelid worms, small clams, and crustaceans. A good example of this type of adaptation can be found in the various sand crabs of the family Hippidae, which populate many beaches around the world. These animals have a short body with limbs highly modified to dig quickly into wet sand. As soon as they are freed from the substrate by a passing wave or they leave by moving to the surface, they are able to reburrow quickly before the water motion can carry them offshore (Fig. 6.29*C*). Small clams of the genus *Donax* also do the same thing, and razor clams of the genus *Siliqua* are extraordinarily fast at reburrowing.

Other adaptations are correlated with the burrowing action of these organisms. For example, some mollusks tend to have very smooth

(*A*)

(*B*)   (*C*)

Figure 6.29 Some sandy beach animals of the Pacific coast of North America. *(A) Tivela stultorum* (Pismo clam). *(B) Olivella biplicata. (C) Blepharipoda occidentalis* (sand crab). (Photos by the author.)

shells that reduce the resistance to burrowing into the sand, while others, as Stanley (1969) has noted, have developed special ridges on the shell to grip the sediment to aid in penetrating into the substrate. Similarly, shore echinoderms such as sand dollars have much reduced spines to allow them to burrow into the sand.

A special adaptation observed by Chia (1973) in small sand dollars (*Dendraster excentricus*) is that they tend to accumulate iron compounds in a special area of their digestive tracts, which serve as a "weight belt" to keep them down in the presence of wave action.

Reproductive problems facing sandy beach organisms include when to reproduce and whether to reproduce repeatedly (iteroparous) or but once (semelparous). A critical factor here is to coordinate gamete production or release of young with the tidal cycle such that the reproductive products are not stranded by the tide or completely consumed by predators. To this end, most sandy beach animals display lunar rhythms of reproductive activity. Many are also iteroparous, though at present we cannot demonstrate that the semelparous forms are at any particular disadvantage in this environment. The mode of development has a latitudinal gradient, with the tropical species having a majority of planktotrophic larvae and the temperate zone species having more lecithotrophic larvae.

A common aspect of many macrofaunal animals is their tendency toward gregariousness, about which we know very little. Perhaps the tendency is due to water movements. On the other hand, it may be biological, as when we see scavengers congregating on large food masses. At present, this is all speculation.

Sandy beach macrofaunal animals are quite vulnerable to predation and many show adaptations to avoid predation. Chief among these is the ability to burrow deeply, as exhibited in the bivalves *Donax* and *Tellina* and the gastropod *Bullia*. Also, the tidal migrations of *Bullia* and the hippid crabs, which tend to keep them in the swash zone, may reduce predation by birds or fishes, which usually cannot operate in these areas. Crustaceans of the upper beach areas often avoid diurnal predators by residing in deep burrows during the day or by hiding under piles of wrack (Fig. 6.30).

A final set of adaptations has to do with the problem of preventing clogging of respiratory surfaces by the resuspended sand. In order to prevent this from occurring, the intake siphons of sandy beach clams are often fitted with various screens, which prevent entrance of sand but allow passage of water. Similarly, in the sand crabs, the antennae held together form a tube to the surface through which water enters the branchial chamber. These are densely clothed with

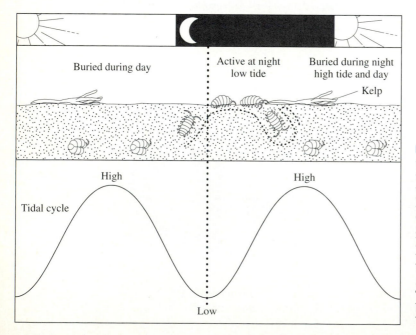

Figure 6.30 Endogenous activity rhythms of the semi-terrestrial isopod *Tylos*, which is active only during low tide at night, when it emerges from its burrow to roam the shore in search of kelp, on which it feeds. (After A. McLachlan, Sandy beach ecology, pp. 321–380. *In:* A. McLachlin and T. Erasmus [eds.], Sandy beaches as ecosystems, Dr. W. Junk, 1983.)

closely spaced hairs designed to prevent entrance of sand.

Types of Organisms

The most conspicuous organisms that are absent from sand beaches and sand flats are the large plants. No macroscopic plants occur on open sand beaches, presumably because there is no way for them to attach and maintain themselves in the wave action. Perennial macroscopic plants are also absent from protected sand flats where the unstable substrate seems also to preclude their persistence, but certain ephemeral algae, such as *Ulva* or *Enteromorpha,* may be seasonally abundant. On open sand beaches the only primary producers are certain benthic diatoms and surf-living phytoplankton. These microflora organisms often exhibit vertical migration either within the sediments or between the sediment and the water column. Migrations may lead to concentrations of these autotrophs into cells that can result in highly productive patches with high rates of productivity. Protected sand flats, however, support a large and diverse microflora of benthic diatoms, dinoflagellates, and blue-green algae, which typically appear as a brownish or greenish film on the sediment surface. The second major group of organisms conspicuous by its absence is the sessile animals such as barnacles and mussels, so dominant on rocky shores. Again, there is no place for them to attach. The sand beach and intertidal flats are dominated by representatives from three invertebrate groups: polychaete worms, bivalve mollusks, and crustaceans. Various combinations of species from these three groups dominate sand beaches and protected sand flats throughout the world. There is a tendency for crustaceans to be more abundant on exposed and tropical beaches and bivalves to be the more abundant on protected and temperate shores. The total number of macrofaunal species of all groups increases with decreasing wave exposure, but the highest biomass comes from exposed beaches.

Feeding Biology

The absence of large multicellular plants and algal mats means that there is very little primary productivity on open sand beaches. Although diatoms are present, the opaqueness of the sand ensures that the diatom population is restricted to the very surface layers. As a result, there are no macroscopic herbivores on a sand beach. Since there is virtually no primary productivity, the animals living on the beach must depend for food upon either the phytoplankton carried in seawater above the beach, the organic debris brought in by the waves, or consuming the other beach animals. These food sources dictate a benthic invertebrate association on open sand beaches dominated by filter feeders, detritus feeders, and scavengers. On protected sand flats the productivity is confined to the microfloral films and ephemeral macroalgae, but even the latter do not appear to be grazed extensively, leaving 90 to 95 percent of these macrophytes to be broken down into detritus. There are relatively few resident carnivores among open-beach animals, because to be such would require active movement across the substrate in search of prey. This is probably not feasible in the face of the heavy wave action. In addition, the highly erratic nature and unpredictability of large food items, either living or dead, mean that specialized feeders are rare and most carnivores and scavengers are opportunists.

Suspension-feeding animals are those that filter particles out of the water column. These particles may be plankton organisms or, particularly on the sand beach, they may be various organic particles resuspended from the bottom by the passing waves. The dominant group of suspension feeders on sand beaches are the bivalve mollusks such as razor clams (*Siliqua, Ensis*), surf clams (*Tivela, Spisula*), and coquinas (*Donax*). On protected sand flats suspension feeders include small bivalves (*Gemma*) and amphipods (Haustoriidae; Fig. 6.31).

Open sand beaches tend to have lesser amounts of organic detritus than protected sand flats, but enough debris from various sources finds its way to the beach to be a reliable source of food for certain organisms. Since this detrital material is often carried up and down the beach, suspended in the wave wash, rather than being deposited on the bottom, the mechanisms employed by detritus feeders on the open beach are often different from those employed by detritus feeders on protected shores.

The sand crabs of the family Hippidae, such as *Emerita analoga* and *Blepharipoda occidentalis* of the Pacific coast of America, employ a

Figure 6.31 Some representative New England sand flat benthic invertebrates, indicating general life habits. (Not to scale.) Surface deposit feeders: (A) *Spiophanes bombyx* (spionid polychaete); (B) *Saccoglossus kowalewskyi* (protochordate); (E) *Pygospio elegans* (spionid polychaete). Burrowing deposit feeders: (C) *Aricidea* sp. (paraonid polychaete); (D) oligochaete; (H) *Exogone hebes* (syllid polychaete); (I) *Scoloplos* spp. (orbiniid polychaete); (J) *Nephtys* spp. (nephtyid polychaete). Suspension feeders: (G) *Gemma gemma* (venerid bivalve); (F) *Protohaustorius deichmannae* (haustorid amphipod); (K) *Acanthohaustorius millsi* (haustorid amphipod). Conveyor-belt deposit feeder: (L) *Clymenella torquata* (maldanid polychaete). (From R. Whitlatch, The ecology of New England tidal flats: A community profile, U.S. Fish & Wildlife Serv., 1982.)

unique mechanism to trap debris in the wave wash (Fig. 6.29). MacGinitie and MacGinitie (1949) observed that *Emerita* orient so that their heads point shoreward. They remain completely buried as the incoming wave passes over them shoreward. After passage, as the water begins to run down the beach, they stick out their very large second antennae. These are thickly clothed in hairs forming a net that, when spread into the backwash, intercept all particles within a certain size range. The particles are then wiped off by the mouth and ingested. As the tide rises and falls, the position of the wave-wash front also moves. These crabs take advantage of this to maximize their feeding time by being moved up and down the beach with the tide. Movement is effected by surfacing as a wave rolls in, and thus being carried up some distance by the wave, and then quickly reburrowing. In a similar manner,

as the tide recedes, they emerge and roll down the beach before reburrowing and commencing feeding.

A similar method of feeding is employed by the gastropod mollusk *Olivella columellaris* on the coast of Central America. In this case, a mucous net is produced by the animal and held across the wave backwash. This net acts like a filter and captures small particles. Since the animal feeds by consuming the entire net with the particles it has caught, it must produce a new net at frequent intervals.

Most sand beaches also include one or more polychaetes, which burrow through the sand, ingesting the sand and digesting out the organic particles. Much like earthworms on land, however, they are rarely abundant on open sand beaches because the amount of organic material is less than in protected shores.

Sand dollars, one of the echinoderms common on sandy beaches, are generally detritus-feeding animals. These animals usually burrow through the sand. As they do so, the finer organic particles, but not the sand grains, fall down between the short spines, are trapped in mucus, and carried to the mouth. Some species, such as the Pacific coast *Dendraster excentricus,* according to Timko (1976), have even modified this basic feeding process to remove suspended particles by projecting part of the body up into the water column (Plate 39).

On protected sand flats the numbers of detritus feeders are greater both in terms of numbers of species and individuals. By far the dominant group is the polychaete worms. Two types of deposit feeders are recognized, surface deposit feeders and burrowing deposit feeders (Fig. 6.31). Most deposit feeders appear to be generalized feeders whose diet is limited by the size of the particle they can ingest. Surface deposit feeders usually live in vertical tubes (spionid and terebellid polychaetes) or burrow in the surface layer (amphipods). Burrowing deposit feeders consist primarily of polychaete worms.

Carnivores are rare, particularly on open sand beaches, but the few found usually are polychaete worms, large crabs, or gastropod mollusks. The polychaetes, such as species of the genera *Nephtys* and *Glycera,* actively burrow through the sand and seize small prey items with their eversible proboscis. The carnivorous gastropods are represented most commonly by burrowing snails of the families Naticidae (moon snails) and Nassariidae (dog whelks). Moon snails burrow slowly just below the surface layers of the sand in search of bivalve mollusks, which they consume by boring holes through the shell and eating out the contents. Nassariids of the genus *Bullia* are probably the best known of the larger open-sand gastropods. They are common on beaches all around Africa and as far east as India. They are extreme opportunists, consuming virtually all animal material available, live or dead. They detect food with their chemical sense organ, the osphradium, and move quickly to it by crawling or surfing. Once on the food, they attach via proboscis or foot, to avoid being separated from it by the continual movement of the food by the swash. On protected sand flats epifaunal carnivorous crabs such as *Callinectes* and *Carcinus* are common predators of various infaunal invertebrates. The horseshoe crab, *Limulus,* may be seasonally abundant. A number of transient predators also occur. At low tide, various shorebirds actively feed on beach organisms, and at high tide, fishes invade the area, of which the rays, skates, and flatfishes are the most significant as predators of the benthic invertebrate communities (Fig. 6.32).

Community Organization on Open Sand Beaches

Since there are few organisms visible on the surface of the intertidal sand beaches, these areas do not display the surface zonation patterns so obvious on rocky shores. Furthermore, since the various physical factors such as desiccation and temperature that act to limit organisms on rocky shores are of far less importance to organisms living in the sand, it might be

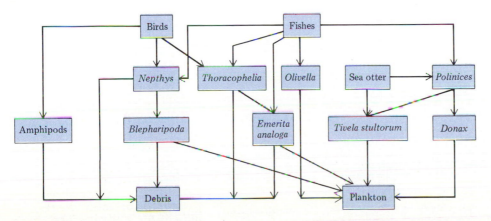

Figure 6.32 Generalized food web for a sand beach in California.

suspected that zonation of organisms would be reduced or absent. In fact, there is a zonation of sandy shore organisms, but it is neither as clearly defined nor as well understood as that for rocky shores. The lack of clear zonation patterns is due partly to the habit of some dominant organisms of moving up and down the beach for feeding and partly to the dearth of studies on the determinants of distribution patterns in this area. Certainly, one of the major contributors to the lack of understanding is the inability of researchers to set up field experiments in this area to determine the results of excluding certain organisms. Wave action precludes such work.

From the few data that are available, it is possible to establish a three-zone division of most sandy beaches that follows quite closely the universal scheme proposed by the Stephensons, mentioned earlier. The highest parts of the sandy beach, corresponding to the supralittoral fringe, are usually inhabited by talitrid amphipod crustaceans (beach hoppers) in the temperate zone and by the fast-moving ghost crabs (*Ocypode*) in the warm temperate and tropics. Both groups excavate burrows and are scavengers. The broad midlittoral area, which corresponds to the area inhabited by barnacles and mussels on rocky shores, is much more variable.

One group found here is isopods of the family Cirolanidae. Also coming into this zone when it is the area of wave wash are the various sand crabs and those other animals that feed in a similar way (see previous section). The lowest zone, the infralittoral fringe or surf zone, is inhabited by the greatest number of species, including the large surf clams (*Tivela, Spisula*), sand dollars (*Dendraster*), various polychaete worms, a host of crustaceans, and the larger carnivorous snails (*Natica, Polinices, Bullia*). There is some tendency for this lowermost zone to be characterized in the tropics by hippid crabs and in the temperate zone by haustorid and other amphipods (Figs. 6.33, 6.34, 6.35).

In contrast to rocky shores where the fauna is sedentary or fixed, sandy beaches are characterized by tidal migrations of many of the macrofauna. Therefore, any zonation pattern established by a researcher reflects the faunal distribution at the time the samples were taken. Most macrofaunal migrations are simple movements of the animals that serve to maintain them in the swash zone, where feeding conditions are good and predation pressure low. These tidal migrations tend to compress the zonation pattern at high tide, when macrofaunal

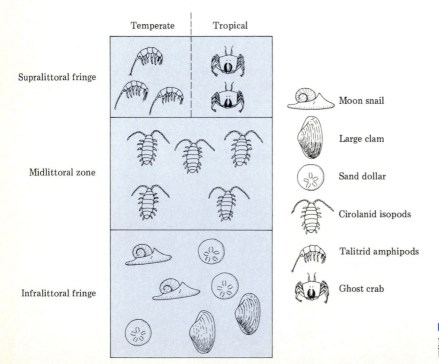

Figure 6.33   Generalized patterns of organism distributions for sandy beaches.

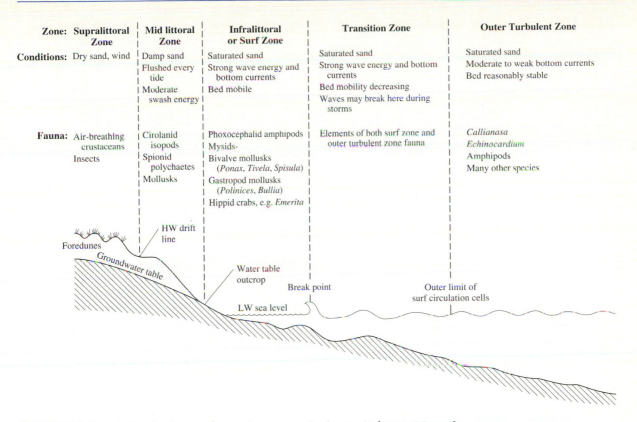

Zone:	Supralittoral Zone	Mid littoral Zone	Infralittoral or Surf Zone	Transition Zone	Outer Turbulent Zone
Conditions:	Dry sand, wind	Damp sand Flushed every tide Moderate swash energy	Saturated sand Strong wave energy and bottom currents Bed mobile	Saturated sand Strong wave energy and bottom currents Bed mobility decreasing Waves may break here during storms	Saturated sand Moderate to weak bottom currents Bed reasonably stable
Fauna:	Air-breathing crustaceans Insects	Cirolanid isopods Spionid polychaetes Mollusks	Phoxocephalid amphipods Mysids Bivalve mollusks (*Ponax, Tivela, Spisula*) Gastropod mollusks (*Polinices, Bullia*) Hippid crabs, e.g. *Emerita*	Elements of both surf zone and outer turbulent zone fauna	*Callianasa* *Echinocardium* Amphipods Many other species

Figure 6.34 Generalized scheme of zonation on sandy shores. (After A. McLachlan, Sandy beach ecology, pp. 321–380. *In:* A. McLachlan and T. Erasmus [eds.], Sandy beaches as ecosystems, Dr. W. Junk, 1983.)

organisms are concentrated, and spread them at low tide.

In addition to tidal changes in faunal composition on open sand beaches, there are seasonal changes. The most dramatic of these changes are those seen on the beaches of India where Ansell et al. (1972) and McLusky et al. (1975) have shown that most of the fauna disappears from the intertidal during the monsoon. Another seasonal change, occurring mainly on temperate zone beaches, is that of onshore and offshore movements in winter, which may be an adaptation to avoid winter storms. Long-term changes may also occur and may be dramatic in the case of periodic mass mortalities.

In direct contrast to rocky shores, where experimental work has demonstrated the importance of physical factors, competition, and predation in determining the observed zonation patterns, the explanations for the establishment of the pattern in open sandy beaches have not been subjected to rigorous experimental analy-

sis. As a result, we know little of the factors or their interactions that establish the sandy beach distribution patterns.

The extremely crowded conditions for organisms observed on rocky shores are not found on sand beaches. Indeed, the fauna is relatively sparse and certainly not occupying all available space. Also, the three-dimensional nature of the sedimentary substrate, coupled with the fact that many macrofaunal species burrow into it, certainly must minimize interference competition. It is also difficult to conceive of organisms overgrowing others in this environment, where animals move freely vertically in the substrate as well as horizontally on the surface. This immediately suggests that competition for space is not a major contributor to the observed distribution pattern. Competition for food also appears negligible, since the sparse populations and abundant plankton make competition unlikely. The fact that most sand-beach animals seem to be opportunistic feeders also suggests competition

Figure 6.35 Zonation of the numerically dominant invertebrate genera on a central California intertidal sandy beach. *Orchestoidea, Paraphoxus* (Crustacea, Amphipoda); *Excirolana* (Crustacea, Isopoda); *Archaeomysis* (Crustacea, Mysidacea); *Emerita, Blepharipoda* (Crustacea, Decapoda); *Nepthys* (Annelida, Polychaeta); *Tivela* (Mollusca, Bivalvia). (Not to scale.)

is not important. Similarly, there are relatively few indigenous invertebrate predators, and it seems unlikely that they are responsible for any major distributional pattern. This is supported by the fact that, in the few occasions that predators have been experimentally excluded from sediments, the diversity has increased, not decreased, as in rocky shores. The effect, however, of large vertebrate predators such as birds and fishes remains to be investigated.

Little is known of the infralittoral or surf zone fish fauna and its potential effects on open beach community structure. This is likely to remain true due to the danger and difficulty of sampling in the surf zone. What we do know is that these fish assemblages are dynamic, have few resident species, and are variable in time and space. Most species are opportunistic feeders.

A similar situation prevails for birds. Here, again, our lack of knowledge stems primarily from a lack of direct studies and experimental manipulations. Indirect work, particularly in South Africa, suggests, however, that birds may be significant predators, removing from 10 to 49 percent of the macrobenthic invertebrate production.

We can summarize by saying that what little evidence there is suggests that both competition and predation decrease in importance as exposure of the beach increases, thus leaving the physical factors—primarily wave action, particle size, and beach slope—as probable major contributors to the patterns observed.

In the absence of indigenous large plants, open sand beaches have food sources that are mainly derived from sources in the surrounding sea. The major determinants of the amount of food on the beach are the proximity and size of the food source and the characteristics of the surf zone delivering the food to the beach. The primary food types that are available on open sand beaches include particulate organic matter, carrion, dead large algae, dissolved organic material, plankton, and resident benthic microflora.

Open sand macrofaunal food chains generally include filter feeders, detritus feeders, scavengers, and carnivores. Filter feeders are more abundant on open sand beaches and detritus feeders on the more protected beaches. Scavengers and carnivores are rarely abundant. Fishes, birds, and certain large crustaceans dominate the top of open sand food chains (Fig. 6.36). In general, open sand beaches have a gradient in which terrestrial predators (mainly birds) dominate the upper shore zones and marine predators

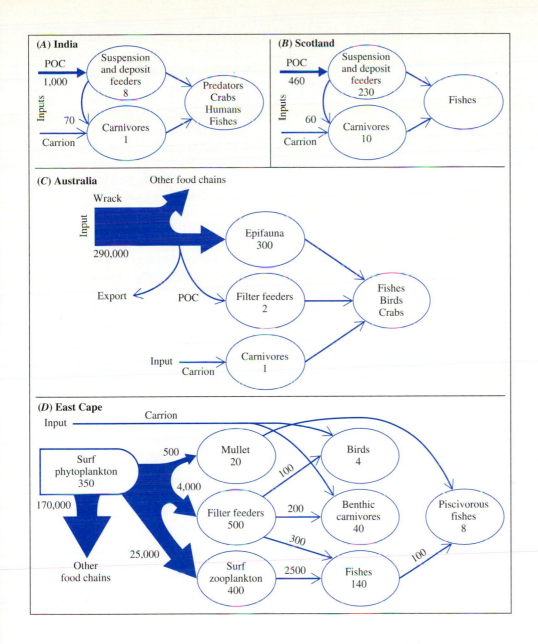

Figure 6.36 Macroscopic food chains and carbon flow in several different open sandy beach systems. *(A)* India. This is an example of a low-energy, mostly reflective beach, with negligible surf and no drift line. Macrofauna are dominated by particulate feeders, with crabs, fish, and humans as the main predators. *(B)* Scotland. An example of a low-energy dissipative beach, with the macrofauna dominated by particulate feeders, but with the dominant predators being fishes, particularly juvenile flatfishes. *(C)* Western Australia. Example of a low-energy, reflective beach without a surf zone, but with large inputs of dead algae (= wrack). The fauna here consists of large numbers of surface dwellers that consume wrack, small populations of hippid crabs and donacid bivalves in the midlittoral, and oxypodid crabs at the drift line. Predators are few. *(D)* South Africa, East Cape. Example of a high-energy beach tending toward the dissipative type and characterized by large numbers of surf diatoms supporting a rich fauna and more complex food chains. (All values are given in grams of carbon per meter per year.) (Modified from A. C. Brown and A. McLachlan, Ecology of sandy shores, Elsevier, 1990.)

(fish) the lower shores. These distributions are summarized in Table 6.1.

Community Organization on Protected Sand Flats

In contrast to the situation on open sand beaches, protected sand flats are accessible for ecological investigations, including various experimental manipulations. As a result, we know more about the structuring of the communities of sand flats. Both physical and biological factors are important.

Physical factors play a significant role in the organization of sand flat communities in a number of ways. In the first and most obvious way, the grain size of the sand may set limits for the organisms that can inhabit the area.

There is a tendency to have a complementary distribution of communities dominated by suspension feeders and deposit feeders due to the aforementioned trophic group amensalism (see p. 172), and to the tendency of deposit feeders to be more abundant in fine sediments, where the fine particles have more organic material and more nutrients for them. Deposit feeders usually dominate fine sands and muds, whereas suspension feeders are most abundant in clean coarse sand. This seems to hold true for the intertidal sand as well as for the subtidal, as described in Chapter 5.

Other direct effects are the result of seasonal changes in the weather. For example, in the New England intertidal sand flats a number of dominant epifaunal invertebrates are affected by the seasonal changes in the water temperature. As the water temperature declines in the fall, Whitlatch (1982) reports that many crustacean species migrate to deeper water, where they become inactive over winter. The same is true for the dominant gastropod, the mud snail *Ilynassa obsoleta*. Since *Ilynassa* has an impact both on algal mats and on sediment stability, and subsequently on infaunal abundance, this seasonal

Table 6.1   MAJOR BEACH TYPES BY CLIMATE ZONE, CONDITIONS, AND MACROFAUNAL COMPOSITION

Type	Latitude	Predominant food sources	Macrofaunal biomass	Dominant trophic group	Predators
High turnover, particulate-based food chain	Tropical	Particulates	Low to moderate	Filter feeders	Invertebrates, fishes
Low turnover, particulate-based food chain	Temperate	Particulates	Low to high	Filter feeders	Fishes, birds, invertebrates
Macrodebris-based food chain	Temperate	Stranded macrophytes	Moderate	Scavengers, grazers	Birds
Sediment-based food chains in sheltered beaches	Temperate or tropical	Deposited detritus microorganisms and benthic microflora	Low to high	Deposit feeders	Birds, fishes, invertebrates
Carrion-based food chains in extremely exposed steep beaches	Temperate or tropical	Stranded carrion	Low	Scavengers, predators	Birds, crabs

Source: From McLachlan, 1983.

migration can have signifiant effects on community structure (see Chapter 5, p. 172). In addition, particularly in high latitudes, the formation of sea ice and its scouring effect on the intertidal flats create changes in the community (see Chapter 5 on polar communities).

The biological factors that act to structure sand flat communities are primarily predation, competition, and disturbance. Of the three, current information suggests that either predation or predation in combination with disturbance is especially significant.

There are a number of sand flat epifaunal predators that may have significant effects on the community. For example, on the East Coast the moon snail *Polinices duplicatus* is an active predator of bivalves, particularly the soft shell clam *Mya arenaria*. Wiltse (1980) has demonstrated that when *Polinices* is excluded from areas, not only does the clam population rise, but so do the populations of other infauna. Thus, this snail affects community structure not only by feeding on clams, but also by disturbing the rest of the infauna by its digging activity as it plows through the substrate. Similar results have been reported by Peterson and Peterson (1979), who caged out predators in North Carolina tide flats. Another group of predators on the East Coast are various large crabs, such as *Carcinus maenas* in New England and *Callinectes sapidus* in the Middle Atlantic states. Virnstein (1977) has demonstrated that *Callinectes* is a voracious predator that also digs in the sediment; exclusion of the crab results in a great increase in infaunal density. A seasonal predator on eastern seaboard tidal flats is the horseshoe crab *Limulus polyphemus.* This animal digs distinctive pits when searching for infaunal prey, and Woodin (1978) has demonstrated that this also significantly reduces other infaunal organisms.

Other large East Coast invertebrate predators include several species of whelks (*Busycon* spp.) and tulip shells (*Fasciolaria*). The whelks, in particular, are devastating to the sand flat clam populations because they can consume all sizes. Unlike the barnacles and some mussels of the rocky intertidal, the clams have no refuge in size.

Despite the massive effects of the larger predators, there are areas that act as refuges from predation and disturbance by these animals. Woodin (1978), for example, has demonstrated that the tube-forming polychaete worm *Diopatra*

cuprea of the eastern United States intertidal sand flats, by forming upright tubes in the sand, effectively deters both *Limulus* and *Callinectes* to the extent that the infaunal abundances are greater in and around the tubes of *Diopatra.* All of the above strongly suggest that predation, and its accompanying disturbance, is the main biological structuring force in intertidal sand flats.

Another significant biological process determining community structure is, surprisingly, deposit feeding. In North Carolina one of the more abundant deposit feeders is the hemichordate worm *Balanoglossus aurantiacus.* This animal, which Peterson and Peterson (1979) have called a "funnel feeder," digs a U-shaped burrow and then collapses the head end. Thus, sediment continues to fall down to form a funnel-shaped depression directed at the head of the worm. The worm then ingests the sediment, consuming the smaller infaunal organisms and causing the death of those that it does not digest. The densities of these large worms and their continual reworking of the sand mean that they keep numbers of infaunal organisms low, as well as excluding potential competitors. In North Carolina the effect of *Balanoglossus* is heightened by the presence of yet another sediment processor in the form of the sea cucumber *Leptosynapta tenuis,* which operates at the sediment surface.

Adult-larval interactions probably also have an effect on the composition of sand flat communities. These interactions should probably be categorized as predation since the interaction consists of deposit or suspension feeders ingesting the settling or newly settled larvae. Woodin (1976) suggests that when adults are in dense assemblages such interactions could be important, but we currently lack any experimental studies to document this.

The role of competition in structuring sand flat communities seems minor. In direct contrast to the results of predator removal in the rocky intertidal, predator removal in soft substrates does not lead to increased competition and competitive exclusion. This is not to say that competition does not occur, for competition for space in sediments has been found by Levinton (1977) and Peterson (1977). The competition, however, does not seem to be a structuring force that determines a community makeup.

Another interesting contrast with rocky intertidal areas is provided by Peterson (1991), who notes that the major predators on rocky shores are the sedentary, slow-moving starfish and mollusks that are limited in effectiveness to the lower parts of the intertidal zone by their lack of mobility and need for coverage by water. In contrast, protected sand flats and mud flats are dominated by highly mobile predators, such as crabs, fishes, and birds (see subsequent section) that have feeding times less affected by the tides and that are capable of ranging throughout the intertidal.

We can summarize this section by stating that the macrofauna of intertidal sand flats tends to display rather high temporal and spatial variability. The total numbers of species and individuals may vary by several orders of magnitude within and between years. This high variability is due to the high predation pressure and biological disturbance that control the infaunal community structure to the extent that the majority of infaunal species tend to be opportunistic in their life history features. This, in turn, promotes wide variability.

MUDDY SHORES

Whereas sharp boundaries exist between rocky shore and sandy shore, such that they are easily defined and recognizable, a similar situation does not prevail when trying to differentiate between sand flats and muddy shores. Indeed, it is not possible to draw sharp boundaries between the two, because as shores become more protected from wave action, they tend to become finer grained and accumulate more organic matter, and thus become more "muddy." Sand and mud shores are therefore the opposite ends of a continuum, the sand beaches having larger grain sizes and the muddy areas the finest grain sizes.

Since sharp boundaries do not exist and one blends into the other along a gradient of increasing protection from wave action, the fauna and flora also show a change from organisms typical of open sand beaches to those typical of muddy shores along the same gradient.

This section will discuss the features characteristic of typical muddy shores; that is, those at the opposite end of the continuum from the open sand beaches. It should be borne in mind,

however, that various transitional communities exist that will show mixtures of organisms from both extremes. Since muddy shores are also composed of sediments, similar factors will apply here as did with sandy shores and will not be repeated.

Muddy substrates are also characteristic of estuaries and salt marshes (Chapter 8). Indeed, most of the muddy shores of the world are associated with estuaries and similar embayments. Hence, many of the types of organisms and adaptations are similar between the two areas. To avoid the problem of overlap in discussing these two areas, this section will treat the infaunal organisms of mud flats, their adaptations, feeding biology, and trophic structure. These considerations will, of course, apply to the estuarine infauna of Chapter 8. The role of other organisms, productivity, and ecological relationships are taken up in Chapter 8.

Physical Factors

The major contrast with open sand beaches is that muddy shores cannot develop in the presence of significant *wave action.* Therefore, muddy shores are restricted to intertidal areas completely protected from open-ocean wave activity. Muddy shores are best developed where there is a source of fine-grained sediment particles. Muddy shores are located in various partially enclosed bays, lagoons, harbors, and especially estuaries (see Chapter 8 for further discussion).

Since these areas are developed where water movement is minimal, the slope of mud shores tends to be much flatter than that observed for sand beaches. This is why these areas are often referred to as "mud flats." Furthermore, these areas are more stable than sand substrates and more conducive to the establishment of permanent burrows.

The very fine particle size, coupled with the very flat angle of repose of these sediments, means that water in the sediments does not drain away and is held within the substrate. This long retention time for water, together with a very poor interchange of the interstitial water with the seawater above, and a high internal bacterial population usually result in complete depletion of the oxygen in the sediments below the first few centimeters of the surface. *Anaerobic* conditions thus prevail within the sediment, and this is one of the most important characteristics of the muddy shore that serve to separate it from sand flats.

Between the upper aerobic layer and the lower anaerobic layer there occurs a transition zone called the *redox potential discontinuity* (RPD) layer. This zone is characterized by a rapid change from a positive redox potential (Eh), as measured by an electrode, to a negative redox potential. Below the RPD zone sediments are anaerobic and decomposition of organic compounds is by anaerobic bacteria, whereas above the RPD zone decomposition is by aerobic means. The RPD layer itself is usually characterized by a gray color of the sediments; the oxidized layer above is usually brown or yellow; and the anaerobic layer is black (Fig. 6.37). The RPD zone is significant biologically for several reasons. First, reduced compounds diffuse upward from below and, as soon as oxygen is available, bacteria oxidize these compounds; the oxidized end products—including CO_2, NO_3, and SO_4—in turn are incorporated into bacterial biomass and form the basis of new food chains. Some compounds, however, diffuse downward below the RPD zone and are then utilized by the anaerobic bacteria. These bacteria, in turn, pro-

duce more reduced compounds, which complete the cycle and also in the process release phosphate, which is important for new plant growth. Also present in the RPD zone are the chemoautotrophic bacteria, which by oxidizing the reduced compounds and fixing CO_2 produce more organic material (see following sections). Little wonder then that the RPD zone has high numbers of organisms inhabiting it.

Muddy shores tend to accumulate organic material, which means that there is an abundant potential food supply for the resident organisms; but the abundant small organic particles "raining" down on the mud flat also have the potential to clog respiratory surfaces.

Adaptations of Organisms

As we noted for sandy shores, the surface of muddy shores is also often rather barren, since few animals inhabit the surface of the mud flat. Most organisms inhabiting these areas show an adaptation to burrowing into and through the soft substrate or else inhabit permanent tubes in the substrate. In contrast to the open sand beaches, however, the presence of the organisms in the mud flat is advertised on the surface by

Figure 6.37 Diagrammatic representation of the physical and chemical characteristics of sediments across the redox discontinuity layer and the biological processes occurring in each. (From various sources.)

the presence of various holes of differing sizes and shapes. Thus, one of the primary adaptations of organisms in mud flats is the ability to burrow into the substrate or to form permanent tubes. The latter is related to the increased stability of the fine sediments.

A second major adaptation concerns the anaerobic conditions that prevail in the substrate. If organisms are to survive while burrowed into the substrate, they must either be adapted to live under anaerobic conditions or else must have some way of bringing the overlying surface water with its oxygen supply down to them. Since most multicellular organisms cannot survive in the absence of oxygen, the latter is the more common adaptation. It is to obtain this oxygen-rich surface water and food that the various burrows, holes, and tubes appear on the surface of the mud flat.

Although most mud flat organisms are intolerant of completely anaerobic conditions, many have adaptations that permit them to exist at lower oxygen tensions than similar forms that live on open sand beaches. The most common adaptation to low oxygen supply is the development of carriers (e.g., hemoglobin) that will continue to pick up oxygen at concentrations well below that of similar pigments in other organisms (Fig. 6.38). Others use glycogen stores for anaerobic metabolism during periods when there is little or no oxygen.

Since wave action is essentially absent on these mud flats, there is no necessity for development of either rapid burrowing or heavy bodies to maintain positions, as is the case on open sandy shores.

Types of Organisms

In contrast to sand beaches, muddy shores often develop a substantial growth of various plants. On the bare mud flats, the most abundant plants are diatoms, which live in the surface layers of mud and often give a brownish color to the surface at low tide. Other plants include large macroalgae such as species of *Gracilaria* (red algae), *Ulva,* and *Enteromorpha* (green algae). These large algae often undergo seasonal cycles of abundance, becoming common in the warmer months and virtually disappearing in colder months. Other areas, particularly the lowest tidal levels, may be covered with a growth of various seagrasses such as the genus *Zostera* (see Chapter 5 for a discussion of these areas). As a result of the occurrence of these various primary producers, there is substantial primary productivity in the mud flats.

Mud flats contain large numbers of bacteria, which feed on the abundant organic matter. Bacteria are the only abundant organisms found in the anaerobic layers of the mud shore and constitute a significant biomass. Among the dominant bacteria inhabiting this area are a number capable of utilizing the potential energy of the various reduced chemical compounds abundant here. These *chemosynthetic* or *sulfur bacteria* obtain energy through the oxidation of a number of reduced sulfur compounds, such as

Figure 6.38 Oxygen dissociation curves for the blood of *Arenicola* (A) and *Nepthys* (NV). (Modified from A. E. Brafield, Life in sandy shores, 1978. From J. D. Jones, J. Exp. Biol. 32:110–125, 1955.)

various sulfides (for example, H_2S). These organisms are thus primary producers of organic matter analogous to green plants. They produce organic matter using energy obtained from the oxidation of the reduced sulfur compounds, whereas plants produce organic matter using energy obtained from sunlight.

Since these autotrophic bacteria are located in the RPD layer of the mud, mud flats are unique among marine environments in that they have two separate layers in which primary productivity occurs: the surface, where diatoms, algae, and marine grasses carry on photosynthesis; and a deep layer, where bacteria conduct chemosynthesis.

The dominant macrofaunal groups on muddy shores are the same as those encountered on sand beaches and sand flats—namely, various polychaete worms, bivalve mollusks, and small and large crustaceans, but of quite different sorts.

Feeding Biology and Trophic Structure

Because of the greater amount of organic matter present in and on muddy shores and because of increased productivity due to both bacteria and plants, there is vastly more food available on muddy shores than on sand beaches. This permits more large organisms to live on mud shores and, indeed, mud flats may be very densely populated.

The dominant feeding types on mud flats are deposit feeders and suspension feeders, a situation somewhat different from that on open sand beaches. Deposit feeders are particularly abundant due to the large amount of organic material and the large populations of bacteria in the sediments. Deposit-feeding polychaetes are represented by genera such as *Arenicola* and *Capitella*. Both feed by burrowing through the substrate, ingesting it and digesting out the organic matter (or bacteria), and passing out the undigested material through the anus. *Arenicola* spp., however, form a U-shaped burrow in which one arm of the U is a permanently open shaft to the surface, while the other arm is filled with sediment that the worm ingests (Fig. 6.39). The worm consumes sediment from the filled shaft, passes it through its gut, and then moves up in the open burrow to defecate at the surface, leaving characteristic mounds. Capitellids, on the other hand, form no permanent tubes, but move like aquatic earthworms through the surface layers of the substrate, ingesting it. Other deposit feeders include the surface-feeding terebellid polychaete worms and the burrowing hemichordates.

Deposit-feeding bivalves are also common in these areas. Temperate-zone mud flats often have large numbers of small tellinid clams of the genus *Macoma* or *Scrobicularia*. These bivalves have siphons that are separated. Each clam lies buried in the substrate with its siphons extended to the surface. There, the intake siphon is moved over the surface like a vacuum cleaner, ingesting organic particles and bringing them down to the clam for digestion (Fig. 6.40).

Suspension feeders include various other species of clams, some crustaceans, and numbers of

Head shaft   Tail shaft

Fecal mass

Gallery

Figure 6.39 Two common polychaete worms of mud flats: *Arenicola* (right) in its U-shaped burrow, and *Capitella* (left) burrowing through the substrate.

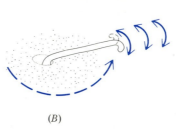

Figure 6.40 *(A) Macoma nasuta* in the substrate. *(B) Macoma nasuta* feeding with its in-current siphon. (Modified from A. E. Brafield, Life in sandy shores, Studies in biology no. 89, Edward Arnold, 1978.)

polychaetes. Suspension-feeding mechanisms are not different from those observed in other areas. There is, however, some question as to how exclusively these forms are feeding on suspended plankton. The fine sediment particles are easily stirred up on mud flats and remain resuspended in the water. Hence, suspension feeders may possibly take in amounts of these particles in addition to various plankton organisms. Perhaps most of the suspension feeders on mud flats also partially take in resuspended sediments and hence are actually feeding on both deposited and suspended material. In general, deposit feeders are more common in fine-grained shores and suspension feeders become more abundant in coarser sediments, where there is little organic matter (Fig. 6.41).

The major carnivores on mud shores are often the fishes, which feed when the tide is in, and the birds, feeding when the tide is out (Fig. 6.42). Indigenous mud flat predators include a few polychaete worms such as *Glycera* spp., moon snails (*Polinices, Natica*), nemertean worms, and crabs.

Despite the relative abundance of plant material on muddy shores, there are very few herbivores, and most plant material finds its way into the food chains only after it has been broken down into small pieces to enter the deposit-feeding food web as detritus.

Thus, the trophic structure of a mud flat is often built up from two bases: a detritus-bacteria base and one based on plants. The detritus base web is derived from plants and other organic

sources and includes the bacteria that live upon the detrital particles. Detrital particles can thus either be taken in directly by the large macrofaunal invertebrates or else be consumed by bacteria. An additional bacterial component at the base of this trophic pyramid is the sulfur bacteria.

Bacteria are consumed commonly by various nematode worms, which thus occupy a position similar to that of herbivores in other areas. Bacteria are also consumed by deposit feeders as they ingest the organic particles on which the bacteria are found. Nematodes and deposit feeders are, in turn, consumed by various carnivores, including predatory invertebrates (moon snails, polychaetes = *Glycera*) as well as birds and fishes (Fig. 6.43).

A second food web is based primarily on the microscopic diatoms as the autotrophic base. The diatoms are consumed by several different polychaetes, mollusks, and crustaceans (Fig. 6.43). These, in turn, are consumed by the large predatory birds and fishes.

Zonation and Community Structure

Little information exists regarding the zonation of mud flats. The very gentle slope of these areas means that the intertidal is often very extensive, more so than rocky or sandy shores. The upper area, the supralittoral fringe, is often inhabited by various species of crabs, many of which burrow into the substrate. The very extensive

Figure 6.41 Some representative New England mud flat benthic invertebrates. Suspension feeder: (B) *Mya arenaria* (soft-shelled clam). Surface deposit feeders: (A) *Polydora ligni* (spionid polychaete); (D) *Corophium* spp. (gammaridean amphipod); (H) *Hydrobia totteni* (hydrobid gastropod); (I) *Streblospio benedicti* (spionid polychaete); (J) *Macoma balthica* (bivalve); (L) *Ilyanassa obsoleta* (mud snail). Burrowing omnivore: (C) *Nereis virens* (nereid polychaete). Burrowing deposit feeders: (E) *Tharyx* sp. (cirratulid polychaete); (F) *Lumbrineris tenuis* (lumbrinerid polychaete); (G) *Heteromastus filiformis* (capitellid polychaete); (K) oligochaete. (From R. Whitlatch, The ecology of New England tidal flats: A community profile, U.S. Fish & Wildlife Serv., 1982.)

midlittoral area is the home of most of the common species of clams and polychaetes. There is no sharp boundary with the infralittoral fringe, and similar organisms are encountered.

The role of physical factors and biological factors in structuring these communities has been studied in a number of places. Structuring forces are similar to those acting in sand flats.

Figure 6.42 Birds feeding on a mud flat. (Photo courtesy of Bruce Stewart.)

275

Figure 6.43 Generalized food web of a muddy shore. (From S. K. Eltringham, Life in mud and sand, Crane Russak, 1971.)

Various biological factors seem to be significant, with predation being perhaps the most important. Since many of the predators are similar to those described for sand flats and since these areas grade into each other, it is not necessary to repeat that material here. Suffice it to say that both sand and mud flats seem as of now to be similarly organized.

INTERTIDAL FISHES

Whereas a great deal of research has been concentrated on the ecology of invertebrates and plants of the intertidal zone on all three types of shores, there are relatively few accounts of the fishes of these areas or of the role they may play in community organization as grazers or predators. Since fishes are often present in considerable numbers in the intertidal and since it is known that they have a significant effect on other communities such as coral reefs (see Chapter 9) and in various subtidal communities (see Chapter 5), it is reasonable to assume that future research may elucidate their role here as well.

Most rocky intertidal fishes, because of the turbulent environment, are of small size. The body shape is usually compressed and elongate (Blenniidae, Pholidae) or depressed (Cottidae, Gobiesocidae), which allows them to inhabit holes, tubes, crevices, or depressions for protection against both desiccation and wave action (Fig. 6.44). Most also lack swim bladders and are closely associated with the substrate. Many of these fishes are adapted to withstand greater ranges of salinity and temperature than their subtidal relatives. A few are even adapted to spend some time out of the water (see Chapter 9 on *Periophthalmus*).

Most rocky intertidal fishes in the temperate zone have been shown to be visual carnivores, again suggesting a potentially significant role in intertidal community organization.

The life history patterns of the few species that have been investigated are all largely similar. Eggs are demersal and laid on stones, rocks, or submerged vegetation. Often, the eggs are guarded by the male. The eggs hatch after a few weeks into planktonic larvae. The planktonic period varies, depending on the species. It may be as long as two months. During this period, the larvae gradually acquire the adult features and, at the end, become demersal. Life spans of the adults are generally short, from two to ten years, and sexual maturity occurs in the first or second year.

Figure 6.44 Examples of rocky intertidal fishes from the four dominant families: *(A)* Blenniidae (blennies). *(B)* Gobiidae (gobies). *(C)* Gobiesocidae (clingfishes). *(D)* Cottidae (cottids). (Modified from R. N. Gibson, The biology and behavior of littoral fish, Oceanogr. Mar. Biol. Ann. Rev. 7:367–410.)

On mud and sand flats, skates and rays as well as various species of flatfishes are significant predators of infaunal organisms. Rays are particularly common on intertidal flats. Their method of foraging, by excavating depressions or pits, means that they cause widespread mortality among the infaunal organisms. Their effect extends beyond the species they consume. Digging uncovers many small infaunal species that are either killed by the disturbance or consumed by the smaller fishes that follow the rays. Flatfishes also may prey on various infaunal invertebrates, but their role in community structure seems more species-specific. Some have a minimal effect because they feed on few important infaunal species, whereas others reduce both abundance and diversity of species in a community. For example, Virnstein (1977) demonstrated in Chesapeake Bay that the hogchoker (*Trinectes maculatus*) had a minimal effect on the infauna, whereas the spot (*Leiostomus xanthurus*) significantly reduced the infaunal numbers. Some intertidal fishes are migratory, moving either tidally, diurnally, or seasonally.

BIRDS

At low tide, a considerable variety of birds are often associated with the intertidal zone. They are particularly conspicuous on intertidal sand and mud flats and much less so in rocky intertidal areas. Sandy and muddy intertidal flats are exploited by a large number of different shorebird species (Fig. 6.45). These are mainly predators of infaunal invertebrates and, given their abundance and high food requirements, one might wonder how so many predatory species can coexist in a single flat without severely depleting the infauna or undergoing severe competition. There are several explanations. In the temperate zone, most of these birds are migratory and are only present during certain portions of the year. While they are present, they may locally deplete the infauna, but it has time to recover during the periods when they are absent. Another explanation resides in the observation that tidal flats are not uniform in community structure, as we have seen. Different infaunal associations are found in different areas due to differences in grain size and due to various succession patterns that follow disturbances. The infauna is also segregated by depth in the sediment, and perhaps by tide level as well. There exists, in other words, a variety of different microhabitats. If the bird species differ in food preference or microhabitat preference, then the various species may be separated in space or time, thereby reducing competition and the chance that resources will become depleted. This seems to be the case. For example, Harrington and Schneider (1978) have shown that sanderlings prefer sandy substrates, while dowitchers

Figure 6.45 New England tidal flat birds showing general feeding positions of the various groups. (From R. Whitlatch, The ecology of New England tidal flats: A community profile. U.S. Fish & Wildlife Serv., 1982.)

are more abundant on mud. Similarly, temporal segregation has been demonstrated among sanderlings and semipalmated plovers. Sanderlings characteristically follow the water's edge as the tide recedes, whereas semipalmated plovers forage in the middle intertidal zone (Harrington et al., 1974). Evans et al. (1979) have shown that competition between knots and dunlins is reduced because knots eat mollusks, whereas dunlins eat polychaetes. Morphological features of the birds, with respect to different leg and bill lengths, also influence where the birds are able to feed. Finally, behavior patterns dictate what food resources may be exploited.

Given that birds are present and abundant and that they do consume intertidal invertebrates, how important are they in structuring communities? It seems, on the basis of most studies, that they can be very important. Evans et al. (1979), for example, reported that wintering shorebirds were responsible for removing 90 percent of the *Hydrobia* snail population and 80 percent of the nereid polychaetes from an English mud flat. On the East Coast of the United States, Harrington and Schneider (1978) reported that the whole group of foraging shorebirds eliminated 50 to 70 percent of the invertebrate population. On the Pacific coast of North America, Quammen (1984) demonstrated that shorebirds had a significant effect on the densities of infaunal invertebrates on mud flat areas but little or no effect on sand flat invertebrate densities.

A special case may exist on "bird islands," those isolated small islands on which birds congregate in great numbers for breeding. In such situations, as Bosman and Hockey (1986) have noted, the greater nutrient input to the intertidal via bird guano leads to enhanced algal production and may, if there are predators of the herbivorous molluscan grazers, such as oystercatchers, lead to an intertidal dominated by foliose algae.

We can summarize by saying that bird predation may be intense and significant, particularly at certain times, and certainly plays an important role in structuring tidal flat communities (Fig. 6.46).

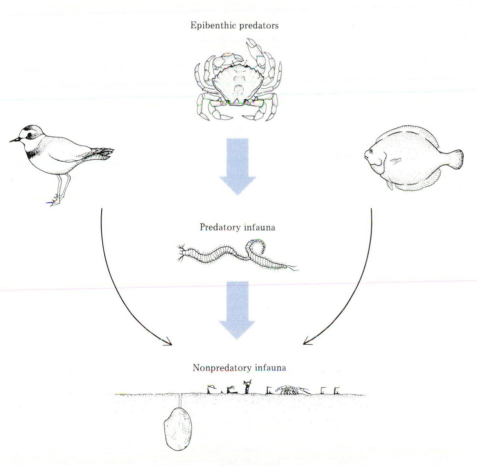

Figure 6.46 Diagrammatic representation of the interactions between various epibenthic predators and the infauna of sedimentary shores. Broad arrows indicate relations that may be particularly important. (From W. G. Ambrose, Role of predatory infauna in structuring marine soft bottom communities, Mar. Ecol. Prog. Ser. 17:102–115, 1984.)

BIBLIOGRAPHY

Ansell, A. D., P. Sivades, B. Narayanan, and A. Trevallion. 1972. The ecology of two sand beaches in southwest India. II. Observations on the populations of *Donax incarnatus* and *Donax spiculum.* Mar. Biol. 17:318–322.

Bosman, A. L., and P. A. R. Hockey. 1986. Seabird guano as a determinant of rocky intertidal community structure. Mar. Ecol. Prog. Ser. 32(2–3):247–257.

Boyle, P. R. 1969. The survival of osmotic stress by *Sypharochiton pelliserpentis.* Biol. Bull. 136:154–165.

Brafield, A. E. 1978. Life in sandy shores. Studies in biology no. 89. Edward Arnold, London. 60 pp.

Bridges, C. A. 1988. Respiratory adaptations in intertidal fishes. Amer. Zool. 28:79–96.

Brown, A. C., and A. McLachlan. 1990. Ecology of sandy shores. Elsevier, N.Y. 328 pp.

Carefoot, T. 1977. Pacific seashores, a guide to intertidal ecology. University of Washington Press, Seattle. 208 pp.

Castenholz, R. W. 1961. The effect of grazing on marine littoral diatom populations. Ecology 42:783–794.

Chia, F. S. 1973. Sand dollar: A weight belt for the juvenile. Science 181:73–74.

Connell, J. H. 1961. The influence of interspecific competition and other factors on the distribution of the barnacle *Chthalamus stellatus.* Ecology 42:710–723.

Connell, J. H. 1970. A predator-prey system in the marine intertidal region. I, *Balanus glandula* and several predatory species of *Thais.* Ecol. Monogr. 40:49–78.

Connell, J. H. 1975. Some mechanisms producing structure in natural communities: A model and evidence from field experiments, pp. 460–490. *In:* Cody, M. L., and J. M. Diamond (eds.). Ecology and evolution of communities. Harvard University Press, Cambridge, Mass.

Davis, P. S. 1969. Physiological ecology of *Patella.* III, Desiccation effects. Mar. Biol. Assoc. U.K. 49:291–304.

Dayton, P. K. 1971. Competition, disturbance and community organization: The provision of and subsequent utilization of space in a rocky intertidal community. Ecol. Monogr. 41:351–389.

Dayton, P. 1973. Dispersion, dispersal and persistence of the annual intertidal alga *Postelsia palmaeformis* Ruprecht. Ecology 54(2):433–438.

Dayton, P. K. 1975. Experimental evaluation of ecological dominance in a rocky intertidal algal community. Ecol. Monogr. 45:137–159.

Denley, E. J., and A. J. Underwood. 1979. Experiments on factors influencing settlement, survival and growth of two species of barnacles in New South Wales. J. Exp. Mar. Biol. Ecol. 36:269–293.

Doty, M. S. 1946. Critical tide factors that are correlated with the vertical distribution of marine algae and other organisms along the Pacific coast. Ecology 27:315–328.

Eltringham, S. K. 1971. Life in mud and sand. Crane Russak, N.Y. 218 pp.

Evans, P. R., D. M. Henderson, T. J. Knights, and M. W. Pienkowski. 1979. Short term effects of reclamation of part of Seal Sands, Teesmouth, on wintering waders and shelduck. I, Shorebird diets, invertebrate densities and the impact of predation on the invertebrates. Oecologia 41:183–206.

Foster, B. A. 1971. On the determinants of the upper unit of intertidal distribution of barnacles (Crustacea, Cirripedia). J. Anim. Ecol. 40:33–48.

Francis, L. 1973. Intraspecific aggression and its effect on the distribution of *Anthopleura elegantissima* and some related anemones. Biol. Bull. 144:73–92.

Frank, P. W. 1965. The biodemography of an intertidal snail population. Ecology 46:831–844.

Gaines, S. D., and J. Lubchenco. 1982. A unified approach to marine plant-herbivore interactions. II. Biogeography. Ann. Rev. Ecol. Sys. 13:111–138.

Gibson, R. N. 1969. The biology and behavior of littoral fish. Oceanogr. Mar. Biol. Ann. Rev. 7:367–410.

Harrington, B. A., S. K. Groves, and N. T. Houghton. 1974. Season progress report. Massachusetts shorebird studies. Contract 14-16-008-687 U.S. Fish & Wildlife Service, Manomet, Mass.

Harrington, B. A., and D. C. Schneider. 1978. Studies of shorebirds at an autumn migration stopover area. Final Report for U.S. Fish & Wildlife Service, Migratory Bird and Habitat Research Laboratory, Laurel, Md.

Haven, S. B. 1971. Niche differences in the intertidal limpets *Acmaea scabra* and *Acmaea digitalis* (Gastropoda) in central California. The Veliger 13:231–248.

Hawkins, S. J. 1981. The influence of season and barnacles on the algal colonization of *Patella vulgata* exclusion areas. J. Mar. Biol. Assoc. U.K. 61:1–15.

Hawkins, S. J., and R. G. Hartnoll. 1983. Grazing of intertidal algae by marine invertebrates. Oceanogr. Mar. Biol. Ann. Rev. 21:195–282.

Hay, C. 1979. Some factors affecting the upper limit of the southern bell kelp *Durvillaea antarctica* (Chamisso) Hariot on the New Zealand shores. J. Roy. Soc. N.Z. 9:279–289.

Hay, M. E. 1981. The functional morphology of turf forming seaweeds: Persistence in stressful marine habitats. Ecology 62:739–750.

Hedgpeth, J. (ed.). 1957. Treatise on marine ecology and paleoecology. Vol. 1, Ecology. Geol. Soc. America Memoir 67. 1296 pp.

Himmelman, J. H., and D. H. Steele. 1971. Food and predators of the green sea urchin *Strongylocentrotus droebachiensis* in Newfoundland waters. Mar. Biol. 9:315–322.

Jones, N. S. 1948. Observations and experiments on the biology of *Patella vulgata* at Port St. Mary, Isle of Man. Liverpool Biol. Soc. Proc. Trans. 56:60–77.

Jones, N. S., and J. M. Kain. 1967. Subtidal algal colonization following the removal of *Echinus*. Helg. Wiss. Meeresunt. 15:460–466.

Kanwisher, J. 1957. Freezing and drying in intertidal algae. Biol. Bull. 113:275–285.

Keough, M. J., and A. J. Butler. 1979. The role of asteroid predators in the organization of a sessile community on pier pilings. Mar. Biol. 51:167–177.

Kozloff, E. N. 1983. Seashore life of the northern Pacific Coast. University of Washington Press, Seattle. 370 pp.

Levinton, J. S. 1977. Ecology of shallow water deposit feeding communities, pp. 191–227. *In:* Coull, B. C. (ed.). Ecology of the marine benthos. University of South Carolina Press, Columbia.

Lewis, J. R. 1964. The ecology of rocky shores. The English Universities Press, Ltd., London. 323 pp.

Littler, M. M., and D. S. Littler. 1980. The evolution of thallus form and survival strategies in benthic marine macroalgae: Field and laboratory tests of a functional form model. Amer. Nat. 116:25–44.

Lubchenco, J. 1978. Plant species diversity in a marine intertidal community: Importance of herbivore food preference and algal competitive abilities. Amer. Nat. 112(983):23–39.

Lubchenco, J. 1980. Algal zonation in the New England rocky intertidal community: An experimental analysis. Ecology 61:333–344.

Lubchenco, J., and B. Menge. 1978. Community development and persistence in a low rocky intertidal zone. Ecol. Monogr. 59:67–94.

Lubchenco, J., and S. D. Gaines. 1981. A unified approach to marine plant-herbivore interactions. I. Populations and communities. Ann. Rev. Ecol. Sys. 12:405–437.

Lubchenco, J., B. A. Menge, S. D. Garrity, P. J. Lubchenco, L. R. Ashkenas, S. D. Gaines, R. Emlet, J. Lucas, and S. Strauss. 1984. Structure, persistence, and role of consumers in a tropical intertidal community (Taboguilla Island, Bay of Panama). J. Exp. Mar. Biol. Ecol. 78:23–73.

MacGinitie, G. E., and N. MacGinitie. 1949. Natural history of marine animals. McGraw-Hill, N.Y. 473 pp.

Mann, K. H. 1972. Macrophyte production and detritus food chains in coastal waters. Mem. Inst. Ital. Idrobiol. 29(suppl.):353–383.

McLachlan, A. 1983. Sandy beach ecology, pp. 321–380. *In:* McLachlan, A., and T. Erasmus (eds.). Sandy beaches as ecosystems. Dr. W. Junk, The Hague. 757 pp.

McLusky, D. S., S. A. Nair, A. Stirling, and R. Bharoava. 1975. The ecology of a central west Indian beach with particular reference to *Donax incarnatus*. Mar. Biol. 30:267–270.

Menge, B. A. 1976. Organization of the New England rocky intertidal community: Role of predation, competition and environmental heterogeneity. Ecol. Monogr. 46(4):355–393.

Menge, B. A. 1991. Relative importance of recruitment and other causes of variation in rocky intertidal community structure. J. Exp. Mar. Biol. Ecol. 146:69–100.

Menge, B. A., J. Lubchenco, L. R. Ashkenas, and F. Ramsey. 1986. Experimental separation of the effects of consumers on sessile prey in the low zone of a rocky shore in the Bay of Panama: Direct and indirect consequences of food web complexity. J. Exp. Mar. Biol. Ecol. 106:225–269.

Moran, M. J. 1980. The ecology and effects on prey of the predatory gastropod, *Morula marginalba*. Ph.D. dissertation, University of Sydney.

Newell, R. C. 1970. Biology of intertidal animals. Elsevier, N.Y. 555 pp.

Newell, R. C. 1979. Biology of intertidal animals. Marine Ecological Surveys Ltd., Faversham, Kent, U.K. 781 pp.

Paine, R. T. 1966. Food web complexity and species diversity. Amer. Nat. 100:65–75.

Paine, R. T. 1974. Intertidal community structure: Experimental studies on the relationship between a dominant competitor and its principal predator. Oecologia 15:93–120.

Paine, R. T. 1979. Disaster, catastrophe and local persistence of the sea palm *Postelsia palmaeformis.* Science 205:685–687.

Paine, R. T. 1988. Habitat suitability and local population persistence of the sea palm *Postelsia palmaeformis.* Ecology 69(6):1784–1794.

Paine, R. T., J. C. Castillo, and J. Cancimo. 1985. Perturbation and recovery patterns of starfish dominated intertidal assemblages in Chile, New Zealand and Washington State. Amer. Nat. 125(6):679–691.

Peterson, C. H. 1977. Competitive organization of the soft bottom macrobenthic communities of southern California lagoons. Mar. Biol. 43:343–359.

Peterson, C. H. 1991. Intertidal zonation of marine invertebrates in sand and mud. Amer. Sci. 79(3):236–249.

Peterson, C. H., and N. Peterson. 1979. The ecology of intertidal flats of North Carolina: A community profile. U.S. Fish & Wildlife Service, Office of Biological Services. 73 pp.

Petraitis, P. S. 1987. Factors organizing rocky intertidal communities of New England: Herbivory and predation in sheltered bays. J. Exp. Mar. Biol. Ecol. 109:117–136.

Quammen, M. 1984. Predation by shorebirds, fish and crabs on invertebrates in intertidal mudflats: An experimental test. Ecology 65:529–537.

Ricketts, E., and J. Calvin. 1985. Between Pacific tides, 5th ed., rev. by D. W. Phillips. Stanford University Press. 652 pp.

Segal, E., and P. Dehnel. 1962. Osmotic behavior in an intertidal limpet, *Acmaea limatula.* Biol. Bull. 122:417–430.

Sousa, W. 1980. Experimental investigations of disturbance and ecological succession in a rocky intertidal algal community. Ecol. Monogr. 49:227–254.

Southward, A. J. 1965. Life on the seashore. Harvard University Press, Cambridge, Mass. 151 pp.

Southward, A. J., and C. E. Southward. 1978. Recolonization of rocky shores in Cornwall after use of toxic dispersants to clean up the *Torrey Canyon* spill. J. Fish. Res. Bd. Canada 35:682–706.

Stanley, S. 1969. Bivalve mollusk burrowing aided by discordant shell ornamentation. Science 166:634–635.

Steneck, R. S. 1982. A limpet-coralline alga association: Adaptations and defenses between a selective herbivore and its prey. Ecology 63:507–522.

Stephenson, T. A., and A. Stephenson. 1949. The universal features of zonation between tide-marks on rocky coasts. J. Ecol. 37:289–305.

Stimson, J. 1970. Territorial behavior in the owl limpet, *Lottia gigantea.* Ecology 51:113–118.

Timko, P. 1976. Sand dollars as suspension feeders: A new description of feeding in *Dendraster excentricus.* Biol. Bull. 151(1):247–259.

Underwood, A. J. 1980. The effects of grazing by gastropods and physical factors on the upper limits of distribution of intertidal macroalgae. Oecologia 46:201–213.

Underwood, A. J., and E. J. Denley. 1984. Paradigms, explanations and generalizations in models for the structure of intertidal communities on rocky shores, pp. 151–180. *In:* Strong, D. R., D. Simberloff, G. Abele, and A. B. Thistle (eds.). Ecological communities: Conceptual issues and the evidence. Princeton University Press, Princeton, N.J.

Underwood, A. J., and P. Jernakoff. 1981. Effects of interactions between algae and grazing gastropods on the structure of a low shore intertidal algal community. Oecologia 48:221–233.

Vadas, R. L. 1968. The ecology of *Agarum* and the kelp community. Ph.D. dissertation, University of Washington, Seattle.

Virnstein, R. W. 1977. The importance of predation by crabs and fishes on benthic infauna in Chesapeake Bay. Ecology 58:1199–1217.

Voss, G. L. 1976. Seashore life of Florida and the Caribbean. E. A. Seeman, Miami. 168 pp.

Whitlatch, R. 1982. The ecology of New England tidal flats: A community profile. U.S. Fish & Wildlife Service, Biological Services Program. 125 pp.

Wiltse, W. I. 1980. Effects of *Polinices duplicatus* (Gastropoda: Naticidae) on infaunal community structure at Barnstable Harbor, Massachusetts. Mar. Biol. 56:301–310.

Woodin, S. A. 1976. Adult-larval interactions in dense infaunal assemblages: Patterns of abundance. J. Mar. Res. 34:25–41.

Woodin, S. A. 1978. Refuges, disturbance and community structure: A marine soft bottom example. Ecology 59:274–284.

Chapter 7

Meiofauna

*I*n addition to the large macrofaunal organisms that inhabit sedimentary shores, there is another world that exists here, comprising those organisms that occupy the microspaces between adjacent particles or live on the individual particles. These are the *meiofaunal* or *interstitial* organisms. The existence of this unique association of organisms was not realized by biologists until the twentieth century, when European scientists began investigations into these areas. Since then, many scientists throughout the world have made studies of this unique habitat. The first studies were primarily taxonomic, to determine what kinds of organisms were present and to name them. More recent ecological studies have given us a better understanding of the peculiar conditions under which these animals live.

An interstitial fauna and flora assemblage has been found in intertidal and subtidal sedimentary substrates throughout the world, in both fresh water and salt water. Major interstitial groups have been found by Coull et al. (1977) down to water depths as great as 5000 m. Studies of these associations are, however, unequally distributed. The marine interstitial fauna is better known than the freshwater interstitial fauna, and the organisms and associations of the European seas are the best known of the various geographical areas. The Pacific coast of North America is practically unknown and unstudied with respect to interstitial fauna, whereas the Atlantic coast of North America has been the site of numerous studies.

This chapter will investigate the principles that govern meiofaunal life in marine sediments.

Definitions

As with any discipline or subdiscipline, there are a few technical terms that are in general use. *Interstitial* is a general term that refers to the spaces between the sedimentary particles and is also used as a synonym to refer to the organisms living there. *Meiofauna* is the preferred term to refer to the organisms that live interstitially. It includes all animals that pass through a 0.5-mm screen but are retained on a 62-μm screen. A synonym is *meiobenthos*. Since the term meiofauna simply defines a size class, but does not specify a spatial location or how the organisms move within the habitat, and since the interstitial space is a function of the size of the particles, a few other terms have been used to refine the classification and indicate location. *Endobenthic* is a term used now to refer to meiofaunal-sized organisms that move within the sediment by displacing particles (i.e., they are bigger than the interstitial space). *Mesobenthic* refers to meiofaunal organisms living and moving within the interstitial space, and *epibenthic* refers to those meiofauna living at the sediment-water interface. In this scheme mesobenthic is virtually equivalent to meiofauna. In this chapter we deal primarily with meiobenthos operating as mesobenthic animals, but in very fine sediments typically meiobenthic organisms such as nematodes, kinorhynchs, copepods, and others will displace sediment grains as they move, thus operating as endobenthic animals.

ENVIRONMENTAL CHARACTERISTICS

The conditions that affect the meiofauna are somewhat different from those that affect the macrofauna in the same areas. Perhaps the most important factor influencing the presence, absence, and types of interstitial organisms is *grain size*. Grain size is of paramount importance in determining the amount of interstitial space available for habitation. The coarser the grain size, the greater the volume of interstitial space and hence the larger the interstitial organisms that can inhabit the area. Conversely, the finer the grain size, the lesser the space available and the smaller the organisms must be to inhabit the

area. Grain size may thus act as a definite barrier to the movement of meiofaunal organisms. Certain organisms will be confined to sediments of certain grain size or areas of a given environment simply because they are too large to fit between the grains of adjacent areas and therefore cannot disperse through the whole area. Meiofaunal organisms show a definite zonation based on grain size. For example, sediments in which the median particle size is less than 125 μm are dominated by burrowing organisms such as kinorhynchs, while sediments with particle sizes above 125 μm are dominated by forms that move in the interstices.

Grain size is also of great importance because it controls the ability of the sediment to retain and circulate water. Meiofaunal organisms are aquatic and therefore require the presence of water in the interstices of the substrate in order to survive (Fig. 7.1). If the grain size becomes too coarse, when the tide ebbs, the water will not be held in the sand by capillary action, but will drain away, leaving only a very thin layer coating the grains. On the other hand, fine-grained sediments are able to hold considerable water in the interstices through capillary action.

Circulation of water through the pore spaces in the sediment is important because this water movement is responsible for renewing the oxygen supply. Circulation is best in coarse-grained sediments and is reduced in fine-grained areas. If the grain size becomes really fine, as in mud shores, the circulation practically ceases, and the result is the creation of anaerobic layers in the sediment (see discussion of RPD layer in Chapter 6). The maximum number of meiofauna have been found where sand grain diameters were between 0.175 and 0.275 mm. Fenchel (1978) reports that the interstitial fauna tends to disappear when the median grain size diameter drops below 0.1 mm.

In addition to grain size, the mineral nature of the grains is also of importance in determining the composition of the meiobenthos. Different associations of organisms may be found in siliceous sand as opposed to carbonate sand.

As suggested above, *oxygen* is an important factor in this environment. Practically all marine sediments have an oxidized layer at the surface, beneath which lies a completely anoxic layer of different chemical composition. Thus,

Water (% of wet weight)

Salinity

2 ‰

< 2 ‰

4 ‰

16 ‰
15 ‰
14 ‰
13 ‰

☐ < 5%

☐ 5–10%

☐ 10–15%

☐ 15–20%

■ 20% (saturation)

Oxygen availability (g $O_2 10^{-7} cm^{-2} min^{-1}$)

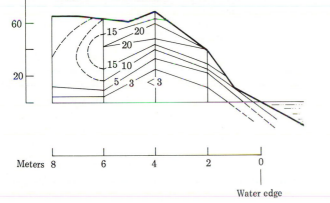

60

20

15   20
20
15 10
5   3   < 3

Meters 8   6   4   2   0

Water edge

Figure 7.1 A transect of a Danish sand beach showing gradients of water, salinity, and oxygen at low tide. (From Fenchel et al., Vertical and horizontal distribution of the metazoan microfauna and of some physical factors on a sandy beach in the northern part of the Oresund, Ophelia 4: 227–243.)

meiofaunal organisms living below a certain depth will encounter oxygen-free conditions. The thickness of the oxygenated layer depends on a number of factors such as grain size, amount of organic material, water turbulence, and bacterial metabolism.

Temperature is another environmental factor of importance in determining presence or absence and distribution of meiofaunal organisms. Temperature ranges are most extreme in intertidal beaches and in the uppermost layers of the beach. Temperature changes are minimal in subtidal sediments and in intertidal sediments below 10–15 cm in depth (Fig. 7.2). The surface layer of the sand on a beach acts as an insulator

to the lower layers, effectively dampening any significant temperature changes. Within the surface layers, however, the temperature may change markedly, depending on the air temperature, the effect of wind and rain, the amount of sunlight hitting the surface, and the temperature of the seawater. In cold temperate and polar regions, the temperature may be low enough to freeze the upper layers. Some meiofaunal organisms have apparently adapted to survive being frozen (e.g., turbellarian *Coronhelmis lutheri*, gastrotrich *Turbanella hyalina*, oligochaete *Marionina southerni*, and harpacticoid copepod *Parastenocaris phyllura*). Because of the range of temperatures that may occur in the upper layers,

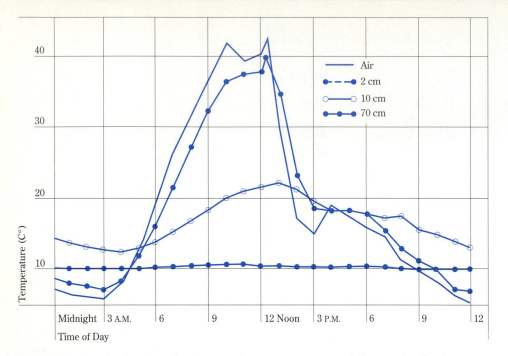

Figure 7.2 Graph showing the change in temperature at different depths in a sand beach in Denmark during a spring day. (From B. O. Jansson, Diurnal and annual variations of temperature and salinity of interstitial water in sandy beaches, Ophelia 4:173–201, 1967.)

the fauna inhabiting these areas may be different from those of lower layers and includes those with a considerable tolerance for temperature change (= eurythermal).

Another factor acting on meiobenthos, particularly of intertidal beaches, is *salinity.* Reduced salinities may occur in intertidal beaches as a result of freshwater runoff over the beach during low tide or heavy rainfall. As noted in the last chapter, such salinity changes are usually confined to the upper layers of the beach because the lower layers retain, through capillary action, elevated levels of salt water. Since fresh water is less dense than salt water, fresh water cannot penetrate below that point where marine water is held by the capillary action. This means that only the uppermost layers are subject to salinity changes. Organisms inhabiting these layers are usually adapted to tolerate considerable salinity changes (e.g., ~ 15 %).

A final environmental factor is that of *wave action.* Wave action acts both intertidally and subtidally, but is most dramatic intertidally.

Breaking and moving waves have the effect of resuspending the upper layers of sand and, seasonally, of completely removing or depositing large amounts of sand. Whenever the beach is thus churned up, the whole array of internal space is rearranged and the organisms themselves are moved about, a process that has profound effects on organisms dependent on staying in the space between grains. It means that, for those organisms, the interstitial space is being constantly rearranged and they are in constant danger of being thrown out into the open water where the risk of predation is very much greater. Special adaptations for coping with this risk are known (see pp. 293–295).

Before leaving the realm of environmental factors, some mention should be made of *light.* Light rarely penetrates below 5–15 mm into sediments. This means that microscopic plants should generally be restricted to the very surface layers. However, diatoms have been found at depths considerably below this, raising speculation as to how they are managing to survive.

Most of the meioflora and microflora are attached to the surface of the sand grains and form an important food source for a variety of meiofaunal grazers.

COMPOSITION OF THE INTERSTITIAL ASSEMBLAGES

The types of organisms constituting the meiofauna are drawn from a very broad range of invertebrate phyla, but some are represented by one or a few species, whereas others are abundant both in terms of individuals and species. Their small size invites comparison with plankton organisms. In general, meiofaunal organisms have a size range similar to that of some of the smaller mesoplankton and the microplankton (see Chapter 2).

Those invertebrate phyla that normally have small bodies are, of course, preadapted to live in the small spaces between sand grains and are represented by many species and individuals. The protozoan phylum Ciliophora is represented abundantly by a large number of species, which are very diverse in form and, interestingly enough, often larger than the metazoan animals found with them (Fig. 7.3). The class Turbellaria, of the phylum Platyhelminthes, is also abundantly represented by numerous small, flat, and elongate worms (Fig. 7.4). Most abundant of all, perhaps, are the ubiquitous roundworms, the nematodes (Fig. 7.5).

(A)   (B)   (C)   (D)   (E)

Figure 7.3 Types of interstitial ciliate protozoans, phylum Ciliophora. *(A) Loxophyllum vermiforme* with a flattened body. *(B) Tracheloraphis remanei* with a cylindrical body. *(C) Remanella faurei* with a flattened elongated body. *(D) Remanella caudata* with an elongated body and tail. *(E) Geleia gigas* with a threadlike body. (From B. Swedmark, The interstitial fauna of marine sand, Biol. Rev. 39:1–42. Copyright © 1964 by Cambridge Philosophical Society; reprinted with permission of Cambridge University Press.)

Figure 7.4 Examples of some turbellarian flatworms from an interstitial marine sand beach in Florida. *(A)* A member of the family Macrostomidae. *(B)* A member of the family Kalyptorhynchidae. *(C) Polystylophora* sp. *(D) Proschizo-rhynchus* sp. *(E) Cicerina* sp. *(F) Nematoplana* sp. (From L. Bush, Characteristics of interstitial sand turbellaria: The significance of body elongation, muscular development, and adhesive organs. Trans. Amer. Microsc. Soc. 87[2]:244–251, 1968.)

Figure 7.5 Some interstitial nematodes. *(A) Paramonohystera wieseri. (B) Cyttaronema reticulatum. (C) Tricoma hopperi. (D) Epsilonema.* (From various sources.)

Several inconspicuous invertebrate phyla are well represented in the meiofauna, probably because they are normally of small size. These include the phyla Gastrotricha, Kinorhyncha, Tardigrada, and Rotifera. The Gastrotricha are particularly numerous in marine sediments and Rotifera in fresh water (Fig. 7.6).

The worms of the phylum Annelida with their elongated shape are well suited to the in-terstitial environment and are abundant. Representatives in the meiofauna are from both the classes Oligochaeta and Polychaeta (Fig. 7.7).

Crustaceans are represented primarily by an abundance of harpacticoid copepods and ostra-cods and a few other small groups such as Mystacocarida (Fig. 7.8).

Phyla that normally consist of large organ-

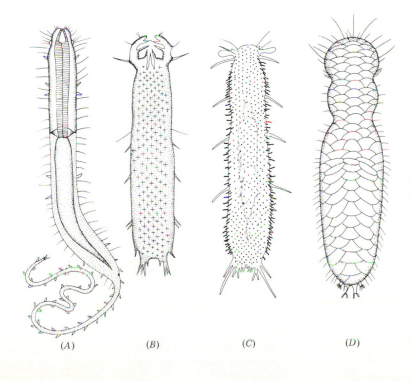

Figure 7.6 Some interstitial gastrotrichs. *(A) Urodasys viviparus. (B) Pseudostomella roscovita. (C) Thaumastoderma heideri. (D) Diplodasys ankeli.* (From B. Swedmark, The interstitial fauna of marine sand, Biol. Rev. 39:1–42. Copyright © 1964 by Cambridge Philosophical Society; reprinted with permission of Cambridge University Press.)

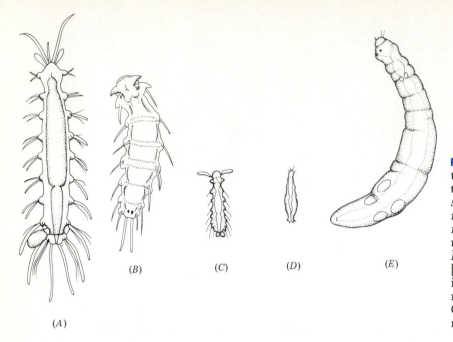

Figure 7.7 Some representative interstitial annelids of the class Polychaeta. *(A) Mesonerilla intermedia. (B) Paranerilla limnicola. (C) Nerillidium simplex. (D) Diurodrilus sp. (E) Trilobodrilus sp.* (From N. C. Hulings [ed.], Proceedings of the first international conference on meiofauna, Smithsonian Contributions to Zoology no. 76, 1971.)

isms or of sedentary or sessile forms are very poorly represented in the meiobenthos. In the case of large-bodied phyla, this is certainly due to the body size, which does not allow them to fit into the small spaces. These groups are, therefore, represented only by a few rather aberrant forms. Examples of such phyla are Echinodermata and Cnidaria (Fig. 7.9). Phyla of sessile or

Figure 7.8 Some crustacea of the interstitial fauna. *(A) Derocheilocaris remanei* (Mystacocarida). *(B) Cylindropsyllis laevis* (Copepoda). (From B. Swedmark, The interstitial fauna of marine sand, Biol. Rev. 39: 1–42. Copyright © 1964 by Cambridge Philosophical Society; reprinted with permission of Cambridge University Press.)

Figure 7.9 Representative interstitial Mollusca, Cnidaria, Echinodermata, and Brachiopoda. *(A) Caecum glabrum* (Mollusca, Gastropoda). *(B) Hedylopsis brambeli* (Mollusca, Gastropoda). *(C) Gwynia capsula* (Brachiopoda). *(D) Labidoplax buskii* (Echinodermata). *(E) Stylocoronella riedli* (Cnidaria, Scyphozoa). *(F) Psammohydra nanna* (Cnidaria, Hydrozoa). (From N. C. Hulings [ed.], Proceedings of the first international conference on meiofauna, Smithsonian Contributions to Zoology no. 76, 1971.)

sedentary habit seem to have few representatives, because the dynamic nature of the substrate makes it difficult to exist if one is permanently fixed in one place. Moreover, most of these phyla are suspension feeders, depending on filtering organisms out of the water. In the interstitial environment, there is relatively little water to filter and very few organisms in the water, as most are attached to, or adhere to, the particles. The sessile groups that are represented by a very few species include the bryozoans and ascidians (Fig. 7.10).

Absent from the interstitial fauna are the following phyla: Phoronida, Echiura, Pogonophora, Porifera, Ctenophora, Hemichordata, and Chaetognatha.

(A)

(B)

Figure 7.10 Sample interstitial Bryozoa and Tardigrada. *(A) Monobryozoon ambulans* (Bryozoa). *(B) Batillipes pennaki* (Tardigrada). *(A,* after N. C. Hulings [ed.], Proceedings of the first international conference on meiofauna, Smithsonian Contributions to Zoology no. 76, 1971. *B,* after L. W. Pollock, Distribution and dynamics of interstitial Tardigrada at Woods Hole, Massachusetts, U.S.A., Ophelia 7[2]: 145–165, 1970.)

SAMPLING AND EXTRACTING MEIOFAUNA

Because interstitial organisms are very small and occupy a unique habitat, special techniques are needed to obtain samples. A great variety of sampling and extracting techniques have been employed by meiobenthologists over the years, and none is really standard. Most techniques are, however, simple and may be grouped into general classes.

Because of the small size of the organisms, only a small sample of the substrate is required.

This is most conveniently obtained by taking a *core.* Coring devices are usually hollow plastic or metal cylinders that may be pushed into the sand by hand, closed off at one or both ends, and removed from the beach (Fig. 7.11). The enclosed sedimentary column may then be extruded and the organisms extracted. It is also possible to take a small volume or core from a larger grab or dredge sample that may be taken for macrofauna. Since the meiofauna exists throughout a number of vertical layers of the sand, it is well to take a core that extends some distance vertically into the substrate. The majority of

Figure 7.11 A simple coring device for obtaining samples of interstitial organisms. (Photo by the author.)

the meiofaunal individuals will, however, be concentrated in the top 10–15 cm, so a depth of 25–30 cm is adequate to sample all but a few forms.

The diameter of the coring device is not critical, but those in common use range in size from 2–4 cm. Larger sizes produce such large samples that the extraction process is prolonged and difficult.

Once the sample is in hand, it is necessary to extract the organisms. Meiobenthic animals are reluctant to leave the sediment and must be removed by special methods in order to be studied. A number of different extraction techniques have been used over the years, and each gives somewhat different results. It should be noted that the fauna may be extracted alive or preserved. In most cases, it is preferable to extract the animals alive, because live animals are much more easily seen and because groups like flatworms and gastrotrichs are virtually unidentifiable when preserved.

The simplest extraction is to scoop up a bucket of sand and let it stagnate until organisms are driven up to the surface by declining oxygen concentration. They may then be collected by pouring off the surface water through a 62-μm screen. Another simple technique is to treat a sediment sample with fresh water, to osmotically shock the animals to release from the particles, and then to pour off the water through a 62-μm screen. Perhaps the most used technique is to treat the sample with an isotonic solution of $MgCl_2$. This will anesthetize the animals and they can subsequently be collected by pouring the $MgCl_2$ solution through a fine net. Even more efficient is the method of *elutriation.* In this technique, a sediment sample is introduced into a container with openings at the top and bottom. The sediment is then constantly stirred up by water entering from below. As the sediment stirs, the organisms are dislodged and carried over and out with the water through the upper opening. This water is filtered on a very fine mesh, which then catches the organisms (Fig. 7.12). With this technique, one must remember to use only filtered water at the input. Otherwise, the sample will be contaminated by plankton organisms. This technique can be used with either living or preserved material.

Figure 7.12 An elutriation apparatus for separating interstitial fauna from the sand. (Photo by the author.)

ADAPTATIONS

Most of the morphological adaptations observed in the interstitial fauna appear to be correlated most closely with a few aspects of the environment, which can be reviewed here. The environment is a dynamic one in which the individual sediment grains of the upper layers are constantly being resuspended and deposited, thus constantly changing the location and amount of interstitial space. In order to survive here, meiofaunal organisms must remain in these spaces in the sediment and must avoid both being crushed by moving sand particles and being thrown out into the plankton, where their survival would be reduced.

Perhaps the most obvious adaptation to this environment is that of *size*. All meiofaunal organisms are very small. Furthermore, in this fauna, one finds the smallest representatives of most large phyla. For example, mollusks, echinoderms, and annelids are usually fairly large animals. All have representatives in the meiobenthos, but they are remarkable for their small size. The holothurian *Leptosynapta minuta* is 2 mm; the gastropod *Caecum glabrum* is 2 mm (Fig. 7.9); the polychaete *Diurodrilus minimus* is but 350 μm (Fig. 7.7); and the hydroid *Psammohydra nanna* but 1 mm (Fig. 7.9). In the case of the ciliate protozoans, however, the size is generally larger than that of ciliates living in other environments. At the same time, however, these larger ciliates are considerably flattened and elongated, which allows them to fit more readily in the interstices (Fig. 7.3). In this case, larger really means longer.

Another adaptation concerns the *shape* of the body. Most of the meiofaunal organisms have elongated or vermiform bodies, even in those organisms that are normally not vermiform (Fig. 7.3). Another type of body shape observed is represented by organisms that are very flat. Flat organisms fit into the narrow spaces well, and they also give a greater surface area for hanging on to the grains. This latter adaptation has survival value in that it permits the organism to stay on the sediment grain during resuspension.

Coincident with small body size, there is a reduction in the complexity and number of the body organ systems in metazoan animals. For example, in the polychaetes, the complex pharyngeal apparatus found in some macrofaunal polychaetes often disappears, as do the parapodial gills and even the kidneys (nephridia). In the meiofaunal hydroids, the number of tentacles is reduced from that observed in the forms not living interstitially.

One might expect that organisms living in an environment in which the substrate was often rearranged would be at risk of being crushed as sediment grains banged into each other. As a response to that, many meiofaunal animals have developed various types of reinforcement to their body walls. Skeletons of spicules are found in ciliates such as *Remanella,* turbellarian flatworms such as *Acanthomacrostomum spiculiferum,* and the acochlidiacean mollusks *Rhodope* and *Hedylopsis.* Gastrotrichs, on the other hand, have an armor of scales, while the nematodes

depend on a thick, heavy cuticle. Those delicate animals lacking a protective armor seem to have developed great abilities to extend and contract the body quickly, which also permits escape from crushing.

Perhaps because the environment is so dynamic, we find that most of the inhabitants are free moving. This is true even for groups that are normally all sessile, such as the Bryozoa and Ascidiacea. It is possible, however, that since all the common ways of extracting the meiofauna depend on the fauna leaving the substrate, there may be other sessile forms existing that we have not yet observed.

Two final types of adaptation may be noted, and both are in response to the necessity of the meiofaunal organisms' remaining in the sediment and not being thrown out during the resuspension process. Many species of various phyla, such as flatworms and gastrotrichs, have adhesive organs to fix them to the sediment grains. Crustaceans and tardigrades effect the same ability using hooks or claws on their limbs. *Statocysts,* organs that detect gravity and hence differentiate up and down, are also common. Such sense organs enable any organism to determine quickly which way to move during resuspension and thus return to the sediment rather than up into the open water.

Life History

As yet, we know relatively little of the life histories of many meiofaunal organisms. There are really two groups of interstitial organisms, a *temporary meiofauna,* which consists of the newly settled juveniles of the macrofaunal organisms, and a *permanent meiofauna,* consisting of those organisms that spend their entire lives in the interstices of the sediment. The temporary meiofaunal organisms are not considered here.

Since the permanent meiofaunal organisms are so small in size, the number of gametes that they can produce at any one time is also limited. In contrast to the macrofauna, which often produce thousands or hundreds of thousands of eggs, the production of eggs by meiobenthic organisms is almost always below 100 per individual and usually ranges between one and ten! When the number of eggs is that small, a species cannot afford to lose many or there will be no next generation. As a result, these animals have evolved a number of adaptations to ensure survival of the few propagules. First, there are adaptations that ensure that fertilization will occur. One way is to have copulation in which the sperm are directly transferred to the female. Such is the case with harpacticoid copepods. Another way is to package all the sperm in a single unit called a *spermatophore* and attach it to the female where it will be available to fertilize the eggs as they issue. Such a situation occurs in certain of the polychaete worms and the acochlidiacean mollusks. Finally, some species are hermaphroditic, having both male and female systems in the same individual and thus ensuring fertilization. This situation occurs in gastrotrichs, some hydroids (*Otohydra*), and some polychaetes (*Protodrilus*).

Once fertilization has occurred, it is important that the embryos be protected from loss. Usually, macrobenthic organisms produce larvae that enter the plankton in order to obtain maximum dispersal. However, the plankton is a notoriously dangerous place, with the result that most of the embryos are destroyed via predation. As a consequence, to compensate, the benthic animals using this route must produce large numbers of larvae. Since the meiofauna cannot produce large numbers, it follows that it would not be a good survival strategy to put larvae in the plankton. In fact, this is what is observed. Very few meiobenthic animals have planktonic larvae, and those that do are the ones that produce the largest numbers of eggs. Most meiofaunal organisms have instead brood protection. In such a case, the eggs are kept with the female until the young have reached a sufficient size to be self-sustaining. That selection pressure for brood protection is high in this environment can be deduced by observing that brood protection occurs here in groups of invertebrates in which it is otherwise virtually unknown, such as cnidarians (*Otohydra*) and gastrotrichs (*Urodasys viviparus*).

For those few that do not have brood protection, additional safeguards are found. Eggs that are spawned are either sticky so as to quickly attach to sand, or else are confined in cocoons that are sticky. In any case, the development is usually direct; that is, the egg hatches as a juvenile and does not have a free-swimming larva or a larva is produced that is nonpelagic and remains in the interstices of the sediment.

Seasonal reproductive periods have been observed in many species. Studies by Gerlach (1971) suggest that generation times of interstitial organisms range from a few days to over a year. The average number of generations per year is three.

ECOLOGY

On a worldwide basis, Coull and Bell (1979) report that the average number of meiofaunal organisms per square meter is 10^6 and the biomass between 1 and 2 g/m^2. Both numbers of organisms and biomass are highest in intertidal areas and decrease with increasing depth in the oceans. The most numerically abundant taxa in all environments are nematodes and harpacticoid copepods.

The composition of the meiofauna varies both vertically and horizontally in the substrate. The factors creating this zonation are the aforementioned grain size differences and physiochemical factors, particularly temperature, oxygen, and salinity. Characteristic patterns of species assemblages are correlated with these physical factors. Most meiofaunal organisms are concentrated in the uppermost sediment layers, particularly in muddy sediments where Coull and Bell (1979) found that as much as 94 percent of all meiofauna were in the uppermost 1 cm. In sand beaches meiofauna have been found as deep as 90 cm. Subtidally, perhaps 95 percent of the meiobenthos occur in the upper 7 cm. The variability in vertical extent seems to be related to the oxygen concentration, most taxa being present only in aerobic sediments. There is often a definite vertical zonation (Fig. 7.13). Vertical zonation, however, may also be due to predation. Coull et al. (1989) demonstrated that vertical cores taken where the fish *Leiostomus xanthurus* was feeding showed that the meiofauna was significantly reduced in the upper 2 cm.

There is a specialized faunal assemblage that is confined to the anoxic layers, the *thiobios.* Characteristic of this assemblage are ciliates of the orders Trichostomatida, Heterotrichida, and Odontostomatida, of which the latter seems to contain only anaerobic species and is confined to

Oxidized zone

Reduced zone

Sulfide zone

this area. Zooflagellates, certain nematodes, turbellarians, gnathostomulids, rotifers, and gastrotrichs are also present, but it is not known whether these latter forms are permanent members of the community or transients from the upper oxidized zones.

Vertical migration of the organisms also occurs. This migration is usually triggered by temperature or salinity changes and is especially prominent in beaches in the temperate zone. Thus, the same organisms are often found deeper during the winter months and shallower during the summer. Strong tidal or wave action following storms may also trigger a migration, which may occur on a much shorter time scale, on the order of a few days or a tidal cycle.

There are also seasonal changes in the abundance of meiofaunal organisms. Certain species are more abundant at one season than another (Fig. 7.14). For example, on Danish beaches, Muus (1967) found that copepods were at a minimum in winter and peaked in numbers in the summer. Fenchel (1978) has noted that in temperate shallow-water areas, there is an annual change in the densities of the dominant groups, with copepods succeeded by oligochaetes, nematodes, ostracods, and turbellarians. This temporal succession appears to be regulated by primary productivity and predation. Not only do abundances vary seasonally, but so does the species composition, with different species becoming abundant at different times of the year. This is probably correlated with changes in the physiochemical factors mentioned above, but may also be due to migration.

Meiofauna seem to be sensitive to human-induced or natural perturbations. Studies by Coull and Palmer (1984) have recorded an immediate decrease in abundance and diversity of constituent meiofauna following a disturbance. In such perturbations, the crustaceans seem to be the most affected and the nematodes the least affected. Recovery rates from perturbations appear to be variable, taking from only hours up to months. Because of their rapid response to a disturbance, large numbers, stationary life habits, short generation time, and intimate association with sediments, meiofaunal organisms would seem to be ideal organisms to use as indicators of pollution. However, the great difficulty in identifying these tiny animals and the general lack of experts seem to preclude any widespread use for this purpose.

Meiofaunal organisms are patchily distributed horizontally, even when the sediment grain distribution is homogeneous. We are ignorant of the causes of such patchiness, but various workers have suggested several reasons. Prominent among these suggestions is the idea that food sources are patchy, which in turn leads to concentration of the meiofauna. Others have suggested that the occurrence of macrofaunal organisms or their constructions, such as burrows and tubes, increases the heterogeneity of the environment and the patchiness of distribution. Finally, since most meiofaunal organisms lack a larval dispersal mechanism, reproduction may result in aggregations of juveniles.

There is little information on the long-term variability of meiofauna. The few studies have produced conflicting results. The earliest, a 9-year study by McIntyre and Murison (1973), suggested that annual variations in meiofauna abundance were small, whereas Coull and Bell

Figure 7.13 Vertical zonation of interstitial animals in a Scandanavian beach. (1) Microdasyoid, *Urodasys;* (2) gastrotrich, *Chaetonotus;* (3) nematode; (4) turbellarian, *Coelogynopora;* (5) tardigrades; (6) harpacticoid, *Cylindropsyllis;* (7) archiannelid, *Nemillidion;* (8) hydroid, *Halammohydra.* Cilates: (9) *Pseudoprorodon;* (10) *Helicoprorodon;* (11) *Tracheloraphis;* (12) *Loxophyllum;* (13) *Litonotus;* (14) *Dilepthus;* (15) *Lynchella;* (16) *Chlamydodon;* (17) *Pleuronema;* (18) *Frontonia;* (19) *Blepharisma;* (20) *Condylostoma;* (21) *Diophrys;* (22) *Strombidium;* (23) *Aspidisca;* (24) *Paraspathidum;* (25) *Mesodinium;* (26) *Remanella;* (27) *Kentrophorus;* (28) *Cardiostomum;* (29) *Homalozoon;* (30) *Lacrymaria;* (31) *Geleia;* (32) *Sonderia;* (33) *Metopus;* (34) *Caenomorpha;* (35) *Saprodinium;* (36) *Myelostoma;* (37) *Parablepharism.*

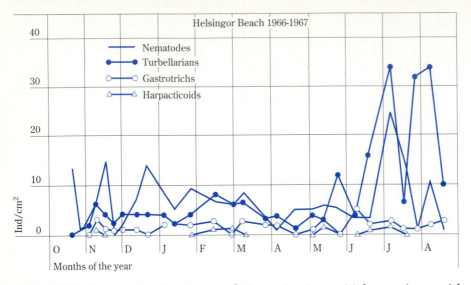

Figure 7.14 Changes in abundance of the major interstitial organisms with season in a Danish beach. (From T. Fenchel, The ecology of marine microbenthos. IV, Structure and function of the benthic ecosystem: Its chemical and physical factors and the microfaunal communities with special reference to the ciliated protozoa. Ophelia 6:1–182, 1969.)

(1979) in a 5-year investigation came to the opposite conclusion. More recently, Coull (1985a) in an 11-year study of two sites, one sand and one mud, demonstrated that the variability in the mud site was twice that of the sand, and the year-to-year variability was greater than the seasonality, but no long-term cycles were found.

Trophic relationships among the components of the interstitial fauna are not well known. It has been determined that certain flatfishes such as *Pleuronectes platessa, Platichthys flesus, Microchirus boscanion,* and *Limanda limanda* in European waters actively prey upon meiofauna when they are young and recently metamorphosed. McIntyre (1969) reports that their stomachs have been found to contain numbers of harpacticoid copepods, ostracods, oligochaetes, and nematodes. Recently, Nelson and Coull (1989) have demonstrated that the juveniles of the fish *Leiostomus xanthurus* are selective in preying preferentially on the meiofaunal copepods in contrast to the more abundant nematodes. Other macrofaunal predators on meiofauna include shrimps, larger polychaetes, and hydroids. Macrofaunal deposit feeders such as polychaetes and holothurians must also destroy

numbers of meiofauna existing on the particles that they consume.

In addition to macrofaunal predators, there are interstitial predators, including certain nematodes, turbellarians, some tardigrades, and hydroids such as *Protohydra.* The latter is a particularly voracious feeder on copepods and nematodes, according to Muus (1966). The result of predation by both meiofaunal and macrofaunal predators is to reduce the populations of other interstitial organisms, but it is not known how extensive this reduction is.

In addition to predators, the interstitial fauna contains three other feeding types: herbivores, detritus feeders, and suspension feeders. While some turbellarians have been shown to be predators, others appear to be scavengers, feeding on dead meiofaunal and macrofaunal organisms. Harpacticoid copepods seem to be primarily herbivores, feeding on the few attached diatoms or else feeding on detritus. Ostracods also ingest diatoms. Detritus feeders include gastrotrichs, some nematodes, and archiannelid polychaetes. Suspension feeders are, as noted before, quite rare, due to lack of an interstitial plankton.

A considerable degree of food specialization appears to exist among the meiofauna. Fenchel

(1968) studied the interstitial ciliates and demonstrated a surprising degree of food specificity for various bacterial and algal taxa. In addition, he showed that in certain habitats, whole groups of species subdivided their food resource according to size. A similar food niche separation has been reported by Tietjen and Lee (1977) for nematodes.

What controls meiofaunal communities? Again we lack sufficient data to answer. Coull (1985b), in a study of two sites in South Carolina over an 11-year period, suggested that in fine-grained sediment (mud) the component organisms were controlled by predators, while in the nearby sand habitat the organism abundances were controlled by physical factors.

Meiofauna—Macrofauna Interactions

What is the significance of the meiofauna to the general ecology of benthic sedimentary systems and what, if any, is its relationship to the macrofauna? These are questions that have generated considerable interest and conflicting opinions over the years.

It appears that the average meiofaunal biomass is from one-tenth to one-fifth of that of the resident macrofauna. In certain areas, such as the intertidal sand and mud flats and in the deep sea, the ratio is closer to one to one (Platt, 1981). This suggests that in these latter areas meiofauna plays an important role in energetics. Because of their small size, higher metabolic activity, and high turnover rate, productivity of meiofaunal organisms per unit of biomass is greater than that of the macrofauna. Therefore, even in systems dominated by macrofauna, total organic meiofaunal production approaches that of the macrofauna. Clearly, then, the meiofauna represents an important reservoir of organic material. Furthermore, meiofaunal animals stimulate bacterial production in the sediments by their metabolic secretions.

What is the ultimate fate of this production? Is it merely recycled within the meiofaunal system or is there a linkage with the macrofauna and higher trophic levels? Earlier workers, such as McIntyre and Murison (1973), thought the meiobenthic system was separate and that its production was unavailable to higher trophic

levels of the macrofauna. However, it now appears likely that these two systems are linked to some degree. For example, nonselective deposit feeders of the macrofauna cannot avoid consuming meiofauna, and if the digestion is equally nonselective, then meiofauna must contribute to their nutrition. Similarly, resuspension of bottom particles, which in turn are captured by large suspension feeders, would lead to a similar macrofaunal ingestion. Macrofaunal predators that selectively pick meiofaunal organisms seem rare, but Feller and Kaczynski (1975) have demonstrated that juvenile salmon are predators on meiobenthic copepods, and Bell and Coull (1978) have demonstrated that the grass shrimp *Palaemonetes pugio* not only feeds on meiofaunal organisms, but can control meiofaunal population numbers and community structure. Other fishes, particularly young stages of flatfishes, gobies, and mullet, have large quantities of meiofaunal organisms in their guts. Thus the meiofaunal organisms probably fit into the macrofaunal food web, such as is illustrated in Fig. 7.15.

The extent to which meiofauna enters the macrofaunal food web seems to vary depending on the sediment type. It is likely that meiofauna is more significant as food to macrofauna in muddier sediments than in sands. This is due in part to the fact that in muds the meiofaunal organisms are all concentrated in the topmost layers, where they are more accessible, whereas in sand they may be much deeper and relatively less accessible to many macrofaunal organisms.

Since the macrofauna do interact with the meiofauna, what effects do they have on meiofaunal community structure? This is little known at the present, but the aforementioned study by Bell and Coull (1978) has demonstrated that, in at least one case, meiofaunal numbers, particularly of the slower-moving wormlike meiobenthos, could be significantly reduced, suggesting that a predator-controlled system is in operation.

Species Richness, Distribution, and Dispersal

One of the more remarkable features of the meiofauna is the species richness. Fenchel et al.

Figure 7.15 Hypothetical benthic food web, illustrating possible connections between the meiofauna and macrofauna. (From H. M. Platt, Meiofaunal dynamics and the origin of the metazoa, pp. 207–216. *In:* P. L. Foray [ed.], The evolving biosphere, Cambridge University Press, 1981.)

(1967) found as many as 70 species of meiofaunal organisms in 50 cm² of sand, and they regularly obtained 30 to 50 species of ciliate protozoan alone. Even more striking is the discovery by Gerlach (1965) that the Arctic interstitial fauna is rich in contrast to the impoverished macrofauna at that latitude. Based on the limited studies available now, it appears that there is no gradient in species richness in this fauna moving from polar regions to the tropics. This is in direct contrast to the common distribution pattern of most macrofaunal organisms, where lower latitudes tend to have greater richness than higher latitudes.

Zoogeographically, the meiofauna have a pattern in which the common genera are often cosmopolitan, but the species are not. As a result, several different zoogeographical regions or provinces can be recognized.

Because most meiofaunal organisms lack a motile larval phase, their dispersal seems to be mainly by the transport of adults by means of waterfowl, by rafting on floating substrates (including icebergs), and by turbulent water (Gerlach, 1977).

BIBLIOGRAPHY

Bell, S. S., and B. C. Coull. 1978. Field evidence that shrimp predation regulates meiofauna. Oecologia 35:141–148.

Bush, L. 1968. Characteristics of interstitial sand turbellaria: The significance of body elongation, muscular development and adhesive organs. Trans. Amer. Microsc. Soc. 87:244–251.

Coull, B. C., 1973. Estuarine meiofauna: A review: Trophic relationships and microbial interactions, pp. 499–512. *In:* Stevenson, L. H., and R. R. Colwell (eds.). Estuarine microbial ecology. University of South Carolina Press, Columbia.

Coull, B. C. 1985a. Long-term variability of estuarine meiobenthos: An 11-year study. Mar. Ecol. Prog. Ser. 24:205–218.

Coull, B. C. 1985b. The use of long-term biological data to generate testable hypotheses. Estuaries 8:84–92.

Coull, B. C., R. L. Ellison, J. W. Fleeger, R. P. Higgins, W. D. Hope, W. D. Hummon, R. M. Rieger, W. E. Sterrer, H. Thiel, and J. H. Tietjen. 1977. Quantitative estimates of the meiofauna from the deep sea off North Carolina, U.S.A. Mar. Biol. 39:233–240.

Coull, B. C., and S. S. Bell. 1979. Perspectives of marine meiofaunal ecology, pp. 189–216. *In:* Livingston, R. J. (ed.). Ecological processes in coastal and marine systems. Plenum, N.Y.

Coull, B. C., and M. A. Palmer. 1984. Field experimentation in meiofaunal ecology. Hydrobiologia 19:1–19.

Coull, B. C., M. A. Palmer, and P. E. Meyers. 1989. Controls on vertical distribution of meiobenthos in the mud: Field and flume studies with juvenile fish. Mar. Ecol. Prog. Ser. 55:133–139.

Feller, R. J., and V. W. Kaczynski. 1975. Size selective predation by juvenile chum salmon (*Oncorhynchus keta*) on epibenthic prey in Puget Sound. J. Fish. Res. Bd. Canada 32:1419–1429.

Fenchel, T. 1968. The ecology of marine microbenthos. II. The food of marine benthic ciliates. Ophelia 5:73–121.

Fenchel, T. 1969. The ecology of marine meiobenthos. IV. Structure and function of the benthic ecosystem, its chemical and physical factors and the microfaunal communities with special reference to the ciliated protozoa. Ophelia 6:1–182.

Fenchel, T. 1978. Ecology of micro- and meiobenthos. Ann. Rev. Ecol. Systematics 9:99–121.

Fenchel, T., B. O. Jansson, and W. von Thun. 1967. Vertical and horizontal distribution of the metazoan microfauna and of some physical factors on a sandy beach in the northern part of the Oresund. Ophelia 4:227–243.

Gerlach, S. A. 1965. Uber die Fauna in der Gezeitenzone von Spitzbergen. Bot. Bothab. Acta Univ. Gothab. 3:3–23.

Gerlach, S. A. 1971. On the importance of marine meiofauna for benthos communities. Oecologia 6:176–190.

Gerlach, S. A. 1977. Means of meiofauna dispersal. Mikrofauna Meeresboden 61:89–103.

Higgins, R. P., and H. Thiel (eds.). 1988. Introduction to the study of meiofauna. Smithsonian Institution Press, Washington, D.C. 488 pp.

Hulings, N. C. (ed.). 1971. Proceedings of the first international conference on meiofauna. Smithsonian Contributions to Zoology no. 76. 205 pp.

Hulings, N. C., and J. S. Gray (eds.). 1971. A manual for the study of meiofauna. Smithsonian Contributions to Zoology no. 78. 84 pp.

McIntyre, A. D. 1969. Ecology of marine meiobenthos. Biol. Rev. 44:245–290.

McIntyre, A. D., and D. J. Murison. 1973. The meiofauna of a flatfish nursery ground. J. Mar. Biol. Assoc. U.K. 53:93–118.

Muus, B. J. 1967. The fauna of Danish estuaries and lagoons. Meddr. Danm. Fisk.-og Havunders 5:1–316.

Muus, K. 1966. Notes on the biology of *Protohydra leuckarti* Greef (Hydroidea, Protohydridae). Ophelia 3:141–150.

Nelson, A. L., and B. C. Coull. 1989. Selection of meiobenthic prey by juvenile spot (Pisces): An experimental study. Mar. Ecol. Prog. Ser. 53:51–57.

Platt, H. M. 1981. Meiofaunal dynamics and the origin of the metazoa, pp. 207–216. *In:* Foray, P. L. (ed.). The evolving biosphere. British Museum (Natural History) and Cambridge University Press, U.K.

Pollock, L. W. 1970. Distribution and dynamics of interstitial Tardigrada at Woods Hole, Massachusetts, U.S.A. Ophelia 7:145–165.

Sterrer, W., and P. Ax (eds.). 1977. The meiofauna species in time and space. Akademie der Wissenschaften und der Literatur, Mainz, West Germany. 316 pp.

Swedmark, B. 1964. The interstitial fauna of marine sand. Biol. Rev. 39:1–42.

Tietjen, J. H., and J. J. Lee. 1977. Feeding behavior of marine nematodes, pp. 21–35. *In:* Coull, B. C. (ed.). Ecology of marine benthos. University of South Carolina Press, Columbia.

Vernberg, W., and B. C. Coull. 1981. Meiofauna, pp. 147–177. *In:* Vernberg, F. J., and W. B. Vernberg (eds.). Functional adaptations of marine organisms. Academic Press, N.Y.

Estuaries and Salt Marshes

Whereas the intertidal zone is the place where terrestrial and the marine habitats meet, an estuary is a place where fresh water and salt water meet. It acts as a transition zone between the two aquatic ecosystems. Estuaries have been and remain in close association with humans, because many of the major cities of the world are established on estuaries. In contrast to other ecotones or transitional areas, estuaries are depauperate in terms of numbers of species of permanent organisms. In this chapter, we shall discuss the various special physical and chemical conditions in an estuary and how these act to create a rigorous environment in which relatively few species can persist.

TYPES OF ESTUARIES

There are many definitions for an estuary. This diversity of definitions is due in part to the fact that several geomorphological features of coastlines, such as lagoons, sloughs, fjords, and other shallow embayments, are often considered estuaries. A simple definition is that an *estuary* is a partially enclosed coastal embayment where fresh and sea water meet and mix. This definition implies the free connection of the sea with the freshwater source, at least during a part of the year. This definition excludes permanently isolated coastal water impoundments, as well as isolated brackish or saline bodies of water such as the Caspian Sea, Aral Sea, and Great Salt Lake.

As a result of the geomorphology of an estuary, the geological history of the area, and the prevailing climatic conditions, there may be different estuarine types, each displaying somewhat different physical and chemical conditions. These may be grouped into a few basic types (Fig. 8.1). Perhaps the most common type of estuary is the *coastal plain estuary*. Coastal plain estuaries were formed at the end of the last ice age when the rising sea level invaded low-lying coastal river valleys. Estuaries such as the Chesapeake Bay and the mouths of the Delaware and Hudson rivers in the United States and the Cornwall and Devon estuaries in Great Britain are examples of this type. Similar to this is the *tectonic estuary*. In this class of estuary, the sea reinvades the land due to subsidence of the land, not as a result of a rising sea level. A good example is San Francisco Bay. A third type of estuary is the *semienclosed bay* or *lagoon*. Here, sandbars build up parallel to the coastline and partially cut off the waters behind them from the sea. This creates a shallow lagoon behind the sandbars, which collects the freshwater discharge from the land. The water in such lagoons varies in salinity, depending on the climatic conditions, whether or not any major river flows into the lagoon, and the extent to which the bars restrict seawater access. Such estuaries are common in North Carolina and along the Texas and Florida Gulf coasts, in northwestern Europe (the Netherlands), and parts of Australia. A final type category of estuary is the *fjord*. These are valleys that have been deepened by glacial action and are then invaded by the sea. They are characterized by a shallow sill at the mouth that greatly restricts water interchange between the deeper waters of the fjord and the sea. Often, these deeper waters are stagnant because of lack of circulation. Fjords are abundant on the coasts of Norway, Chile, Scotland, Alaska, New Zealand, and British Columbia.

Estuaries may be classified in yet another way, depending on the way the salinity gradients are formed. In most estuaries, there is a gradient in salinity from full seawater (33–37 ‰) at the mouth to fresh water at the upper reaches. Fresh water is less dense than seawater and, where the two meet, the fresh water will "float" on the seawater. Mixing will occur where the two come in contact, but the extent of the mixing varies with many other environmental factors, including basin shape, tide, river flow, and rainfall. In estuaries where there is substantial freshwater outflow and reduced evaporation (typical temperate zone estuaries), the fresh water will move out over the top of the salt water, mixing with it near the surface and reducing the salinity, leaving the deeper waters more saline. In such a situation, a cross section of the estuary will show isohalines (lines of equal salinity), which extend upstream at the bottom (Fig. 8.2). At any given point on the estuary, a vertical column of water will have highest salinity at or near the bottom and lowest at or near the surface. This is a *positive estuary* or *salt wedge estuary*. These estuaries are also called *river-dominated* or *stratified*. Such estuaries form a continuum—from those with little mixing and very prominent salt wedges, through those with partial mixing and lesser salt wedges, to *homogeneous, marine-dominated,* and *neutral* estuaries where either complete mixing or having an evaporation rate equal to the freshwater inflow gives similar salinities vertically from surface to bottom at any point. Where an estuary fits in this continuum depends not only on the amount of mixing of the water masses, but also on the tidal regime, the geometry of the estuarine basin, and the river flow. The tidal regime and river flow may be further altered seasonally. In general, positive estuaries are common, while neutral estuaries are rare. Galveston Bay, Texas, and Alligator Harbor, Florida, are examples of neutral estuaries. In desert climates where the amount of freshwater input to the estuary is small and the rate of evaporation high, a *negative* or *evaporite estuary* results. In a negative estuary, the incoming salt water enters at the surface and is somewhat diluted by mixing with the small amount of fresh water. The high evaporation rate, however, causes this surface water to become hypersaline. Hypersaline water is more dense than seawater, sinks to the bottom, and moves out of the estuary as a bottom current. A salinity profile of such an estuary is the reverse of the positive

(A)

Figure 8.1 Types of estuaries. *(A)* Bay or lagoon lying behind a barrier beach; the major lagoons of the North Carolina coast. *(B)* Tectonic estuary; San Francisco Bay. *(C)* Fjord; Naeroy Fjord in Norway. (A, after C. Peterson and N. M. Peterson, The ecology of intertidal flats of North Carolina: A community profile, U.S. Fish & Wildlife Serv. FWS/OBS–79/39, 1979. B, after T. J. Conomos [ed.], San Francisco Bay, the urbanized estuary, 1979, courtesy of Pacific Division—American Association for the Advancement of Science. C, photo by the author.)

(B)

(C)

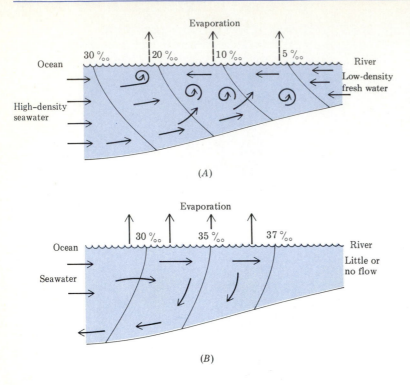

(A)

(B)

Figure 8.2 Positive and negative estuaries. *(A)* Positive (salt wedge) estuary. Freshwater runoff is significant throughout the year and greater than evaporation. Seawater enters along the bottom and gradually mixes with outward-flowing fresh water. *(B)* Negative estuary. Freshwater flow is diminished or absent during part of the year. Seawater enters along the surface. Evaporation is greater than runoff, so the salinity increases as one moves up the estuary. The hypersaline water sinks and flows out below the incoming seawater.

estuary, with highest values at the bottom and lowest at the top (Fig. 8.2). A final category is the *seasonal* or *intermittent* estuary. These estuaries are formed in areas where there is a marked wet and dry season (Mediterranean climate). In the rainy season they have fresh water coming down and are open to the ocean. In the dry season they have no freshwater input, may become dry or stagnant, and are often cut off from the ocean by seasonal sand bars. Salinity in these estuaries varies not spatially but temporally.

PHYSICAL CHARACTERISTICS OF ESTUARIES

The physical-chemical regime of estuaries is one with large variations in many parameters, which often create a stressful environment for organisms. It is probably due to such stresses that the number of large species living in an estuarine area is small in comparison to other marine habitats.

Salinity

The dominant feature of the estuarine environment is the fluctuation in salinity. By definition,

a salinity gradient exists at some time in an estuary, but the pattern of that gradient varies seasonally, with the topography of the estuary, with the tides, and with the amount of fresh water. We have already discussed the pattern of salinity distribution in salt wedge, partially mixed, and negative estuaries (see above). There are, however, yet other factors that act to alter salinity patterns. The tide itself is one such force. Where the tidal range is significant, high tide acts to drive the salt water further up the estuary, displacing the isohalines upstream. Low tides, by contrast, displace the isohalines downstream. The result is that there is a certain area of the estuary that is subject to a salinity regime that changes with each tide (Fig. 8.3). This is the area of an estuary having maximum salinity fluctuation. The time scale of these salinity fluctuations is such that the salinity range over a 6- to 12-hour period equals or exceeds the entire annual range of salinities for some areas, even within an estuary.

A second force is the *Coriolis effect*. The rotation of the earth has the effect of deflecting flowing water. In the Northern Hemisphere, this effect deflects outflowing fresh water to the right as one looks down the estuary toward the sea. Salt water flowing into the estuary from the

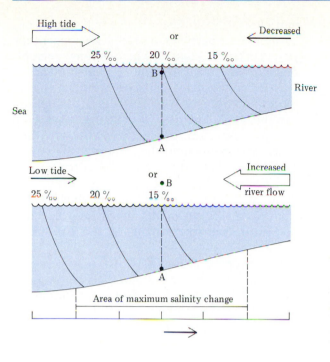

Figure 8.3 Change of salinity in an estuary with change in tide level or change in river discharge. Point A is on the bottom and B is intertidal. Because point B is covered only at high tide, it is inundated only with high-salinity water and is unaffected by low-salinity water at low tide. Point A, in contrast, is covered by water of different salinity at different tidal levels.

ocean is also displaced to the right looking from the sea toward the estuary. The opposite is true in the Southern Hemisphere. As a result, two points, each on opposite sides of the estuary equidistant from the mouth, may well have predictably different salinities (Fig. 8.4).

Seasonal changes in salinity in the estuary are usually the result of seasonal changes in evaporation, freshwater flow, or both. In areas where freshwater discharge is reduced or absent for a part of the year, higher salinities (= salt wedge) may be found further upstream. With the onset of increased freshwater flow, the salinity gradients are moved downstream toward the mouth. Therefore, at different seasons of the year, a given point in the estuary may experience different salinities.

Thus far, we have considered only the salinity changes in the water column itself. In the substrate, a different situation may prevail. Since estuarine substrates are sand and mud, water is held in the interstices between the particles. This "interstitial" water originates from the overlying water. The interstitial water changes in salinity much more slowly than the overlying water because of the slow interchange between the two. The interstitial water and the surrounding mud and sand are therefore "buffered" with respect to the overlying water. Organisms dwelling within the substrate are thus subject to less drastic salinity changes than organisms in the water column at the same point. This buffering effect is probably most effective in the lower intertidal zone, but not in the high intertidal. Variability in interstitial salinity tends to increase with tidal height. This is because interstitial soil water at higher elevations has more opportunity for dilution during rains and is subject to more evaporation in dry weather (Fig. 8.5).

Substrate

Most estuaries are dominated by muddy substrates, which are often very soft. These are derived from sediments carried into the estuary by both seawater and fresh water. Wind (aeolian) transport of larger sand particles into the estuary is often significant in certain areas, particularly for coastal lagoons behind barrier beaches. In the case of fresh water, rivers and streams carry silt particles in suspension. When these suspended particles reach and mix with seawater in the estuary, the presence of the various ions in the seawater causes the silt particles to flocculate, thus creating larger, heavier particles, and settle out, forming the characteristic mud bottom. Seawater also carries a considerable amount of suspended material.

Figure 8.4 Result of Coriolis effect on salinity in the Chesapeake Bay estuary. Note the deflection of the isohalines to the right in this Northern Hemisphere estuary. Circled numbers refer to salinity values in parts per thousand. (After D. S. McLusky, Ecology of estuaries, Heinemann, 1971.)

When it enters an estuary, the sheltered conditions reduce the water motion that has been responsible for keeping the various particles in suspension. As a result, the particles settle out, contributing to the formation of the mud or sand substrate. The relative importance of freshwater-borne or marine-borne particles to the development of the muddy substrate varies from estuary to estuary and also geographically, but in North America and Europe the main sedimentary source is the sea.

The deposition of particles is also controlled by currents and the size of the particle. Large particles settle out faster than small, and strong currents keep particles in suspension longer than slow currents. Hence, where strong currents prevail, the substrate will be coarse (sand or gravel), as only large particles are permitted to settle out; whereas where waters are calm and currents weak, fine silt will settle out. Thus, both seawater and fresh water tend to drop their coarse sediments first, the former at the mouth of the estuary and the latter in the upper reaches or the river itself, leaving the area of mixing to be dominated by fine silt (= mud), which results from decreased water movement and flocculation from the intermixing of the two water masses.

Catastrophic events may also play a significant role. Massive storms and their accompanying floods may produce large depositions or removal of sediment in estuaries, causing severe mortality of organisms. For example, Zedler (1982) reported 4 cm of silt deposited during a 1980 flood on the Tijuana estuary in southern California, and 40 percent of the low-tide volume of Magu Lagoon was lost during the same floods. Similarly, on the east coast, Andrews (1973) reported that the massive flooding of fresh water into Chesapeake Bay following hurricane Agnes in 1972 caused severe reduction in salinity and catastrophic losses of organisms.

The ecological conditions within the mud are similar to those already noted for muddy shores (see pp. 270–271). Indeed, many of the mud shores throughout the world are found in estuarine situations.

Among the particles that settle out in the estuary are many of organic origin derived from both the terrestrial and marine environments. As a result, the substrate that accumulates is very rich in organic material. This material may, in turn, serve as a large food reservoir for estuarine organisms. The large surface area relative to volume of the very small particles means that there is a very large area for bacteria to inhabit.

Temperature

Water temperatures in estuaries are more variable than in the nearby coastal waters. In part, this is due to the fact that there is usually a

Figure 8.5 Comparison of salinity fluctuations in the water column with that interstitially in the bottom mud. Data from the Pocasset River estuary in Massachusetts. (After P. C. Mangelsdorf, Jr., Salinity measurements in estuaries, pp. 71–79. *In:* G. H. Lauff [ed.], Estuaries pub no. 83 AAAS, 1967.)

smaller volume of water in an estuary and a larger surface area, and therefore, it heats up and cools down more rapidly under prevailing atmospheric conditions (fjords, being deep and with a large volume, do not show this). Another reason for the variation is freshwater input. Fresh water in rivers and streams is much more subject to seasonal temperature change than seawater. Rivers in the temperate zones are colder in winter and warmer in summer than adjacent seawater. When these fresh waters enter the estuary and mix with the seawater, they alter the temperature. As a result, estuarine waters are colder in winter and warmer in summer than surrounding coastal waters. The time scale is of interest in that with the changing tides, a given point in the estuary will show large temperature variation as a function of the difference between seawater and river water temperature.

Because the fresh waters show the greatest temperature ranges, it follows that as one progresses up the estuary, the temperature range on an annual basis becomes greater. Similarly, the temperature range is least at the entrance to the estuary where mixing with fresh water is minimal. On a short-term basis, the central area of an estuary may show the greatest change with changing tide.

Temperature also varies vertically. The surface waters have the greatest temperature range and the deeper waters the smallest. In salt wedge estuaries, this vertical temperature difference also reflects the fact that surface waters are dominated by fresh water, whereas the deeper waters may be completely or predominantly marine.

Wave Action and Currents

Estuaries are surrounded by land on three sides. This means that the water distance over which a wind can blow to create waves is minimal, at least with respect to the oceans. Since the height of waves depends on the "fetch" or the open-water distance over which the wind can blow, it follows that a small area of water can generate only small waves. The shallow depth of the water in most estuaries also precludes development of large waves. The narrowness of the mouth of the estuary, coupled with the shallow bottom, combine to dissipate rapidly the effects of any waves entering the estuary from the sea. As a result of these processes, estuaries are generally places of calm water. Reduced wave action also promotes the deposition of fine sediments, allowing the development of rooted plants, which in turn adds to the stabilizing of these sediments.

Currents in estuaries are caused primarily by tidal action and river flow. Currents are generally confined to channels, but within these channels, velocities up to several knots can occur. The highest velocities occur in the middle of channels, where the frictional resistance from the bottom and side banks is lowest. Although the estuary is an overall place of sediment deposition as noted above, the channels where currents are concentrated are often areas of marked erosion. Whenever currents change position, new channels may be quickly eroded and old channels filled. This is particularly true in the intertidal areas where a natural cycle of erosion and deposition occurs. However, in most estuaries, deposition exceeds erosion so there is a net accumulation of silt. In intermittent estuaries, where the mouth is often closed off from the sea during the dry part of the year, water movement is severely reduced at that time, leading to stagnation, reduced oxygen content, algal blooms, and fish kills.

For most estuaries, there is a continual input of fresh water at the head. A given amount of this fresh water moves down the estuary, mixing to a greater or lesser degree with seawater. A volume of this water of sufficient size is eventually discharged from the estuary or evaporated to compensate for the next volume introduced at the head. The time interval that is required for a given water mass of fresh water to be discharged from the estuary is the *flushing time.* This time interval can be a measure of the stability of the estuarine system. Long flushing times are important to the maintenance of estuarine plankton communities.

Turbidity

Because of the great number of particles in suspension in the water of estuaries, at least at certain times of the year, the turbidity of the water is high. Highest turbidities occur during the times of maximum river flow. Turbidity is generally at a minimum near the mouth, where full seawater occurs, and increases with distance inland. The more an estuary approaches a lagoon type, the more the turbidity is a function of the plankton concentration or wind speed (producing resuspension).

The major ecological effect of turbidity is a marked decrease in the penetration of light. This, in turn, decreases photosynthesis by phytoplankton and benthic plants, thereby reducing productivity. Under conditions of severe turbidity, phytoplankton production may be negligible, and the major production of organic matter is by emergent marsh plants (see pp. 328–330).

Oxygen

The regular influx of fresh and salt water into the estuary coupled with the shallowness, turbulence, and wind mixing usually means that there is an ample supply of oxygen in the water column. Since the solubility of oxygen in water decreases with increased temperature and salinity, the precise amount of oxygen in the water will vary as those parameters vary. In salt wedge estuaries, or any deep estuary during the summer when a thermocline can develop and where there is a vertical salinity stratification, there is often little interchange between the oxygen-rich surface waters and the deeper layers. This isolation of the deep waters from interchange with an oxygen source, coupled with high biological activity and slow renewal or flushing rate, may deplete the oxygen of the bottom waters.

Oxygen is severely depleted in the substrate. The high organic content and high bacterial populations of the sediments exert a large oxygen demand on the interstitial water. Since the fine particle size of the sediments restricts the interchange of interstitial water with the water column above, oxygen is quickly depleted. Estuarine sediments are, therefore, anoxic below the first few centimeters unless they have large particle size or large numbers of burrowing animals such as the ghost shrimp *Callianassa* and the hemichordate worm *Balanoglossus,* which by their activities oxygenate lower sediment layers.

THE BIOTA OF ESTUARIES

Faunal Composition

There are three components of the fauna of estuaries: marine, freshwater, and brackish water or estuarine.

The marine component of the fauna is the largest in terms of numbers of species and includes two subgroups. The *stenohaline* marine animals are typical marine forms that either are unable, or barely able, to tolerate salinity changes. This component is usually restricted to the mouths of estuaries where salinity is generally

25 ‰ or above. These animals are often the same species found in the open sea. *Euryhaline marine animals* compose the second subgroup. These are typical marine animals that are capable of tolerating varying amounts of salinity reduction below 30 ‰. Such species are capable of penetrating varying distances up the estuary. Most tolerate salinities down to 15 to 18 ‰, with a few hardy species tolerating levels down to 5 ‰ (Fig. 8.6).

The brackish water or *true estuarine* component comprises those species that are found in the middle reaches of the estuary in salinities between 5 ‰ and 18 ‰, but that are not found in fresh water or in full seawater. Examples of this component include the polychaete *Nereis diversicolor*, various oysters (*Crassostrea, Ostrea*) and clams (*Scrobicularia plana, Macoma balthica, Rangia flexuosa*), certain small gastropods (*Hydrobia*), crabs (*Callinectes*), and shrimps (*Palaemonetes*; Fig. 8.7). It may be, however, that certain of these estuarine genera are limited in a seaward direction not by physiological tolerances but by biological interactions such as competition and predation. Hence this component may not be easily defined.

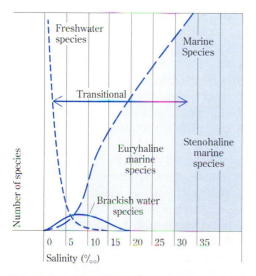

Figure 8.6 Numbers of species in each of the three major components of the fauna of estuaries and their distribution with salinity. (After D. S. McLusky, Ecology of estuaries, Heinemann, 1971.)

The final component is derived from fresh water. These animals usually cannot tolerate salinities much above 5 ‰ and are restricted to the upper reaches of the estuary (Fig. 8.6).

There is also what we might term a transitional component. This includes those organisms such as migratory fishes that pass through the estuary on their way to breeding grounds, either in fresh or salt water. Common examples are the salmon (*Salmo, Oncorhynchus*) and the eels (*Anguilla*). Also included here are those forms that spend only a part of their life histories in the estuary. Usually it is the juvenile stage, the adults being found at sea. Good examples of this latter group are the various shrimps of the family Penaeidae (*Penaeus setiferus, P. aztecus, P. duorarum*), which form the basis of the Gulf of Mexico shrimp fishery. The young occur in estuaries. The transitional fauna also includes those forms that enter the estuary only to feed and includes many birds and fishes.

The number of species of organisms inhabiting estuarine systems, as noted by Barnes (1974), is generally significantly lower than the number inhabiting nearby marine or freshwater habitats. This is probably partly due to the inability of freshwater organisms to tolerate the increased salinities and marine organisms to tolerate the decreased salinities of the estuary.

The true estuarine organisms are derived primarily from marine stocks and not fresh water. This is similar to the situation in the other transitional zone, the intertidal, which is also populated mainly with marine organisms, not terrestrial forms. In contrast to the intertidal, however, the number of true estuarine species is very small, and the middle reaches of estuaries are depauperate (Fig. 8.6).

Because the marine animals are able to tolerate a greater reduction in salinity than the freshwater species are able to endure salinity increases, and because the true estuarine organisms are primarily derived from marine stocks, the majority of the estuary is inhabited by marine animals.

Why are there so few estuarine species? The most common explanation is that the fluctuating environmental conditions, mainly salinity, are of such magnitude that only a few species have been able to evolve the necessary physiological

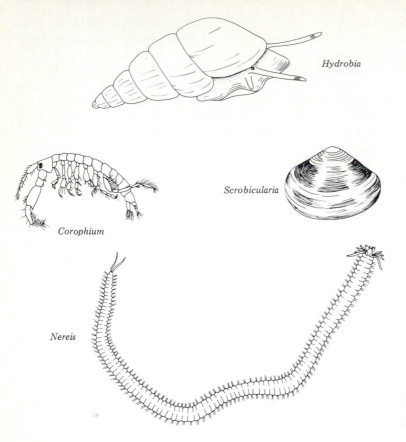

Hydrobia

Corophium

Scrobicularia

Nereis

Figure 8.7 Some typical estuarine animals. (Not to scale.)

specializations to enable them to exist there. Another explanation is that estuaries have not existed long enough in geological time to permit a complete estuarine fauna to develop. It should be noted that these two hypotheses are not mutually exclusive, so both could be operative. Yet a final reason may be that estuarine areas have little topographic diversity, being mainly broad expanses of mud. There are fewer niches and, therefore, fewer species. It is not possible at this time to say if one, all, or none of the above are responsible for the fact that there are so few species, but apparently euryhalinity is a trait not easily acquired.

Estuarine Vegetation

The macroflora of estuaries is also limited. Most of the permanently submerged portions of estuaries consist of mud substrates unsuitable for macroalgal attachment. In addition, the highly turbid water restricts light penetration to a narrow upper layer. As a result, the deeper layers of the estuary are often barren of plant life. The uppermost layers of water and the intertidal zone have a limited number of plants. In the lower reaches of the estuary at and below mean low water, there may be beds of seagrasses (*Zostera, Thalassia, Cymodocea*). These beds are considered as subtidal communities in Chapter 5.

The intertidal mud flats are inhabited by a limited number of green algal species. Common genera include *Ulva, Enteromorpha, Chaetomorpha,* and *Cladophora.* These are often seasonally abundant, disappearing during certain times of the year (Fig. 8.8).

Estuarine mud flats often have an abundant diatom flora. In fact, as Lackey (1967) noted, the benthic diatoms are more abundant in estuaries than their planktonic relatives. Many are motile and undergo a rhythmic migratory pattern, moving up to the surface or down into the mud

Cladophora sericea

Chaetomorpha linum

Enteromorpha intestinalis

Figure 8.8 Typical estuarine algae. (Not to scale.)

depending on the illumination. Estuarine muds also provide a suitable habitat for filamentous blue-green algae of several types. These algae may form mats up to 1 cm thick on the mud surface and on the bases of various emergent vascular plants.

The highly turbid water in estuaries means that by far the dominant vegetation in terms of biomass is emergent plants. These are generally long-lived flowering plants that root in the upper intertidal and form the characteristic *salt marsh*

communities that fringe estuaries throughout the temperate zones of the world. The dominant genera include *Spartina* and *Salicornia* (see subsequent sections of this chapter for discussion). In the tropics, another community, the mangrove forest, is found (see Chapter 9 for a discussion).

A final component to mention is the bacteria. Both the water and the mud of estuaries are extremely rich in bacteria because of the abundance of organic matter to decompose. Estuarine waters have been shown by Wood (1965) to contain hundreds of times more bacteria than seawater, and the upper layers of mud more than a thousand times more bacteria than the overlying water. Densities of bacteria in estuarine muds of 100 to 400 million per gram have been reported.

Estuarine Plankton

The estuarine plankton is also reduced in number of species. It thus follows the same trend as observed for the macrofauna and macrovegetation. Diatoms frequently dominate the phytoplankton, but dinoflagellates may achieve dominance during the warmer months and may remain dominant in some estuaries at all times. Dominant diatom genera include *Skeletonema, Asterionella, Chaetoceros, Nitzchia, Thalassionema,* and *Melosira.* Abundant dinoflagellate genera include *Gymnodinium, Gonyaulax, Peridinium,* and *Ceratium.* The phytoplankton may also be temporarily enriched by resuspension of characteristic bottom-dwelling diatoms. High turbidity and rapid flushing may restrict phytoplankton numbers and productivity in some estuaries. However, where turbidity is low and flushing time long, diverse populations and relatively high productivity may result. Consequently, depending on prevailing conditions, estuaries may differ considerably among themselves with respect to both phytoplankton numbers and productivity.

Zooplankton in estuaries mirror the phytoplankton in being limited in species composition. The species composition also varies, both seasonally and with salinity gradients up the estuary. The few true estuarine zooplankters occur in larger, more stable estuaries, where salinity gradients are less variable; shallow, rapidly flushed estuaries are inhabited mainly by a typical marine zooplankton assemblage carried in and out with the tide. Characteristic estuarine zooplankters include species of the copepod genera *Eurytemora, Acartia, Pseudodiaptomus,* and *Centropages;* certain mysids such as species of the genera *Neomysis, Praunus,* and *Mesopodopsis;* and certain amphipods such as species of *Gammarus.* The estuarine zooplankton average about 1 ml/m^3 displacement volume, or some-

what greater than the concentration in adjacent coastal waters.

ADAPTATIONS OF ESTUARINE ORGANISMS

The variable nature of the estuarine habitat, especially defined by fluctuating salinities and temperatures, makes this a particularly stressful and rigorous habitat. In order for organisms to survive and successfully colonize this area, they must possess certain adaptations.

Morphological Adaptations

Few morphological adaptations can be recognized among estuarine organisms that can be attributed simply to living under conditions of fluctuating temperature and salinity. Most are simply the result of adaptations to a given habitat such as burrowing into mud. (See Chapter 6 for a discussion of adaptations of organisms living in mud.) Mud-dwelling organisms, whether estuarine or not, often have, for example, fine fringes of hair or setae, which guard the entrances to respiratory chambers so as to prevent the clogging of those respiratory surfaces by the silt particles. Such a situation prevails for estuarine crabs and many bivalve mollusks.

Other morphological changes in estuarine organisms, reported by Remane and Schlieper (1971), include a generally smaller body size than relatives living in full seawater and a reduced number of vertebrae among fishes. Reproduction is also affected. Marine species in an estuary often have a lowered reproductive rate and lowered fecundity; freshwater species may be partially sterile.

There are also certain morphological adaptations among vascular plants living in estuaries and salt marshes. All appear to be relatively small, perhaps the result of long-term selection stemming from water stress. All seem to have a special tissue, the *aerenchyma,* that brings oxygen down to the roots, otherwise embedded in anoxic mud. This tissue may also serve to bring oxygen indirectly to animals in the mud. Many plants also have mechanisms for dealing with salinity. Some, such as *Spartina alterniflora,* have glands for eliminating salt and thus maintaining a stable water balance. Others have extensive stores of carbohydrates in the roots to provide a sugar source to cope with salinity fluctuations. Many also have high lignin content, which may be an adaptation to add strength to the plant to cope with high internal salinity. Abundant stomata and thin cuticles are suggestive of high transpiration rates.

Physiological Adaptations

The dominant adaptations that are required for continued estuarine life are those associated with maintaining the ionic balance of the body fluids in the face of fluctuating external salinities. *Osmosis* is the name given to the physical process in which water passes through a semipermeable membrane, separating two fluids of different salt concentration, moving from the area of lower to higher salt concentration. The ability to control the concentration of salts or water in internal fluids is called *osmoregulation.* Most marine organisms do not have the ability to control their internal salt content and are *osmoconformers.* Their ability to penetrate estuaries is thus limited to their tolerance for changes in their internal fluids. *Osmoregulators* are those organisms that have physiological mechanisms to control the salt content of their internal fluids. Most estuarine animals either are osmoregulators or else they possess the ability to function with fluctuating internal salt concentrations.

Penetrating into an estuary means encountering water with lowered salinity. Since the internal salt concentration of marine species is higher than that of the estuarine water, water tends to move across membranes into their bodies in an attempt to equalize the concentrations. Regulation means excreting the excess water without losing salts or excreting water and salts and replacing lost salts with active uptake of ions from the environment. For freshwater animals, which move from a more dilute to a less dilute medium when entering estuaries, the reverse is true.

Osmoregulatory ability is found in several taxa of estuarine animals, primarily polychaete worms, mollusks, and crustaceans. The field of

osmoregulation has been the subject of intensive research for many years, resulting in a large literature. Detailed coverage is beyond the scope of this introductory book, but students interested in more in-depth studies may refer to the literature cited at the end of this chapter. Suffice it to say that osmoregulation can operate in three ways: (1) animals may move water; (2) they may move ions; or (3) they may adjust the internal water-ion balance. In advanced invertebrates and vertebrates, the osmoregulatory organ is generally the kidney, where excess water is excreted and needed ions resorbed. Among many invertebrates gills are the most common osmoregulatory structures. However, special cells may also exist in other parts of the body, particularly for taking up or removing certain ions.

In certain soft-bodied forms, such as polychaete worms, an osmoregulatory mechanism is developed but is relatively slow to respond. These animals are able to tolerate wide ranges of internal ion concentrations, at least for certain periods of time. *Nereis diversicolor,* for example, when placed in dilute seawater takes up water due to osmosis. After some time, however, it begins to osmoregulate and, according to McLusky (1971), the internal water content falls to near that found when it was in full seawater (Fig. 8.9). This osmoregulation not only is delayed, but does not come into effect until the salinity of the external environment falls below a certain level. That is, *N. diversicolor* remains an osmoconformer as salinity decreases and regulates only when the external salinity falls below a certain level (Fig. 8.9).

Bivalve mollusks are generally poor osmoregulators and respond to drastic salinity decreases by closing up in their shells to prevent excessive dilution of internal fluids with water. Certain gastropods, such as *Hydrobia,* do have limited osmoregulatory ability, and the members of the genus are found in enormous densities in northern European estuaries, where Muus (1963) reports they live in salinities down to 5 ‰ or lower.

Among higher crustaceans, such as crabs, osmoregulation is well developed. Greatly restricted body permeability due to the exoskeleton and the pronounced ability to regulate ion concentrations of internal fluids are prob-

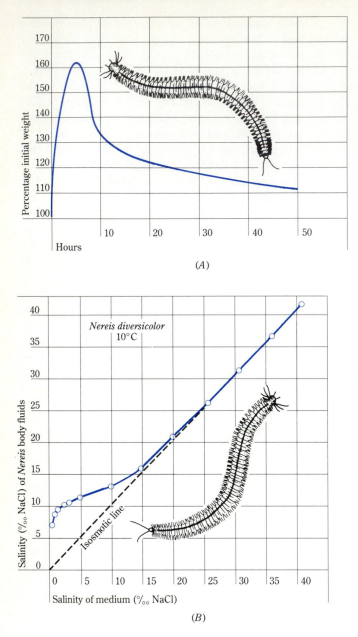

(A)

(B)

Figure 8.9 Changes in the body fluids of *Nereis diversicolor* with change in salinity. *(A)* Weight changes occurring after the animals are transferred to 20 percent seawater. Initial weight increase is due to the take-up of water by osmosis. The weight decrease occurs when the animal osmoregulates. *(B)* Osmotic concentration of the body fluids in relation to salinity change. The straight line on the graph indicates osmoconformity. *(A, B,* after D. S. McLusky, Ecology of estuaries, Heinemann, 1971.)

able reasons for their success in estuaries. Among vertebrates, osmoregulation is primarily through water removal by the excretory organs coupled with ion uptake from the environment to make up for unavoidable losses in water removal. In invertebrates other mechanisms may be involved. For example, the cells of crustaceans are regulated by changes in the concentration of amino acids. This "intracellular osmotic regulation" results from movements of free amino acids in and out of individual cells to equalize the osmotic concentration in the cells with the extracellular bathing fluid, thus preventing excessive water buildup and possible rupture of the cells (Fig. 8.10).

As Kinne (1963) has noted, there appears to be some correlation between temperature and the ability to osmoregulate. In the tropics where

Figure 8.10 Graph showing the change in salinity of the blood of the crab *Australoplax tridentata* with change in the salinity of the external medium. (Modified from D. S. McLusky, Ecology of estuaries, Heinemann, 1971.)

water temperatures are higher and the difference in temperature between fresh and salt water is minimal, there are considerably more estuarine species, and stenohaline marine species penetrate further upstream than they do in the temperate zone estuaries. The reasons for this phenomenon are not as yet clearly understood.

Vascular plants growing under estuarine conditions also display a set of physiological adaptations to osmotic stress. These rooted plants must maintain high concentrations of osmotically active substances in order to counter the water-retaining capacity of the surrounding medium. Generally, when the osmotic potential (= increase in ion concentration) of the water in the soil around the roots increases, the plants respond by increasing the ion concentration of their tissues. Such an osmotic adjustment is the plant equivalent of osmoregulation in animals. In saline environments, salts are continuously transported into the plant, and in order to prevent salts from reaching a lethal level, most plants have mechanisms to get rid of excess salts. Such mechanisms may include salt-secreting glands, leaching by rains of salts deposited on external surfaces by transpiration, and the shedding of leaves loaded with excess salts. Nonsucculent halophytes tend to lose salts by some sort

of secretion, whereas succulents release salts by shedding leaves.

Behavioral Adaptations

A certain number of behavioral adaptations are common to many estuarine organisms. Among the invertebrates one such adaptation is burrowing into the mud. Although this adaptation is certainly not one that is exclusive to estuarine animals, being common to many invertebrates in soft sediments throughout the world's oceans, it has two beneficial effects for an estuarine animal. First, in the case of those species with limited or imperfect osmoregulatory abilities, existence in the mud means exposure to the interstitial water, which has much less salinity and temperature variation than the open water above. It is thus a means to achieve reduced salinity and temperature changes. Second, burying in the substrate means that these species are less likely to be consumed by surface or water-dwelling predators such as birds, fishes, or crabs.

Many adult estuarine crabs are able to live successfully in low salinities due to their developed osmoregulatory abilities. However, their eggs and young often lack these regulatory abilities. Therefore, many exhibit a specific migratory pattern, moving from the estuary to the adjacent sea for the breeding season. One example of this type of behavior is the crab *Eriocheir sinensis,* which has been introduced into Europe from its native China. Although adults live hundreds of miles up into rivers, *E. sinensis* migrates back to the sea to breed and to spend the early part of larval life. Similarly the blue crab *Callinectes sapidus* lives as an adult in estuaries of the American East and Gulf coasts, including Chesapeake Bay. The females migrate to waters of higher salinity to hatch the young. Once the larval stages are passed, the young crabs migrate back up the estuary (Fig. 8.11).

On the other hand, certain other species take advantage of the great amount of food in the estuary and the relative paucity of predators to use the estuary as a nursery for their young. Most of these species are fishes. Examples include mullet (*Mugil* sp.), striped bass (*Roccus saxatilis*), and flounder (*Platichthys flesus*),

Figure 8.11 The life cycle of the blue crab *Callinectes sapidus* in estuaries of the Atlantic coast of the United States.

which enter the estuary as juveniles and migrate back to the sea when mature. Various shrimp (*Penaeus*) also use the estuary as a nursery.

ECOLOGY OF ESTUARIES

Having considered the major physical-chemical features of estuaries, the common organisms, and the major adaptations of the organisms, it is now possible to consider certain aspects of the ecology of these transition areas.

Productivity, Organic Matter, and Food Sources

If we ignore for now the salt marsh, the primary productivity of estuaries resides in the phytoplankton, the benthic diatoms, seagrasses, and the various algal mats. Of these, the benthic diatoms and algal mats appear to be significant. When all these sources are considered together, however, primary productivity by algae has traditionally been considered to be very low. Yet estuaries are areas that have large amounts of organic material, large numbers of organisms, and high secondary productivity. If algal primary productivity is low, where does this organic material originate? Classically, it has been considered that the major source of primary productivity in estuaries resided not in the estuaries at all, but in the emergent plants of the salt

marsh surrounding the estuary (see pp. 328–331). In an oft-cited study by Teal (1962) on a Georgia salt marsh, it was found that there was a net productivity of 6850 kcal/m^2/year, whereas the various algae contributed only 1600 kcal/m^2/year. Algae may, however, be more important in some areas such as the Pacific coast of North America, where Zedler et al. (1978) have shown that they contribute significantly to primary productivity. In addition, studies by Haines (1979) in the Georgia marshes have suggested that the detritus derived from benthic and planktonic algae forms the bulk of the organic detritus. Haines found salt marsh plant detritus to be largely accumulated and consumed within the marsh itself, leaving the algal detritus as the major material exported. In this scenario, the role of the marsh is primarily as a nursery for fishes, crabs, and shrimps, which are later exported to coastal waters.

Primary productivity in and around the estuary, however, is not the only source of organic material. As we noted previously, estuaries act as sinks for organic material brought down by rivers and in from the sea. Thus, organic matter arrives from these sources as well. It is difficult to assess the role of primary productivity within the estuarine system in contributing to the total organic production for several reasons. In the first place, there are few herbivores that feed directly on the plants themselves. Therefore, most of the plant material must be broken down to detritus before entering the various food webs. This process of breakdown involves bacterial action, which is an additional complicating factor. Also, once the plant material has become detritus, it is not easily distinguished from other organic detritus brought into the system by the river and the sea. Thus, of the great amounts of organic material that move through the estuary, some fraction is produced in the estuary itself (autochthonous), but some is transported in, the result of primary production elsewhere (allochthonous), including transport from the surrounding salt marsh. Currently, the relative amount contributed by each source is under debate.

McLusky (1989) has considered the various sources of "food" in the form of carbon in the temperate estuaries of Europe and North America. From these studies he has separated estuaries into two major types that are at the opposite ends of a spectrum and named for their primary geographical locations. In the *European* type (e.g., Britain, the Netherlands) the estuary is dominated by large mud flats, relatively barren of large plants, but with large populations of surface benthic diatoms, which produce most of the primary productivity along with water-column phytoplankton. These estuaries derive most of their energy sources from outside the estuary in the form of organic matter imported from the sea or river. They are net recipients of energy. Their ability to support large populations of consumers stems not from their productivity but from the fact that they are sinks for imported energy. At the opposite extreme is the *American* estuary (e.g., Louisiana, Florida, Georgia), dominated by extensive stands of *Spartina* grass and other salt marsh plants surrounding the less extensive mud flats. Phytoplankton and benthic diatoms are still significant primary producers, but are overshadowed by the huge productivity of *Spartina* and the salt marsh plants. This type of estuary produces an excess of carbon, which is then exported to the surrounding sea, usually as detritus.

The detritus forms a substrate for a rich bacterial and algal growth, which, in turn, serves as an important food source for various suspension- and detritus-feeding animals. When one refers, therefore, to the detritus food source of estuaries, *detritus* is used in the widest sense to refer to organic particles, bacteria, algae, and even associated protozoans. The net effect of this detritus is to form an accumulation of food for estuarine organisms. As an example of the amount of food, Odum and de la Cruz (1967) noted that open seawater contains 1–3 mg of dry organic matter per liter, whereas drainage waters in estuaries may contain up to 110 mg dry weight organic matter per liter.

Food Webs

Low primary productivity in the water column, few herbivores, and the presence of large amounts of detritus have traditionally suggested that the basis of the food web in estuaries is detritus. This claim is currently under debate among scientists, but it is a convenient starting point.

To some extent to consider detritus as the basis of estuarine food webs is misleading, for it implies that detritivores actually digest the organic particles. In fact, they most likely are digesting the bacteria and other microorganisms on the particles and excreting the particles without further degradation. This should be kept in mind in the following discussion.

Detritus may be consumed directly, either while in suspension in the water mass by various suspension-feeding benthic invertebrates, or more commonly by direct consumption of the material in or on the substrate. Suspension feeders, such as the clams *Cardium, Mya,* and *Mercenaria,* may obtain a small portion of their food from suspended detritus along with phytoplankton and zooplankton. Direct detritus consumers (= deposit feeders) include certain clams such as *Scrobicularia* and *Macoma,* which have separated siphons and feed by "vacuuming" up the detrital particles from the substrate via their inhalant siphons (Fig. 6.40).

Polychaete worms are very abundant in estuaries and are represented by many detritus feeders. Members of the families Spionidae, Terebellidae, and Ampharetidae spread out long tentacles over the surface of the mud and collect and transport the organic particles via ciliary-mucous tracts. Still others, such as the families Capitellidae and Arenicolidae, are deposit feeders, ingesting the substrate directly, digesting out the organic material and bacteria during passage through their guts. Bacteria may be consumed directly by small protozoans and nematodes, both of which occur in large numbers in the upper layers of the mud.

Others, such as amphipods of the genus *Corophium,* mechanically sort particles using their appendages, while small gastropods such as *Hydrobia* and *Batillaria* rasp up the surface of the mud (Fig. 8.12). Such sorting out or digesting out of the sediments leads to a reworking of the substrate.

The large numbers of deposit feeders and suspension feeders are, in turn, consumed by a number of predators, both vertebrate and invertebrate. The main invertebrate predators are various species of crabs and shrimps; certain polychaetes, such as the genera *Glycera, Nephtys,* and perhaps some *Nereis;* and various predatory gastropod mollusks, such as *Polinices, Aglaja, Chelidonura,* and *Busycon.*

The dominant predators, however, are various species of fishes and birds. In European estuaries, for example, Hartley (1940) found that the plaice (*Pleuronectes platessa*) feeds on polychaetes, while the flounder (*Platichthys flesus*) feeds on mysids, shrimps, and amphipods. In Louisiana estuaries, Darnell (1961) found the three most important larger invertebrate species eaten by fishes to be the clam *Rangia cuneata* and two crabs, *Rithropanopeus*

harrisii and *Callinectes sapidus.* Different fish species tend to concentrate on different types of animals, and feeding habits change with age. A common pattern in estuarine fishes is to progress from eating zooplankton, to detritus, to macroinvertebrates, to even other fishes. Much of the productivity of estuaries that ends up in fishes is ultimately lost to the estuarine system when the fishes move offshore as adults.

Estuarine birds include many ducks and geese as well as a host of shorebirds, wading birds, and gulls and terns. The majority of the shorebirds feed upon the infaunal invertebrates of the mud flats, which they obtain by probing at low tide (see also the discussion in Chapter 6, pp. 277–279). The variety of bill lengths among these birds enables the different species to feed at different levels in the mud (Fig. 8.13). They follow the rise and fall of the tide during daylight hours to obtain their food. Observations and analyses of stomachs of these birds suggest that they are capable of consuming large numbers of invertebrates. Studies by Drinnan (1957), Goss-Custard (1969), and Prater (1972) have reported that populations of oystercatchers (*Haematopus ostralegus*) consume 315 *Cardium,* that redshanks (*Tringa totanus*) consume 40,000 *Corophium,* and that knots (*Calidrus canutus*) consume 730 *Macoma* per day. The effect of this

Figure 8.12 Mud snails, *Batillaria attramentaria,* on the surface of the mud in a California estuary. (Photo by the author.)

Figure 8.13 Diagram showing bill length of some common European shorebirds in relation to the depth of some common estuarine invertebrates. (After J. Green, The biology of estuarine animals, University of Washington Press, 1968.)

predation on the populations of infaunal invertebrates is not well known, but various scientists have suggested that birds take between 4 and 20 percent of the populations.

In recent years, there has been considerable interest in conducting field experiments that exclude large predators from the soft-bottom invertebrate communities of estuaries to determine the effect of predators on these communities. In summing up the results of these experimental manipulations, Peterson (1979) has noted that, when such systems are freed from predation, the usual results are for the density of benthic organisms to increase, but without a corresponding tendency toward competitive exclusion by certain dominants. This is in marked contrast to the results we have seen on rocky shores, where predator removal has led to a competitive dominant occupying all space, forcing other species out. Why should such a different result occur here? Peterson (1979) has suggested several reasons. The initial one is simply that such experiments have not been conducted for a sufficient length of time. A second is that interference competition, which operates in the rocky intertidal, cannot operate in these soft muds because of the inability of the organisms to obtain a purchase from which to push away or crush another organism and because of the ability of the organisms to exploit the third dimension (= depth) of the soft substrate. Competitive exclusion by overgrowth, common on rocky shores, is also precluded because of high sediment mobility, which excludes those species best able to employ this mechanism—namely, colonial or clonal organisms. It may also be possible that various adult-larval interactions are of sufficient intensity and development that they can hold the community at a density below its carrying capacity and hence prevent competitive exclusion without the necessity of predation (see also the discussion of intertidal mud flats in Chapter 6).

The estuarine food web can be summarized as having a number of pathways of energy flow generated primarily from a detrital base, and by a somewhat incomplete use of energy, allowing export to other areas primarily by means of fishes and birds. The incidence of omnivory is generally higher than in other systems. It is the abundance of food and the relative lack of predators that has permitted estuaries to be nurseries exploited by the juveniles of many animals that spend their adult lives elsewhere. It is also the reason for their use as a feeding ground by migratory adults of many species of birds and fishes. The complex interactions of the estuarine food web are summarized in simplified form in Figs. 8.14 and 8.15.

Plankton Cycles

Although the nutrient levels in estuaries are high, they are not well balanced; nitrogen is often low and may even occasionally be limiting to estuarine phytoplankton. As we observed in the open sea (Chapter 2), there is an annual cycle of phytoplankton in estuaries also. In the temperate zones, low phytoplankton populations usually occur in late fall and winter due to reduced light and high turbidity, resulting from high river runoff and turbulence. This is followed by a later winter diatom bloom. This bloom is terminated in late spring, often not by zooplankton grazing, but probably by depletion

of nitrogen sources, leading to the accumulation of diatoms on the mud surface. Populations remain low in the summer due to low nutrient levels and grazing, but there may be occasional blooms of dinoflagellates, leading to "red tide" situations. A large fall bloom occurs in some estuaries that rivals the spring bloom, whereas in others it is absent or reduced.

When one considers the relative abundance of nutrients, the phytoplankton in estuaries have a relatively low annual rate of production. Mc-Lusky (1989) suggests that this apparently stems from three factors: first, the shallow water, the depths of which are much less than the optimal depth for producing maximum photosynthesis; second, the turbidity of the water, which limits light penetration; and, third, the continual removal of the phytoplankton, which results from the rapid flushing rate of the estuary.

The zooplankton of estuaries may also be limited or reduced. This limitation is either the result of turbidity, which by limiting the phytoplankton limits the food available to the zooplankton, or the result of currents. Currents

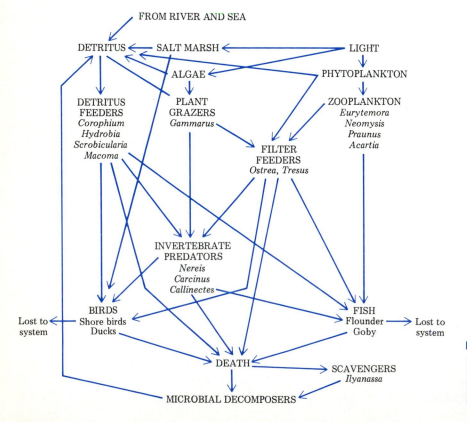

Figure 8.14 A generalized estuarine food web. (Modified from D. S. McLusky, Ecology of estuaries, Heinemann, 1971.)

Figure 8.15 The food web of a typical estuary showing some of the major aquatic trophic groupings. (After Day et al., Estuarine ecology, Wiley, 1989.)

carry the zooplankton out to sea. Elimination by currents can be reduced by the zooplankton remaining close to the bottom.

The zooplankton cycle in estuaries is quite variable and, as Deevey (1960) found, may even change drastically from year to year. It does not closely follow the phytoplankton cycle. Zooplankton peak in number later in the spring than the phytoplankton, and their numbers remain high throughout the summer, declining through the fall to winter lows. Zooplankton in estuaries appear to be consuming only 50 to 60 percent of the net phytoplankton production. This leaves a significant portion available to the bottom suspension feeders; indeed, in shallow estuaries, the suspension-feeding benthos competes directly with the zooplankton for food, a situation not present in offshore waters.

SALT MARSHES

Bordering temperate and subpolar estuaries and protected marine shores and embayments throughout the world are special plant associations known as salt marshes (Fig. 8.16). These marshes form a transition area between aquatic and terrestrial ecosystems.

Definition and Characterization

Salt marshes are communities of emergent herbs, grasses, or low shrubs rooted in soils that are alternately inundated and drained by tidal action. They occur mainly at the higher tidal levels in areas of protected water and most often in association with estuaries. Since the dominant plants are emergent flowering plants, they invade only the shallowest intertidal areas. They are *halophytes,* meaning that they grow in soils with a high salt content. The dominant plants are herbaceous angiosperms. Because the upper portions of the plants are above water even during periods of high tide, this association has both terrestrial and aquatic components.

Environmental Characteristics

The salt marsh is a rigorous environment that shows wide variations in several environmental factors. It shares with the estuary a fluctuating *salinity* that results from the interaction of river

Figure 8.16 The distribution of salt marshes in the world.

flow and marine water, but because it is intertidal, the salinity variation may be more sudden and more extreme than that experienced in the waters of the estuary itself. A sudden rainstorm at low tide may reduce surface salinities to near zero, whereas the return of the high tide may inundate the marsh with nearly full-strength seawater. It is not uncommon for marshes to experience salinities varying from 20 to 40 ‰ in a single tidal cycle.

Similarly, *temperature* undergoes wide fluctuations. Exposure at low tide opens the marsh and substratum to the extreme air temperatures of the terrestrial environment. Since these marshes occur in the temperate zone, this means air temperatures below freezing in winter and above 30°C in summer. Indeed, mud surface temperatures may vary 10°C during a single day.

The *substrate* is typically mud, essentially similar to the estuarine sediments, and with a high salt content, which results from the perfusion of salt water coupled with a high evaporation rate. Salt-marsh soils are often anaerobic. Within the marsh, the rate of sedimentation is higher in the lower part of the marsh than in the upper.

The various environmental factors influencing the dominant vegetation are summarized in Figure 8.17.

Composition and Distribution

One of the most characteristic general features of salt marshes is that they are species poor. Because of the rigorous environmental conditions and the high soil salt content, relatively few plants and animals are able to inhabit this area. Furthermore, these tolerant organisms show a high degree of taxonomic similarity over a wide geographic area. Dominant plants of salt marshes worldwide are grasses of the genus *Spartina* and species of the genera *Juncus* and *Salicornia* (Fig. 8.18). Characteristic animals of marine origin are various species of crabs (*Uca, Hemigrapsus*), mussels (*Modiolus*), certain snails (*Littorina, Cerithidea, Melampus*), and smaller crustaceans such as amphipods.

The insects represent the major terrestrial animal component of the marsh and may live permanently on the marsh. There are also terrestrial animals, including raccoons, that enter the marsh only for feeding. At high tide, marine and

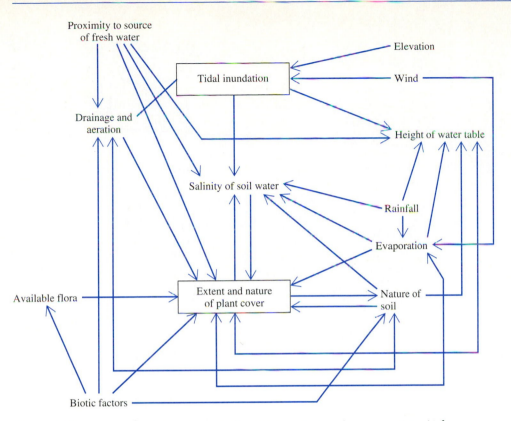

Figure 8.17 Environmental factors influencing salt-marsh vegetation. (After D. S. McLusky, The estuarine ecosystem, 2nd ed., Chapman and Hall, 1989.)

estuarine animals enter the marsh, and at low tide, terrestrial species forage in the marsh (Fig. 8.19).

Despite the taxonomic similarity of the salt-marsh species over broad geographic areas, certain combinations of topographic features and species distributions have produced several different marsh types. In the United States, there are two major groups of salt marshes, those of the Atlantic and Gulf coasts on the one hand and those of the Pacific coast on the other. On the East and Gulf coasts salt marshes are far more extensive than West Coast marshes and occupy large areas of gently sloping coastline which surrounds the numerous broad estuaries and shallow bays associated with the extensive shallow offshore continental shelf. The Pacific coast, by contrast, has few rivers and bays, and steep coastal mountains lead directly to a narrow continental shelf. Consequently, river mouths are narrow and the extent of both estuaries and

salt marshes is restricted. The East and Gulf coast marshes are dominated by various species of *Spartina,* which succeed each other in elevation and geographically. The more poorly developed West Coast marshes are characterized by broad bands of *Salicornia. Spartina* is of lesser importance or, in some cases, absent.

Physiognomy and Zonation

The general appearance of a salt marsh is similar throughout the world. The picture is that of a broad, flat expanse of low herbaceous, grassy, or shrubby plants, all of generally similar form, cut by a dendritic arrangement of channels leading to larger tidal creeks (Fig. 8.20). These, in turn, lead out to mud flats devoid of macrophytes or to the open waters of the estuary. Within the marsh are scattered open areas (pans) and shallow pools. At the upland side of the marsh, a variety of terrestrial shrubs and trees replace the

Spartina foliosa

Salicornia

Juncus gerardi

Figure 8.18 Some dominant salt-marsh emergent plants.

halophytes. In contrast to the rocky intertidal where the majority of organisms are marine, in salt marshes the plants are mainly terrestrial in origin.

A cross section of a typical marsh from the tidal creek to the true terrestrial vegetation reveals a zonation pattern. This pattern varies in detail and species composition from one geographical area to another because of tidal regime, drainage, slope, climate, and other factors, but a basic common zonation pattern exists.

The lowest zone is that of the creek bank and bottom, usually barren of macrophytic vegetation and containing mainly infaunal estuarine and marine animals and many mud snails. The next zone above comprises the creekside marsh, a border zone between the creek bank and the main part of the marsh. The next zone is the

main part of the marsh, which may be variously subdivided. On the East and Gulf coasts, the subdivision is into a lower, tall *Spartina* zone and an upper, short *Spartina* or *Juncus* zone. The final marsh zone is a *Salicornia* zone, often with various other plants (Fig. 8.21). On the West Coast, marshes tend to have a narrow fringe of *Spartina* in the lower zone, followed by a broad expanse of *Salicornia* occupying most of the midintertidal zone. The highest intertidal area is occupied by a more diverse assemblage, of which *Jaumea, Distichlis,* and *Limonium* are common (Fig. 8.22). In contrast to intertidal rocky shores, where the reverse is true, Adam (1990) notes that competition among the various plant species tends to be the factor governing the uppermost zonal boundaries of plants in salt marshes, while the lower limits are set by

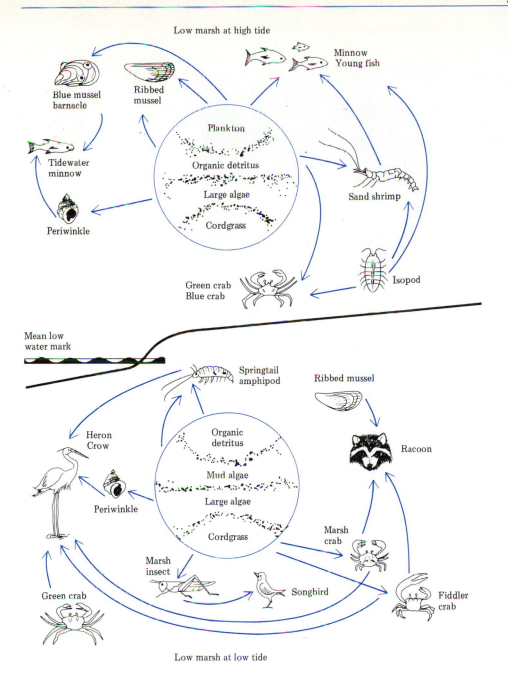

Low marsh at high tide

Blue mussel
barnacle

Ribbed
mussel

Minnow
Young fish

Tidewater
minnow

Plankton

Organic detritus

Large algae

Cordgrass

Sand shrimp

Periwinkle

Green crab
Blue crab

Isopod

Mean low
water mark

Springtail
amphipod

Ribbed mussel

Heron
Crow

Organic
detritus

Racoon

Mud algae

Periwinkle

Large algae

Cordgrass

Marsh
crab

Marsh
insect

Songbird

Fiddler
crab

Green crab

Low marsh at low tide

Figure 8.19 Characteristic animals present in a salt marsh at low and high tides on the Atlantic coast of North America.

physiological tolerances for submersion in salt water (for comparison see pp. 232–234 in Chapter 6).

Although the physical development of a marsh is a constant process resulting from accre- tion of sediment, there is little evidence that a predictable or directional succession will follow this accretion. Instead, as Adam (1990) notes, there appear to be long periods of vegetation stability punctuated by brief periods of rapid

Figure 8.20 A tidal creek penetrating into a coastal salt marsh on the Pacific coast of North America. (Photo by the author.)

change. Thus, it seems more appropriate to think of salt marshes as steady-state systems in which various small temporal changes may occur, but with no overall directional change for the whole marsh. Marshes undoubtedly do change over time, but now it is not possible, according to Adam, to predict what the changed system will be like nor to determine if it will repeat a historical pattern or sequence. This, in turn, suggests that, when changes in a marsh do occur, they are the result of changes in the physical environment rather than the result of biological interactions.

Productivity

Primary productivity within the marsh is carried out by the emergent marsh plants and the various microalgae that occur on the surface of the marsh plants and the mud. In contrast to the water and mud flats of estuaries, productivity is very high in marshes, especially for the rooted emergent plants.

Although salt marshes have been extolled as one of the most productive of all plant communities in many introductory texts, it is now clear that both the various methods and assumptions employed to obtain the primary production figures often have been flawed or have suffered from various methodological limitations that make them suspect or have given rise to widely varying estimates. Therefore, the figures presented here may well be revised in the future and

are only presented to give a feel for the relative production among different areas.

It is most convenient to consider primary productivity in two groups: vascular emergent plants and various algae. The most complete data on productivity exist for the vascular plants and especially for the East Coast marshes. The major primary producers in these marshes are the various species of *Spartina.* Teal (1962) has shown Georgia marshes to have an average net productivity of 1600 g C/m²/year. This decreases to the north, such that in New Jersey Good (1965) reports only 325 g C/m²/year. On the Gulf Coast marshes the primary productivity of vascular plants ranges from 300 to 400 to more than 3000 g C/m²/year according to Turner (1976). On the Pacific coast, Atwater et al. (1979) report San Francisco marshes at 50 to 1500 g C/m²/year of net production (compare with Table 2.2, p. 67), and Seliskar and Gallagher (1983) have given estimates of from 100 to 1000 g C/m²/year for the marshes of the Pacific Northwest.

Algal productivity estimates are fewer, more variable, and, for many areas, simply unavailable. On the Atlantic coast, studies by Teal (1962), Gallagher and Daiber (1974), and Van Raalte et al. (1976) have provided productivity estimates from 20 to 25 percent of that of the emergent vegetation. On the Pacific coast Zedler (1982) has shown, however, that algal productivity in southern California marshes is 40 to 60 percent of that of the emergent vegetation. In the Pacific

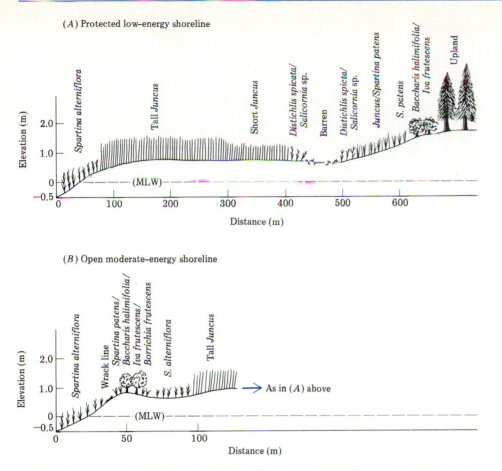

Figure 8.21 Generalized diagrams of Gulf coast salt marshes on (*A*) protected low-energy shorelines and (*B*) open moderate-energy shorelines. (From J. P. Stout, The ecology of irregularly flooded salt marshes of the northeastern Gulf of Mexico: A community profile, U.S. Fish & Wildlife Serv. 85 [7.1], 1984.)

Northwest, Seliskar and Gallagher (1983) estimate algal production to be considerably less, 100 g C/m²/year.

Zedler (1982) has suggested that the reduced productivity of vascular plants and enhanced productivity of algae observed in southern California marshes, but not in the East Coast marshes, may be due to the fact that these California marshes exist surrounding intermittent estuaries where the soils are hypersaline during the summer growth period. The hypersalinity reduces the growth of the vascular plants, leading to a relatively open canopy; hence more light penetrates to the substrate, enhancing algal production. Zedler also points out that this system may be more energy efficient, since algae can be consumed directly, whereas the vascular plants must be attacked by decomposers first, thereby adding another step to the food chain and reducing the total energy available to higher trophic levels.

Because there are few herbivores, most of the productivity of the marsh is not consumed directly. In the North Temperate Zone the major natural grazers on the emergent salt-marsh vegetation are the various ducks and geese. These waterfowl may have a significant effect where they occur in large numbers (see pp. 198–200). Why there should be a dearth of herbivores is not known, but since marsh vascular plants have high salt content and low nutritional value, it may be that they are unpalatable, difficult to break down, or both. The majority of the vascular plant production, perhaps 90 percent of it, is

Figure 8.22 Generalized zonation of a Pacific coast salt marsh based on the marshes of San Francisco Bay.

channeled into detritus and is either decomposed in situ or washed out of the marsh. The dead plants are broken down by bacteria on the surface of the mud or in the water. During this process the food value of the plant parts is actually increased as the indigestible cellulose is broken down into digestible carbohydrates, and the protein content rises as the microorganism biomass increases on the pieces of decomposing plant. On the other hand, the algal mats are often subject to heavier grazing pressure, probably in part because of their higher nutritional value and easier digestibility. For example, Pace et al. (1979) excluded the snail *Ilyanassa obsoletus* and found a rapid increase of the algal microflora.

Over the years there has developed a prevailing view among biologists that, because of the high productivity of marshes and because the productivity is little utilized or even underutilized by resident animals, salt marshes serve as an outwelling nutrition source that sustains the productivity of estuarine and near-shore waters, particularly fisheries. This concept was widely supported by the statement of Teal (1962) that 45 percent of marsh production was removed, permitting that fraction to be used to support organisms outside the marsh. Since that time additional studies and the use of new methodologies have led to a considerable modification of this view. According to Nixon (1980), there now seems to be agreement that salt marshes do export a certain amount of dissolved and particulate organic matter, but the magnitude of this exported material and its importance to productivity in adjacent waters are certainly variable,

depending on a number of biological and geographical features. For example, Nixon (1980) showed that, where marsh area is large relative to near-shore water, the organic supplement from the marsh may well amount to about 50 percent of the open-water phytoplankton production. Such areas would be like those found off Georgia. In other cases, such as Chesapeake Bay and most New England marshes, the organic supplement is 10 percent or less. Furthermore, there seems to be no evidence that the exported carbon ends up providing any significantly greater production of fish or shellfish than is found in coastal areas without marshes. We do not know why the export of organic matter does not lead to increased production at the higher trophic levels, but Nixon suggests it may be due to factors such as rapid sedimenting, the fact that the outwelled particles are of little nutritive value, or that the carbon is rapidly respired by bacteria.

Marshes also apparently export dissolved nitrogen and phosphorus, but according to Nixon (1980), these nutrients do not appear to have significant effects on primary productivity in near-shore waters, since primary productivity is not any higher in areas where marshes occur than in areas where they do not.

On the Pacific coast, the marshes are small and adjacent to a nutrient-rich, upwelling ocean. Therefore, any outwelling of carbon or nutrients would be insignificant to ocean productivity in these areas.

We can summarize this section by saying that our understanding of the relationships between salt marshes and estuaries and near-shore coastal waters has advanced in the past 20 years, but these are complex systems and our knowledge is still very incomplete.

Interactions and Food Webs

The salt marsh is an ecotone between estuarine and terrestrial systems. There are numerous interactions between the animals of both areas and the marsh plants and their environment.

In general, salt-marsh animals find themselves in a stressful environment of widely varying physical conditions. Few animals possess tolerance limits that include the range of temperature and salinity experienced in a marsh; therefore, marsh animals tend to evolve strategies that minimize the environmental variations to which they are exposed. As a result, there are more aquatic species found in low areas seldom exposed to the air and more terrestrial forms in areas above the effects of the highest tides. The fewest species occur in the intermediate area.

Marsh animals minimize the environmental changes in a variety of ways. Crabs commonly burrow into the mud, by which they escape desiccation and surround themselves with soil water of higher salinity. Coffee bean snails (*Melampus*) and certain insects escape inundation at high tide by climbing up the marsh plants. This may also be a strategy to avoid predators. A great many animals simply migrate in and out of the marsh with the tide—aquatic animals in at high tide and terrestrial ones at low tide.

The marsh plants themselves are beneficial to the animals in a number of ways (Fig. 8.23). They may provide cover that reduces predation. The various small invertebrates, such as crabs, snails, and insects, are less subject to predation from birds and fish if they are beneath the canopy of marsh plants than if they are exposed on the open mud flat. For example, Willason (1980) found that shelter of the marsh was important in determining survival of the crab *Pachygrapsus crassipes,* and Boland (1981) showed that most predatory birds utilized open mud flats rather than marsh areas. In southern California evidence exists that the native horn-snail (*Cerithidea californica*) is more heavily preyed on in open mud flats than in areas within the marsh. The plants also act to stabilize the mud and provide a substrate to which certain animals may attach. Without the presence of *Spartina,* for example, the mussel *Geukensia* would have nothing to which to attach its byssal threads; barnacles would be unable to exist due to lack of a surface to cement to; and the terrestrial insects and pulmonate snails would have no vertical substrate to climb to enable them to escape inundation at high tide. The marsh plants also act indirectly by reducing wave and current action, which might otherwise remove or destroy some animals. They also reduce light penetration to the mud surface and ameliorate the temperatures on the mud surface, thus decreasing the range of variation to which the animals are exposed.

In turn, the animals have effects on the plants. The direct effects are minimal because only a few relatively rare grasshoppers and leafhoppers actually consume the living plants; and these do not appear to have any significant effect on the plants. The burrowing activities of various crabs allow aeration of an otherwise stagnant mud and may benefit the roots of the marsh plants. On the other hand, burrowing may lead to harmful cropping of the roots. Burrowing by crabs and isopods such as *Sphaeroma* may also lead to bank erosion and loss of parts of the marsh as they collapse into the tidal creeks. Such destruction of marshland has been reported by Josselyn (1983) for San Francisco Bay. Invertebrate fecal matter may provide nutrients to the marsh plants, but this remains to be tested. Certain snails, such as *Melampus* and *Littorina,* graze the algae that settle on the stems of the marsh halophytes and may enhance their growth by eliminating or reducing this algal cover.

Competitive interactions between animals and plants in the salt marsh have been little studied and remain a fertile area for future work. The few studies that exist suggest that this process may be of considerable importance. For example, Race (1982) has demonstrated that in San Francisco Bay the introduced snail *Ilyanassa obsoletus* has eliminated the native snail *Cerithidea californica* from mud flats. The latter is now found only in the upper portions of the marsh, where it is more tolerant of desiccation than *Ilyanassa.* Similarly, Sliger (1982) found that the two common marsh crabs *Pachygrapsus crassipes* and *Hemigrapsus oregonensis* were segregated vertically. *Pachygrapsus* appeared to be the dominant crab and could eliminate *Hemi-*

Figure 8.23 Microhabitats in the salt marsh: (1) underneath the overhanging margin of a pan; (2) permanent water; (3) water surface; (4) organic mud on bottom of pan; (5) roots; (6) plant stems; (7) leaves; (8) flowers; (9) seeds; (10) soil surface under relatively open vegetation; (11) soil surface under dense vegetation; (12) major rooting zone. (After P. Adam, Saltmarsh ecology, Cambridge University Press, 1990.)

grapsus from higher tidal areas. But *Hemigrapsus* had a refuge in the lower intertidal areas because of its ability to tolerate silty water. Its gills are not so easily clogged by silt as are those of *Pachygrapsus.*

Marshes also experience a number of disturbances, both human induced and natural. Natural disturbances, such as those caused by wave or current action or by biological agents, including predatory activity, burrowing or excavating, and accumulations of drift plant and animal debris, among others, will lead to patchy distributions of organisms in the marsh. Human-induced disturbances—for example, diking, dredging, channeling water, pollution, introduction of exotic species, particularly large grazers such as cattle, and construction of structures—often have a more profound effect on the marshes. They can lead to large-scale changes not only in the composition of a marsh but also can threaten its very existence. For example, San Francisco Bay, the largest estuarine salt-marsh system on the Pacific coast has, according to Nichols et al. (1986), been reduced from an areal extent of 2200 km^2 in 1850 to less than 125 km^2 today. All of this is due to the effects of humans. In addition, deliberate and inadvertent introduc-

tions of exotic species have led to the current condition in which nearly all common macroinvertebrate species in the bay are alien (Nichols et al., 1986). Of the major estuaries of the United States, San Francisco Bay is now considered to be the one that has been most changed by human activity.

In some cases, however, human-induced disturbances have given us better insights on how marshes operate. For example, Valiela et al. (1975) investigated the results of placing sewage on marshes on the East Coast. They discovered that the growth of *Spartina* was stimulated to such an extent that other plants were eliminated from the area. They also discovered that the *Spartina* plants accumulated high levels of toxic metals, such as cadmium and lead.

The food webs of salt marshes are based primarily on the consumption of detritus (and associated bacteria) and perhaps various algae. Since most of the detrital consumers distinguish only by particle size, a certain amount of live benthic algae is also consumed. These detrital consumers plus the few insects feeding directly on the marsh plants are, in turn, consumed by a variety of both aquatic and terrestrial carnivores (Fig. 8.24).

Figure 8.24 The food web of an Atlantic coast salt marsh. (After J. Teal, Energy flow in the salt marsh ecosystem of Georgia, Ecology 43:614–624, 1962. Copyright © 1962 by the Ecological Society of America.)

BIBLIOGRAPHY

Adam, P. 1990. Saltmarsh ecology. Cambridge University Press, N.Y. 461 pp.

Andrews, J. D. 1973. Effects of tropical storm Agnes on epifaunal invertebrates in Virginia estuaries. Chesapeake Sci. 14:223–234.

Atwater, B. F., S. G. Conard, J. N. Dowden, C. W. Hedel, R. L. McDonald, and W. Savage. 1979. History, land forms and vegetation of the estuary's tidal marshes, pp. 347–386. *In:* Conomos, J. (ed.). San Francisco Bay, the urbanized estuary. Pacific Division AAAS, San Francisco. 493 pp.

Barnes, R. S. K. 1967. The osmotic behaviour of a number of grapsoid crabs with respect to their differential penetration of an estuarine system. J. Exp. Biol. 47:535–551.

Barnes, R. S. K. 1974. Estuarine biology. Studies in biology no. 49. Edward Arnold, London. 76 pp. (Distributed in U.S. by University Park Press, Baltimore.)

Boland, J. M. 1981. Seasonal abundance, habitat utilization, feeding strategies and interspecific competition within a wintering shorebird community and their possible relationships with the latitudinal distribution of shorebird species. M. S. thesis, South Dakota State University.

Chapman, V. J. (ed.). 1977. Ecosystems of the world. 1, Wet coastal ecosystems. Elsevier, N.Y. 428 pp.

Conomos, J. (ed.). 1979. San Francisco Bay, the urbanized estuary. Pacific Division AAAS, San Francisco. 493 pp.

Darnell, R. 1961. Trophic spectrum of an estuarine community based on studies of Lake Ponchartrain, Louisiana. Ecology 45(3):553–568.

Day, J. W., C. A. S. Hall, W. M. Kemp, and A. Yañez-Arancibia. 1989. Estuarine ecology. Wiley, N.Y. 558 pp.

Deevey, G. B. 1960. The zooplankton of the surface waters of the Delaware Bay region. Bull. Bing. Ocean. Coll. 17(2):54–86.

Drinnan, R. E. 1957. The winter feeding of the oyster catcher (*Haematopus ostralegus*) on the edible cockle (*Cardium edlule*). J. Anim. Ecol. 26:441–469.

Gallagher, J. L., and F. C. Daiber. 1974. Primary production of edaphic algal communities in a Delaware salt marsh. Limn. Oceanogr. 19:390–395.

Good, R. E. 1965. Salt marsh vegetation, Cape May, New Jersey. N.J. Acad. Sci. Bull. 10(1):1–11.

Goss-Custard, J. D. 1969. The winter feeding ecology of the redshank *Tringa totanus*. Ibis 111:338–356.

Green, J. 1968. The biology of estuarine animals. University of Washington Press, Seattle. 401 pp.

Haines, E. B. 1979. Interactions between Georgia salt marshes and coastal waters: A changing paradigm, pp. 35–46. *In:* Livingston, R. J. (ed.). Ecological processes in coastal and marine systems. Plenum, N.Y.

Hartley, P. H. T. 1940. The saltash tuck-net fishery and the ecology of some estuarine fishes. J. Mar. Biol. Assoc. 24:1–68.

Josselyn, M. 1983. The ecology of San Francisco Bay tidal marshes: A community profile. U.S. Fish & Wildlife Service Division of Biological Services FWS/OBS-83/23. 102 pp.

Kinne, O. 1963. The effects of temperature and salinity on marine and brackish water animals. Oceanogr. Mar. Biol. Ann. Rev. 1:301–340.

Lackey, J. 1967. The microbiota of estuaries and their roles, pp. 291–302. *In:* Lauff, G. H. (ed.). Estuaries. Publ. no. 83, AAAS, Washington, D.C. 757 pp.

Lauff, G. H. (ed.). 1967. Estuaries. Publ. no. 83, AAAS, Washington, D.C. 757 pp.

Long, S. P., and C. F. Mason, 1983. Saltmarsh ecology. Blackie and Son, Ltd., London. 160 pp.

McLusky, D. S. 1971. Ecology of estuaries. Heinemann, London. 144 pp.

McLusky, D. S. 1981. The estuarine ecosystem. Wiley, N.Y. 150 pp.

McLusky, D. S. 1989. The estuarine ecosystem, 2nd ed. Chapman and Hall, N.Y. 215 pp.

Muus, B. J. 1963. Some Danish Hydrobiidae with a description of a new species, *Hydrobia neglecta.* Proc. Malac. Soc. London 35:131–138.

Nichols, F. H., J. E. Cloern, S. N. Luoma, and D. H. Peterson. 1986. The modification of an estuary. Science 231:567–573.

Nixon, S. W. 1980. Between coastal marshes and coastal waters—A review of twenty years of speculation and research on the role of salt marshes in estuarine productivity and water chemistry, pp. 437–525. *In:* Hamilton, P., and K. B. MacDonald (eds.). Estuarine and wetland processes. Plenum, N.Y.

Odum, H. T., B. J. Copeland, and E. A. McMahan. 1974. Coastal ecosystems of the United States, 4 vols. Vol. I, 533 pp.; Vol. II, 521 pp.; Vol. III, 453 pp.; Vol. IV, 470 pp. (Vol. II has information on marshes and estuaries.) The Conservation Foundation, Washington, D.C.

Odum, E. P., and A. A. de la Cruz. 1967. Particulate detritus in a Georgia salt marsh-estuarine ecosystem, pp. 383–388. *In:* Lauff, G. H. (ed.). Estuaries. Publ. no. 83, AAAS, Washington, D.C. 757 pp.

Pace, M. L., S. Shimmel, and W. M. Darley. 1979. The effect of grazing by a gastropod, *Nassarius obsoletus,* on the benthic microbial community of a salt marsh mudflat. Est. Coast. Mar. Sci. 9:121–134.

Peterson, C. H. 1979. Predation, competitive exclusion and diversity in the soft bottom benthic communities of estuaries and lagoons, pp. 233–264. *In:* Livingston, R. J. (ed.). Ecological processes in coastal and marine systems. Plenum, N.Y.

Peterson, C., and N. M. Peterson, 1979. The ecology of intertidal flats of North Carolina: A community profile. U.S. Fish & Wildlife Service Biological Services Program. 73 pp.

Pomeroy, L. R., and R. G. Wiegert (eds.). 1981. The ecology of a salt marsh. Springer-Verlag, N.Y.

Prater, A. J. 1972. Ecology of Morecambe Bay. III, The food and feeding habits of knot (*Calidris canutus*) in Morecambe Bay. J. Appl. Ecol. 9:179–194.

Race, M. 1982. Competitive displacement and predation between introduced and native mud snails. Oecologia 54:337–347.

Reid, G. K., and R. D. Wood. 1976. Ecology of inland waters and estuaries, 2nd ed. Van Nostrand, N.Y. 485 pp.

Reimold, R. J., and W. H. Queen (eds.). 1974. Ecology of halophytes. Academic Press, N.Y. 605 pp.

Remane, A., and C. Schlieper. 1971. Biology of brackish water, 2nd ed. Schweizerbartsche, Stuttgart.

Seliskar, D. M., and J. L. Gallagher. 1983. The ecology of tidal marshes of the Pacific Northwest Coast: A community profile. U.S. Fish & Wildlife Service Division of Biological Services, Washington, D.C. FWS/085-82/32. 65 pp.

Sliger, M. 1982. Distribution and microhabitat selection of *Hemigrapsus oregonensis* (Dana) and *Pachygrapsus crassipes* (Randall) in Elkhorn Slough, Monterey County, California. M.S. thesis, California State University, Hayward. 64 pp.

Stickney, Robert R. 1984. Estuarine ecology of the southeastern United States and the Gulf of Mexico. Texas A & M University Press, College Station, Texas. 310 pp.

Stout, J. P. 1984. The ecology of irregularly flooded salt marshes of the northeastern Gulf of Mexico: A community profile. U.S. Fish & Wildlife Service Biol. Rep. 85(7.1). 98 pp.

Teal, J. 1962. Energy flow in the salt marsh ecosystem of Georgia. Ecology 43:614–624.

Turner, R. E. 1976. Geographic variations in salt marsh macrophyte production: A review. Contrib. Mar. Sci. 20:47–68.

Valiela, I., J. M. Teal, and W. Sass. 1975. Production and dynamics of salt marsh vegetation and the effects of experimental treatment with sewage sludge: Biomass, production and species composition. J. Appl. Ecol. 12:973–982.

VanEngel, W. A. 1958. The blue crab and its fishery in Chesapeake Bay. Commercial Fisheries Rev. 20(16):6–17.

Van Raalte, C. D., I. Valiela, and J. M. Teal. 1976. Production of epibenthic salt marsh algae: Light and nutrient limitation. Limn. Oceanogr. 21:862–872.

Waisel, Y. 1972. Biology of halophytes. Academic, N.Y. 395 pp.

Willason, S. W. 1980. Factors influencing the biology and coexistence of grapsoid crabs, *Pachygrapsus crassipes* (Randall) and *Hemigrapsus oregonensis* (Dana) in a California salt marsh. M.A. thesis, University of California, Santa Barbara.

Wood, E. J. F. 1965. Marine microbial ecology. Chapman and Hall, London. 243 pp.

Zedler, J. B. 1982. The ecology of southern California coastal salt marshes: A community profile. U.S. Fish & Wildlife Service Biological Services Program FWS/OBS–81/54. 110 pp.

Zedler, J., T. Winfield, and D. Mauriello. 1978. Primary productivity in a southern California estuary. Coastal Zone 3:649–662.

Zobell, C. E., and C. B. Feltham. 1942. The bacterial flora of a marine mud flat as an ecological factor. Ecology 23:69–78.

Tropical Communities

*F*or color, sheer beauty of form and design, and tremendous variety of life, perhaps no natural areas in the world can equal coral reefs (Plates 49, 50). Their beauty has fascinated generations of people, both scientific and lay, down through the years. By the same token, few areas of the marine environment have less initial aesthetic appeal than the dark, mud-wreathed areas known as mangrove forests, where passage is barred by a maze of tangled roots standing above a soft mud surface into which one quickly sinks. Yet these two very different associations are characteristic of vast tropical regions of the world, are unique, and must be considered if one is to understand the functioning of shallow-water areas in the tropics. In this chapter we shall take up first the distribution, zonation, and structure of these assemblages and then attempt

to account for the maintenance and continuation of these communities through a consideration of the component organisms and their interactions.

CORAL REEFS

Over a vast region (millions of square miles) of the tropics, the shallow inshore waters are dominated by the formation of coral reefs and, indeed, they are often used to define the limits of the tropical marine environment (Fig. 9.1). Coral reefs are unique among marine associations or communities in that they are built up entirely by biological activity. The reefs are essentially massive deposits of calcium carbonate that have been produced primarily by corals (phylum Cni-

Figure 9.1 Distribution of coral reefs in the world; 20°C isotherm (dashed line).

daria, class Anthozoa, order Madreporaria = Scleractinia) with minor additions from calcareous algae and other organisms that secrete calcium carbonate. Although corals are found throughout the oceans of the world in polar and temperate waters as well as in the tropics, it is only in the tropics that reefs are developed. This is because there are two different groups of corals, one called *hermatypic* and the other *ahermatypic*. Hermatypic corals are those that produce reefs; ahermatypic do not. Ahermatypic corals are distributed worldwide, but hermatypic corals are found only in the tropical regions. The distinguishing feature between the two is that most hermatypic corals have in their tissues small symbiotic (living together) plant cells called *zooxanthellae*, whereas most ahermatypic corals do not. The role of these plants will be considered later (see pp. 349–350).

Reef Distribution and Limiting Factors

A glance at a map of coral reef distribution in the world immediately shows that nearly all coral reefs are found only in those waters bounded by the 20°C surface isotherm (Fig. 9.1). Hermatypic corals can maintain themselves for periods at temperatures somewhat below this; however, as Wells (1957) notes, no reefs develop where the annual mean minimum temperature is below 18°C. Optimal reef development occurs in waters where the mean annual temperatures are about 23–25°C. Some coral reefs can tolerate temperatures up to about 36–40°C. It may be noted that reefs are reduced or absent from large areas on the west coast of South and Central America and also from the west coast of Africa, both areas of which lie well within the tropical zone. The reason for the great reduction or absence of reefs from these areas is that the west coasts of both of these continents are areas of strong upwelling of cold water, which reduces the temperature of the shallow inshore waters below that required for reef development in many places. Furthermore, both of these coasts have strong, north-flowing cold currents that keep the temperature down, the Peru Current on the South American coast and the Benguela Current off West Africa.

Coral reefs are also limited by *depth*. Coral reefs do not develop in water that is deeper than about 50–70 m. Most reefs grow in depths of 25 m or less. This explains why these structures are restricted to the margin of continents or islands.

The peculiar feature of the development of atolls, which occur rising out of water many kilometers deep, will be discussed later (see pp. 340–342). The reason for the depth restriction has to do with the hermatypic corals' requirement for light. *Light* is one of the most important factors limiting coral reefs. Why? Sufficient light must be available to allow photosynthesis by the symbiotic zooxanthellae in the coral tissue. Without sufficient light, the photosynthetic rate is reduced and, with it, the ability of the corals to secrete calcium carbonate and produce reefs. The compensation point for most reef corals seems to be the depth where light intensity has been reduced to 1 to 2 percent of surface intensity, according to Yentsch (1966).

Another factor that acts to restrict coral reef development is *salinity*. Hermatypic corals are true marine organisms and are intolerant of salinities deviating significantly from that of normal seawater (32–35 $^0/_{00}$). Wherever inshore waters are subject to continuing influxes of fresh water from river discharge so that the salinity is lowered, reefs will be absent. Such is the case along large portions of the Atlantic coast of South America, where the Amazon and Orinoco rivers discharge a huge volume of fresh water, and reefs are absent (Fig. 9.1). On a smaller scale, this occurs in many areas of the tropics where rivers and streams running out to sea cause breaks in reef development. At the other extreme, coral reefs do occur in regions of elevated salinity such as the Persian Gulf, where reefs flourish at 42 $^0/_{00.}$

Often correlated with freshwater runoff is the factor of *sedimentation*. Sediment, both in the water and settling out on the coral reefs, has an adverse effect on the corals. Many corals can remove limited amounts of sediment by trapping it in mucus and carrying it off by ciliary action. Most hermatypic corals, however, cannot withstand heavy sedimentation, which seems to overpower their ciliary-mucus cleansing mechanism and to smother them and clog their feeding structures. Sediment in the water (= turbidity) has an additional adverse effect in that it reduces the light necessary for photosynthesis by the zooxanthellae in the coral tissue. As a result, coral reef development is reduced or eliminated in areas of high turbidity. When this sediment is carried by rivers or streams, the combination of reduced salinity and excess sediment is responsible for the absence of reefs. Hermatypic coral species may vary, however, in their tolerance for sedimentation; few are able to tolerate rather high sedimentation rates, and these are found as isolated colonies in sedimentary areas.

In general, coral reef development is greater in areas that are subject to strong wave action. Coral colonies with their dense, massive skeletons of calcium carbonate are very resistant to damage by wave action. At the same time, the wave action provides a constant source of fresh, oxygenated seawater and prevents sediment from settling on the colony. Wave action is also responsible for renewing the plankton, which serves as food for the coral colony.

Finally, coral reefs are limited in an upward direction by emergence into air. Whereas abundant secretion of mucus may prevent dehydration for a short time, an hour or two, most corals are killed by long exposure to air, thus limiting their upward growth to the level of the lowest tides. During an unusual and extremely low tide series over a five-day period in the Gulf of Aquaba, Loya (1976) found that coral mortality was between 80 and 90 percent, which suggests this factor can be of considerable importance. On very low spring tides, coral reef areas may, however, be exposed for short periods up to a few hours, which does not seem to harm them, at least in the Indo-Pacific region.

All these factors are summarized diagrammatically in Fig. 9.2.

Structure of Corals

Since the dominant members of coral reefs are the corals, it is necessary to have some understanding of their anatomy. Corals are members of the phylum Cnidaria, which includes such diverse forms as jellyfish, hydroids, the freshwater *Hydra,* and sea anemones. Corals and sea anemones are members of the same taxonomic class, Anthozoa, and differ primarily in that corals secrete an external calcium carbonate skeleton, whereas anemones do not. Corals may be colonial or solitary, but almost all hermatypic corals are colonial, with the various individual coral animals or *polyps* occupying little cups or

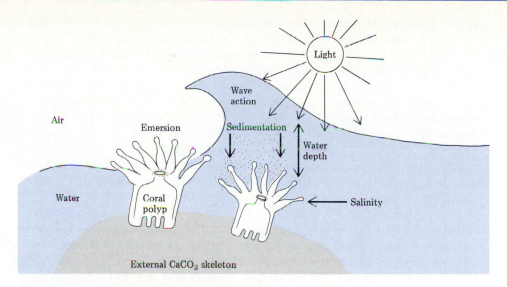

Figure 9.2 Summary of physical factors acting on coral polyps and coral reefs that may act to limit their distribution.

corallites in the massive skeleton (Fig. 9.3; Plates 51 and 52). Each cup or corallite has a series of sharp, bladelike *sclerosepta* rising from the base. The pattern of these septa differs from species to species and is often used as a taxonomic tool to distinguish among coral species. The sclerosepta of the skeletal cup alternate with the internal septa of the gastrovascular cavity of the coral polyp. Each polyp is a three-layered animal with an outer epidermis separated by a fibrous mesoglea from an internal gastrodermis. Around the mouth is a series of tentacles that have batteries of stinging capsules or nematocysts, which the animals use to capture their zooplankton food.

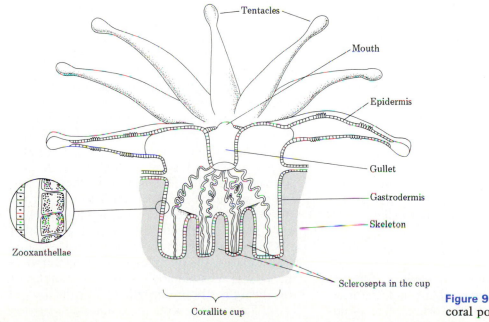

Figure 9.3 Anatomy of a coral polyp.

The symbiotic zooxanthellae, actually dinoflagellates (see Chapter 10), are found intracellularly in the gastrodermal layer. Coral colonies grow by having the polyps bud off new polyps asexually. New colonies are established by the fragmentation of skeletal pieces or through the settlement of a planktonic *planula* larva, which itself is the result of sexual reproduction.

Types of Reefs

Coral reefs occur in many different sizes and shapes, resulting from the particular hydrological and geological situations that recur in different areas of the tropics. This, in turn, has led scientists to classify reefs into various schemes, often generating a bewildering array of names and terms. Recognizing this tendency, but in an attempt to keep things simple and prevent excessive terminology, we choose to limit the number of types to those that are most common. In general, coral reefs are grouped into one of three categories: *atolls, barrier reefs,* and *fringing reefs.* Atolls are usually easily distinguished because they are modified ring-shaped reefs that arise out of very deep water far from land and enclose a lagoon, which itself may contain *lagoon reefs* or *patch reefs* (Fig. 9.4). With few exceptions, atolls are found only in the Indo-Pacific area. Barrier reefs and fringing reefs, on the other hand, tend to grade into each other and are not readily separable. Some scientists would prefer to group them into a single category. Both types occur adjacent to a landmass, with a barrier reef being separated from the landmass by a greater distance and deeper water channel than the fringing reef (Fig. 9.4). Fringing reefs and barrier reefs are common throughout the coral reef zones in all oceans. The largest barrier reef is the Great Barrier Reef of Australia, which stretches for almost 2000 km along the eastern coast of Australia, from near New Guinea to just north of Brisbane. The second largest is that off the Yucatan Peninsula in Mexico and Belize.

Origin of Reefs

Different types of reefs and reefs in different oceans may have diverse origins and histories. The greatest interest in the origin of reefs has been with respect to atolls. For many years, humans speculated as to how such reefs could develop out of such deep water, miles from the nearest emergent land. This interest was heightened when it was discovered that reef corals could not live deeper than 50–70 m. This led to the development of several theories concerning the origin of atolls. Only one need be discussed here—that is, the *subsidence* or *compensation theory.* This explanation for the origin of atolls was first promulgated by Charles Darwin following his five-year voyage on the *Beagle,* during which time he had the opportunity to study reefs in several areas. According to Darwin's subsidence theory, atolls have their origin when fringing reefs begin to grow on the shores of newly formed volcanic islands, which have pushed to the surface from deep water. These islands then often begin to subside and, if the subsidence is not too fast, reef growth will keep up with the subsidence, forming next a barrier reef and finally an atoll as the island disappears beneath the sea (Fig. 9.5). When the island has disappeared, continued growth of the corals on the outside keeps the reef at the surface; but on the inside, where the island used to be, quiet water conditions and high sedimentation prevail, which preclude continued vigorous coral growth, and hence, a lagoon develops. This theory thus links all three reef types in an evolutionary sequence, but is not an explanation for all fringing and barrier reef types.

Since the current surface features of atolls give no evidence of a volcanic base, in the years after the development of Darwin's theory other explanations were offered and the whole concept of the origin of atolls became embroiled in the "coral reef problem." If Darwin's theory was correct, it must be assumed that drilling down through the current atoll reefs would yield layer after layer of reef limestone until, finally, volcanic rock was encountered. Such ability to drill to the base of atoll reefs and resolve the problem had to wait until the mid-twentieth century. In 1953, Ladd et al. reported borings at Eniwetok Atoll in the Marshall Islands that penetrated 1283 m of reef limestone and then hit volcanic rock. This was the initial evidence that Darwin's theory was substantially correct. Correctness of this theory has been strengthened by the discovery of flat-topped submerged mountains or *guyots* that, at present, have their tops many hun-

(A)

(B)

Figure 9.4 Aerial photo of an atoll (A) and a barrier reef (B). (Photos by Dr. Richard Mariscal.)

dreds or thousands of meters below the ocean surface, but have on their surface the remains of shallow-water corals. Evidently, these mountains sank too fast for reef growth to keep within the lighted zone.

Although the subsidence theory links together all three reef types in a successional sequence, this does not mean that everywhere barrier reefs and fringing reefs occur today a similar sequence is happening. Indeed, the reasons for the occurrence of barrier and fringing reef types around continental margins and around high, nonvolcanic islands are simply that these areas offer suitable environmental conditions for the growth of reefs and a suitable substrate on which to begin growth. Hence, the extensive reefs around the Indonesian Islands, the Philippines, New Guinea, Fiji, and most of the Caribbean

Figure 9.5 Geological evolution of a coral atoll according to the subsidence hypothesis of Darwin. (After T. Storer, R. Usinger, R. Stebbins, and J. Nybakken, General zoology, 6th ed., McGraw-Hill, 1979. Copyright © 1979; used with the permission of McGraw-Hill Book Company.)

Islands and Florida are there because a suitable substrate in shallow water existed on which they could begin growth. In none of these areas are large land areas subsiding, nor will these reefs ultimately become atolls.

Whereas atolls are often very old structures (for example, the 1283 m of reef limestone at Eniwetok represent reef growth for about 60 million years), other reefs may be much younger. This is particularly true for the reefs in the Atlantic Ocean. They are all geologically very recent, most dating only from the last glacial age, or 10,000 to 15,000 years before the present.

Composition of Reefs

Although the corals are the major organisms that form the basic reef structure, there is a bewildering array of other organisms associated with reefs, such that these areas are perhaps the most diverse and species-rich areas that exist in the marine environment today. Members of practically all phyla and classes may be found on coral reefs. It is not practical here to mention all of these forms, nor do we even know the roles played by many of them, but before considering the ecological structure of reefs, it is worth pointing out some of the more important and abundant groups.

In addition to the stony corals (order Scleractinia), certain other cnidarians contribute strongly to the reef. These include those anthozoan relatives of the corals called gorgonians (order Gorgonacea), commonly called sea fans and sea whips (Fig. 9.6; Plate 50). These organisms have internal skeletons of spicules. They are particularly conspicuous on the Atlantic reefs. The soft corals (order Alcyonacea) are very

common on Indo-Pacific reefs and may be more abundant than the stony corals in some areas of the reef. They are often rare on Atlantic reefs. The hydrocoral *Millepora* (order Hydrocorallina), often called "fire coral" for its powerful nematocysts, is a conspicuous member of Atlantic reefs.

The *coralline algae* are an extremely important group in constructing and maintaining reefs. These red algae precipitate $CaCO_3$ as do the corals, but the algae tend to be encrusting, spreading out in thin layers over the reef, literally cementing the various pieces together. By this action, the various individual calcium carbonate pieces are welded together to form a strong bulwark that resists wave destruction. These plants form the "algal ridge" on reefs, which is the most rapidly calcifying zone on the reef. Still other green algae are nonencrusting, grow erect, and secrete calcium carbonate, but are not coralline. Together with the corals, the coralline and calcareous green algae make up the three dominant calcium carbonate–secreting organisms of the reef. Much of the sand in coral reef systems is derived from breakup of these erect algae, usually of the green alga genus *Halimeda*. Other free-living, noncoralline algae are inconspicuous on reefs, but a significant algal flora also exists just below the surface layers of calcium carbonate in the coral colonies themselves.

Significant contributions to the calcium carbonate deposits on reefs are also made by mollusks of various types. The most conspicuous and important of these are the various giant clams (*Tridacna, Hippopus*), which Salvat has recorded occurring in numbers as great as 200 per square meter in some atoll lagoons in the Tuamotu Archipelago. Generally, McMichael

Figure 9.6 Some of the dominant and conspicuous components of a coral reef.

(1974) records their density at 1 per square meter or less (Plate 55). Giant clams are absent from the Atlantic reefs. Various small gastropods are also abundant on reefs, but usually inconspicuous due to their size or because they are hidden.

Echinoderms, primarily sea urchins, sea cucumbers, starfish, and feather stars (crinoids), are another abundant and conspicuous group on reefs, but their role in reef ecosystems (except for the sea star *Acanthaster* and certain urchins) is incompletely understood (Plate 54).

Although various crustaceans and polychaete worms are very abundant on reefs, they are often inconspicuous, and their various functions in the reef ecosystem are poorly known. Various sponges are common on reefs, but seem to play a minor role in modern reef construction. The only sponges contributing to the construction are the sclerosponges and they are restricted to deeper slopes and in caves or overhangs, where

coral cover and calcification, according to Wilkinson (1983), are low. Siliceous sponges (Demospongiae) may, however, be important in holding coral and rubble together, preventing loss from the reef until it can be fused together by coralline algae. Wilkinson (1983) has also demonstrated that certain of the most common barrier reef sponges have blue-green algae symbionts responsible for net primary productivity.

The final large group, which is very conspicuous on reefs and most certainly contributes to the structure of the reef system, is the various fishes (Plates 56 to 61).

Bacteria are also probably very abundant on reefs and are responsible for the decomposition and quick cycling of organic matter.

Coral Distribution and Reef Zonation

The largest number of reef coral species and genera occur in an area of the Indo-Pacific that includes the Philippine Islands, the Indonesian Archipelago, New Guinea, and northern Australia. Within this area, Crossland (1952) and Wells (1954) list 50 genera and several hundred species. Considering the whole Indo-Pacific region, Wells (1957) lists 80 genera and 700 species. The number of genera and species declines away from this area, both north and south and east and west, but most of the central Indo-Pacific has reefs with from 20 to 40 genera of corals. This is in contrast to the Atlantic Ocean, where Goreau and Wells (1967) report only 36 genera and 62 species are found. A dominant reef-building genus in both oceans, *Acropora,* has 200 species in the Indo-Pacific but only three in the Atlantic, and some important reef-building genera of the Pacific such as *Pocillopora, Pavona,* and *Goniopora* are absent from the Atlantic. Thus, the species richness of Atlantic reefs is much less than that of Pacific reefs. Whether in the Atlantic or Pacific, as one moves away from the centers of the distribution of coral species, the number of species declines and is dramatically reduced at the edges of the coral reef zone. For example, the northernmost Pacific reefs, at Midway Island in the Hawaiian chain, have but seven genera of corals according to Grigg (1983).

Coral reefs are large and complex associations of organisms that have a number of different habitat types all present in the same system. Thus, for example, there are areas of hard substrate on which many sessile organisms attach and that are analogous to rocky shores; at the same time, there are areas of sand, which require a different set of adaptations, much as we saw for soft bottoms (see Chapter 5). Similarly, there are areas of heavy wave action and strong currents and areas of virtual calm, where water movement is minimal. Still other areas may have a lush growth of calcareous green algae (*Halimeda*) mimicking seagrass beds, and in other areas we may have mangroves or seagrass beds. All of these different habitat types are found in a relatively small area. Therefore, one of the reasons for the great diversity of life in coral reefs is the great diversity of habitats. As might be expected, these different habitat types, as well as others not mentioned here, vary in their extent or presence on reefs in different geographic areas; hence, it is difficult to give a pattern of

zonation for reefs that will be generally applicable. On all reefs, gradients in water motion (waves, currents) and in light penetration (depth, turbidity) are the two most significant physical factors determining zonation. When these are coupled with the various biological interactions (competition, predation, grazing), the result is the various zonal patterns. Broad, large-scale zonation patterns are generally the result of physical factors; the small-scale "patchiness" is usually the result of biological interaction. We discuss here the broad-scale zonation patterns imposed by physical factors. Biological factors are taken up subsequently. The pattern of zonation, though most complex, is most consistent for atolls, and it can be given as a general guide.

Reef development and zonation on atolls are strongly influenced by wave action, a product of the prevailing wind, and, therefore, even though these reefs are essentially circular, a cross section shows different zonation, depending on whether the reefs face the prevailing wind and waves or lie in the lee of them. Figure 9.7 is a diagrammatic cross section of a typical atoll, which indicates the major features of zonation.

Beginning on the windward side, the first zone on a reef is the *outer seaward slope,* where living coral begins to become abundant at a depth of about 50 m. Corals in this area are few and often delicate. At about 15 m, there is often a terrace or step in the otherwise steep outer slope; from here to the surface, coral growth is lush. This is also the area where wave action is the most severe, and hence we know little of this area because of its inaccessibility to humans. At the surface of the water is the *windward reef margin,* which is characterized by the development of coral-algal spurs or buttresses extending out into the waves and down in depth to the level of the terrace. These spurs or buttresses alternate irregularly with deep channels or grooves into which water surges onto the reef itself and through which water and associated debris leave the reef (Fig. 9.8). This is the so-called *spur and groove* or *buttress zone.* The sides of the grooves or channels and the spurs themselves support a lush growth of the dominant reef-forming corals such as *Acropora,* and it is here that coral reef accretion occurs most rapidly. This is because environmental conditions are optimal. Immediately behind the buttress zone is a smooth,

Figure 9.7 Diagrammatic cross section of a typical atoll: (1) outer seaward slope, (2) windward reef margin with spur and buttress zone, (3) algal ridge, (4) the reef flat, (5) seaward beach of the island, (6) lagoon beach of the island, (7) lagoon floor, (8) lagoon reefs, (9) leeward reef flat, (10) leeward reef margin, (11) leeward reef slope. (Modified from C. M. Yonge, The biology of coral reefs, Adv. Mar. Biol. 1:209–260, 1963.)

coral-free ridge of encrusting coralline algae called the *algal ridge*. This ridge receives the full impact of the wave energy and, therefore, is virtually free of any organisms except the encrusting coralline algae. The coralline algae present are usually of several genera, including *Porolithon, Hydrolithon,* and *Lithothamnion* or related genera, and hence the origin of the alternate name for this zone, the "Lithothamnion ridge." Algal ridges are not developed on all windward reefs in the Indo-Pacific and are uncommon in the Atlantic. The grooves of the buttress zone, often called "surge channels," penetrate behind the algal ridge as deep channels

Figure 9.8 Generalized sketch of the spur and buttress zone of a windward reef. (AR) Algal ridge; (B) buttresses or spurs; (G) grooves; (SC) surge channels; (T) 18-m terrace. (After C. M. Yonge, The biology of coral reefs, Adv. Mar. Biol. 1:209–260, 1963.)

or often as "blow holes," where the channel has been roofed over by algal or coral growth, leaving only one or a few openings on the inside of the ridge through which water may enter or leave. These "blow holes" may offer a spectacular display under certain wave conditions, as the water is forced into the tubes at the outer slope of the reef and explodes as a geyser on the inside. This whole area, algal ridge and buttress zone, is an extremely dangerous one, since wave action is violent and a step into a blow hole or surge channel could mean disaster.

Behind the algal ridge is a very shallow *reef flat* zone. This is a complex area with gradients in environmental factors such as temperature, turbidity, and exposure across it, from the algal ridge to the shore of the atoll island. These gradients, coupled with differences in depth and substrate type (coral rock, sand), provide a great number of habitats that have caused this zone to be variously subdivided. It is also one of the most species-rich areas of the reef. Near the algal ridge, conditions for coral growth are excellent and a great variety of corals is found. This is another area where active accumulation of reef material occurs. Conditions for coral growth depart from optimal as one approaches the island and, as a consequence, corals become less

and less abundant. Deeper areas of the middle reef flat often experience the development of micro-atolls, which are simply larger colonies of a massive coral, such as some species of *Porites,* in which the uppermost part of the colony has been killed, either by exposure to air or excess sedimentation. This is followed by erosion of the skeleton so that it becomes a small basin filled with water in which other organisms may grow (Fig. 9.9). These micro-atolls support a variety of organisms. Also present in this reef flat area are the giant clams (*Tridacna, Hippopus*).

The reef flat zone ends at the *seaward beach* of the atoll island. On the lagoon side of the island is a narrow beach, followed by the quiet waters of the lagoon. Two major zones can be recognized in the lagoon, the *lagoon reefs* and the *lagoon floor.* Lagoon reefs are found around the perimeter of the lagoon and also exist as patch reefs or pinnacles rising from the lagoon floor to the surface. The depth of the lagoon is usually less than 50 m and, therefore, within the depth range of coral growth. Conditions within the lagoon are not as good for coral growth as on the windward side, because wave action and circulation are not as great and sedimentation is greater. Hence, coral growth is often restricted to areas where conditions are better, leading to the

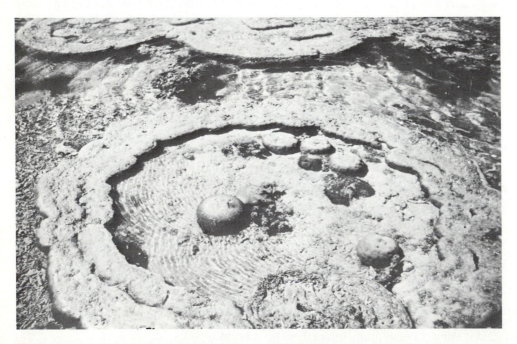

Figure 9.9 A micro-atoll on the Great Barrier Reef. (Photo by the author.)

existence of patch reefs and pinnacles and leaving large areas of the lagoon floor free of coral growth. Sedimentation leads to large areas of sand, and the lagoon floor may have extensive beds of seagrasses (*Thalassia, Cymodocea*) or of green algae such as *Caulerpa* and *Halimeda.*

The leeward side of the lagoon is margined by the *leeward reef flat,* which is generally narrower than the seaward reef flat and separated from the lagoon reefs by a barren boulder zone or by an area of poor coral growth, dominated by isolated colonies of *Porites.* The seaward margin of this flat is similar to the windward flat. At the leeward reef margin, the algal ridge is usually poorly developed or absent and there are no surge channels or buttresses. At the seaward edge, extending down to 15–20 m, coral growth is lush and diverse, dominated by the branching arms of *Acropora.* In addition to the massive corals seen on the seaward reef margin, numerous branching species are found here, due to the absence of violent wave action. Below this, the leeward reef slope is similar to the outer seaward slope.

Fringing reefs and barrier reefs lack the complexity of zones found in atolls and, in effect, can be considered as truncated portions of the atoll zonation scheme. Fringing reefs and barrier reefs on the windward sides of islands have a zonation like that of the windward side of atolls and those on the leeward side similar to the leeward reef of atolls. The major difference between the two is that barrier reefs, especially large ones such as the Great Barrier Reef, may have a very extensive lagoon between the reef flat and the shore of the continent or island and that lagoon may itself develop additional coral reefs.

Atlantic and Indo-Pacific Reefs

As we have already indicated, there are certain differences between the Atlantic and Indo-Pacific reefs, which we will now detail. They include not only structure and zonation features, but also presence, absence, and abundance of various faunal and floral components.

The physical conditions under which reefs grow in both the Atlantic and Pacific are quite similar with respect to temperature, salinity, and turbidity, but the number of coral genera and species is very different between the two areas as described in the preceding section. This reduction in number of species between Pacific and Atlantic reefs is not limited to corals, but also occurs among most of the reef components, including mollusks, fishes, and crustaceans. For example, there are perhaps 5000 species of mollusks on Indo-Pacific reefs, but only 1200 on the Atlantic reefs. Similarly, there are 2000 species of Indo-Pacific reef fish compared to 600 Atlantic reef fish. Atlantic reefs commonly have large numbers of sea fans and whips (gorgonians), whereas these are much less in evidence in Indo-Pacific reefs. Indo-Pacific reefs have large numbers of soft or alcyonacean coral, such as *Sarcophyton* and *Lobophyton,* whereas these are inconspicuous in the Atlantic. Algal ridges and coralline algae were long thought to be of minor significance on Atlantic reefs, but recent work by Adey et al. (1977) and Steneck and Adey (1976) now suggest that they are important. Atlantic reefs also lack the giant clams (*Tridacna, Hippopus*) and the octocorals *Heliopora* and *Tubipora.* There are no anemone fish (*Amphiprion*) in the Atlantic, though the Atlantic does have fishes that associate with anemones. Similarly, crabs and shrimp that guard certain corals against predators are absent from the Atlantic, as is the coral-eating starfish *Acanthaster planci.*

Dominance is also different in the two regions. Atlantic reefs are dominated by *Acropora palmata* (elkhorn coral) and *Millepora complanata,* a hydrocoral. No *Millepora* dominates any Indo-Pacific reef. Rhythms also differ in that most Atlantic corals are nocturnal, but in the Pacific many, such as *Porites, Goniopora,* and *Favia,* are diurnal.

There are also differences in relative abundances of some of the noncoral species. Wilkinson (1987b) has noted that sponge biomass is from 2 to 10 times greater on Caribbean reefs than on the Great Barrier Reef, but few, if any, Caribbean sponges have phototrophic symbionts that are present in many Pacific sponges. Similarly, Sammarco (1987) has noted that echinoids are the major grazers in the Atlantic, while fishes fill that role in the Pacific.

Not only are there differences in species composition and relative abundances between these two areas, there are also differences in age and reef morphology (= construction). Atlantic reefs usually rest on shallow banks or platforms, which are probably the result of erosion by wave action during the Pleistocene period, when sea level was much lower than at present. Therefore, borings through Atlantic reefs do not show the great depth of reef limestone that indicates the great age of the Pacific reefs. Why this great discrepancy in age? This is not known for certain, but two explanations may be offered. In the first place, the Atlantic is a geologically more recent ocean, and hence, there has been less time for reefs to develop. Second, in the ice age the Atlantic may have become too cold for reefs to survive, and hence they died out and were replaced only when the seas warmed up following the retreat of the glaciers. Glynn (1973) has stated that Atlantic reefs, though of more recent origin, do develop massive structural frameworks similar to Pacific reefs and show similar vigorous growth.

Zonation is somewhat different. Atlantic reefs have the buttress zone deeper than the Pacific reefs, and active coral growth extends to 100 m in the Atlantic, but only to 60 m in the Pacific. Atoll islands are often absent in the Atlantic, but reef flats and shallow lagoons are common. Also, the areal coverage by corals differs. In the Atlantic corals cover about 60 percent of the substrate as opposed to 80 to 90 percent in the Pacific.

Finally, there are some subtle differences in the life histories and ecological processes. For example, Sammarco (1987) reports that coral recruitment in the Caribbean, at least for the dominant genus *Acropora,* seems to be primarily asexual via branch breaking, whereas in the Pacific recruitment is by sexually reproduced planulae. Similarly, Vermeij (1987) reports that Pacific invertebrates tend to be more specialized, particularly in dealing with predators and symbiotic relationships, than their Atlantic relatives. Birkeland (1989) has also noted that no asteroid species are influential predators of corals in the Atlantic.

Productivity

Coral reefs are truly oases in a watery desert. As we noted in Chapter 2, tropical marine waters tend to be extremely poor in nutrients, and hence very low in productivity (see pp. 66–67). Yet coral reefs abound with life and all studies done on them have indicated a very high productivity. Most studies—such as those by Kohn and Helfrich (1957), Odum and Odum (1955), Johannes et al. (1972), and Smith and Marsh (1973)—have given gross primary productivity estimates of 1500–5000 g C/m^2/yr. This is in contrast to a gross primary productivity in the open tropical oceans of 18–50 g C/m^2/yr. How can coral reefs maintain such production in such nutrient-poor areas? This is less hard to explain for reefs bordering high islands or continents, where much nutrient material is available from the land and from the shallow inshore waters, than it is for atolls isolated far from land. In the latter case, we do not as yet know all the reasons for the high productivity, but we do know that the nutrient concentrations in the water over the reef differ only slightly from that in water off the reef (Larkham, 1983). This suggests some reasons for maintaining high productivity. In the first place, the amount of plant tissue present on a reef and capable of photosynthesis is very large, much larger than in a similar area of the open ocean. Photosynthetic organisms present on a reef include zooxanthellae, prokaryotic symbionts in sponges, crustose coralline algae, turf algae, large macroalgae, sand algae, seagrasses, and phytoplankton. Each of these components varies in productivity and in amount of areal cover on reefs (Table 9.1), but together they form a significant biomass, one that, according to Odum and Odum (1955), is greater than the animal biomass, even if less conspicuous. Of these various plant components, it appears that the ubiquitous small turf algae are the most significant primary producers (Table 9.1). Even given the potential productivity of this large plant biomass on reefs and the optimal physical conditions, it would not be realized if there were not enough nutrients, the lack of which is the very reason for the low productivity of the open ocean immediately offshore. The secret, apparently, of reefs is their ability to hold tenaciously to all nutrients in the system and to cycle them in very efficient ways similar to those in rain forests, which are also highly productive but exist on nutrient-poor soil. Reefs also act as a sink for any nutrients brought in from the outside. This ensures that nutrients are cycled within the reef system and are not lost to deeper offshore waters. It also means that any plankton from the open sea offshore that impinges on the reef remains there, as do the nutrients brought with them.

Table 9.1 PRIMARY PRODUCTIVITIES OF COMPONENT PLANT TYPES ON CORAL REEFS AND THE DISTRIBUTION OF THESE COMMUNITIES ON REEFS

Plant type	Range of productivity (g C/m²/day)	Approximate area/ cover on reefs %
Benthic algae	0.1–4	0.1–5
Turf algae	1–6	10–50
Zooxanthellae	0.6	10–50
Sand algae	0.1–0.5	10–50
Phytoplankton	0.1–0.5	10–50
Seagrasses	1–7	0–40

Source: From Larkham, 1983.

As an example of the cycling that prevents loss, we can point to the zooxanthellae in the coral tissue. Placing the plants in the coral tissue means that the algae will not be washed from the reef. It also means that any nutrients produced by the coral (NO_3, PO_4) as a result of its metabolism will be available directly to the plant without cycling it outside into the water first, where it might be lost. Similarly, most of the free-living algae on the reef contain calcium carbonate. They are too heavy to be moved easily off the reef; hence, their nutrients remain on the reef to nourish additional generations. Rapid cycling is also aided by the large bacteria populations in coral reefs. These bacteria act quickly on dead material, breaking it down and making available enclosed nutrients. For all these reasons and probably for others yet to be discovered, coral atolls have very high productivity, as do fringing and barrier reefs. However, even given the above, it is as yet far from understood just how such systems are maintained. Much of this explanation can be forthcoming from a more complete understanding of how the various components in the reef ecosystem interact. Of these components, the first and most significant are the corals themselves. They are the subject of our next consideration.

Biology of Hermatypic Corals

Hermatypic corals are the dominant group involved in reef formation and maintenance; therefore, a knowledge of certain of the biological aspects of the group has considerable relevance to our understanding of the ecology of reefs themselves.

Nutrition Reef corals have long been known to be carnivorous animals, similar in that respect to most other members of their phylum. They have tentacles studded with stinging capsules, nematocysts, which are used to sting and capture small plankton organisms. Since they occur in colonies of hundreds or thousands of individuals, and the number of colonies is also huge, the total area for feeding is enormous. In addition to the tentacles and their nematocysts, the outer epidermis of corals is ciliated and produces mucus. This ciliary mucous mechanism is generally used by corals to rid themselves of sediment settling on the surface; in some corals, it has been modified to be used in feeding as well. Those species employing mucous mechanisms to capture plankton tend to have smaller, shorter tentacles than those that feed directly on plankton by snaring them with their tentacles. Food organisms are detected by means of chemoreception.

Since corals exist in such numbers on reefs, and the surrounding ocean is notoriously poor in plankton, how is it possible for these animals to obtain enough food? This is precisely the question that has puzzled generations of coral reef biologists and is only now becoming clear. Rigorous plankton sampling on reefs and just offshore by Johannes et al. (1970), Porter (1974), and others, both during the day and at night, has revealed an abundant plankton population. However, a large fraction of this plankton population is not from the open-ocean area, but is a population indigenous to the reef itself. This indigenous plankton population is composed primarily of demersal meroplankton, which spends the daylight hours on the bottom, rising into the waters over the reef only at dusk. This may also explain why many corals are closed up during the daylight hours, expanding to feed only during the hours of darkness. In spite of this discovery of enhanced abundances of plankton over the reefs, Johannes et al. (1970) have determined that the amount of plankton available to the corals is sufficient to satisfy only 5 to 10 percent of their total food requirements. Where, then, does the remainder of their food come from? Only one source seems possible, and that is the zooxanthellae found in the coral tissues.

The role of zooxanthellae in coral biology has long puzzled biologists. It was early learned that corals were carnivores and that they could not digest plant tissue. Furthermore, early experimental studies showed that, if large tentacled corals were starved or kept in the dark, they would expel their zooxanthellae, but could continue to live without them. This was possible because these particular corals could capture zooplankton and utilize them. These observations suggested that the zooxanthellae were really unnecessary to the coral. More recently, however, Muscatine and Cernichiari (1969) employed radioactive tracers to prove that organic compounds fixed by the zooxanthellae in photosynthesis are transferred to the coral, where they undoubtedly serve as food for the coral. Franzisket (1969) and Johannes (1974) also demonstrated that, if corals are not fed but kept in the light, they gain weight. This was further established in actual field studies by Wellington (1982b). This could happen only if the zooxanthellae furnished food to them. Whereas the role of zooxanthellae in nutrition thus seems proved, we do not know as yet how important this role is for all coral species. Studies by Porter (1976) and Wellington (1982b) suggest that corals are primarily *phototrophic* (meaning dependent on zooxanthellae) but that corals with large polyps and tentacles are less dependent on zooxanthellae for nutrition than other species with smaller polyps.

Growth and Calcification   Although there has been much interest in the growth rate of both corals and coral reefs for many years, it is only in recent years that we have come to have a reasonable understanding of growth in individual corals and reefs. The first requirement for active coral growth is light. As Connell (1973) has shown, if corals are shaded or in some way prevented from obtaining access to the light,

they cease growth and, if light is withheld long enough, they will die. This requirement for light is undoubtedly due to the needs of the zooxanthellae, and it suggests the other important role of zooxanthellae in the life of corals. Zooxanthellae have been found by Goreau (1961) to significantly increase the calcification rate of corals and hence the rate of growth of the coral colonies. Exactly how the zooxanthellae act to increase this rate of skeletal accretion is not fully understood, but it seems as if light enhances oxygen production, which, in turn, stimulates coral metabolism and leads to increased calcium carbonate deposition. This again points up the important role of the zooxanthellae in the coral reef ecosystem, where rapid calcification is needed to maintain the reef against the various destructive forces acting against it.

The most direct way to measure coral growth is to record a particular dimension of the coral and then, after an appropriate time interval, repeat the measurement. Dimensions measured according to Buddemeir and Kinzie (1976) have included length of branch, diameter of the colony, volume, area of the colony, and dry weight. Other means of measuring coral growth have included monitoring the rate of calcification by radioisotope uptake. A rapid and simple way of recording coral growth has been employed by Connell (1973). It is to take photographs of an area of the reef and then measure the colonies later from the developed print (Fig. 9.10).

The rate at which coral colonies grow differs with different species, age of the colonies, and in different areas of a reef, but some generalities can be noted. Young, small colonies tend to grow more rapidly than older, larger colonies, and branched or foliceous corals tend to grow more rapidly than massive (= brain) corals. In growth corals increase their diameters rapidly. *Acropora,* for example, is a foliaceous genus; measurements of the Atlantic species reported in

Figure 9.10 Photographs of the same 1-m quadrat showing the changes in growth and occurrence of coral colonies over an 8-year period on the Great Barrier Reef at Heron Island. For each photograph the square meter frame, divided into 20 × 20 cm squares, was positioned on permanent stakes. Note the great changes that have occurred in growth and in presence of corals over short time intervals. *(A)* 1963. *(B)* 1965. *(C)* 1969. *(D)* 1971. (Photos courtesy of Dr. Joseph Connell, Dept. of Biological Sciences, University of California, Santa Barbara.)

(A) (B)

(C) (D)

Davies (1983) suggest that the species could grow from 2.5 to 26.6 cm/year in length. *Montastrea annularis,* a massive type, had a linear growth from 0.81 to 2.5 cm/year. Despite much recent work that has given us good estimates of growth rates in some species, the fact that there is growth throughout the life of the colony, and that growth rates in the larger, older colonies are variable and difficult to estimate, means that we still do not have any good feel for the age structure of the colonies seen on a reef, nor do we know how long corals can live. What little information there is, primarily that of Connell (1973), suggests that most coral colonies on reefs are ten years of age or younger. However, some of the very large massive corals, such as those forming the aforementioned micro-atolls, may be 100 years old or even older, suggesting that corals can live much longer than the ages generally realized by most colonies on the reef.

It is also difficult to translate the growth rates of individual species into growth rates of a whole reef system. As a result, most estimates of the rate of reef growth have been made by extrapolating from changes in reef topography over a period of years or from age of the reef as determined from geological data such as thickness of reef limestone deposits. These estimates vary quite widely and also vary for reefs in different areas of the world. The range reported by Stoddard (1969) is from 0.2 mm of upward growth per year to perhaps 8 mm.

The growth form of a given coral species may also vary considerably, depending on its location on a reef. In contrast to the same species in shallow water, a coral species at greater depth is thinner and more spindly, probably as a result of less calcification. Wave action tends to force branched species into shapes in which the branches are shorter and stubbier, and currents cause branching forms to have a definite alignment of the branches. This ability of coral species to have different skeletal shapes has resulted in much confusion among coral reef taxonomists, such that different growth forms of the same species have often been named as different species.

Sexual Maturity, Reproduction, and Recruitment Corals perform both sexual and asexual reproduction. Asexual reproduction is generally accomplished by budding off a new individual from the parent, and continued budding is the mechanism of increasing the size of the colony, but generally not for producing new colonies. Highsmith (1982) has, however, reported fragmentation as a means of starting new colonies in some species, and Jackson and Hughes (1985) have shown in their study site in the Caribbean that over a three-year period 38 new colonies of the coral *Leptoseris cucullata* were started by fragmentation but only 16 by larval recruitment. Sexual reproduction results in the formation of a free-swimming *planula* larva, which, when it settles down, initiates development of a new colony (Fig. 9.11). It has now been demonstrated, however, that some planulae are produced asexually so asexual reproduction may be more important than first believed in the establishment of new colonies. This may be especially true if the aforementioned fragmentation proves to be widespread.

From our very limited data, it appears that most corals reach sexual maturity at between 7 and 10 years of age. Corals are primarily hermaphroditic but a few are dioecious.

The time at which corals form gametes depends on the size and age of the parent colony. This means that fragmentation or partial destruction of a colony can affect fecundity of the colony. Corals are either broadcast spawners, releasing eggs and sperm into seawater, or they are brooders, retaining fertilized eggs in the gastrovascular cavity until development has proceeded to the planula stage (Fig. 9.11). For broadcast spawners, spawning is regulated by lunar events, and there are some interesting regional differences. According to Szmant-Froelich (1985), many species on the Great Barrier Reef spawn in late spring to early summer, after the October and November lunar cycles, and are often synchronous, whereas in the Caribbean the major spawning period is longer, lasting from July through September, and nonsynchronous. Among the brooders, release of planulae also appears to be under lunar control, and the cue, according to Szmant-Froelich (1985), appears to be nocturnal illumination.

The planulae are released and swim in the open water for an undetermined time, but generally for only a few days, before settling down and starting a new colony. Since adult corals are

Figure 9.11 Reproduction of corals. *(A)* Adult polyp. *(B)* Planula larva. *(C)* Later planula with developing septa. *(D)* Young polyp after attachment. (Modified from L. Hyman, The invertebrates, Vol. 1, Protozoa through Ctenophora, McGraw-Hill, 1940. Copyright © 1940; used with the permission of McGraw-Hill Book Company.)

fixed in place, the planula larvae are the means of dispersal of the various coral species.

The rate of recruitment of new coral colonies to the reef system is quite variable in time and in space. In both the Great Barrier Reef and in the Caribbean, Sammarco (1985) found that, whereas the genus *Acropora* was the dominant coral genus in both areas, recruits were rare in the Caribbean but among the most abundant in the Great Barrier Reef. There appears, therefore, to be a major difference between the two geographic areas in that there is marked diminishing of recruitment success by broadcast species in the Caribbean. This could, as Sammarco (1985) notes, have implications for reef recovery from catastrophic events. In the Great Barrier Reef system the higher number of recruits and seemingly higher survival rate would lead to faster reef recovery, whereas in the Caribbean recov-

ery would be much slower. A study by Connell (1973) has indicated that the rate of recruitment on the Great Barrier Reef was between 0 and 13 new colonies per square meter per year. The more common coral species were the most abundant of the new colonizers, but there were great differences in species that settled among different areas. It also appeared that mortality was high among the newly settled colonies. Growth rates of newly settled broadcast corals appear to be slower than of those juveniles resulting from brooding corals, according to Babcock (1985).

Species Interactions and Ecology of Reefs

At this point, we have covered the basic biology of the dominant organisms of the reefs—namely, the scleractinian corals—and have outlined the general zonation pattern of reefs and the physical conditions acting on them. Reefs, however, are much more than simple static systems. They are living systems that increase or decrease in size as a result of the complex interactions among various biological and physical forces. We can now consider what we know about how these interacting factors produce the reef as we view it, and also how the interacting forces ensure the persistence of a reef system through time. Since coral reefs are perhaps the most complex systems in the marine environment and since they have not received as much detailed ecological study as, say, rocky shores, our knowledge of this aspect of the coral reef system is not as complete as we would like. However, it is still possible to suggest how some factors interact to maintain the system. The general picture beginning to emerge is that the interactions that account for the reef are similar to those responsible for the patterns we observed on rocky shores.

Competition One of the most obvious features of a coral reef in an area of active coral growth is that there is virtually no open space. The primary space is almost completely covered by the dominant space occupiers—namely, the corals themselves or an algal turf (= all the small, short macroalgae growing up a few centimeters from the substrate). Since all space is occupied and since corals must have access to the light to survive, we might expect competition to occur among corals for space and light.

Such competition does occur among coral colonies. Upright, branching coral colonies grow more rapidly than encrusting or massive corals, and they often extend themselves up and over the encrusting forms, shutting them off from the light. Where this occurs, the part of the encrusting colony in the shade dies. This is *exploitative competition*. Since these branching corals are able to grow faster and shade out the slower-growing massive and encrusting corals, how then do these slow-growing forms manage to persist? There are a number of reasons, but the most intriguing was the initial discovery by Lang (1973) that the slow-growing coral species have the ability to extend out filaments from their gastrovascular cavities that, when they encounter the living tissue of an adjacent coral colony of a different species, digest that tissue, thereby

killing that portion of the adjacent colony (Fig. 9.12). This is *interference competition* and is reminiscent of the activities of anemone clones (see p. 248). Thus, the slow-growing corals can prevent the faster-growing corals from overshadowing them: they maintain space on the reef by simply killing off that portion of the colony about to overgrow them (see Plate 45).

This ability was suggested by Lang as a mechanism to maintain the local diversity of coral species on reefs. Subsequent testing of this "aggressive pecking order" as a means of maintaining corals has, however, revealed that this may be a far more complicated interaction than first believed. For example, in the eastern Pacific, Glynn (1974), in laboratory experiments, found that the massive slow-growing coral *Pavona* was dominant over the faster-growing foliaceous coral *Pocillopora*. This was anticipated from Lang's studies. In the field, however, *Pocillopora* seemed to damage *Pavona* more often. Wellington (1980) showed that in fact during the initial contact between the two genera *Pavona* did extend its mesenterial filaments and partially kill the *Pocillopora*. However, after from 7 to 60 days, the undamaged tentacles of *Pocillopora* polyps at the edge of the area killed by *Pavona* began to elongate to 30 times their normal length and changed their nematocyst batteries to be more powerful. These "sweeper tentacles" then were passively carried by water currents into the *Pavona* colony, where they in turn inflicted damage. This retaliation permitted the *Pocillopora* to repair the original damage and continue growth toward the *Pavona*. The ultimate outcome is that *Pocillopora* wins out in the direct conflict. A similar system of sweeper tentacles has been reported by Richardson et al. (1979) for two corals in the Caribbean, suggesting that this phenomenon is more widespread. The competitive superiority of *Pocillopora* helps to explain why it is the dominant coral on shallow reefs of the eastern Pacific region.

At the shallowest depths on coral reefs, cover by coral or other organisms approaches 100 percent, according to Huston (1985). Under such crowded conditions the rapidly growing, branching corals reduce or eliminate the slower-growing, massive corals by overtopping them, thereby reducing their access to light and water movement. Such exploitative competition may

Figure 9.12 Photo showing the killing of portions of a tabular *Acropora* colony (left) by the brain coral *Ctenella chagius* (right), thereby preventing overgrowth by the faster-growing *Acropora*. (Photo from C. R. C. Sheppard, Interspecific aggression between reef corals with reference to their distribution, Mar. Ecol. Prog. Ser. 1:237–247, 1979.)

explain the dominance of *Pocillopora* at the shallowest depths in the eastern Pacific and *Acropora* in the Caribbean. If then the slow-growing corals cannot gain access to full sunlight, how do they survive? One mechanism, according to Huston (1985), is that they are more "shade tolerant" and grow at greater depths, where reduced competition under the low light conditions permits their survival. This, in turn, leads to increased species diversity with depth. We see this not only in corals but in other members of the reef fauna as well.

Such complicated competitive hierarchies among various coral species, and their change with depth, is but one mechanism that acts to maintain coral diversity and zonation patterns. The degree to which competition between coral species controls community structure on reefs is presently unknown.

Competition between corals and other reef taxa, particularly various fast-growing algal species, is reduced by diverse grazing fishes and invertebrates (see subsequent section), which in turn is advantageous to the corals. Also, whereas algae are competitively superior and overgrow juvenile coral colonies in shallow water, Birkeland (1977) has shown that the corals can maintain their space in deeper waters. The importance of algal grazers to corals is shown by grazer

exclusion experiments that demonstrate, according to Brock (1979), Hatcher (1983), and Sammarco (1980), that the algae are competitively dominant over corals.

Finally, although competition between corals and taxa other than algae seems to be generally uncommon, there is one other set of organisms in which competition for space seems common. That group is what we might call the *cryptofauna,* the smaller colonial encrusting invertebrates such as sponges, bryozoans, and tunicates. As Jackson (1977, 1979) has demonstrated, individual clones of these fast-growing invertebrates often undergo intense competition for space among themselves but also can overgrow solitary or semicolonial corals. Due to indeterminate and rapid asexual growth, they have an advantage over the slower-growing ahermatypic corals. Their competitive abilities are offset, as Jackson (1977) has noted, by their increased susceptibility to predation and disturbance. One effect of the intense space competition on reefs is the tendency of various organisms to utilize the same space in some close association. This is *commensalism.* Commensal relationships are much more common among reef organisms than elsewhere (see Chapter 10 for complete discussion).

Predation   Although large numbers of species of various other invertebrates exist on coral reefs, one sees virtually no invertebrates on the reef except for a few large echinoderms (sea cucumbers, urchins, feather stars; Plate 54) and large mollusks (*Tridacna* and others). The reef seems dominated by the corals and by the abundant fishes. This is because the other invertebrates are all hidden from view. It has been suggested by Bakus (1964) that the reason for this cryptic pattern is the intense predation pressure on the reef. Any soft-bodied invertebrate in the open would be quickly consumed. This can be verified by turning over coral heads and exposing the fauna, which immediately attracts fishes to the area to consume the exposed organisms. Presumably, the larger echinoderms and mollusks are more immune to predation.

The role of predation in determining the structure and composition of coral reefs is not as well studied or understood as it is for rocky intertidal areas, but we do have an increasing body of evidence of its importance. A surprising number of animals feed on coral and hence can be classified as predators. Most of these predators, however, are small in relation to the coral colony. The process of predation in these forms resembles the process of grazing by herbivores, whereby patches of coral polyps are removed but the entire colony is not destroyed. Thus, if not too much tissue is removed, the coral may regrow polyps to cover the area that has been "grazed," very much as grass regrows after grazing. These small predators include a number of gastropod mollusks (families Architectonidae, Epitoniidae, Ovulidae, Muricidae, and Coralliophilidae), at least one nudibranch mollusk (*Phestilla*), amphinomid polychaete worms (*Hermodice*), certain barnacles (*Pyrgoma*), and several crabs (*Mithraculus, Tetralia*). Taken together, these predators do not appear to have a significant effect on coral colonies, nor do they seem able to affect the community structure.

Two other taxa of predators are, however, capable of destroying coral colonies and thus of modifying the structure of reefs. These are the starfish *Acanthaster planci* and various fishes. *Acanthaster planci* is a very large, multiarmed starfish that feeds only on the tissue of living coral (Fig. 9.13). Because of its size, it is capable of destroying an entire colony during feeding. Normally, *Acanthaster* is rare on coral reefs in the Indo-Pacific (not found in the Atlantic), but when present in moderate to large numbers can exert an influence on the composition of reefs. For example, on eastern Pacific reefs, Glynn (1976) found that *Acanthaster* preferred to feed on the abundant, fast-growing *Pocillopora* colonies, but in the field actually fed most of the time on the slow-growing, nonbranching colonies in the deeper parts of the reef. This selective predation on nonpocilloporid corals reduced coral cover and diversity. The reason that the preferred *Pocillopora* was not eaten was that the colonies harbored symbiotic shrimp (*Trapezia, Alpheus*) that repulsed the starfish and protected the colonies.

The second influence on the reef structure by *Acanthaster* comes when conditions bring about very high starfish densities. In recent years, *Acanthaster planci* has undergone a population explosion on certain reefs in the western Pacific such that the normal population density report-

Figure 9.13 Photo showing the large numbers of *Acanthaster planci* on a heavily infested reef in Palau in 1979. (Photo courtesy of Dr. Charles Birkeland.)

ed by Endean (1973) of one to three starfish observed in four to five hours of search has gone to more than 100 starfish seen in 20 minutes of search. The latter figure represents tens of thousands of starfish on each reef (Fig. 9.13). In these cases, the entire reef is destroyed by the voracious starfish (see pp. 369–370), and nearly all corals are consumed.

Two groups of fishes actively graze over coral colonies: those species that consume the coral polyps themselves (= corallivores), such as puffers (Tetraodontidae), filefish (Monacanthidae), triggerfish (Balistidae), and butterfly fish (Chaetodontidae); and a group of multivores (= omnivores) that remove the coral polyps to obtain either the algae in the coral skeleton or various invertebrates that have bored into the skeleton—for instance, surgeonfish (Acanthuridae) and parrot fish (Scaridae; Fig. 9.14).

Figure 9.14 (A) Fishes that actively graze coral colonies: butterfly fish (Chaeto-dontidae), *Chaetodon, Colunula;* parrot fish (Scaridae), *Scarus;* triggerfish (Balis-tidae), *Balistapus;* puffers (Tetraodontidae), *Arothron.* (B) Herbivorous fishes of the reef: triggerfish (Balistidae), *Balistapus, Rhinecanthus;* squirrel fishes (Holo-centridae), *Holocentrus;* surgeonfish (Acanthuridae), *Acanthurus, Ctenochae-tus, Zebrasoma, Naso;* damselfishes (Pomacentridae), *Pomacentrus;* butterfly fishes (Chaetodontidae), *Chaetodon;* puffers (Tetraodontidae), *Arothron;* mullet (Mugilidae), *Mugil;* boxfishes (Ostraciontidae), *Ostracion;* blennies (Blenniidae), *Cirripecies.* (Not to scale.) (Modified from R. W. Hiatt and D. W. Strasburg, Ecological relationships of the fish fauna on coral reefs of the Marshall Islands, Ecol. Monogr. 30:65–127, 1960. Copyright © 1960 by the Ecological Society of America.)

Corallivores that feed by biting off and ingesting portions of the coral skeleton tend to feed on the faster-growing branching species such as *Porites* and *Acropora* in the Atlantic and on *Pocillopora, Acropora,* and *Montipora* in the Pacific. There is evidence that these corallivorous fish can influence community structure. For example, Neudecker (1977) found that transplanted *Pocillopora damicornis* colonies were severely damaged if transplanted to depths of 15–30 m but were unaffected by corallivorous fish at depths of 2–3 m. Similarly, Wellington (1982a) found that *P. damicornis* grew better in damselfish territories where it was protected from corallivorous fish by the aggressive territorial defense of the damselfish. These two cases—plus the observations that *Pocillopora* is rarely found deep on reefs, and that their colonies when caged away from predators can grow faster at 15 m than at 2 m—are strong evidence that corallivorous fish determine community structure. The extent to which this activity restricts corals to certain regions or is responsible for coral distribution patterns is, however, unknown at present. What is certain is that the two families of fishes, Acanthuridae (surgeonfishes) and Scaridae (parrot fishes), that do consume living and dead coral are among the more abundant fishes on reefs; by passing coral skeletons through their guts, they are significant producers of sand and sediment on reefs.

Grazing Aside from the abundant and conspicuous coralline algae on reefs, there is little evidence of other large algae or other plants. If you look closely, however, you will find in many parts of the reef a short algal turf or small clumps of algae. Algae can be major space competitors with corals on reefs, invading and establishing themselves more quickly than corals. Since the algae are not dominant on the reef and obviously do not outcompete the corals, what restrains them? The answer appears to be intense grazing pressure by fishes and also certain invertebrates, primarily sea urchins. Many families of fishes contain herbivorous grazers, including Siganidae and Pomacentridae, as well as the aforementioned Acanthuridae and Scaridae. Continued grazing by these fishes, plus the grazing by abundant sea urchins such as the ubiquitous genus *Diadema,* keep the algae reduced to a

minimum and enhance the survival of coral recruits (Fig. 9.14). As Huston (1985) points out, if the herbivores are either removed from reefs or prevented from grazing by caging, coral is rapidly overgrown and killed by algae.

Grazing by herbivorous fish and urchins directly affects the algae, but only indirectly the corals. The obvious effect of the grazers on the algae is the reduction of the algal biomass. This reduction can be enormous. In the Virgin Islands, when Sammarco et al. (1974) removed all the *Diadema antillarum* from an isolated reef, they had a tenfold increase in algal biomass within four months. There may also be more complex and subtle effects. Grazing, for example, may prevent a single competitively dominant alga from taking over all the space and reducing the diversity. For instance, Sammarco et al. (1974) reported higher algal diversity in grazed patch reefs than in ungrazed reefs in the Virgin Islands; the latter were dominated by a single algal species. The universality of this may be questioned, however, because the same investigator did not find that situation in a similar study in Jamaica.

Grazing can also result in shifts in abundance of groups of algal species due to their differential palatability to the grazers. This is best exemplified by the coralline algae and the fleshy filamentous species. The latter group grows more rapidly and may eliminate the slower-growing corallines if ungrazed. However, the fleshy filamentous algae are much more palatable and so are preferred by the grazers. In areas of heavy grazing, therefore, the coralline algae increase at the expense of the fleshy forms.

The indirect effects of grazing on corals and coral growth produce a more complex situation. Although it is undoubtedly true that herbivores, in general, permit the survival of coral that would otherwise be eliminated by competition with the more rapidly growing fleshy algae, it is also possible that they may have certain deleterious effects. For example, sea urchins seem to be rather unselective grazers in contrast to fishes and may destroy young coral colonies as they graze. The studies done by Sammarco et al. (1974) with *Diadema* suggest that at very high densities this urchin will graze off all organisms, not just algae, and thus prevent coral growth as well. At lower densities the urchin effectively removes algae and permits coral colonies to become established. Presumably, grazing by moderate densities of *Diadema* clears areas for coral planulae to settle into and thus contributes to coral reef maintenance.

Within the tropics it appears that the relative importance of grazing fishes and urchins to the maintenance and structure of reefs varies with human fishing pressure. Fishing removes fishes but, according to Hay (1984), it removes both herbivorous fishes and those that prey on urchins. This means that urchins become relatively more important on reefs that are subject to heavy fishing pressure.

The relative importance of fishes and urchins on reefs has been further elucidated as a result of the catastrophic die off of virtually all of the *Diadema antillarum* in the Caribbean. Beginning at the Atlantic entrance to the Panama Canal and at Barbados, *D. antillarum* populations began to mysteriously disappear starting in 1983. Subsequently, over the next 13 months, the populations died off throughout the Caribbean. According to Lessios (1988), no tropical western Atlantic area was spared, and mortality was 93 to 99 percent. Mortality was followed by nearly complete failure of larval recruitment in 1984.

The immediate response to this loss of a major grazer was the increase in the algal biomass, and this was most dramatic in areas where herbivorous fish had been reduced by fishing pressure. In areas that experienced the largest algal blooms, the fleshy algae increased at the expense of the encrusting coralline forms. In the years since the die off, studies strongly suggest that the relaxing of grazing pressure by *Diadema antillarum* has had a major effect on algal cover, biomass, and composition, but also that this may be much modified if herbivorous fish are abundant to compensate for the urchins.

A special case of the complex effect of grazers on coral is that exemplified by the damselfishes (Pomacentridae). Damselfishes are territorial fishes that graze either selectively or nonselectively on the algae within their territory. They defend their "algal gardens" against other grazing fishes. They may even control the algal species composition in their territory by selective "weeding," according to Ogden and Lobel (1978). Since the algae in the territory are pro-

tected from heavy grazing by other grazers, Potts (1977) found that they grow sufficiently rapidly to kill the small settling corals by overgrowth and shading. Corals are therefore reduced or absent. Damselfish "gardens" may occupy significant portions of reef surfaces. Wellington (1982a) suggests that up to 60 percent of the area of reef flats may be so occupied; hence, they are important contributors to community structure as well as providing refuges for juvenile invertebrates and demersal plankton. All is not, however, as simple as just indicated. Since damselfish protect their territory from all grazing fishes including grazing corallivorous fishes, and given that massive corals recover from fish grazing and grow more slowly than branching foliose corals, the damselfish contribute to the selective dominance of the reef areas by branching corals in the areas they occupy. Thus, according to Kaufman (1977), damselfish in the Caribbean tend to eliminate the massive coral *Montastrea annularis* from the upper parts of the reef where damselfish are most common. This leads to the dominance in the area by the fast-growing coral *Acropora cervicornis*. In the eastern Pacific there is an even more dramatic example of the effects of damselfish on reef structures. Here Glynn and Wellington (1983) have demonstrated that the damselfishes also remove the dominant urchin, *Eucidaris thouarsii,* from their territories, thus contributing to the survival of corals that might otherwise be destroyed by the grazing of the urchin.

Finally, though we have emphasized that heavy algal growth will kill corals, Huston (1985) thinks it may well be that a certain amount of algal growth may promote coral diversity by reducing the interspecific contact and subsequent competition between coral colonies.

Role of Algae in Reef Systems

As we have noted, algae tend to be small and inconspicuous on coral reefs, in marked contrast to the great kelp forests of temperate shores. Yet, algae, through various biological interactions, are ecologically important to the community structure of the reef system. We discuss now some additional, primarily physical, roles for algae.

First, the encrusting red coralline algae such as *Lithothamnion* are important in maintaining the integrity of the reef by constantly cementing together various pieces of calcium carbonate, thus reinforcing the reef framework against destruction by wave action and preventing individual pieces from being carried off the reef. The algal ridge is further responsible for breaking the velocity of the waves and producing calmer conditions, allowing growth of other organisms in the reef flat behind it. The second group of calcareous algae are the greens, of which the genus *Halimeda* is dominant. These algae are significant contributors to the sand found in reefs, particularly in the lagoon area. They, therefore, create a special habitat type.

Certain algae also bore into coral skeletons and thus are responsible for the breakdown of reef structure. Algae also create habitats in and around themselves and may furnish necessary shade from the hot tropical sun for certain organisms.

They, of course, are important as primary producers in the reef system and as a food for various herbivores. These various roles are summarized in Fig. 9.15.

Ecology of Reef Fishes

It is impossible to conceive of a coral reef without the presence of myriads of brilliantly colored fishes gracefully moving among the various corals and associated organisms. Indeed, fishes are certainly the most abundant and conspicuous large organisms that one encounters on a reef (Plates 56 to 61). Because of their abundance and all-pervading presence in all areas of the reef, it seems obvious that they must contribute in significant ways to the reef ecosystem. We have noted some of these important structuring relationships in previous sections and will in this section review other aspects of their ecology.

The species richness of coral reef fishes is very similar to that observed for corals. That is to say, the central Indo-Pacific area of the Philippine and Indonesian archipelagos has the greatest number of species and the number decreases in all directions away from this center (Table 9.2). The Atlantic reefs are also relatively species poor in this comparison. The number of species that have been found on a single reef is really quite remarkable—500 for one reef in the Great Barrier Reef system. How does this great diversity of species come about, and how is it maintained? This is the question that has intrigued coral reef biologists for many years, and it is still generating debate. We can only suggest some of the reasons here; more remain to be discovered.

One of the reasons for the high diversity of species on reefs is the great variety of habitats that exists on reefs. Coral reefs encompass not only coral, but also areas of sand, various caves and crevices, areas of algae, as well as shallow and deep water and different zones progressing across the reef. This great diversity of habitats alone can go far to explain the increased number of fishes. These various habitats and their associated fish inhabitants are summarized diagrammatically in Figs. 9.16 and 9.17.

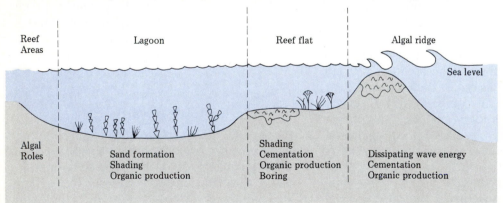

Figure 9.15 The various roles of free-living algae in different zones of a reef. (Modified from M. Doty. *In:* Proceedings of the second international symposium on coral reefs, Vol. 1, Great Barrier Reef Committee, Brisbane, 1974. Used by permission of the Great Barrier Reef Committee.)

Table 9.2   NUMBER OF FISH SPECIES IN SEVERAL CORAL REEF AREAS

Geographical area	Number of fish species
Philippine Islands	2177
New Guinea	1700
Great Barrier Reef	1500
Seychelles Islands	880
Marshall and Mariana islands	669
Bahama Islands	507
Hawaiian Islands	448

Source: After Goldman and Talbot, 1976.

The numerous habitats, however, are not enough to explain the high diversity of coral reef fishes, particularly in local areas. High local fish diversity has led recently to numerous experimental studies attempting to explain how so many species can continue to coexist in a given area. As a result, perhaps three opposing theories of reef fish diversity and community structure have arisen. The more classical view is that coexistence is the result of a high degree of specialization such that each species has a specific set of adaptations that give it the competitive edge in at least one situation on a reef. That is to say, these fishes have narrower ecological niches and hence more species can be accommodated in a given area. This view is partially supported by correlations between fish species diversity and habitat complexity or "rugosity" by Risk (1972) and Luckhurst and Luckhurst (1978). The opposing view, promulgated by Sale (1977, 1980), is the "lottery" hypothesis. This hypothesis rests on the fact that most coral reef fishes generally produce large numbers of larvae

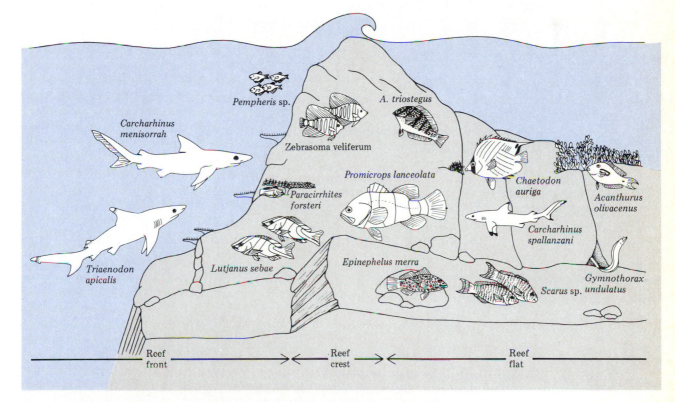

Figure 9.16 Generalized picture of the characteristic fish species and their habitats on a seaward coral reef in the Marshall Islands. Requiem sharks (Carcharinidae), *Triaenodon, Carcharinus;* moray eels (Muraenidae), *Gymnothorax;* hawk fishes (Cirrhitidae), *Paracirrhites;* groupers (Serranidae), *Epinephalus, Promicrops;* snappers (Lutjanidae), *Lutjanus;* surgeonfishes (Acantharidae), *Acanthurus, Zebrasoma;* parrot fishes (Scaridae), *Scarus;* butterfly fishes (Chaetodontidae), *Chaetodon,* sweepers (Pempheridae), *Pempheris.* (Not to scale.) (Modified from R. W. Hiatt and D. W. Strasburg, Ecological relationships of the fish fauna on coral reefs of the Marshall Islands, Ecol. Monogr. 30:65–127, 1960. Copyright © 1960 by the Ecological Society of America.)

Figure 9.17 Fishes associated with individual coral colonies of the *(A)* branching and *(B)* plate type. Butterfly fishes (Chaetodontidae), *Chaetodon, Forcipeger;* damselfishes (Pomacentridae), *Chromis, Dascyllus;* wrasses (Labridae), *Epibulus, Thalassoma, Stethojulis, Gomphosus;* velvet fishes (Caracanthidae), *Caracanthus;* gobies (Gobiidae), *Paragobiodon;* puffers (Tetraodontidae), *Arothron;* filefishes (Monocanthidae), *Oxymonocanthus;* hawk fishes (Cirrhitidae), *Paracirrhites;* triggerfishes (Balistidae), *Balistapus, Rhinecanthus, Abalistes;* squirrel fishes (Holocentridae), *Holocentrus.* (Not to scale.) (Modified from R. W. Hiatt and D. W. Strasburg, Ecological relationships of the fish fauna on coral reefs of the Marshall Islands, Ecol. Monogr. 30:65–127, 1960. Copyright © 1960 by the Ecological Society of America.)

that are dispersed into the plankton; it states further that fishes are not specialized (many similar species having the same requirements) and that there is active competition among the species. Local success and persistence result from chance as to which species of the planktonic larval pool occupies the vacant space. The third view, by Talbot, Russell, and Anderson (1978), is that the fish populations do not reach equilibrium. Predation, catastrophe, and unpredictable recruitment act to ensure that the populations never become large enough to undergo competitive exclusion, because they are kept below the numbers at which resources or food become limiting. Evidence is currently inconclusive as to which of these views, if any, is correct.

Perhaps as a result of the large numbers of species and the partitioning of the habitat, we find that most reef fishes, despite their obvious mobility, are restricted to certain areas of the reef and are very localized. They also do not migrate, and many of the smaller species such as gobies, blennies, and damselfishes are known to defend territories.

One of the most interesting discoveries concerning the fishes of coral reefs is the differences in fishes between day and night. Most people see reefs during the day, when the great majority of the fish species are visible. At night, however, these diurnal fishes seek shelter in the reef and are replaced by a smaller number of nocturnal species not seen at all during the day (Fig. 9.18; Plate 60). Since some of the nocturnal species are ecologically similar to certain diurnal species (Apogonidae, for example, replace Pomacentridae), this is another way of permitting a greater number of species to exist on the reef without competing directly. According to Hobson (1968), all nocturnal species of reef fishes are predaceous.

The feeding relationships of reef fishes are of some interest (Fig. 9.19). The most abundant feeding type on reefs are the carnivores, which may constitute 50 to 70 percent of the fish species. Goldman and Talbot (1976) say that many of these carnivores appear not to be highly specialized to a given food source, but instead are opportunistic, taking whatever is available to them. They also feed upon different prey at different stages in their life cycles. The fact that the majority of the fishes on reefs are generalized

carnivores seems to argue against the aforementioned explanation for the great species richness of reefs as being due to narrower niches (= more specialization). Indeed, this lack of specialization makes it difficult to explain satisfactorily the species richness. This is not to say that no specialized carnivores exist; there are a few. Correlated with the generalized feeding habits of most carnivores, the numbers of true scavenger fishes are few because the carnivores act also as scavengers and simply pick up any recently dead organisms.

Herbivores and coral grazers make up the next largest group of fishes (about 15 percent of the species), and the most important of these are the families Scaridae and Acanthuridae. The remainder of the fishes are generally classed as omnivores or multivores and include representatives from virtually all families of fishes on the reef (Pomacentridae, Chaetodontidae, Pomacanthidae, Monocanthidae, Ostraciontidae, Tetraodontidae). Only a few fishes are zooplankton feeders, and these are mainly small, schooling fishes of the families Pomacentridae, Clupeidae, and Atherinidae.

A side effect of the feeding of some fishes, such as grunts (haemulids), is to enhance the nutrition and growth of corals. Meyer et al. (1983) have demonstrated that schools of certain fishes, which rest by day in coral heads, feed at night on seagrass beds and then return to defecate material rich in nitrogen and phosphorus into the coral heads. This results in faster coral growth.

When one considers that the preponderance of fishes on the reef are unspecialized carnivores, it is easy to understand why most invertebrate organisms are hidden from view, since to expose a soft body on a reef would certainly invite instantaneous attack by some carnivore. How, then, do a few invertebrates, notably the large sea cucumbers, remain abundant lying unprotected on reefs? The answer apparently is that those few that are visible have evolved mechanisms to deter potential predators. The most common is a toxin or poison. According to Bakus (1973), tropical sea cucumbers produce toxic substances that can kill fishes and, in addition, have viscous, sticky strands (cuvierian organs), which can be extruded to literally tie up any potential predator.

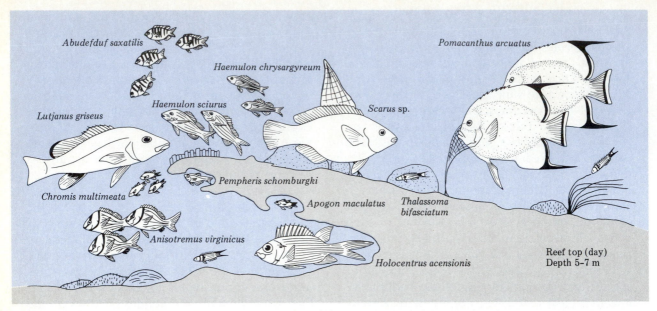

Abudefduf saxatilis

Haemulon chrysargyreum

Pomacanthus arcuatus

Lutjanus griseus

Haemulon sciurus

Scarus sp.

Chromis multimeata

Pempheris schomburgki

Apogon maculatus

Thalassoma bifasciatum

Anisotremus virginicus

Holocentrus acensionis

Reef top (day)
Depth 5–7 m

(A)

Clepticus parrai

Chromis cyanea

Chromis multilineata

Holacanthus tricolor

Haemulon striatum

Lachnolaimus maximus

Scarus spp.

Deep reef (day)
Depth 25–30 m

(B)

(A)

(B)

Figure 9.18 Day and night fish distributions on a Florida coral reef at two different depths. (A) Reef top at 5–7 m. (B) Deep reef at 25–30 m. Damselfishes (Pomacentridae), *Abudefduf, Chromis*; wrasses (Labridae), *Ciepticus, Thalassoma, Lachnolaimus*; parrot fishes (Scaridae), *Scarus*; angelfishes (Chaetodontidae), *Pomacanthus, Holocanthus*; grunts (Pomadasyidae), *Haemulon, Anisotremus*; cardinal fishes (Apogonidae), *Apogon*; squirrel fishes (Holocentridae), *Holocentrus*; sweepers (Pempheridae), *Pempheris*; snappers (Lutjanidae), *Lutjanus*; halfbeaks (Hemiramphidae), *Hemiramphus*. (Data for the drawing based on W. A. Starck and W. P. Davis, Night habits of fishes of Alligator Reef, Florida, Ichthyological Aquar. J. 38[4]:313–356, 1966.)

Figure 9.19 The trophic relationships of coral reef fishes. (After R. H. Lowe-McConnell, Ecology of fishes in tropical waters, Studies in biology no. 76, Edward Arnold, 1977.)

The development of distasteful or toxic substances is not limited to invertebrates. A great many of the fishes also produce toxic substances. These may take the form of venom associated with various spines or of poisonous material simply extruded onto the body surface (*crinotoxin*), or the flesh and internal organs may be toxic. Truly venomous fishes are rare on reefs and are confined mainly to the stonefishes (Synanceiidae) and scorpion fishes (Scorpaenidae), but a large number of fishes have toxic secretions on their outer surfaces, including the abundant parrot fish (Scaridae), wrasses (Labridae), and surgeonfish (Acanthuridae). One explanation for the prevalence of these substances is that they deter predation by the abundant carnivores.

The final category, namely, toxic flesh or internal organs, has caused the most interest among humans, because eating tropical fishes with toxins in the body flesh or organs produces a serious disease called *ciguatera*. Ciguatera has been one of the most mysterious and puzzling diseases found in the tropics. As noted by Banner (1976), the symptoms in humans are primarily neurological, including exhaustion, visual disturbance, "inversion" of senses (hot feels

cold, and vice versa), paralysis, loss of reflexes, and finally, death by respiratory failure. The mystery came originally from several sources: (1) The symptoms are not consistent. (2) The fishes that have the toxin can include virtually any or all of the large reef fishes, ranging from carnivores to herbivores, but the fishes vary in toxicity, both in time and in space, so that a given species may be toxic on one reef but not on another, or the same species may be toxic at one time of the year on a given reef and not at another. (3) There seems to be no repeatable pattern to the occurrence of the toxin.

We now know somewhat more about the disease. Perhaps most significant was the discovery in 1977 of the causative organism, the dinoflagellate *Gambierdiscus toxicus*. This dinoflagellate is scarce on reefs with large amounts of living coral, but proliferates after disturbances have created dead coral surfaces covered by filamentous or calcareous algae. The toxin, which is a polyether, has also been isolated and partially characterized. An efficient and sensitive bioassay employing mosquitoes has also been developed. We are, however, still ignorant of most of the ecological aspects of *G. toxicus* and the factors that will trigger a bloom, as well as of the mechanisms by which the toxin comes to reside in the flesh of certain carnivorous fish. Ciguatera, therefore, still remains of considerable importance to the people of the tropics since the fisheries of many areas cannot be developed until this toxicity problem is solved.

Apparently, coral reef fishes are also subject to occasional epidemics that cause mass mortality. A Caribbean-wide mass die off of fishes was recorded by Williams and Williams (1987–88) to have occurred in August and September of 1980 following the passage of a hurricane. However, this mass mortality event was not studied so the nature of the cause(s) is unknown.

Another of the outstanding characteristics of reef fishes is their color (Plate 59). Why, especially given the great predation pressure, should they be so brightly colored? Biologists cannot agree as yet, and it is likely that color and pattern serve several different functions. One explanation is that the bright colors serve to advertise that the species is toxic or otherwise distasteful so that predators will ignore it (= warning coloration). Another explanation of the colors is that they serve for species recognition. They may also serve to camouflage the species, either by breaking up the shape of the fish or else making it appear as something else. This latter explanation has been especially invoked to explain the dramatic banding patterns seen in some reef fishes (Fig. 9.20; Plate 56).

A final aspect of reef fish ecology concerns the phenomenon of *cleaning behavior*. Cleaning behavior is a specialized form of predation in which certain small fishes (*Labroides* spp.) or shrimps remove various ectoparasites from other, usually larger, fish species. Cleaning behavior is extremely widespread and apparently occurs on all reefs. In this process, the cleaner fishes (or shrimps) often set up "cleaning stations" where they appear to advertise their presence through bright, contrasting colors. The fish to be cleaned comes to the cleaning station area (often a prominent coral head or boulder) and remains motionless as the cleaner moves over its body removing the parasites (Fig. 9.21). The cleaners may even enter the mouth and gill chambers of the fish. A sort of "truce" among predators prevails at the cleaning stations, as the fishes being cleaned are vulnerable to predation. Fish-

Lo vulpinus

Heniochus acuminatus

Figure 9.20 Reef fishes with dramatic banding patterns.

es even "line up" at these stations, awaiting their turn for cleaning.

The importance of this cleaning behavior to the fish population and economy of the reef has not been established. Some biologists believe that, if cleaners are removed from a reef, the fish fauna decreases, moving away or otherwise showing signs of distress. Yet, other biologists have removed cleaners with no noticeable effect on the reef or reef fishes.

We can summarize this section by stating again the importance of fishes to the reef ecosystem. Fishes are important in determining the zonation of various parts of the reef through their grazing activities, which probably keep the algae reduced to a short mat and prevent them from outcompeting the corals. At the same time, their grazing on corals is responsible for the exclusion of certain corals from areas of the reef,

Figure 9.21 A cleaner wrasse (*Labroides*) cleaning the mouth of a larger grouper.

which also contributes to the zonation of reefs. Consumption of coral by fishes also leads to the breakdown of coral and contributes to the sediment budget of a reef. The tremendous number of carnivores among the fishes undoubtedly exerts a great predation pressure on all reef organisms and is the reason for the cryptic nature of most soft-bodied invertebrates and for the development of various toxic passive defense mechanisms among fishes and invertebrates.

Coral Reef Cryptofauna and Bioerosion

The term *cryptofauna*, applied to coral reefs, refers to all invertebrates that inhabit the coral substrates, boring into the calcium carbonate, living in crevices and cracks, and occupying space under coral heads and boulders. According to Hutchings (1983), two groups exist: those that actually bore into the calcium carbonate—the true borers—and those that are unable to bore the coral rock but live in preexisting holes, cracks, and crevices—the "opportunists." True borers of coral rock include sponges, bivalve mollusks, sipunculans, and some polychaete worms. The nestlers or "opportunists" represent most marine invertebrate phyla, but crustaceans and polychaete worms seem to be particularly common.

Colonial cryptofaunal organisms, including sponges, bryozoans, and ascidians, tend to dominate the surface fauna of the undersides of corals, caves, and overhangs. Their ability to reproduce asexually and to overgrow other potential space occupiers probably gives them a competitive edge in these areas.

Very little is known about the structure, recruitment, or succession of these cryptofaunal communities. According to Hutchings (1983), the community seems to be dominated by deposit feeders. These animals feed mainly on organic detritus that has been collected in mucus secreted by the coral and moved off the coral colony. They may play an important part in the coral reef system by using the large amounts of coral-produced mucus and in recycling the organic material and nutrients trapped therein. Cryptofaunal organisms are also an important food source for certain of the reef carnivores such as holocentrid fishes (Plate 60) and certain gastropod mollusks such as *Conus* (Kohn and Nybakken, 1975).

Most certainly the cryptofauna is significant in bioerosion of the reef. The rates of bioerosion due to activities of sponges, echinoids, and grazing mollusks have been assessed and when added to that of various grazing fishes suggest that bioerosion may be the most important of the processes that act to break down coral reefs. For instance, Stearn and Scoffin (1977) reported that on Barbados corals and algae laid down about 160 metric tons of calcium carbonate annually over an area of about 1 hectare. In that same area boring sponges, barnacles, and bivalves annually removed 1.5 metric tons from live corals and 23.5 metric tons from dead coral. Parrot fish removed an additional 1 metric ton and the urchin *Diadema antillarum* removed 163 metric tons. This suggests that the reef is eroding faster than it can be built up. However, a second study by Scoffin et al. (1980) on the same reef suggested a higher calcification rate of 206 tons per year and a total bioerosion rate of 163 tons per year. This points up the variability in the data and suggests that few conclusions can yet be drawn about calcification and bioerosion rates without more study.

A more recent review of bioerosion by Hutchings (1986) has further substantiated this earlier work on the significance of biological agents in reef destruction. Hutchings also noted that, in addition to the major groups of sponges, mollusks, echinoderms, and fishes, erosive agents such as fungi, endolithic algae, polychaetes, and sipunculan worms are present on reefs but that little or nothing is currently known about their relative importance. More important, the rates at which any and all bioerosive organisms remove calcium carbonate from reefs appear to be quite variable, both in time and in space, thus confounding the problem of determining the role of these organisms in the overall ecology of coral reefs.

Catastrophic Mortality and Recovery of Reefs

Although reefs appear to be large and very stable systems, existing, in the case of atolls, for millions of years, they do in fact suffer large-scale destruction from various forces. Perhaps the major source of massive reef mortality is mechanical destruction by severe tropical storms. Hurricanes or typhoons are sufficiently intense that their passage over a reef area often means destruction of large areas of reef. Since most coral reefs lie within the zone most frequently traversed by typhoons or hurricanes, the chance that a reef or a section of a reef will be destroyed or severely damaged is great. Damage in such storms usually is due to coral colonies being uprooted and carried off the reef, thus freeing space for new occupants.

The second major source of catastrophic mortality of reefs, at least in recent times, is the population explosion of the sea star *Acanthaster planci*. Since 1957, when the population explosions were first observed, *A. planci* has caused catastrophic mortality of reefs in a number of places in the western Pacific. The ability of this starfish to destroy large areas of reef is tremendous. In Guam, Chesher (1969) estimated that 90 percent of the coral reef along 38 km of shoreline was destroyed in 2½ years, and on the Great Barrier Reef, Endean (1973) indicated that the bulk of the corals on an 8 km² reef was destroyed in 12 months. More recent surveys (1985–86) reported in Walbran et al. (1989) indicated that 27 percent of the Great Barrier Reef has been affected by *Acanthaster*. When these rates are multiplied by the number of reefs in the Pacific that were found to be infested in the 1960s (including 60 percent of the reefs of the Great Barrier Reef between 16°S and 19°S),

one can understand the concern for the survival of reefs.

What causes these population explosions and are they natural phenomena? These are the two major questions that biologists have asked regarding the "*Acanthaster* problem," and their answers point up how little we really know about how coral reefs function. Although there is no consensus concerning the answer to either question, it now seems as though the explosions are natural phenomena; that similar outbreaks have occurred in the past is evidenced by the findings of aggregations of *Acanthaster* skeletal remains in sediment cores by Frankel (1977) and more recently by Walbran et al. (1989), by the mention in folklore and memories of certain island peoples according to Birkeland (1982), by their occurrences in areas remote from human interference and activities, and by their recurrence in several areas of the Pacific. Whether such explosions are natural to other reef systems is not known. Even if these earlier outbreaks were natural, the recent ones could still have been caused by human factors. Whether natural or human-induced, what is the trigger? Examination of the infested reefs has produced no consistent pattern of human-induced or natural activity that would allow us to isolate a single cause. Many suggested reasons have been proffered, such as various dredging activities that open new space for the *A. planci* juveniles; that some chemical (e.g., pesticides) somehow freed *A. planci* from its normal population controls; that humans removed an important predator, allowing increased survival of juveniles; or that severe tropical storms opened space for juvenile settlement or caused adults to aggregate.

Although no one knows for certain what the cause is, of the above suggestions the two that have received most attention and coverage in the press are the latter two, involving predators and the effects of storms. The predator-removal theory necessitates the presence of predators on *Acanthaster* that normally keep the population in check. Is there such a predator? Yes, there is a large gastropod, *Charonia tritonis*, which does feed on *Acanthaster*. It is also in demand by shell collectors. Endean (1977, 1985) argues that the removal of this relatively uncommon gastropod from the reefs by collectors could have allowed the population explosion to begin, but currently there is little evidence to support this hypothesis. If this explanation is correct, then *C. tritonis* is a "keystone" species in the same sense that *Pisaster ochraceus* was in the rocky intertidal. However, we know so little about the interactions on reefs that this certainly could not have been predicted ahead of time. Another, more acceptable, theory is the "runoff" theory by Birkeland (1982). This hypothesis suggests that juvenile recruitment of *A. planci* is enhanced by a combination of unusually low salinities, high nutrients, and high temperatures. In this hypothesis, the occurrence of a year or years of abnormally heavy rainfall in an area, especially if coupled with human destruction of the native vegetation on adjacent land areas, causes increased runoff, which in turn provides enough nutrients to stimulate phytoplankton blooms to provide food for large numbers of *Acanthaster* larvae. The increased survival of the larvae and their subsequent settlement onto the reefs lead to outbreaks three years later. This model invokes basically a natural explanation that fits the pattern of recent outbreaks, but currently has little evidence to support it. Correlated with this theory is the adult-aggregation theory of Dana et al. (1972), which says that the destruction of reef areas by typhoons causes the starfish to condense into aggregations, which then proceed to attack the remaining living coral en masse.

A third cause of catastrophic mortality is the phenomenon known as *El Niño*. The El Niño is the name given to a naturally recurring periodic phenomenon manifested as a massive influx of nutrient-poor warm water into the usually cold, nutrient-rich surface waters off the coasts of Ecuador and Peru. In fact, El Niño is but part of an interacting set of atmospheric and oceanographic conditions called the "El Niño–Southern Oscillation" (ENSO) that periodically alters the atmospheric pressure, winds, rainfall patterns, ocean currents, and sea level over large areas of the tropical and subtropical Pacific Ocean. These natural ENSO events vary greatly in intensity, areal extent, and ecological effects and are highly unpredictable. The greater the intensity, the greater the change in oceanographic and meteorological conditions, the larger the area affected, and the greater the effect on living organisms. The 1982–83 El Niño, according to

Glynn (1988), was perhaps the strongest event of this century and produced unprecedented disturbances to marine populations and communities. It may have had a periodicity of 1 in 100 to 1 in 250 years. As a result of this event, the ocean surface temperatures in the eastern Pacific rose to 30–32°C, some 2–4°C above normal, and remained there for several months. This prolonged elevation of temperature, coupled with changes in sea level that were up to 44 cm below normal in the eastern Pacific and unusual hurricane activity and hurricane paths, caused widespread death of large tracts of reef corals in the Gulf of Panama, Galápagos Islands, Tahiti, and the Society Islands, as well as in other places throughout much of the tropical Pacific (Lessios, Glynn, and Robertson, 1983). Mortality rates for coral ranged from 50 to 98 percent overall according to Glynn (1988). Particularly devastated were the reefs in the Galápagos, where Glynn (1985) reported 95 percent of corals died, and in the Gulf of Panama, where 70 to 80 percent of the corals were destroyed.

The loss of the corals and coral reef framework in areas such as Tahiti, the Society Islands, and the Galápagos Islands had profound secondary effects. On the one hand, it eliminated habitats for numerous reef-associated species, according to Glynn (1988). Thus, whole fish communities disappeared from areas of dead coral. On the other hand, in the eastern Pacific the recruitment of sea urchins was enhanced and densities of *Diadema* in Panama and *Eucidaris* in the Galápagos rose. Their activities, in turn, caused increased bioerosion, which destroyed the dead reef framework. Even corals that survived the initial mortality by the change in physical factors were subjected to increased predation by corallivorous sea stars when the protective shrimp and crab symbionts died or migrated.

A puzzling event, which has the potential to cause catastrophic coral mortality, is the phenomenon that has come to be known as *coral bleaching*. Corals bleach when they expel the zooxanthellae that normally inhabit their tissues. Following the expelling of the zooxanthellae, corals become white, hence the name of the phenomenon. If the zooxanthellae are absent for a significant time, the coral will die. Coral

Plate 49 Underwater view of a Pacific Ocean coral reef. Note the clear water, lack of unoccupied primary space, and dominance of madreporarian corals. *(Photo courtesy of Drs. Lovell and Libby Langstroth.)*

Plate 50 Underwater photo of the crowding of gorgonians and corals on a coral reef in Palau. *(Photo courtesy of Drs. Lovell and Libby Langstroth.)*

Plate 51 A single coral colony showing the characteristic appearance when the polyps are expanded. *(Photo courtesy of Drs. Lovell and Libby Langstroth.)*

Plate 52 Close-up view of expanded coral polyps. *(Photo courtesy of Drs. Lovell and Libby Langstroth.)*

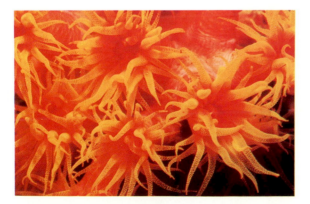

Plate 53 Expanded polyps of *Tubastrea* on a reef in Palau. This small coral does not build reefs. *(Photo courtesy of Drs. Lovell and Libby Langstroth.)*

Plate 54 A crinoid or feather star crawling over gorgonians on a coral reef in Palau. *(Photo courtesy of Drs. Lovell and Libby Langstroth.)*

Plate 55 A giant clam, *Tridacna maxima,* on a Palau coral reef. The patterned mantle tissue is expanded. The mantle tissue contains the symbiotic zooxanthellae, which furnish the clam with a certain portion of its nutrition. *(Photo courtesy of Drs. Lovell and Libby Langstroth.)*

Plate 56 Moorish Idols, *Zanclus canescens,* on a coral reef in Palau. *(Photo courtesy of Drs. Lovell and Libby Langstroth.)*

Plate 58 An angelfish of the genus *Pomacanthus.* *(Photo courtesy of Drs. Lovell and Libby Langstroth.)*

Plate 57 A hawkfish swimming among the branches of a coral on a reef in Palau. *(Photo courtesy of Drs. Lovell and Libby Langstroth.)*

Plate 59 A lionfish, *Pterois volitans,* of the Pacific Ocean coral reefs. *(Photo courtesy of Drs. Lovell and Libby Langstroth.)*

Plate 60 A nocturnal reef fish, *Myripristis murdjan.* Note the large eyes. *(Photo courtesy of Drs. Lovell and Libby Langstroth.)*

Plate 61 An anemone fish of the genus *Amphiprion* nestled in the tentacles of its anemone. *(Photo courtesy of Drs. Lovell and Libby Langstroth.)*

Plate 63 Commensal barnacles, *Cryptolepas rhachianectes,* and "whale lice," *Cyamus scammoni,* on the skin of a California gray whale, *Eschrichtius robustus.* *(Photo by the author.)*

Plate 62 Mangrove seedlings coming up among the stilt roots of the adult trees. Lizard Island, Australia. *(Photo by the author.)*

Plate 64 Commensal slipper shells, *Crepidula adunca,* adhering to the black shell of the predatory gastropod *Mitra idae.* *(Photo courtesy of Drs. Lovell and Libby Langstroth.)*

bleaching events have been reported for many reefs in both the Caribbean and the Indo-Pacific in the decade of the 1980s. Usually the bleaching affects most, but often not all, coral species and is most severe in shallow-water areas. Widespread severe bleaching events occurred in the Caribbean in 1987–88 and in the eastern Pacific following the previously described El Niño in 1982–83. A recent bleaching event has been reported by Goreau (1990) for Jamaica.

The cause of bleaching is unclear, but most scientists tend to attribute it to stress, particularly stress induced by water temperatures of 30°C or above. Bleaching events appear to be growing more common, and concern has been raised that the pattern of bleaching events may be an early warning of the onset of a global warming (see Chapter 11). If bleachings continue to increase, there is concern for the integrity of many reefs because either the length of the events or the closeness of events will be such that corals will not have time to recover, and large areas may die.

The activities of humans can directly cause catastrophic mortality on reefs through dredging and pollution. The reefs in Kaneohe Bay, Hawaii, for example, were destroyed through sewage pollution. During World War II, the military closed off the lagoon at Palmyra Atoll, which resulted in the death of all lagoon reefs. In certain areas of India, coral is actively mined for use in building, and this is also causing catastrophic mortality. In the Philippines, dynamiting for fishes is a major cause of destruction.

When catastrophic mortality occurs, how long does it take a reef to recover? Again, we are ignorant, but the little information we have on recolonization of reefs destroyed by hurricanes suggests that it may take as long as 25 to 30 years before the reef has completely recovered. Estimates of recovery time from *Acanthaster* devastation range from 7 to 40 years. In the case of the 1982–83 El Niño, Glynn (1985) feels that some areas may take 100 or more years to recover. The reasons for the widely varying estimates of recovery time are due to the fact that recovery depends on several factors, including the extent of initial destruction, the nearness of a source of larvae for recolonization, and favorable water currents and conditions for establishment of the larvae. Recovery of some reefs may be inhibited by the mats of algae and soft corals, which quickly take over the dead coral skeletons and thus do not offer suitable substrates for the recolonizing reef corals.

MANGROVE FORESTS

Mangrove forest or *mangal* is a general term used to describe a variety of tropical inshore communities dominated by several species of peculiar trees or shrubs that have the ability to grow in salt water. These "mangroves" are flowering terrestrial plants that have reinvaded the fringes of the sea. The term *mangrove* refers to the individual plants, whereas mangrove forest, mangrove swamp, tidal forest, and mangal refer to the whole community or association dominated by these plants. Walsh (1974) reports that from 60 to 75 percent of the coastline of the earth's tropical regions is lined with mangroves, so their ecological importance is clear. Recognizable descriptions of mangal associations were made by the Greeks as early as 325 B.C. In this section, we shall cover the basic ecological factors governing the existence of these communities.

Structure and Adaptations

The mangrove forest comprises trees and shrubs belonging to some 12 genera of flowering plants in 8 different families (Table 9.3). The most important or dominant genera appear to be *Rhizophora, Avicennia, Bruguiera,* and *Sonneratia*. Mangroves share a number of characteristic features that allow them to live in shallow marine waters. They are, first of all, shallow rooted, with roots that spread widely or else with

Table 9.3   THE GENERA OF MANGROVES

Avicennia	Aegiceras
Suaeda	Aegialitis
Laguncularia	Rhizophora
Lumnitzera	Bruguiera
Conocarpus	Ceriops
Xylocarpus	Sonneratia

Source: After Lugo and Snedaker, 1974.

peculiar prop roots that stem from the trunk or branches (Fig. 9.22). The shallow roots often send up extensions, called *pneumatophores,* to the surface of the substrate that allow the roots to receive oxygen in the otherwise anoxic mud in which these trees grow (Fig. 9.23). The leaves are tough and succulent and have internal water storage tissue and elevated salt concentrations. Some mangroves have salt glands, which help to maintain the osmotic balance by secreting salt (see also salt glands in seabirds in Chapter 3). Others exclude salt and separate fresh water from salt water at the roots by a "reverse osmosis" process. Nearly all mangroves share two reproductive strategies, according to Rabinowitz (1978). These are: vivipary, or uninterrupted development of the embryo following fertilization such that there is no "resting" or true "seed" stage; and dispersal by water. The latter is particularly well developed in the genus *Rhizophora* and may serve as a model. The seed, while still on the parent plant, germinates and begins to grow into a seedling without any intervening resting stage. During this time the seedling elongates and the weight distribution changes, so it becomes heavier at the outer, free end (Fig. 9.24). Eventually, this seedling is dropped from the parent plant and, because of the weight distribution, floats upright in the water. It then is carried by the water currents until it enters water shallow enough for its root end to strike the bottom. When this happens, it puts out roots to anchor itself and continues growth into a tree. The advantages of this system of reproduction are obvious for a plant living on the fringe of the sea. Having seeds able to float means it can be dispersed by water currents, while the fact that it floats upright with much of

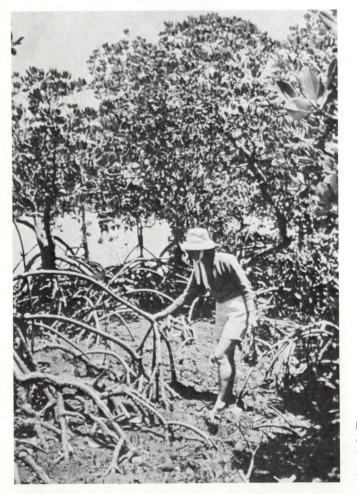

Figure 9.22 Mangrove forest showing typical root system and roots descending from the branches. (Photo by Dr. Richard Mariscal.)

Figure 9.23 Pneumatophores on the roots of a mangrove tree. (Photo courtesy of Wards Natural Science Establishment.)

the seed below the water means that, when water shallow enough for the mangrove to grow in is reached, the seed will ground itself (Plate 62).

Distribution

Mangrove forests are distributed throughout the tropical and subtropical oceans of the world. They are able to grow only on shores that are sheltered from wave action; otherwise, the seeds would never be able to ground properly and put down roots. These shores may be directly along the lee sides of islands or island chains or on islands or landmasses behind protecting offshore coral reefs. They are particularly well developed in estuarine portions of the tropics, where they reach their greatest areal extent.

Mangroves occur over a larger geographical area than coral reefs and may be found in areas well outside the tropics, such as along the west coast of central and northern South America and Africa, where reefs are rare or absent, and as far south as the north island of New Zealand and nearly to the mouth of the Rio de la Plata in Argentina (Fig. 9.25). These mangrove forests may also penetrate some distance upstream along the banks of rivers (as far as 300 km along the Fly River in New Guinea; see McNae, 1968). Mangroves are usually absent from atolls and isolated high islands like Hawaii. (Mangroves were introduced to Hawaii in 1902.)

Physical Conditions of Mangrove Forests

Since mangroves can establish themselves only where there is no significant wave action, the first physical condition noted in mangrove areas is that the water motion is minimal. This lack of vigorous water motion has in its turn significant effects. Slow water movement means that fine sediment particles tend to settle out and accumulate on the bottom. The result is an accumulation of mud, and hence, the substrate in mangrove swamps is usually mud. In this sense, then, these areas are much like the previously discussed muddy shores, where poor interstitial circulation and high bacterial numbers lead to anoxic conditions. Perhaps this is also why mangroves have such shallow roots and pneumatophores. We should note that mangroves have the ability to grow on many types of substrates, so it

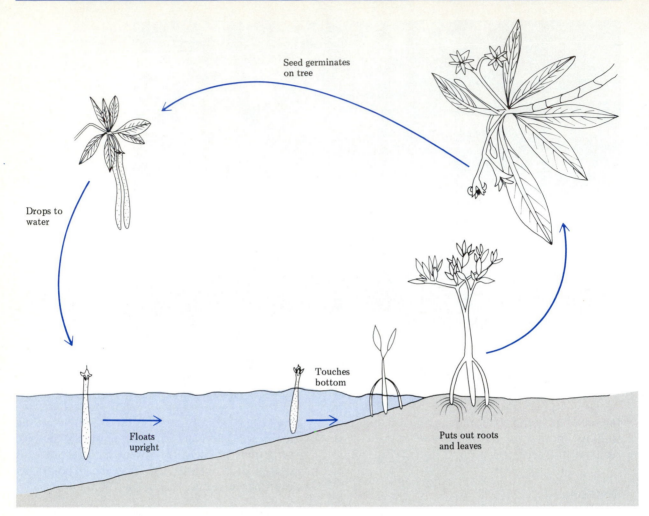

Seed germinates
on tree

Drops to
water

Touches
bottom

Floats
upright

Puts out roots
and leaves

Figure 9.24 The life cycle of a typical mangrove tree.

is not unusual to see small islands of mangroves even on coral rock or sand. However, the extensive mangrove forests appear to flourish only on muds or fine-grained sediments (Odum et al., 1982).

The initial slow water movement in mangrove forests is further enhanced by the mangroves themselves. As we noted above, many mangroves have peculiar prop (stilt) roots, which extend downward from the trunk and branches. These roots are often so numerous that they form an impenetrable tangle between the surface of the mud and the surface of the water. The result of the presence of these dense root systems is to decrease the water movement even more, and thus the very finest particles settle out around the roots of the mangroves, creating ever accumulating layers of sediment. Once dropped,

the sediment is usually not picked up again and transported out. The result of this trapping is the rather rapid accumulation of sediment within the mangrove association. Mangrove forests are very significant producers of new coastal land at the expense of the sea. The rate at which this occurs is really quite remarkable. McNae (1968) notes, for example, that the great Venetian traveler Marco Polo visited a port city called Palembang in Sumatra in 1292. At that time, the city was directly on the sea. At present it lies 50 km inland! This translates into an accumulation of land at the rate of 73 m/year over the last 675 years. In other areas, the rate may be even faster (200 m/year in the Bodri Delta of east Java). This creation of land means that, as one progresses through a mangrove association from seaward side to landward edge, one will see a

Figure 9.25 Distribution of mangrove forests in the world.

progressive change from marine to terrestrial conditions. This is probably also the main reason for the zonation observed in mangrove forests (see below). The soils of mangrove forests are characterized by low oxygen and high salt content. They are fine grained, with a high organic content.

Although mangroves seem to be facultative halophytes and are capable of growing in freshwater areas, mangrove forests do not develop in strictly freshwater environments. This is probably due, as Kuenzler (1974) notes, to the fact that they are not good competitors and the salinity is important in reducing competition from freshwater and terrestrial vascular plants.

The final physical factor to be considered is the tide. The tidal range and type of tide vary across the geographical range of the mangroves. Mangrove forests develop only in shallow water and intertidal areas and are thus strongly influenced by the tides. Whereas tidal action is not a direct requirement to the mangrove according to Odum et al. (1982), it is indirectly important in a number of ways. First of all, tidal action helps exclude other competitively superior vascular plants, permitting mangroves to exist. Second, tides bring salt water up the estuaries, permitting

mangroves to penetrate inland. Tides also transport nutrients into and export material out of mangrove forests. Finally, the action of the tides prevents soil salinities from reaching lethal levels in areas of high evaporation, and they aid in dispersing propagules. Perhaps these are the reasons that mangrove communities reach their greatest development where there are large tidal fluctuations. It is the tide and its vertical range that determine the periodicity of the inundation of the forests. This periodicity of inundation seems to be of considerable importance in determining what kinds of mangrove associations develop in an area and may be responsible for the different types of zonation observed.

Zonation

Up until now, most studies on mangrove forests have been descriptive works in which the emphasis has been on describing the changes in vegetation across the associations, from seaward edge to true terrestrial communities. As might be expected, each of these studies has produced differing schemes that reflect the differing underlying conditions (such as the aforementioned tidal factors). No one has suggested a truly universal scheme such as that proposed for rocky shores (see Chapter 6), but a general scheme of wide applicability for mangrove forests of the Indo-Pacific region may be suggested.

The seaward area of most Pacific mangrove forests is dominated by one or more species of *Avicennia*. This *Avicennia* fringe is usually narrow because *Avicennia* seedlings do not grow well under conditions of shade or heavy silt-

ation, which prevail in the interior of the forest. Associates in this zone and growing seaward of it are trees of the genus *Sonneratia*, which grow where they experience daily wetting (Fig. 9.26).

Behind the *Avicennia* fringe lies the *Rhizophora zone*, dominated by one or more species of *Rhizophora*. These are the trees that most characterize mangrove communities, because

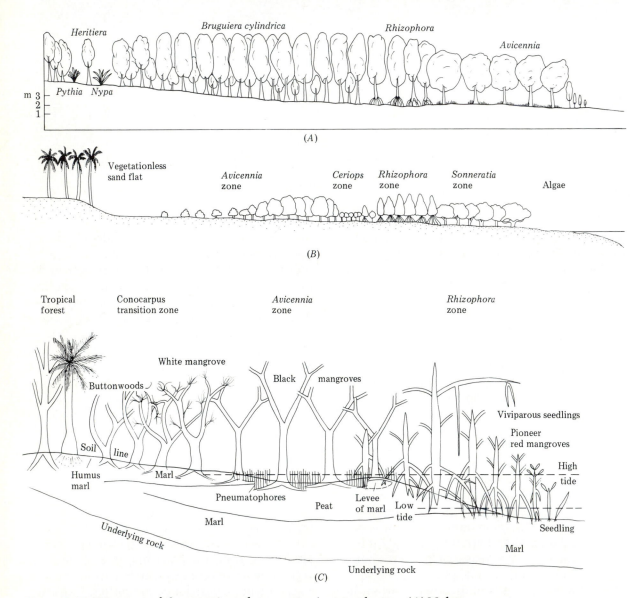

Figure 9.26 Diagrams of the zonation of mangroves in several areas. *(A)* Malaya. *(B)* East Africa. *(C)* Florida. (*A, B,* after V. J. Chapman [ed.], Ecosystems of the world, I. Wet coastal ecosystems, Elsevier, 1977. *C,* after J. H. Davis, The ecology and geologic role of mangroves in Florida, in Papers from Tortugas Laboratory XXXII, Carnegie Institution of Washington pub. 517, 1940.)

they have the arching stilt roots that make these areas such an impenetrable maze for humans. Species of *Rhizophora* are often quite tall and develop across a broad area of the intertidal, from levels flooded at all high tides to areas flooded only on highest spring tides.

As one progresses landward, the next zone is the *Bruguiera* zone. Trees of the genus *Bruguiera* develop on more heavy sediment (clays) at the level of high water of spring tides.

The final zone of the mangrove forest, which is sometimes present, is the *Ceriops* zone, an association of small shrubs. This is a variable zone and may, in fact, be merged with the trees of the *Bruguiera* zone.

In the Americas, the zonation of mangrove associations is less pronounced, in part due to the lesser number of mangrove species. For example, many Indo-Pacific regions have 30 to 40 mangrove species present, but Florida has just three: red mangrove (*Rhizophora mangle*), black mangrove (*Avicennia germinans*), and white mangrove (*Laguncularia racemosa*). Another species, buttonwood (*Conocarpus erectus*), though not a mangrove, does often occur between mangroves and terrestrial vegetation. Zonation in the Americas, according to West (1977), places the *Rhizophora* mangal zone at the seaward edge. This zone is followed inland by a broad band of *Avicennia* and then a belt of *Laguncularia* or *Conocarpus* (Fig. 9.26).

It should also be remembered that the above are general schemes and not all mangrove forests will correspond to them. In fact, in some areas the association may be considerably abbreviated or represented only by a few individuals. This is particularly true at the limits of the distribution of the mangroves. The zonation may also be interrupted where local conditions are such that evaporation of water from the soils makes them hypersaline. Hypersalinity tends to kill off mangroves, creating bare areas. Full development of mangrove forests is thus found in areas of high rainfall or in areas where rivers furnish enough fresh water to preclude the development of hypersaline conditions.

Zonation may also be limited by tidal action. Wherever the tidal range is small, the intertidal zone is also restricted, as are the mangrove forests. The most extensive forests are developed on shores that have a substantial vertical tidal range.

Associated Organisms

Mangrove communities are unique: due to the vertical extent of the trees, true terrestrial organisms can occupy the upper levels, while simultaneously true marine animals occupy the bases. Mangrove forests, then, form a strange mixture of marine and terrestrial organisms and have been suggested as a pathway from land to sea, and vice versa.

Terrestrial organisms seem to show no special adaptation for life in the mangrove forest, since they spend their lives out of reach of the marine waters in the upper reaches of the trees, though they may forage on the marine animals at low tide. There is a large variety of terrestrial arthropods in the mangrove forest. It includes herbivorous and carnivorous insects, but their significance to the mangrove trees is little known.

The marine organisms are of two types: those that inhabit the hard substrate represented by the numerous stilt roots of the mangroves and those that occupy the mud. Mangrove associations are different from those of muddy shores primarily because of the vast hard surface area of roots available to organisms, which is absent from typical mud shores (see pp. 270–271).

The dominant groups of marine animals in the mangrove forests are mollusks, certain crustaceans, and some peculiar fishes. Mollusks are represented by a number of snails, one group of which generally lives on the roots and trunks of the mangrove trees (Littorinidae) and the other on the mud at the base of the roots, comprising mainly detritus feeders (Ellobiidae and Potamididae). Little is known of the contribution of these snails to the mangrove forest. A second group of mollusks includes the bivalves. Oysters are the dominant bivalves. They attach to the roots of the mangroves, where they form a significant biomass (Fig. 9.27).

Mangrove forests are also inhabited by numerous large-sized crabs and shrimps. These animals excavate burrows in the soft substrate and include such common genera as *Uca,* the fiddler crabs; *Cardisoma,* the tropical land crabs; and various ghost crabs (*Dotilla, Cleistostoma*). The mangrove crab *Aratus pisonii* is extremely

Littorina

Ascidia

Crassostrea

Aratus

Stenopus

Stichopus

Sabellastarte

Bartholomea

Arboreal fauna, e.g., insects, tree-snakes, iguanas (of terrestrial origin)

Mangrove fauna

High tide

Marine hard-bottom fauna (see symbols)

Marina fauna visitors at high tide, e.g., fishes (*Mugil sphenoides*), crabs (*Callinectes*), prawns (*Penaeus*)

Low tide

Marine soft-bottom fauna, e.g., *Uca*, gobies *Crassostrea*

← Marine → ← Mangrove fauna →

Figure 9.27 Representation of the macrofauna of a mangrove forest showing the vertical distribution and ecological relationships. (Modified from H. Friedrich, Marine biology, University of Washington Press, 1969.)

abundant in Caribbean mangroves, where it lives on the roots just above the waterline. Also abundant in Caribbean mangroves is *Goniopsis cruentata,* which lives among the roots. It does not construct burrows but may take refuge in the burrows dug by other crabs. These crabs usually are specialized to feed on the detrital particles found in the mud. Generally, they separate the organic detrital particles from the nonorganic matter by filtering the substrate through a set of fine hairs around the mouth. These crabs also show varying degrees of adaptation toward a more terrestrial mode of life. This usually expresses itself in a vascularization of the walls of the gill chambers so that they become more "lunglike." As we noted previously, mangrove

associations are transition areas between land and sea, and the crab fauna living in mangroves certainly reflect this in their partial adaptation to air breathing.

The burrows of these crabs, as well as those of certain shrimps such as *Upogebia* and *Thalassina,* serve several functions. They serve as refuges from predation, as a breeding place, and as an aid to feeding. However, they also serve the mangrove community in that they allow oxygen to enter more deeply into the substrate, thus ameliorating anoxic conditions. The mangrove crab *Aratus pisonii* is not a particle feeder. It appears to be an omnivore that has a diet that includes mangrove leaves. Thus, it is a potentially significant herbivore.

The large numbers of marine invertebrates inhabiting the prop roots form an association that is poorly understood and little studied. The few studies done suggest that some intriguing interactions occur among the epifaunal organisms and that they have a corresponding significance to the mangroves. For example, Perry (1988), working on the red mangrove *Rhizophora mangle* in Costa Rica, experimentally demonstrated some interesting ecological interactions among the mangrove, the epifauna on the mangrove roots, and various benthic and water-column predators. In this system the isopod *Sphaeroma peruvianum* bores into the roots of the mangrove, while barnacles (*Balanus* spp.) settle on the surface of the roots. Both act to cause a decrease in root growth rate of 50 percent and 30 percent, respectively, as well as a decrease of root production of over 50 percent in both cases. However, not all roots are equally affected. Those roots not in contact with the substratum are most affected, whereas those in contact suffer considerably less. The reason for this discrepancy is that those roots in contact with the substrate are accessible to a benthic predator, the hermit crab *Clibarnarius panamaensis*, which climbs the roots at high tide to feed on barnacles. According to Perry, the absence of the isopod from certain roots can also be attributed to predation, as demonstrated by predator-exclusion work. However, the exact predator remains unknown. A somewhat similar situation may occur with respect to algae. Littler et al. have found that in mangroves in Belize fleshy algae grow on roots that do not intersect the bottom but not on those that do. Their experiment demonstrated that the hanging roots offered protection from benthic herbivores such as urchins.

According to Sutherland (1980), who studied the same red mangrove species in Venezuela, these epifaunal organisms that occur on mangrove roots seem to be long lived and resistant to invasion by other organisms. They also show little or no evidence of succession, a given root remaining with the same species array for many years. In contrast to Perry, Sutherland found predation had little effect on community development. These conflicting studies suggest that our understanding of how species interact in mangroves is still very poor.

Mangrove roots are not the only parts of the plant that are subject to consumption. Wada and Wower (1989) demonstrated that the ocypodid crab *Macrophthalmus quadratus* grazes on the pneumatophores of the mangrove *Sonneratia alba* and removes epiphytic algae and bark. Onuf et al. (1977) found that mangroves in Florida responded to nutrient enrichment derived from the guano of a bird-breeding colony by higher growth rates compared to nonenriched areas, but that the nutrient-enriched mangroves also had increased herbivory on the leaves by terrestrial herbivores. Underwood and Barrett (1990) have experimentally demonstrated a relationship between the root-dwelling oyster *Crassostrea commercialis*, which resides on the roots of the mangrove *Avicennia marina*, and the presence of the littorinid snail *Bembicium auratum*; the oysters increase the distribution, abundance, and size of the *Bembicium*. Underwood and Barrett ascribe this positive relationship to the ability of the oyster to provide a refuge from predation for the littorinid, thus protecting them from crustaceans and fish.

Mangrove areas also serve as nursery grounds for penaeid shrimps and fishes such as mullet, which may spend the early part of their life cycles in these areas before moving offshore. The prop root area of mangroves is known to be a nursery for juvenile spiny lobsters (*Panulirus argus*) in the Caribbean. They may spend up to two years in this area.

Conspicuous by their size and abundance in the water and on the mud in Indo-Pacific mangrove forests are the small, big-eyed fishes of the genus *Periophthalmus* and relatives. These fishes, collectively called "mud skippers," are remarkable because they spend most of their time out of water, crawling around on the mud flats or even climbing the mangrove roots (Fig. 9.28)! In fact, they act much like frogs or toads. These fishes create burrows in the mud, which serve for refuge and for breeding. They are quite fast, moving over the exposed mud flats by "walking" on their strong pectoral fins or by using a series of "skipping" motions or "bounds," in which the tail and caudal fins provide the thrust.

Aside from their peculiar unfishlike habit of crawling around on the mud flats, the other characteristic feature of these fishes is the modification of the eyes. The eyes are set high on the

Figure 9.28 Fishes of the genus *Periophthalmus* in the mud and on the roots of the mangroves. (From various sources.)

head and are arranged so that they focus best in air, not in the water. When mud skippers are swimming, then, the eyes protrude onto the surface, and hence these fishes look for all the world like frogs. A final noteworthy adaptation is in the respiratory system. The gills are often reduced, and aerial respiration is accomplished by vascularized sacs in the mouth cavity and gill chambers.

Succession and Mortality

Mangrove associations are subject to mortality from a number of different natural and human-induced causes. They exist under rather delicately balanced conditions involving a somewhat predictable steady sedimentation rate, minimal water movement, a certain tidal regime, and water and soil of a certain salinity. Any change that upsets this balance produces corresponding changes in the mangrove community. If the

changes are slow enough, there may be a gradual change or *succession*.

Davis (1940), working in Florida, was the first to suggest a connection between zonation, succession, and changes in physical factors. In this now classical view, Davis argued that the mangrove zonal patterns were equivalent to seral stages in succession. In his scheme, the seaward zone of red mangroves is the "pioneer" or initial seral state. Farther landward, the black and white mangrove zones were regarded as progressively later stages of the successional sequence, leading eventually to terrestrial communities. This view, however, is not universally accepted. Other workers, such as Thom (1975), have demonstrated that mangrove zonation seems to be mostly a response to external physical factors rather than a successional sequence induced by the plants themselves. This view is supported by Rabinowitz (1975), who demonstrated in transplant experiments that various species of mangroves could grow well within any zone and that physical factors alone were not responsible for their absence. It now appears that zonation of mangroves is not controlled directly by physical factors but by an interplay of these factors with interspecific competition and propagule dispersion and sorting. Unless there is heavy sedimentation, once a mangrove area reaches some sort of equilibrium state, change tends to occur only if there is some sort of disturbance. That is, there is no further "succession" induced simply by the plants. Disturbances may be small (lightning strikes, fires, or droughts) or they may be large (sea level changes or hurricanes). Such perturbations seem to set zonation back to an earlier state. Evidence from Florida, according to Odum et al. (1982), suggests that hurricane perturbations are frequent enough on a regular basis to keep mangrove systems in a cyclical pattern of succession rather than in the classic facilitation sequence that would continually produce additional terrestrial forests as the climax stage. In this newer view, the role of mangroves in building land at the expense of the sea is seen as more passive than active, and various physical processes are the dominant forces determining the rate of growth or recession of mangrove shorelines. Mangroves in this scenario have the role of stabilizing sediments, the origin and major deposition of which are due to physical processes.

Perhaps the greatest cause of large-scale mortality is typhoons and hurricanes. These violent storms destroy large areas of mangrove forests by uprooting trees, by massive sedimentation, or by altered salinity of water and soil. Hurricane Donna, which struck Everglades National Park in 1960, caused a mortality ranging from 25 to 75 percent of 100,000 acres of mangroves in the park. Recovery from such massive mortality has a similar time span for mangroves as for coral reefs, and estimates of recovery time are about 20 to 25 years (Fig. 9.29).

Another suggested source of mortality is the small isopod *Sphaeroma terebrans*. This isopod burrows into the prop roots of red mangroves, and the boring, plus secondary decomposition by bacteria and fungi, usually severs the prop

Figure 9.29 Photo of dead mangroves destroyed in Everglades National Park by the hurricane of 1960. (Photo by the author.)

root at the level of mean high tide. Rehm and Humm (1973) were the first to describe this phenomenon and suggested that this prop root destruction left the underlying sediment free to erode away, thus allowing the shallow-rooted trees to fall into the water. In this way the whole forest would slowly be cut down and destroyed. Since that time other workers have taken issue with this "eco-catastrophe" view. For example, Estevez and Simon (1975) have shown that the infestations of the isopod are patchy and limited in extent. They also demonstrated that burrowing was related to salinity changes. Simberloff et al. (1978) suggested that since the boring activity stimulated the mangrove to produce more root growth and branching, which is beneficial to the tree, the isopod has a positive effect. We thus find ourselves in the peculiar position of having dramatically opposed views concerning this problem, which remains unresolved.

Most other massive mortality is due to human activity. Perhaps the greatest recent destruction has been as a result of herbicide spraying on the mangrove forests of Vietnam during the war. According to Tschirley (1969), perhaps 100,000 hectares of these forests were defoliated and destroyed by herbicides during the war. It is not known how long recovery of these forests will take, or whether they will regenerate at all in view of the extreme sensitivity of mangroves to residual herbicides.

Mangrove forests have also been destroyed by filling, by dredging, and by channelizing waters. Historically, mangrove forests probably occurred along the Persian Gulf, from which they are now absent due to cutting by early humans for firewood and boat building. In the tropics today, the mangroves are exploited by humans mainly for firewood, but may be partially destroyed in order to build ponds for fish and shrimp culture or for salt production; mangroves have also been filled for construction.

Mangrove forests have demonstrably important values to humans. They are the nurseries for penaeid shrimp, spiny lobsters, blue crabs, and various fishes such as the mullet. The whole Tortugas pink shrimp industry of Florida, an 11-million-pound-per-year harvest, is highly dependent on the mangrove system. Mangroves also shelter a number of endangered species such as the American crocodile, the brown pelican,

and the Atlantic Ridley sea turtle. Finally, their ability to stabilize land in the face of tropical storms may be important in protecting human habitations.

BIBLIOGRAPHY

Adey, W. H., I. G. MacIntyre, and R. Stuckenrath. 1977. Relict barrier system off St. Croix: Its implications with respect to late Cenozoic coral reef development in the Western Atlantic. Proc. 3rd Int. Coral Reef Symp. 2:15–22.

Babcock, R. C. 1985. Growth and mortality in juvenile corals (*Goniastrea, Platygyra* and *Acropora*): The first year. Proc. 5th Int. Coral Reef Symp. 4:355–360.

Bakus, G. 1964. The effects of fish-grazing on invertebrate evolution in shallow tropical waters. Occ. paper no. 27, Allan Hancock Foundation. 29 pp.

Bakus, G. 1973. The biology and ecology of tropical holothurians, pp. 325–367. *In:* Jones, O. A., and R. Endean (eds.). Biology and geology of coral reefs. Vol. II, Biology 1. Academic Press, N.Y.

Banner, A. H. 1976. Ciguatera: A disease from coral reef fish, pp. 177–213. *In:* Jones, O. A., and R. Endean (eds.). Biology and geology of coral reefs. Vol. III, Biology 2. Academic Press, N.Y.

Birkeland, C. 1977. The importance of rate of biomass accumulation in early successional stages of benthic communities to the survival of coral recruits. Proc. 3rd Int. Coral Reef Symp. Vol. 1, Biology. pp. 15–21.

Birkeland, C. 1982. Terrestrial runoff as a cause of outbreaks of *Acanthaster planci* (Echinodermata: Asteroidea). Mar. Biol. 69:175–185.

Birkeland, C. 1989. The influence of echinoderms on coral reef communities, pp. 1–80. *In:* Jangoux, M., and J. M. Lawrence (eds.). Echinoderm studies. Vol. 3. A. A. Balkema, Rotterdam.

Brock, R. E. 1979. An experimental study on the effects of grazing by parrot fishes and the role of refuges in benthic community structure. Mar. Biol. 51:381–388.

Buddemeir, R. W., and R. A. Kinzie III. 1976. Coral growth. Oceanogr. Mar. Biol. Ann. Rev. 14:183–225.

Chapman, V. J. (ed.). 1977. Ecostystems of the world. Vol. 1, Wet coastal ecosystems. Elsevier, N.Y. 428 pp.

Chesher, R. H. 1969. Destruction of Pacific corals by the sea star *Acanthaster planci*. Science 165:280–283.

Connell, J. 1973. Population ecology of reef building coral, pp. 205–245. *In:* Jones, O. A., and R. Endean (eds.). Biology and geology of coral reefs. Vol. II, Biology 1. Academic Press, N.Y.

Crossland, C. 1952. Madreporaria, Hydrocorallinae, Heliopora, and Tubipora. Sci. Repts. Great Barrier Reef Exped. 1928–29. 6:85–257.

Dana, T. F., W. A. Newman, and E. W. Fager. 1972. *Acanthaster* aggregations: Interpreted as primarily responses to natural phenomena. Pac. Sci. 26:355–372.

Davies, P. J. 1983. Reef growth, pp. 69–106. *In:* Barnes, D. J. (ed.). Perspectives on coral reefs. AIMS Contribution no. 200.

Davis, J. H., Jr. 1940. The ecology and geologic role of mangroves in Florida. Carnegie Inst. Wash. Publ. 517. Tortugas Lab. Pap. 32:303–412.

Endean, R. 1973. Population explosions of *Acanthaster planci* and associated destruction of hermatypic corals in the Indo-West Pacific region, pp. 389–438. *In:* Jones, O. A., and R. Endean (eds.). Biology and geology of coral reefs. Vol. II, Biology 1. Academic Press, N.Y.

Endean, R. 1977. *Acanthaster planci* infestations of reefs of the Great Barrier Reef. Proc. 3rd Int. Coral Reef Symp. 1:185–191.

Endean, R. 1985. Ecocatastrophe on The Great Barrier Reef. Proc. 5th Int. Coral Reef Symp. 5:309–314.

Estevez, E. D., and J. L. Simon. 1975. Systematics and ecology of *Sphaeroma* (Crustacea, Isopoda) in the mangrove habitat of Florida, pp. 286–304. *In:* Walsh, G., S. Snedaker, and H. Teas (eds.). Proceedings of the International Symposium on the Biology and Management of Mangroves. University of Florida, Gainesville.

Fadlallah, Y. H. 1983. Sexual reproduction, development and larval biology in scleractinian corals. A review. Coral Reefs 2:129–150.

Frankel, E. 1977. Evidence from the Great Barrier Reef of ancient *Acanthaster* aggregations. Atoll Res. Bull. 220:75–93.

Franzisket, L. 1969. Riffcorallen können autotroph leben. Naturwissenschaften 56:144.

Ghiold, J. 1990. White death—The fate of a deserted coral. New Scien. 126(1719):46.

Glynn, P. W. 1972. Observations on the ecology of the Caribbean and Pacific coasts of Panama, pp. 13–20. *In:* Jones, M. L. (ed.). The Panamic biota: Some observations prior to a sea level canal. Bull. Biol. Soc. Wash. 2:13–30.

Glynn P. 1973. Aspects of the ecology of coral reefs in the western Atlantic region, pp. 271–324. *In:* Jones, O. A., and R. Endean (eds.). Biology and geology of coral reefs. Vol. II, Biology 1. Academic Press, N.Y.

Glynn, P. W. 1974. Rolling stones among the Scleractinia: Mobile coraliths in the Gulf of Panama. Proc. 2nd Int. Coral Reef Symp. 2:183–198.

Glynn, P. 1976. Some physical and biological determinants of coral community structure in the eastern Pacific. Ecol. Monogr. 46:431–456.

Glynn, P. W. 1984. Widespread coral mortality and the 1982–83 El Niño warming event. Envir. Conserv. 11:133–146.

Glynn, P. W. 1985. Corallivore population sizes and feeding effects following El Niño (1982–1983) associated coral mortality in Panama. Proc. 5th Int. Coral Reef Symp. 2:149.

Glynn, P. W. 1988. El Niño–Southern Oscillation 1982–1983: Nearshore population, community and ecosystem responses. Ann. Rev. Ecol. Sys. 19:309–345.

Glynn, P. W., and G. M. Wellington. 1983. Corals and coral reefs of The Galápagos Islands. University of California Press, Berkeley. 330 pp.

Glynn, P., G. M. Wellington, and C. Birkeland. 1979. Coral reef growth in the Galápagos: Limitations by sea urchins. Science 203:47–48.

Goldman, B., and F. Talbot. 1976. Aspects of the ecology of coral reef fishes, pp. 125–154. *In:* Jones, O. A., and R. Endean (eds.). Biology and geology of coral reefs. Vol. III, Biology 2. Academic Press, N.Y.

Goreau, T. F. 1961. On the relation of calcification to primary production in reef building organisms, pp. 269–285. *In:* Lenhoff, H. M., and W. F. Loomis (eds.). The biology of Hydra, etc. University of Miami Press, Coral Gables, Fla.

Goreau, T. F., N. I. Goreau, and T. J. Goreau. 1979. Corals and coral reefs. Sci. Amer. 241(2):124–136.

Goreau, T. F., and J. W. Wells. 1967. The shallow-water Scleractinia of Jamaica: Revised list of species and their vertical distribution range. Bull. Mar. Sci. 17(2):442–453.

Goreau, T. J. 1990. Coral bleaching in Jamaica. Nature 343:417.

Grigg, R. W. 1983. Community structure, succession and development of coral reefs in Hawaii. Mar. Ecol. Progr. Ser. 11:1–14.

Hatcher, B. G. 1983. Grazing in coral reef ecosystems, pp. 169–179. *In:* Barnes, D. J. (ed.). Perspectives on coral reefs. AIMS, Townsville, Australia.

Hay, M. E. 1984. Patterns of fish and urchin grazing on Caribbean coral reefs: Are previous results typical? Ecology 65:446–454.

Hiatt, R. W., and D. W. Strasburg. 1960. Ecological relationship of the fish fauna on coral reefs of the Marshall Islands. Ecol. Monogr. 30:65–127.

Highsmith, R. C. 1982. Reproduction by fragmentation in corals. Mar. Ecol. Progr. Ser. 7:207–226.

Hobson, E. S. 1968. Predatory behavior of some shore fishes in the Gulf of California. Bur. Sports Fish. and Wildlife (U.S.), Res. Rept. 73. 92 pp.

Huston, M. A. 1985. Patterns of species diversity on coral reefs. Ann. Rev. Ecol. Sys. 16:149–177.

Hutchings, P. A. 1986. Biological destruction of coral reefs: A review. Coral Reefs 4:239–252.

Hutchings, P. W. 1983. Cryptofaunal communities of coral reefs, pp. 200–208. _In:_ Barnes, D. J. (ed.). Perspectives on coral reefs. AIMS, Townsville, Australia.

Jackson, J. B. C. 1977. Competition on marine hard substrata: The adaptive significance of solitary and colonial strategies. Amer. Nat. 111:743–767.

Jackson, J. B. C. 1979. Overgrowth competition between encrusting cheilostome ectoprocts in a Jamaican cryptic reef environment. J. Animal Ecol. 48:805–823.

Jackson, J. B. C., and T. P. Hughes. 1985. Adaptive strategies of coral-reef invertebrates. Amer. Sci. 73:265–274.

Johannes, R. E. 1974. Sources of nutritional energy for reef corals. Proc. 2nd Int. Coral Reef Symp. 1:133–137.

Johannes, R. E., N. T. Kuenzel, and S. L. Coles. 1970. The role of zooplankton in the nutrition of some scleractian corals. Limnol. and Oceanogr. 15:579–586.

Johannes, R. E., et al. 1972. The metabolism of some coral reef communities: A team study of nutrient and energy flux at Eniwetok atoll. Bioscience 22:541–543.

Jones, O. A., and R. Endean (eds.). 1973. Biology and geology of coral reefs. Vol. II, Biology 1. Academic Press, N.Y. 480 pp.

Jones, O. A., and R. Endean (eds.). 1976. Biology and geology of coral reefs. Vol. III, Biology 2. Academic Press, N.Y. 435 pp.

Kaufman, L. 1977. The three spot damselfish: Effects on benthic biota of Caribbean coral reefs. Proc. 3rd Int. Coral Reef Symp. 1:559–564.

Kohn, A., and P. Helfrich. 1957. Primary productivity of a Hawaiian coral reef. Limnol. Oceanogr. 2(3):241–251.

Kohn, A., and J. Nybakken. 1975. Ecology of _Conus_ on eastern Indian Ocean fringing reefs: Diversity of species and resource utilization. Mar. Biol. 29:211–234.

Kuenzler, E. J. 1974. Mangrove swamp systems, pp. 346–371. _In:_ Odum, H. T., B. J. Copeland, and E. A. McMahon (eds.). Coastal ecological systems. Vol. 1. Conservation Foundation, Washington, D.C.

Ladd, H. S., E. Ingerson, R. C. Townshend, M. Russell, and H. K. Stephenson. 1953. Drilling on Eniwetok Atoll, Marshall Islands. Bull. Amer. Assoc. Petrol. Geol. 37:2257–2280.

Lang, J. 1973. Interspecific aggression by scleractinian corals, 2. Why the race is not only to the swift. Bull. Mar. Sci. 23(2):260–279.

Larkham, A. W. D. 1983. The primary productivity of plant communities on coral reefs, pp. 221–230. _In:_ Barnes, D. J. (ed.). Perspectives on coral reefs. AIMS, Townsville, Australia.

Lessios, H. A. 1988. Mass mortality of _Diadema antillarum_ in the Caribbean: What have we learned? Ann. Rev. Ecol. Sys. 19:371–394.

Lessios, H. A., P. W. Glynn, and D. R. Robertson. 1983. Mass mortalities of coral reef organisms. Science 222:715.

Lobel, P. S. 1980. Herbivory by damsel fishes and their role in coral reef community ecology. Bull. Mar. Sci. 30:273–289.

Lowe-McConnell, R. H. 1977. Ecology of fishes in tropical waters. Studies in biology no. 76, Edward Arnold, London. 64 pp.

Loya, Y. 1976. Recolonization of Red Sea corals affected by natural catastrophes and man-made perturbations. Ecology 57(2):278–289.

Luckhurst, B. E., and K. Luckhurst. 1978. Analysis of the influence of substrate variables on coral reef fish communities. Mar. Biol. 49:317–323.

Lugo, A. E., and S. C. Snedaker. 1974. The ecology of mangroves. Ann. Rev. Ecol. Sys. 5:39–64.

McMichael, D. F. 1974. Growth rate, population size and mantle coloration in the small giant clam _Tridacna maxima_ (Roding) at One Tree Island, Capricorn Group, Queensland. Proc. 2nd Int. Coral Reef Symp. 1:241–254.

McNae, W. 1968. A general account of the fauna and flora of mangrove swamps and forests in the Indo-West Pacific region. Adv. Mar. Biol. 6:73–270.

Meyer, J. L., E. T. Schultz, and G. S. Helfman. 1983. Schools: An asset to corals. Science 220:1047–1048.

Motoda, S. 1940. The environment and life of massive reef coral, _Goniastrea aspera_ Verrill, inhabiting the reef flat in Palao. Palao Tropical Biol. Sta. Studies 2:61–104.

Muscatine, L., and E. Cernichiari. 1969. Assimilation of photosynthetic products of zooxanthellae by a reef coral. Biol. Bull. 137:506–523.

Neudecker, S. 1977. Transplant experiments to test the effect of fish grazing on coral distribution. Proc. 3rd. Int. Coral Reef Symp. 1:317–323.

Newel, N. D. 1972. The evolution of reefs. Sci. Amer. 226(6):54–65.

Odum, H. T., B. J. Copeland, and E. A. McMahan (eds.). 1974. Coastal ecological systems of the United States. Vol. I: B, Natural tropical ecosystems of high diversity. Conservation Foundation, Washington, D.C., pp. 346–514.

Odum, H. T., and E. P. Odum. 1955. Trophic structure and productivity of a windward coral reef community on Eniwetok Atoll. Ecol. Monogr. 25:291–320.

Odum, W. E., C. C. McIvor, and T. J. Smith III. 1982. The ecology of the mangroves of South Florida: A community profile. U.S. Fish & Wildlife Service Office of Biological Services FWS/OBS-81/24. 144 pp.

Ogden, J. C., and P. S. Lobel. 1978. The role of herbivorous fishes and urchins in coral reef communities. Envir. Biol. Fishes 3:49–63.

Onuf, C. P., J. M. Teal, and I. Valiela. 1977. Interactions of nutrients, plant growth and herbivory in a mangrove ecosystem. Ecology 58:514–526.

Pearson, R. G. 1975. Coral reefs, unpredictable climatic factors and *Acanthaster.* Crown-of-Thorns starfish seminar proceedings, pp. 131–134. Aust. Govt. Pub. Ser., Canberra.

Perry, D. M. 1984. Direct and indirect regulatory effects by root fauna on the growth rate of prop roots in the red mangrove. Amer. Zool. 24(3):126.

Perry, D. M. 1988. Effects of associated fauna on growth and productivity in the red mangrove. Ecology 69(4):1064–1075.

Porter, J. 1972. Predation by *Acanthaster* and its effect on coral species diversity. Amer. Nat. 106(950):487–492.

Porter, J. 1974. Zooplankton feeding by the Caribbean reef-building coral *Montastrea cavernosa.* Proc. 2nd Int. Coral Reef Symp. 1:111–125.

Porter, J. W. 1976. Autotrophy, heterotrophy and resource partitioning in Caribbean reef building corals. Amer. Nat. 110:731–742.

Potts, D. C. 1977. Suppression of coral populations by filamentous algae within damselfish territories. J. Exp. Mar. Biol. Ecol. 28:207–216.

Rabinowitz, D. 1975. Planting experiments in mangrove swamps of Panama, pp. 385–393. *In:* Walsh, G., S. Snedaker, and H. Teas (eds.). Proceedings of the International Symposium on the Biology and Management of Mangroves. University of Florida, Gainesville.

Rabinowitz, D. 1978. Dispersal properties of mangrove propagules. Biotropica 10:47–57.

Rehm, A. E., and H. J. Humm. 1973. *Sphaeroma terebrans:* A threat to the mangroves of southeastern Florida. Science 182:173–174.

Richardson, C. A., P. Dustan, and J. C. Lang. 1979. Maintenance of living space by sweeper tentacles of *Montastrea cavernosa,* a Caribbean reef coral. Mar. Biol. 55:181–186.

Risk, M. J. 1972. Fish diversity on a coral reef in the Virgin Islands. Atoll Res. Bull. 153:1–6.

Sale, P. F. 1977. Maintenance of high diversity in coral reef fish communities. Amer. Nat. 111:337–359.

Sale, P. F. 1980. The ecology of fishes on coral reefs. Oceanogr. Mar. Biol. Ann. Rev. 18:367–421.

Sammarco, P. W. 1980. *Diadema* and its relationship to coral spat mortality: Grazing, competition and biological disturbance. J. Exp. Mar. Biol. Ecol. 45:245–272.

Sammarco, P. W. 1985. The Great Barrier Reef versus the Caribbean: Comparisons of grazers, coral recruitment patterns, and reef recovery. Proc. 5th Int. Coral Reef Symp. 4:391–397.

Sammarco, P. W. 1987. A comparison of some ecological processes on coral reefs of the Caribbean and the Great Barrier Reef. Unesco Rept. Mar. Sci. no. 46, pp. 127–166.

Sammarco, P. W., J. S. Levinton, and J. C. Ogden. 1974. Grazing and control of coral reef community structure by *Diadema antillarum* Philippi (Echinodermata: Echinoidea): A preliminary study. J. Mar. Res. 32(1):47–53.

Scoffin, T. P., C. W. Stearn, D. Boucher, P. Frydl, C. M. Hawkins, I. G. Hunter, and J. K. MacGeachy. 1980. Calcium carbonate budget of a fringing reef on the west coast of Barbados. II, Erosion sediments and internal structure. Bull. Mar. Sci. 30:475–528.

Simberloff, D., B. J. Brown, and S. Lowrie. 1978. Isopod and insect root borers may benefit Florida mangroves. Science 201:630–632.

Smith, S. V., and J. A. Marsh. 1973. Organic carbon production and consumption on the windward reef flat of Eniwetok atoll. Limn. Oceanogr. 18:953–961.

Stearn, C. W., and T. P. Scoffin. 1977. Carbonate budget of a fringing reef, Barbados. Proc. 3rd Int. Coral Reef Symp. 2:471–476.

Stehli, F. G., and J. W. Wells. 1971. Diversity and age patterns in hermatypic corals. Sys. Zool. 20:115–126.

Steneck, R. S., and W. H. Adey. 1976. The role of environment in control of morphology in *Lithophyllum congestum,* a Caribbean algal ridge builder. Botan. Mar. 19:197–215.

Stoddard, D. R. 1969. Ecology and morphology of recent coral reefs. Biol. Rev. 44(4):433–498.

Sutherland, J. P. 1980. Dynamics of the epibenthic community of roots of the mangle, *Rhizophora mangle,* at Bahia de Buche, Venezuela. Mar. Biol. 58:75–84.

Szmant-Froelich, A. 1985. Reproduction and recruitment of corals: Conclusion. Proc. 5th Int. Coral Reef Symp. 4:399–400.

Talbot, F. H., B. C. Russell, and G. R. V. Anderson. 1978. Coral fish communities: Unstable, high-diversity systems? Ecol. Monogr. 48:425–440.

Thom, B. G. 1975. Mangrove ecology from a geomorphic viewpoint, pp. 469–481. *In:* Walsh, G., S. Snedaker, and H. Teas (eds.). Proceedings of the International Symposium on the Biology and Management of Mangroves. University of Florida, Gainesville.

Tschirley, F. H. 1969. Defoliation in Vietnam. Science 163:779–786.

Underwood, A. J., and G. Barrett. 1990. Experiments on the influence of oysters on the distribution, abundance and sizes of the gastropod *Bembicum auratum* in a mangrove swamp in New South Wales, Australia. J. Exp. Mar. Biol. Ecol. 137(1):25–45.

Vaughn, T. W. 1915. The geologic significance of the growth rate of the Floridian and Bahamian shoal-water corals. J. Wash. Acad. Sci. 5:591–600.

Vermeij, G. J. 1987. Interocean differences in architecture and ecology: The effects of history and productivity. Unesco Rept. Mar. Sci. no. 46, pp. 105–126.

Victor, B. C. 1983. Recruitment and population dynamics of coral reef fish. Science 219:419–420.

Wada, K., and D. Wower. 1989. Foraging on mangrove pneumatophores by ocypodid crabs. J. Exp. Mar. Biol. Ecol. 134(2):89–100.

Walbran, P. D., R. A. Henderson, A. J. Timothy Jull, and M. J. Head. 1989. Evidence from sediments of long term *Acanthaster planci* predation on corals of the Great Barrier Reef. Science 245:847–850.

Walsh, G. E. 1974. Mangroves, a review, pp. 51–174. *In:* Reimold, R. J., and W. J. Queen (eds.). Ecology of halophytes. Academic Press, N.Y.

Wellington, G. W. 1980. Reversal of digestive interactions between Pacific reef corals: Mediation by sweeper tentacles. Oecologia 47:340–343.

Wellington, G. W. 1982a. Depth zonation of corals in the Gulf of Panama: Control and facilitation by resident reef fishes. Ecol. Monogr. 52:223–241.

Wellington, G. W. 1982b. An experimental analysis of the effects of light and zooplankton on coral zonation. Oecologia 52:311–320.

Wells, J. W. 1954. Recent corals of the Marshall Islands, Bikini and nearby atolls. U.S. Geol. Survey Paper 260–I:385–486.

Wells, J. W. 1957. Coral reefs, pp. 609–631. *In:* The treatise on marine ecology and paleoecology. Vol. I, Ecology. Memoir 67, Geol. Soc. of Amer.

West, R. C. 1977. Tidal salt marsh and mangal formations of middle and south America, pp. 193–213. *In:* Chapman, V. J. (ed.). Wet coastal ecosystems. Elsevier, N.Y.

Wilkinson, C. 1983. Net primary productivity in coral reef sponges. Science 219:410–411.

Wilkinson, C. R. 1987a. Interocean differences in size and nutrition of coral reef sponges. Science 236:1654–1657.

Wilkinson, C. R. 1987b. Sponge biomass as an indication of reef productivity in two oceans. Unesco Rept. Mar. Sci. no. 46, pp. 99–104.

Williams, E. H., and L. B. Williams. 1987–88. Caribbean marine mass mortality: A problem with a solution. Oceanus 30(4):69–75.

Yentsch, C. S. 1966. Primary production, pp. 722–725. *In:* Fairbridge, R. W. (ed.). The encyclopedia of oceanography. Van Nostrand, N.Y.

Yonge, C. M. 1963. The biology of coral reefs. Adv. Mar. Biol. 1:209–260.

Chapter 10

Symbiotic Relationships

One of the most striking features of marine organisms is the large number of close associations that can be observed among many different species. These are a series of associations that are not predator-prey or herbivore-plant relationships in which one member consumes the other. These close relationships between unlike species generally seem to be either unharmful to either member or, more likely, to be beneficial to one or both. *Symbiosis,* the name given to such associations, means an interrelationship between two different species. Symbiotic relationships are found in the terrestrial environment as well as in the aquatic, but appear to be disproportionately common, extensive, and well developed in the marine environment, such that they

merit separate coverage in a book such as this.

Definitions and Coverage

Symbiotic relationships cover a broad spectrum of associations from random, casual, or facultative associations through more and more obligatory groupings that benefit one or both members, finally to those that are parasitic. Such differences in the degree of association have led to the subdivision of symbiosis into more narrowly defined groupings. The term *commensalism* is often used to refer to an association that is clearly to the advantage of one member while not harming the other member. *Inquilinism* is a special subdivision of commensalism, in which an animal lives in the home of another, or in its digestive tract, without being parasitic. *Mutualism* is that form of symbiosis in which two species associate together for their mutual benefit. In a mutualistic relationship, the partners are often called *symbionts.* In a commensal or inquiline relationship, however, the partner gaining advantage is called the *commensal* and the other the *host.*

Parasitism generally refers to an association in which one species lives in or upon another and draws nourishment from that species at the expense of, or to the detriment of, the other. In other words, it is an association in which the advantage is solely to one member at the expense of the other. Parasitism is an association that seems to be extremely common in both marine and terrestrial environments. It will not be discussed in this chapter, since it contains few features that are unique to the marine environment.

Basically, there are two broad groupings of nonparasitic symbiotic associations in the sea: those between algal cells and various invertebrate animals and those between various animals, both vertebrate and invertebrate. We shall consider these in that order. A final category is that between various bacteria and animals, which has become particularly well publicized through the discoveries of organisms living in deep-sea thermal-vent systems. These symbioses were covered in Chapter 4.

SYMBIOSIS OF ALGAE AND ANIMALS

All symbiotic relationships in the sea between plants and animals are between unicellular algae or their parts and a wide variety of marine invertebrate animals. These symbiotic relationships have been shown to be most common in tropical waters, but they also are prevalent in temperate oceans. They do, however, appear to be absent, or virtually so, from polar waters. Furthermore, the associations are, for obvious reasons, restricted to the shallow subtidal or intertidal areas or to the uppermost layers of the pelagic realm where sufficient light for photosynthesis is present.

Types and Composition of the Associations

There are basically two types of symbiotic association between algae and invertebrates. The more common is to have the entire functioning algal cell associated with the invertebrate animal. The second is to have only the functioning chloroplasts from the algal cells incorporated into the tissues of the invertebrate body.

The algal cell symbionts have been typically classified into groups on the basis of their color. *Zooxanthellae* is the name given to those cells appearing brown, golden, or brownish yellow in color, and *zoochlorellae* to those that appear green. A third, smaller group appears blue or bluish green and has been called *cyanellae* (Table 10.1). These color groups, however, do not distinguish among the actual algal species involved. Zooxanthellae, the most common in the seas of the world, are primarily species of dinoflagellates (see p. 40), but include a few diatoms and cryptophyceans, haptophyceans, and chrysophyceans (Fig. 10.1). As Schoenberg and Trench (1976) have noted, the most common zooxanthellae species is the dinoflagellate (Dinophyceae) *Symbiodinium* (= *Gymnodinium*) microadriaticum. The zoochlorellae are much less common in the sea, but dominate symbiotic associations in fresh water. Among the marine associations, zoochlorellae include the algae *Platymonas* (= *Tetraselmis*) *convolutae* and *Pedinomonas* of the Prasinophyceae and *Chlorella* and *Chlamydomonas* of the Chlorophyceae. Cyanellae are all blue-green algae and include the genera *Aphanocapsa* and *Phormidium,* and probably others as yet unidentified (Table 10.1). They are most common as symbionts in sponges and planktonic diatoms.

The algae all occur inside the bodies of the animals, and most are found either within vacuoles inside individual tissue cells or else in various body spaces between or within tissue layers. Generally, the algal cells are restricted to

Table 10.1   SUMMARY OF THE TYPES OF ASSOCIATIONS BETWEEN ALGAE
             AND MARINE INVERTEBRATES

Algal symbiont group	Algal taxon	Invertebrate animal host taxa
Zooxanthellae	Dinophyceae	Protista, Porifera, Cnidaria, Platyhelminthes, Mollusca
	Haptophyceae	Protista
	Chrysophyceae	Protista
	Bacillariophyceae	Platyhelminthes (*Convoluta convoluta*)
	Cryptophyceae	Protozoa
Zoochlorellae	Prasinophyceae	Platyhelminthes (*Convoluta roscoffensis*), Protista
	(?) Chlorophyceae	Cnidaria, Protista
Cyanellae	Cyanophyceae	Porifera, Protista, Echiura
Prochlorophyta	*Prochloron*	Urochordata
Chloroplasts	Chlorophyceae	Mollusca (Sacoglossa)

Source: Modified from D. C. Smith, Symbiosis of algae with invertebrates. Oxford Biology Reader No. 43, 16 pp. Copyright © 1973 by Oxford University Press. Reprinted with permission from Carolina Biological Supply Co.

certain tissues or areas of the host, and they grow and reproduce within the invertebrate without undergoing digestion by the host.

Among those associations that involve only the chloroplasts and the invertebrate body, the chloroplasts are usually derived from the cells of larger green algae (*Codium, Caulerpa, Cladophora, Bryopsis*), which are first ingested by the animal in feeding and subsequently transferred from the digestive tract of the animal to other tissues.

Algal symbiosis, either cellular or chloroplast, is widespread among various invertebrate groups. Symbiotic associations occur in Pro-

Figure 10.1 Symbiotic zooxanthellae (dinoflagellates) from the anemone *Anthopleura xanthogrammica.* (Photomicrograph by the author.)

tista, Porifera, Cnidaria, Platyhelminthes, Mollusca, and Echiura. Within these phyla, Mc-Laughlin and Zahl (1966) report about 130 genera having algal symbionts. It is most common among Protista and Cnidaria.

Although most algal symbionts can be readily identified in their hosts, it has been recently discovered that a number of protistans previously considered photosynthetic, such as the ciliate *Mesodinium rubrum* and the dinoflagellates *Peridinium balticum* and *Kryptoperidinium foliaceum,* are in fact symbiotic associations in which the algal cells have lost their cellular integrity such that they do not appear as separate from the host body. These associations are examples of what has been called *cryptic symbiosis.*

Origin of the Association

Although we will probably never know the exact origin of algal-invertebrate symbiotic relationships, it is possible to outline some suggestions. In all the phyla mentioned above that contain symbiotic algae or chloroplasts, the final phase of digestion is intracellular, where individual cells take up particles from the stomach. It is thus possible to conceive of the origin of this association through the ingestion by cells of the digestive tract of either intact algal cells or chloroplasts, taken in by the animal during feeding. If, then, the algal cells were resistant to digestive action or the animal lacked enzymes to digest plant cellulose, the basis for a new association would be laid. This association could then evolve into a more obligatory relationship, provided that the new association conveyed an enhanced survival value to both symbionts. Since these associations, in fact, are common, they must have some selective value (see pp. 395–396).

Distribution of Algae–Invertebrate Associations

Although there is neither space nor time to list and discuss all the species in which algal symbionts occur, it is of use to discuss briefly the various groups of organisms in which this association is prevalent.

Beginning with protozoans, symbiotic relationships are found in all of the epipelagic planktonic Radiolaria, where the zooxanthellae occur in the outermost frothy layer (Fig. 10.2). They also are found in a number of planktonic Foraminifera (*Globigerinoides*) and in all the larger benthic species (*Elphidium, Nonion, Heterostegina*) and even in marine ciliates (*Paraeuplotes, Trichodina*).

Symbiosis in sponges is particularly important in species that inhabit coral reefs. The association is usually with blue-green algae. According to Wilkinson (1983), sponges with such symbionts constitute 80 percent of the sponge species and biomass on certain reefs of the Great Barrier Reef. Zooxanthellae occur in *Cliona* and cyanellae in *Demospongia.*

Among the Cnidaria, symbiosis becomes extremely common, especially among tropical cnidarians. Virtually all tropical, shallow-water anemones, soft corals, sea fans, whips, and stony corals have symbiotic zooxanthellae in their tissues. Even certain of the tropical jellyfish (*Cassiopeia*) contain zooxanthellae. In temperate seas, the incidence is not as great, though many anemones such as *Anthopleura* retain the symbionts (Plate 42).

A peculiar symbiosis occurs in some compound ascidians (*Didemnum, Trididemnum*) where an unusual prokaryotic unicellular alga called *Prochloron,* which combines features of both blue-green and green algae, occurs embedded in the test or in the cloacal system.

Whereas a few symbiotic relationships occur in Ctenophora (*Beröe*), Annelida (*Eunice*), and Echinodermata (*Ophioglypha*), the remaining major reservoirs of symbiotic relationships are found in the Platyhelminthes and Mollusca. Among marine flatworms, the classic case is that of the genus *Convoluta,* which has been extensively investigated. Symbiotic relationships in the Mollusca are generally restricted to certain marine gastropods of the order Sacoglossa, where the animals retain only the chloroplasts, and to the bivalve family Tridacnidae and the genera *Corculum* and *Fragum.*

Modifications Resulting from the Association

The symbiotic association between the algae and the invertebrates is generally a very close one (mutualistic), which has resulted in rather signif-

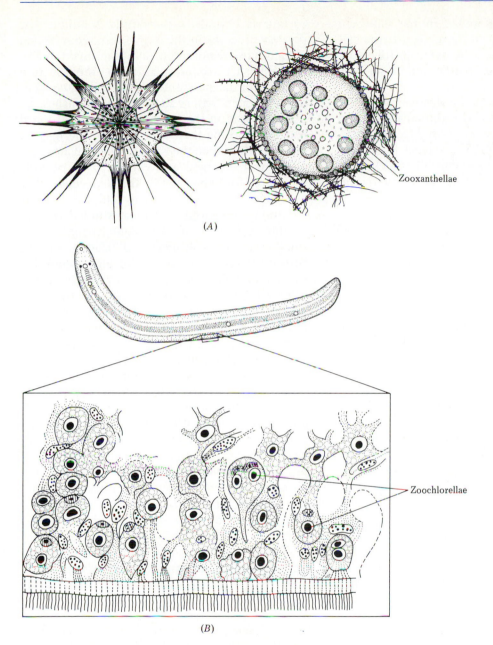

(A)

(B)

Figure 10.2 Zooxanthellae and zoochlorellae in various invertebrate taxa. *(A)* Zooxanthellae in the radiolarian *Acanthometra pellucida* (left) and *Sphaerozoum acuferum* (right). *(B)* Zoochlorellae in the acoel flatworm *Convoluta roscoffensis.* (*A, B,* after J. A. Nicol, The biology of marine animals, Wiley, 1960.)

icant anatomical and physiological changes in both the algal cells and the various invertebrate hosts.

Perhaps the most profound changes that occur as a result of the association are found in the algal cells. Most marine symbiotic algae are dinoflagellates. The symbionts, however, have lost their locomotory flagellae and the characteristic grooves around their body. Furthermore, they have cell walls that are much reduced in thickness. The zoochlorellae in the flatworm *Convoluta* have lost even more, in that the cell wall disappears, as does the light-sensitive stigmata. These cells become little more than bags containing chloroplasts.

If the above algal cells are removed from the host and grown outside the animal in culture, they will develop the characteristic flagellae, cell walls, and other organs of a typical free-living form. It is thus apparent that all the changes observed are a direct result of the symbiotic association (Fig. 10.3).

In the case of the animals, the changes vary with the type of organism and with the degree of interdependence established between the symbionts. The single universal modification is that all of these invertebrates live in shallow water, where they can obtain adequate light so that the algae can carry on photosynthesis.

Perhaps the least amount of modification occurs in the lower invertebrates, such as protozoans and Porifera, where the algae occur as simple inclusions in the cytoplasm of the animal. Among sponges, those with symbiotic blue-green algae enhance their light-gathering ability according to Wilkinson (1983) by morphological flattening, thereby increasing the area available to intercept light.

Among the Cnidaria, definite modifications begin to become apparent. The algae occur in marine Cnidaria in the innermost of the cell layers, the gastrodermis (endodermis). The numbers of zooxanthellae in different corals vary, and those that have the most seem to have reduced tentacle size, indicating that they are less able to capture zooplankton food. Among certain soft corals of the family Xeniidae, the digestive regions of the animal are reduced and the animals are not responsive to animal food. In the jellyfish *Cassiopeia,* a striking modification is behavioral. These animals, rather than swimming in open waters as do most jellyfishes, lie upside down on the bottom in shallow tropical waters, exposing their oral arms to the light in order to illuminate their algae. The oral arms are also much enlarged and expanded to provide more area for habitation by the algal cells. Truly a strange way of life for a jellyfish!

Among flatworms, the most studied case of symbiosis is that of *Convoluta roscoffensis,* a small animal inhabiting sand beaches on the Brittany coast of France. The alga inhabiting *Convoluta* is *Platymonas* (= *Tetraselmis*) *convolutae,* which also occurs free-living in the same area. *Convoluta* exhibit perhaps the most profound behavioral and life history modifications in response to the alga, but little in the way of anatomical changes (Fig. 10.2).

Convoluta roscoffensis live burrowed into the sand of beaches near the upper reaches of the tidal zone. Whenever the tide is in, the animals are found buried. When the tide recedes, the animals move up onto the surface of the sand, where they spread out to expose their symbionts to the light. When the tide begins to return, the vibrations trigger a reburrowing response so that the animals are safely under the sand again before the rising tidal waters are able to sweep them out. Even more significantly, as Smith (1973) notes, these animals apparently do not feed as adults. Furthermore, if the young worms do not ingest the algal symbionts upon hatching,

Granule

Chloroplast

Nucleus

(A)

Chloroplast

Transverse flagellum

Nucleus

Longitudinal flagellum

(B)

Figure 10.3 Symbiotic *(A)* and free-living *(B)* zooxanthellae cells.

they will not complete development and will die even if they feed.

It is among the mollusks, however, that we find the most dramatic changes of all resulting from the association. Although there are thousands of species of mollusks in the seas of the world (the second largest phylum after arthropods), symbiotic associations with algal cells have developed in only nine species. All are bivalve mollusks, seven in a single family, the Tridacnidae, which includes the giant clams. Six of the species are in the genus *Tridacna* itself and the remaining one in the genus *Hippopus*. The remaining two species, *Corculum cardissa* and *Fragum fragum*, are in the family Cardiidae.

All tridacnid clams are distributed only in the Old World tropics of the Indo-Pacific. They also include the largest bivalve mollusk in existence: *Tridacna gigas*, which has been recorded by Rosewater (1965) to reach 1.2 m in length and 263 kg in weight. The other species are more modest in size, but all are still large in comparison with most other bivalve species.

These tridacnid clams are inhabitants of coral reef areas, where they are found abundantly in the shallow, sunlit water (Fig. 10.4). They usually have a very uncharacteristic position for a bivalve mollusk, in that they either lie on the surface of the bottom or bore into coral or coral rock with the opening between the valves facing up toward the surface of the water. The valves usually gape widely, and within this opening can be seen an extensive, brightly colored tissue layer. It is in this tissue layer that the symbiotic zooxanthellae are found. The brightly colored tissue exposed within the gape is siphonal tissue; the color results from the interaction of various pigments deposited there (Plate 55). The reason for the bright colors is that they act to protect the tissues of the clam from the damaging effects of sunlight, while passing enough light to allow the zooxanthellae to photosynthesize.

What is truly remarkable about tridacnids, however, is the tremendous change that their bodies have undergone from that of a typical clam in order to accommodate this symbiotic association. Bivalves generally rest with their foot either embedded in the substrate or held against it. In this position, a normal clam has its hinge uppermost. Obviously, this would not do for a tridacnid, since such a position would not permit the tissue and zooxanthellae to be illuminated. The result, as Yonge (1975) has described, is that the entire tridacnid has undergone a tremendous rotation with respect to the foot, so that the hinge comes to lie on the underside next to the foot and the opening of the shell faces upward (Fig. 10.5). At the same time, the si-

Figure 10.4 A giant clam, *Tridacna gigas*, with siphonal tissue expanded. (Photo by the author.)

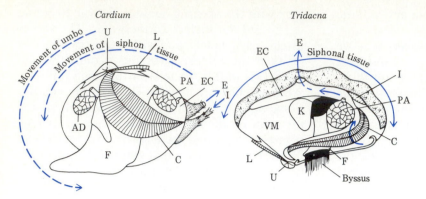

Figure 10.5 Comparison of the body orientation and internal anatomy of a typical clam, *Cardium*, with that of *Tridacna*. (AD) Anterior adductor; (C) ctenidium; (E) exhalent current; (EC) exhalent chamber; (F) foot; (I) inhalent current entering inhalent aperture; (K) kidney; (L) ligament; (PA) posterior adductor; (U) umbo; (VM) visceral mass. (Modified from C. M. Yonge, Symbiosis, ch. 15. *In:* J. E. Hedgpeth [ed.], The treatise on marine ecology and paleoecology, Vol. I, Ecology, Memoir 67, Geol. Soc. of Amer., 1957.)

phons and siphonal tissue underwent an expansion and grew and extended themselves, covering the length of the upward-facing opening and providing the expanded area for occupation by the zooxanthellae. As a result of this rotation and expansion, one of the tridacnid's shell-closing muscles was lost. Because of these profound anatomical changes, the tridacnids are markedly different in body orientation from any other bivalves. These changes can be attributed only to the association with the symbiotic algae.

Symbiotic relationships with chloroplasts have been reported by Greene (1970) in the marine gastropods of the order Sacoglossa (*Elysia, Placida, Placobranchus;* Fig. 10.6). Animals of this order regularly consume the contents of

Figure 10.6 *Elysia hedgpethi*, a sacoglossan mollusk that contains chloroplasts obtained from the green algae *Bryopsis corticulans* or *Codium fragile*. (Photo by the author.)

algal cells, usually of the order Caulerpales. Algae of this order are particularly suitable for these suctorian molluscan feeders because they have partially syncytial (= no cell walls) construction. Some species of sacoglossans have evolved the ability to retain the chloroplasts in a functioning position on the dorsal surface. The only modifications so far observed are that they do not feed very often, and the tropical forms provide a screen of light-absorbing material above the chloroplasts to cut down the light intensity.

Value of the Association

Since symbiotic associations with algae are so common among invertebrates, and since they often result in profound anatomical or behavioral changes in the partners, it seems only natural to assume that they must have some positive value to each of the partners. What might be the value of such associations?

We have already partially answered this question with respect to corals in Chapter 9, where we saw that the corals obtain food materials (see pp. 349–350) from the zooxanthellae and also that the zooxanthellae enhance the ability of corals to lay down calcium carbonate (see pp. 350–352). In turn, the zooxanthellae in corals receive nutrients in the form of nitrates and phosphates produced in the metabolic processes of the coral, but otherwise rare in external waters.

Among most invertebrates that have a symbiotic association with algal cells, the cells retain their integrity and are not digested by the animal to obtain nutrients. Thus, the nutrients that pass are in the form of chemical compounds. Energy-containing molecules, such as glycerol, produced by the zooxanthellae in photosynthesis, pass to the animal, and nitrates and phosphates, needed nutrients for the algae, pass from the invertebrate to the algal cell.

A similar situation also occurs between the sacoglossan mollusks and the symbiotic chloroplasts, where Greene (1970) has shown translocation of organic material from chloroplast to animal. However, in those sacoglossans thus far investigated, the duration of life of the chloroplasts appears to be much shorter and the animals must periodically replenish the supply by ingesting the contents of algal cells.

With respect to the giant clams, however, Fankboner (1971) has suggested that, in addition to reciprocal transference of nutrients, the animals may also actively digest the zooxanthellae cells. This would seem to be counterproductive, especially considering the great anatomical changes that these animals have undergone in order to provide properly for their algal guests. Yet, apparently, the clams can discern between healthy zooxanthellae cells and those that are degenerating or senile. As a result, the clam transports only the senile or degenerating cells from the blood spaces of the outer mantle, where they photosynthesize, into the deeper tissues, where they are consumed by phagocytic blood cells. In this manner, then, the clams retain the symbiotic relationship, and at the same time they cull the unfit algal cells to obtain additional nutrients otherwise lost when the cell dies.

Other values may result from the association. As we have noted, most of these associations are tropical, and these waters are typically low in plant nutrients. It may well be that the algae benefit by the association through access to a larger and more reliable source of nutrients in the form of the metabolic products of the animal (NO_2, PO_4, CO_2) than they would obtain from the open water. Among the benthic Foraminifera the partnership probably accounts for the 20- to 100-fold increase in the calcification rate observed between those with symbionts and those without them.

Tropical waters also are lower in oxygen than temperate waters, because water at higher temperatures holds less oxygen. Since the photosynthetic process produces oxygen, it may be that the animals, especially in the crowded conditions of the shallow waters of the tropics, gain through additional amounts of oxygen produced by the symbiotic algae. Zooxanthellae also gain protection from actinic damage of the intense solar radiation of the tropics by living within the tissues of various marine animals.

In the symbiotic relationship between cyanobacteria and sponges, Wilkinson (1983) has demonstrated that the cyanobacteria are photosynthetically active and has estimated that 5 to 12 percent of the photosynthetically produced carbon is translocated to the host sponge. That this additional energy source is significant has been demonstrated by Wilkinson and Vacelet (1979), who have shown faster growth rates of sponges in light than in the dark. It has also been suggested that the cyanobacteria may protect host-sponge cells from high light intensity.

Establishment and Transmission of Algal Symbionts

Whereas each generation of sacoglossans must actually feed on algae to obtain its supply of symbiotic chloroplasts, perpetuation of the symbiotic association with zooxanthellae from generation to generation of hosts is accomplished differently among the different groups. Basically, two major routes are available. Either the zooxanthellae cells must be passed on directly from the parent to the eggs or larvae, or each new generation must reinfect itself anew from algal cells in the surrounding environment.

Passage of algal symbionts directly to the next generation via the egg or larvae seems to be the method used by most Cnidaria. In brooding corals, for example, the zooxanthellae enter the eggs or larvae at some time prior to their release from the parent. Among broadcast spawners, however, the zooxanthellae do not make their appearance, according to Babcock (1985), until 5 to 10 days after the larvae have settled. It is also assumed, though without much evidence, that the giant clams transmit the zooxanthellae through the eggs. It has been suggested that the clams receive them from the surrounding corals as well.

For other hosts, the perpetuation of a symbiotic relationship depends on the reinfection of each generation. In such cases, the algal symbionts must be obtained anew from the environment. Among protistans and Porifera, this is the common means of transmission. Each new generation obtains the algal cells through ingestion of the free-living form of the symbiont. Surprisingly, this is also the case in *Convoluta roscoffensis,* where the association is a highly dependent one. In this case, the algae-free larvae ingest the free-living *Platymonas* (= *Tetraselmis*) *convolutae.* In order to ensure that the larvae will be infected, however, a chemical has evolved that attracts the alga to the egg cases of *Convoluta,* so that when the young hatch and begin to feed, the algae are present. Under such conditions, the new generation is virtually assured of reinfection.

A final possibility for transmission is the ingestion by the potential host of food organisms themselves with symbiotic algae. Although this has not as yet been shown to be important, it remains an area requiring study.

Little is known about the transmission of cyanobacteria to succeeding generations of sponges, but cyanobacteria have been found in the eggs of some sponges, suggesting that these algae may be transmitted directly.

SYMBIOSIS AMONG ANIMALS

We are concerned here with those special associations in which members of different species are regularly associated with each other in nonparasitic relationships. Such symbiotic associations are widespread in the sea, primarily in the crowded reaches of the epipelagic, intertidal, and shallow subtidal zones. As with algal symbiosis, such relationships appear to be somewhat more common or spectacular in the tropics, but they are certainly also common in temperate seas.

Types of Associations

Symbiotic relationships among marine animals cover a broader spectrum than the strictly mutualistic associations we have seen among plants and chloroplasts and marine invertebrates. The simplest type of associations are commensal, wherein the "guest" lives on another organism "host" or in or upon some construction of that organism, such as a tube or burrow. Such associations are similar to epiphytic relationships seen among terrestrial plants, wherein the epiphyte simply uses the other plant as a substrate without actually taking sustenance from it. In these cases, the commensal usually gains in some measurable way from the association and the host is not seriously inconvenienced.

Marine commensals that live on or upon other invertebrates are called *epizoites.* Those that live inside other animals but are not parasites are called *endozoites.* Epizoites are extremely abundant in marine waters and many are probably not true commensals; the relationship is the result of organisms that normally settle on the substrate but instead settle at random on the outside of a slow-moving or sessile invertebrate. These will not be considered further here. Other epizoites are highly specific, and a symbiotic relationship is certainly the case, with the "guest" somehow seeking out the correct "host." These symbiotic epizoites are the most abundant group of commensals and are spread among the protozoans and the phyla Cnidaria, Entoprocta, Annelida, Arthropoda, and Mollusca.

Many invertebrates harbor specialized ciliate protistans, either on their external surfaces, internally in the digestive tract, or among the gills. For example, the vorticellid ciliate *Ellobiophyra donacis* is restricted to the gills of the clam *Donax vittatus* (Fig. 10.7), while the collared ciliate *Lobochona prorates* is found on the telson of the isopod crustacean *Limnoria tripunctata.* Many other examples are also known. Endozoic commensal ciliates are common as well, especially in the digestive tracts of larger animals. Thus, for example, Beers (1961) found that the large green sea urchin of the cold temperate waters of North America, *Strongylocentrotus droebachiensis,* has as many as seven species of ciliate protozoans in its gut, which are found nowhere else and die if removed.

Among the Cnidaria, the class Hydrozoa furnishes most of the cases of epizoites. The best studied is the case of two species of *Probisodactyla,* two-tentacled hydroids, which always occur on the rim of the tubes of polychaete worms of the genera *Potamilla* and *Schizobranchia* in the North Pacific Ocean. Similarly, along the Pacific, Atlantic, and Gulf coasts, the hydroid genus *Clytia* occurs on the clams of the genus *Donax.* Hydroids of the genera *Hydractinia* and *Podocoryne* are well-known ectocommensals on shells occupied by hermit crabs and on the snail *Cantharus,* respectively. Other hydroids are epizoic on gorgonians, pennatulids, and ascidians, but the closeness of the association is not always known.

Turbellarian flatworms are often endozoic in the digestive tracts of larger marine invertebrates and in the mantle cavity of various mollusks. Thus, the polyclad *Notoplana ovalis* occurs in the mantle cavity of the limpet *Patella oculis,* while the triclad *Nexilis epichitonius* occurs in the mantle cavity of the common Pacific coast chiton *Mopalia hindsii.* The rhabdocoels

Syndesmis dendrastorum and *S. franciscanus* are endozoites in the guts of the sand dollar, *Dendraster excentricus,* and the sea urchins *Strongylocentrotus franciscanus, S. purpuratus,* and *S. droebachiensis,* respectively.

Bryozoans offer two remarkable examples of symbiosis. Larvae of *Hipporidra* seek out a tiny shell occupied by a young hermit crab. The bryozoan colony grows in such a way that it forms a chamber that enlarges continuously with the aperture of the snail shell and keeps pace with the size increase of the hermit crab, so that the crab never needs to change shells. The growth forms of the bryozoan take on peculiar spindle shapes that are known to Gulf coast beachcombers as "Texas longhorns." A second bryozoan, *Triticella elongata,* is commonly found on the crab *Pinnixa,* which is commensal in the tubes of the polychaete *Chaetopterus.* The phylum Entoprocta is a small, little-known group, most of which, as Nielsen (1964) has documented, are epizoic commensals on other marine invertebrates, particularly polychaete worms, where they inhabit respiratory current areas (Fig. 10.8).

Among the Annelida, a number of polychaete worms are epizoites with definite host specificity. The best studied is the hesionid *Ophiodromus pugettensis,* which occurs on the ambulacral grooves of the bat star, *Asterina* (= *Patiria*) *miniata,* on the Pacific coast (Fig. 10.9). However, the whole group of so-called "scale worms" (= Polynoidae) are all very common epizoites on various other echinoderms and mollusks.

Among the class Crustacea, the epizoic forms occur mainly within three groups: copepods, amphipods, and decapod crabs. Copepods associated with other marine animals are usually highly evolved and modified parasites, but there are a few that are simply commensal, scampering over the outer surfaces of various invertebrates. Such examples include various species of the genus *Hemicyclops,* which move on the surface of other, larger crustaceans. Still others, like *Paranthessius,* are found in the mantle cavity of various bivalves (*Tresus, Protothaca, Saxidomus*). One of the most specialized epizoic associations is that among several species of barnacles and various large whales. Two families of barnacles are found only on whales and several are species-specific. Thus, *Cryptolepas rachianectes*

(A)

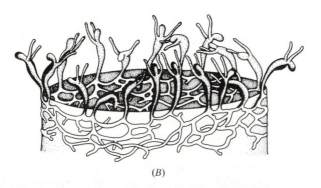

(B)

Figure 10.7 Some epizooites. *(A) Elliobiophyra donacis* on the gills of the bivalve mollusk *Donax.* *(B) Proboscidactyla stellata* on the tube of *Branchiomma vesiculosum.* *(A,* after S. M. Henry [ed.], Symbiosis, Vol. 1, Associations of microorganisms, plants and marine organisms, Academic Press, 1966. *B,* after J. A. Colin Nicol, The biology of marine animals, Wiley, 1960.)

(B)

(A)

Figure 10.8 Epizoic entoprocta, *Loxosomella fauveli,* on the polychaete *Aphrodite aculeata. (A)* Position of the entoprocts in the respiratory channels of the worm. *(B)* Enlarged view of a single *Loxosomella fauveli* individual. (Modified from C. Nielsen, Studies on Danish entoprocta, Ophelia 1[1]:1–76.)

occurs only on the California gray whale, *Eschrichtius gibbosus,* and *Conchoderma auritum* attaches to *Coronula diadema,* which, in turn, is attached to the skin of whales (Fig. 10.10; Plate 63). A very curious form of epizoic relationship occurs between the amphipods of the suborder Hyperiidea and the large jellyfish. Hyperiids are often epizoic or endozoic on jellyfish. Also epizoic on jellyfish are certain juveniles of various benthic crab species (e.g., *Cancer jordani;* Plate 1).

Most mollusk epizoites are found among certain groups of clams. Generally, these are small clams that attach via a foot or byssus to the outside of various other invertebrates. Thus, *Montacuta ferruginosa* is found on the heart urchin, *Echinocardium cordatum; Orobitella*

rugifera, attached via a byssus, is found on the underside of the burrowing shrimp *Upogebia stellata;* and *Mysella pedroana* is seen on the legs of the sand crab *Blepharipoda occidentalis.* Among the gastropods found on other invertebrates the genus *Crepidula* is one of the most common (Plate 64).

The above is certainly not an exhaustive list of the various epizoic and endozoic symbiotic associations, but gives an idea of the extent and diversity of such associations. Many, many more could be listed, but would not serve the purpose of this text.

A second type of symbiotic association is that of organisms associated with animals forming a tube or burrow. In this case, the commensal or guest occupies the tube or burrow constructed

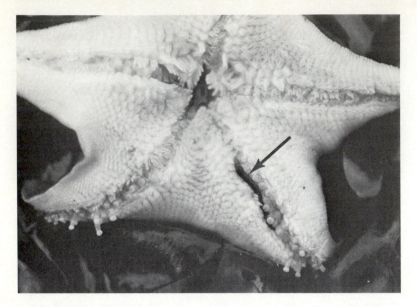

Figure 10.9 The scale worm *Ophiodromus pugettensis* on the bat star *Patiria miniata*. Arrow points to worm. (Photo by the author.)

by the host, but is not necessarily in close association with the body of the host. These commensals use the tube or burrow as a refuge from predation. They may or may not also tap the food source of the host. This type of symbiotic relationship ranges from one simply fortuitous, as in the case of an animal diving into any convenient burrow to avoid a predator, to examples of obligatory associations, where the commensals are not found other than in such tubes or burrows and are dependent on them for continued existence. In the former category are

Figure 10.10 The whale barnacle *Cryptolepas rachianectes* and the cyamid amphipod *Cyamus scammoni* on the skin of the California gray whale, *Eschrichtius gibbosus*. (Photo by the author.)

certain small gobiid fishes, such as *Clevelandia ios,* which take refuge in any available tube or burrow but emerge to feed. Obligate tube dwellers include many small crabs of the family Pinnotheridae, known as "pea crabs." Many species of this family are known throughout the world, where they inhabit the tubes and burrows of polychaete annelids and other invertebrates, sometimes in species-specific associations.

One of the best known examples of symbiosis with tube dwellers is that of the echiurid worm *Urechis caupo* on the Pacific coast of North America. This animal creates a permanent U-shaped burrow in which can be found four or more commensals (Fig. 10.11). At the upper end is the fish *Clevelandia ios,* which merely uses the tube as a refuge. Further in the tube and closely associated with the worm are the scale worm *Hesperonoe adventor* and the pinnotherid crab *Scleroplax granulata. Hesperonoe adventor* seems restricted to *Urechis* burrows and lives against the worm, snatching food from it. Although the crabs are restricted to tubes, they may be found in the burrows of other animals as well. Another commensal is the small clam *Cryptomya californica,* which has very short siphons and inserts them into the burrow of *Urechis,* thus enabling it to live lower in the substrate than the length of its siphons would normally permit. Occasionally, the shrimp *Betaeus longidactylus* may inhabit the burrow.

A similar situation prevails with the tube-building polychaete *Chaetopterus.* This genus is virtually cosmopolitan, building tough parchment tubes. These tubes are also inhabited by various species of pinnotherid crabs and scale worms, the species differing depending on the geographical location.

Another type of association concerns those invertebrates that live in the mantle cavities of various mollusks. This appears to be a very common site for commensals to occupy. Thus, the nemertine *Malacobdella grossa* lives in the mantle cavity of various clams on the Pacific coast. Pinnotherid crabs occupy the mantle cavity of numerous bivalves, including oysters, where they are considered a pest. Polychaete worms inhabit mantle cavities of chitons and limpets. In the tropics, shrimps of the family Palaemonidae are often found in mantle cavities

Figure 10.11 The echiuran worm *Urechis caupo,* to the left, and some of the symbionts that inhabit its burrow. Shown are the pea crab *Scleroplax granulata,* the scale worm *Hesperonoe adventor,* and the long-fingered shrimp *Betaeus longidactylus.* (Photo by Dr. Richard Mariscal.)

of large bivalves such as *Pinna, Atrina,* and *Tridacna.*

The remaining types of associations are those that appear to be mainly mutualistic rather than strictly commensal, as the aforementioned. The organisms involved display varying levels of behavioral and physiological modifications for the association.

One of the most obvious but not well understood associations occurs between various crabs and sea anemones. Many hermit crabs all over the world have anemones attached to their shells, and in some genera, such as *Dardanus,* the presence of anemones seems to be universal. Still other crabs, such as *Hepatus, Munidopagurus macrocheles,* and *Stenocionops furcata,* are found with anemones attached directly to their backs. The ultimate symbiotic relationship appears to be that evolved among a few crabs (*Lybia, Polydactylus*), which carry anemones on their chelae and actively use them for defense or food capture (Fig. 10.12). Interestingly, most of the anemones involved in these associations are of three genera, *Calliactis, Paracalliactis,* and *Adamsia,* species of which are rarely found elsewhere than with crabs. These associations must be mutualistic, because whenever the crab changes shells or molts its carapace, the anemone is usually also transferred. This is effected by the manipulatory movements of the crab, but requires a degree of cooperation on the part of the anemone in loosening its grip on the shell or carapace and allowing itself to be physically moved.

Another type of association is that between other Cnidaria and fishes. Throughout the Indo–West Pacific in the coral reef areas, there is a striking association of small fishes of the genera *Amphiprion, Premnas,* and *Dascyllus* with various large anemones. These fishes are able to live nestled among the tentacles of the anemones by preventing discharge of the formidable nematocysts of the anemone tentacles (Fig. 10.13; Plate 61). Other fishes are not able to prevent this discharge, and the anemones can kill and eat fishes of a similar size. A similar association seems to exist between the dangerous siphonophore, the Portuguese man-of-war (*Physalia*), and the little fish *Nomeus gronovii,* which swims among the tentacles bearing the powerful nematocysts (Fig. 10.17). Still other juvenile fishes often congregate under the bells of the large scyphozoan jellyfishes (*Cyanea, Chrysaora*), where they presumably obtain protection from predation.

A similar situation prevails between fishes and the long-spined tropical sea urchin, *Diadema.* These urchins are circumtropical in shallow water and have enormously long, thin spines, which easily penetrate flesh. At least two tropical fishes, *Aeoliscus strigatus* and *Diademichthys deversor,* have adapted themselves to live

(A)   (B)

Figure 10.12 Symbiotic relationships among crabs and anemones. *(A)* The anemones *Calliactis tricolor* perched on a hermit crab shell. *(B)* The crab *Lybia tessellata* carrying two anemones of the genus *Triactis* in its chelae. (Redrawn from photos.)

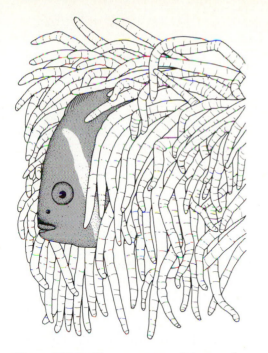

Figure 10.13 The anemone fish *Amphiprion* sp. and its host anemone *Radianthus* sp. (Redrawn from photos.)

among these spines, presumably for protection (Fig. 10.14). They also look like sea urchin spines and so are camouflaged.

There is also the association of fishes with other fishes, usually large, predaceous fishes. Pilot fishes (*Naucrates ductor*) and remoras (*Echeneis remora*) are always found with other, larger fishes, other marine vertebrates such as turtles, or even any moving, inanimate objects (Fig. 10.15).

The final type of association is that known as "cleaning behavior." As noted in Chapter 9, this is an association in which various species of fishes and shrimps actively attract large fishes to themselves for the purpose of cleaning them of various ectoparasites (see pp. 367–368 for a discussion).

Origin of the Association

With such a diverse array of relationships as we have noted, it does not seem likely that there is any single explanation for the origin of all the different types of associations. However, one factor common to all associations is that they have arisen in those areas of the ocean that are the most crowded with life. It would, therefore, appear reasonable that epizoites in particular could have become established due to the great competition for space among the settling invertebrate larvae. As a result, many settled upon other invertebrates and this then later evolved, in some cases, into more obligatory relationships in which those individuals that were commensal gained some advantage over the free-living forms. This is borne out by noting that many epizoites or burrow-inhabiting species are not strongly attracted to any given host species, while others are highly specific. Similarly, for many of the tube and burrow inhabitants, and those found in the mantle cavity of various mollusks, we can probably trace the origin of the relationship to the search for space. After all, tubes, burrows, and mantle cavities really represent a certain amount of potentially occupiable, semi-enclosed space.

The initial association of crabs and anemones may also well have begun when anemones began to take up residence on shells that were already occupied by hermit crabs. Alternatively, the association could have arisen out of the habit of various crabs that pick up materials from the bottom and apply them to their carapaces to aid in camouflage.

Many of the commensals that are epizoites or burrow- or tube-dwellers, especially the polychaete worms, tend to be from groups that show strong *thigmotaxis;* that is, they respond positively to stimulation by contact with other, solid bodies. Such a behavioral response would naturally preadapt them to move into tubes or burrows or into various cavities in the bodies of larger animals. Once so established, selection would favor the development of chemosensory clues enabling these animals to find such burrows, tubes, or cavities, and more specific associations would follow.

Among the fishes and Cnidarians, it seems that the various types of symbiotic relationships have their origin in the protection from larger predators that such relationships confer. It does not appear that this relationship stems from crowding, especially between epipelagic fishes

Figure 10.14 *Diadema*, the long-spined sea urchin, with three shrimp fish, *Aeoliscus strigatus*, and two cling fish, *Diademichthys deversor*, sheltering among the spines. (After R. V. Gotto, Marine animals, partnerships and other associations, Elsevier, 1969.)

and Cnidarians. Provided that the symbionts can avoid having the nematocysts discharged, the tentacles of anemones and jellyfishes with their nematocysts represent a secure, formidable barrier to a large predator. However, in the case of large sharks, pilot fishes, and remoras, the association probably originated in the search for food in the epipelagic. Pilot fishes and remoras probably were originally attracted by the amounts of food that were dropped in the feeding by the larger fishes, and they could have managed to establish themselves either by judiciously avoiding falling prey themselves to the sharks, or by being small enough to

(A)   *(B)*

Figure 10.15 A remora, *Echeneis* sp., showing the suckerlike dorsal fin. *(A)* Dorsal view. *(B)* Side view. (Photos by the author.)

not be recognized as food by the larger fishes.

Cleaning behavior is a very complex relationship involving many fish species, and its origin and ecological functions have been confounded by various alternate hypotheses and factual inaccuracies, as Gorlick et al. (1978) have discussed.

Modifications Due to the Association

As with the algal-animal symbiosis, the animal-animal relationships can result in anatomical, physiological, and behavioral modifications to one or both of the partners.

The fewest modifications occur among the epizoites, especially those that are facultative epizoites and are found also in nonsymbiotic situations. Oftentimes, these animals have no anatomical modifications attributable to the symbiotic association, but those that are found with specific hosts usually have developed the means for *recognition* of the host from a distance. Such ability to recognize hosts is even more highly developed among obligate symbionts of various types. What is the basis for such recognition? Investigators such as Davenport (1950) have shown that it is most likely a chemical, unique to the host, that is released into the water. The commensal has, in turn, developed chemosensory receptors that detect the chemical and enable it to "home" in on its host. If such symbionts are tested in the laboratory in an apparatus designed to give them a "choice," they show a remarkable ability to choose the "host" on which they are found from among other closely related forms. Such attraction by chemical means has been demonstrated for several scale worms (Polynoidae) commensal with various starfish species; for the peculiar fishes, *Carapus,* which live in sea cucumbers; and for the polynoid *Hesperonoe adventor* and its host, *Urechis caupo.*

Yet another modification of commensals, and one that also aids in host recognition, is the ability of the larvae or young to select the appropriate substrate. In many epizoite species that settle onto their hosts from the plankton, metamorphosis and subsequent development do not occur unless the larva is in contact with the appropriate host. The substrate discrimination abilities of some of these symbionts are truly remarkable. Some can apparently recognize substrate surface texture at the molecular level! It is not surprising, then, that whale barnacles can settle not only on whale skin, but only on certain species of whales. However, in this case, it is likely that a chemical unique to the host is also present. Similarly, the clam *Modiolaria* is able to recognize the structure of tunicin, the major chemical component of the external layer of sea squirts, with which it lives.

A basic modification of tube-dwelling symbionts is that they are often smaller or thinner and flatter than their free-living relatives. Thus, the pinnotherids, or "pea" crabs, are among the smallest of the various marine crabs. The shrimps that inhabit the mantle cavities of bivalve mollusks are also among the smallest of the shrimps. Similarly, the scale worms are often very flattened in comparison with other polychaetes. The fishes, such as *Clevelandia ios,* that are commensal in tubes or in the internal spaces of clams and sea cucumbers, *Apogonichthys* and *Carapus,* respectively, are very thin in order that they may fit into these restricted areas. The pearl fishes, *Carapus,* show perhaps the greatest modifications in that they have lost their scales and pelvic fins and have shifted the anal opening far forward under the head. This latter change apparently is to ensure that defecation will occur outside the body of the sea cucumber. Furthermore, as Arnold (1957) has shown, these fishes show behavioral modification in that they enter into the cucumber tail first (Fig. 10.16)! Such extensive modifications suggest a long history of adaptation.

Among those anemones found associated with crabs, a number of behavioral modifications occur. In the first place, the anemones must respond positively to manipulations by the crab when it rushes to transfer the anemone to another shell or from its shed carapace to a new one. This it does by releasing its grip on the old shell so that the crab can remove it with its claws. This requires a higher degree of nervous integration than we have come to believe could be accomplished by the extremely primitive, brainless nervous system of the anemone. We are currently ignorant of an understanding of this process among those crabs that carry anemones in their chelae. The anemones again show a behavioral adaptation in that they allow themselves to be

Figure 10.16 A pearl fish, *Carapus*, entering tail first into the cloacal opening of its holothurian host. (After R. V. Gotto, Marine animals, partnerships and other associations, Elsevier, 1969.)

carried in the chelae and do not close up. In these cases, however, the crab also shows changes in that the chelae are modified to carry the anemones rather than gather food. Behaviorally, the crab extends the anemones when threatened or when gathering food; it uses its second pair of legs to transport food to the mouth, rather than the chelae, since the latter carry the anemones. Perhaps the most drastic changes undergone by the anemone member of these crab-anemone systems is that exhibited by *Adamsia palliata,* which lives with the hermit crab, *Eupagurus prideauxi.* Here, the anemone steadily expands its basal disc to completely encircle the vulnerable abdomen of the hermit crab, thus eliminating the need for the crab to change its shell as it grows. The anemone then expands to allow for the growth of the crab, meanwhile orienting itself so that the oral disc and mouth are ventral to the crab, thus enabling it to obtain food from the crab.

The greatest morphological changes among fishes associated with other fishes occur in the remoras, where the animals have the dorsal fin modified to form a large sucker, which enables them to remain attached to their hosts (Fig. 10.15). Pilot fishes (*Naucrates*) and the anemone fishes (*Amphiprion*) have no special morphological modifications. Among anemone fishes, however, there are definite behavioral and biochemical modifications. Studies have shown that anemone fishes of the genus *Amphiprion* must undergo a period of "acclimatization" to an anemone before they can, with impunity, dive into its tentacles. During this period, the fish is not capable of preventing nematocyst discharge by the anemone. During the acclimatization, Schlichter (1976) has demonstrated that the fish goes through a series of first brief and then longer encounters with the anemone, during which time the fish coats itself with anemone mucus, thus tricking the anemone to discern the fish as itself. Hence, there is no nematocyst discharge from the anemone. When this coating is complete, the acclimatization period is over and the fish is able to dive in among the tentacles without causing nematocyst discharge.

In addition to this fascinating coating process, anemone fishes have additional behavioral modifications. In contrast to most fishes, which swim away when approached by humans or other large potential predators, anemone fishes swim out of their anemones toward the potential predator! This serves to attract the predator to them, at which time, they do an abrupt about-face and dive into the anemone. Presumably, the behavior would tend to make an unwary potential predator chase them and subsequently be caught and consumed by the anemone. In order to further attract attention to themselves, anemone fish also have vivid color patterns, contrasting markedly with their background (Plate 61).

In contrast to the anemone fishes, which develop immunity to the nematocyst of their host, the fishes (*Nomeus gronovii*) associated with the Portuguese man-of-war do not acclimate, nor do they ever obtain complete immunity from the nematocysts (Fig. 10.17). They instead appear to spend their lives playing a perpetual game of Russian roulette with the tentacles of the siphonophore, constantly moving so as to avoid contact with the lethal nematocyst batteries! The same is true for the juveniles of various fishes that shelter below the umbrellas of the large jellyfish.

Among cleaning fishes, certain morphological modifications can be noted, such as pointed,

Figure 10.17 Portuguese man-of-war, *Physalia*, with the fish *Nomeus gronovii* sheltering among the tentacles. (After J. A. C. Nicol, The biology of marine animals, Wiley, 1960.)

narrow snouts and forcepslike teeth, adaptations for picking up the small ectoparasites. Both cleaning shrimps and fishes also tend to have very bright and contrasting colors, which make them stand out against the background and advertise their presence. Certain cleaning shrimps, notably the Pederson shrimp, *Periclimenes pedersoni,* have extraordinarily long antennae, which they wave about to further advertise their presence to the fishes.

Cleaning-behavior relationships also impose behavioral modifications on the fishes that are being cleaned. These animals must seek out and present themselves at the "cleaning station" and then remain motionless while the cleaner moves over their bodies. They must also open their mouths and spread their gills and opercula to allow the cleaner to enter the mouth and gill area. Of course, at the same time, the larger fish being cleaned must refrain from sudden inhalations, which might accidentally ingest the small cleaner!

Value of the Association

These symbiotic associations have obvious value to one or both of the members, and the relative value of the relationship to each varies depending on the association.

The major value of the symbiotic relationship to the epizoites would seem to be with the epizoite rather than the host organism. For some epizoites, the value of the association lies in the fact that the host represents the only available suitable substrate in an otherwise unsuitable area. For example, the hydroid *Clytia bakeri,* which is epizoic on the clam *Donax gouldi* in sand areas in southern California, could not live in the area unless it had the clam to settle upon, as it cannot exist on sand. There is, however, no obvious value to the clam. Other epizoites presumably fix themselves on the host to take advantage of feeding or respiratory currents produced by the host. This seems to be particularly common. The two-tentacled hydroids of the genus *Proboscidactyla* are fixed on the rims of the tubes of their annelid hosts to take advantage of the feeding tentacles and to remove food particles from the host. Similarly, the various epizoic entoprocts are usually found where there are respiratory currents. In this latter case, they do not steal food directly from the host, however, but filter out the food from the moving current. Perhaps most epizoites benefit by the protection afforded them by the larger host.

Thus, the various flatworms occurring in the guts and mantle cavities of their larger hosts are protected from predation and, in the case of intertidal forms, from possible desiccation. Surely, the hyperiid amphipods found on the large scyphozoan jellyfish would be much more subject to predation if free living in the plankton. Finally, the epizoite may also gain the advantage of movement. This advantage may be of particular significance if the epizoite is itself sessile and the host freely moving.

Commensals that inhabit tubes gain primarily protection from predation. This, for example, is the main value to the goby *Clevelandia ios* in the tubes of *Urechis caupo.* Other commensals, however, such as the scale worms (Polynoidae), gain not only protection but also food as they snatch morsels from their tube-dwelling hosts. Commensals in the mantle cavities of bivalve mollusks benefit from the protection of the bivalve shells, but also are able to employ the feeding and respiratory currents of the clam for their own needs. In a few cases, these symbionts may tend toward parasitism in that they may elect to consume pieces of the gills or mantle of their bivalve host.

Among the fishes associated with large jellyfishes, the Portuguese man-of-war, and the large sharks, it seems apparent that the whole value of the association is to the fishes in the form of protection from predation in an area (the epipelagic zone) notorious for great predation pressure. Pilot fishes and remoras are also protected from predation by their association with the large predaceous sharks, but they may also obtain food when the shark is feeding. The fishes associated with sea urchins and living in the respiratory trees of sea cucumbers are protected by their association, but their presence presumably has no correlative advantage to their host.

Among the mutualistic associations, the values of the association are more pronounced and, as the term applies, extend to the host as well. In the various crab and anemone associations, the crabs, as hosts, gain the advantage of the nematocyst batteries of the anemones, either as protection against predators or else as an offensive weapon to capture food. The anemones, on the other hand, while still sessile, gain the advantage of movement; some, such as those on hermit crabs, may also be able to obtain food fragments lost during the feeding process of the crab. A similar situation prevails between anemone fish and their anemones. The fishes gain mainly protection, but they benefit their partners by attracting prey to within reach of the tentacles and also by picking up pieces of food outside the reach of the anemone and depositing them in the anemone's mouth. Fautin (1991) has demonstrated that the fish protect the anemone from predation by predatory fishes. The fish may also benefit the anemone by producing nutrients in the form of wastes that the anemone takes up.

In the case of "cleaning behavior," the cleaner organisms' gain is in the form of food, whereas the large fishes cleaned presumably enjoy an increased measure of health due to the removal of the various ectoparasites.

Luminescent Bacteria

A final category of symbiosis remains that does not fit well with the others and, hence, its consideration here. That is the curious relationship that has developed between various marine animals and luminescent bacteria. In these relationships, the bacteria are usually confined to a cavity in the body of the larger animal near the outer surface. This cavity is usually connected to the outside through an opening of some sort.

Such symbiotic relationships are confined to three groups of marine animals: fishes, squids, and pelagic tunicates. The relationship is most common among fishes and squids inhabiting the mesopelagic zone (see pp. 144–146), but is still less common among these groups than is bioluminescence produced intrinsically without bacteria. The symbiotic bacteria are species of the genera *Vibrio* or *Photobacterium.* The fish or squid of each generation apparently acquire their luminescent bacteria from the environment, where they are widespread. The various suggested values for the production of light by mesopelagic animals have been discussed in Chapter 4.

The relationship of bacteria and the fishes or squids is a mutualistic one in which the bacteria obtain food from the larger animal. In turn, the fishes or squid use the light produced by the bacteria for various defensive or offensive purposes, as discussed in Chapter 4.

The light produced by bacteria is usually continuous, and as a result, the fishes and squids often develop rather elaborate modifications to control the light. These anatomical developments can include the production of a reflecting layer behind the cavity to reflect the light outward, often lenses to focus and concentrate the light, and most often, a screen or shade on the outside that can be raised or lowered to either "turn on" or "turn off" the light. The bacteria show no corresponding changes in behavior or morphology as a result of the association.

The origin of these elaborate associations is not well understood, and the bacteria apparently are not passed from one generation of fishes or squid through the eggs. It would appear that each generation somehow must reinfect its light organs from the external environment.

BIBLIOGRAPHY

Arnold, D. C. 1957. Further studies on the behavior of the fish *Carapus acus*. Publ. Staz. Zool., Naples 23:91.

Babcock, R. C. 1985. Growth and mortality in juvenile corals (*Goniastrea, Platygyra,* and *Acropora*): The first year. Proc. 5th Int. Coral Reef Sym. *4:* 355–360.

Beers, C. D. 1961. The obligate commensal ciliates of *Strongylocentrotus drobachiensis:* Occurrence and division in urchins of diverse ages; survival in sea water in relation to infectivity. Biol. Bull. 121:69–81.

Cheng, T. (ed.). 1971. Aspects of the biology of symbiosis. University Park Press, Baltimore. 327 pp.

Dales, R. P. 1957. Commensalism, Ch. 15. *In:* Hedgpeth, J. E. (ed.). The treatise on marine ecology and paleoecology. Vol. I, Ecology. Memoir 67, Geol. Soc. of Amer.

Davenport, D. 1950. Studies in the physiology of commensalism. I, The polynoid genus *Arctonoe.* Biol. Bull. 98(2):81–93.

Fankboner, P. V. 1971. Intracellular digestion of symbiotic zooxanthellae by host amoebocytes in giant clams (Bivalvia: Tridacnidae) with a note on the nutritional role of the hypertrophied siphonal epidermis. Biol. Bull. 141:222–234.

Fautin, D. G. 1991. The anemone fish symbiosis: What is known and what is not. Symbiosis 10:23–46.

Gorlick, D. L., P. D. Atkins, and G. S. Losey. 1978. Cleaning stations as water holes, garbage dumps and sites for evolution of reciprocal altruism. Amer. Nat. 112(984):341–353.

Gotto, R. V. 1969. Marine animals, partnerships and other associations. Elsevier, N.Y. 96 pp.

Greene, R. W. 1970. Symbiosis in sacoglossan opisthobranchs: Symbiosis with algal chloroplasts; translocation of photosynthetic products from chloroplast to host tissue. Malacologia 10(2):357–380.

Henry, S. M. 1966. Symbiosis. Vol. I, Associations of microorganisms, plants and marine organisms. Academic Press, N.Y. 478 pp.

McLaughlin, J. J. A., and P. Zahl. 1966. Endozoic algae, pp. 257–297. *In:* Henry, S. M. (ed.). Symbiosis. Vol. I, Associations of microorganisms, plants and marine organisms. Academic Press, N.Y.

Nicol, J. A. C. 1960. The biology of marine animals. Wiley, N.Y. 707 pp.

Nielsen, C. 1964. Studies on Danish Entoprocta. Ophelia 1(1):1–76.

Rosewater, J. 1965. The family Tridacnidae in the Indo-Pacific. Indo-Pacific Mollusca 1(6):347–396.

Schlichter, D. 1976. Macromolecular mimicry: Substances released by sea anemones and their role in the protection of anemone fish, pp. 438–441. *In:* Mackie, G. O. (ed.). Coelenterate ecology and behavior. Plenum, N.Y.

Schoenberg, D. A., and R. K. Trench. 1976. Specificity of symbioses between marine cnidarians and zooxanthellae, pp. 423–432. *In:* Mackie, G. O. (ed). Coelenterate ecology and behavior. Plenum, N.Y.

Smith, D. C. 1973. Symbiosis of algae with invertebrates. Oxford biology reader no. 43. Oxford University Press, London. 16 pp.

Smith, D. C., and A. E. Douglas. 1987. The biology of symbiosis. Edward Arnold, Baltimore. 302 pp.

Vernberg, W. B. 1974. Symbiosis in the sea. The Belle W. Baruch Library in Marine Science no. 2. University of South Carolina Press, Columbia. 276 pp.

Wilkinson, C. R. 1983. Phylogeny of bacterial and cyanobacterial symbionts, pp. 993–1002. *In:* Schwemmler, W., and H. E. A. Schenk (eds.). Endocytobiology, endosymbiosis and cell biology. Vol. II. Walter de Gruyter, Berlin.

Wilkinson, C. R., and J. Vacelet. 1979. Transplantation of marine sponges to different conditions of light and current. J. Exp. Mar. Biol. Ecol. 37:91–104.

Yonge, C. M. 1957, Symbiosis. Ch. 15. *In:* Hedgpeth, J. E. (ed.). The treatise on marine ecology and paleoecology. Vol. I, Ecology. Memoir 67, Geol. Soc. of Amer.

Yonge, C. M. 1975. Giant clams. Sci. Amer. 232:96–105.

Human Impact on the Sea

*T*he oceans are truly the last frontier for human exploration and exploitation on this planet. For years, the vastness and volume of the oceans, coupled with the fact that humans are terrestrial organisms naturally ill equipped to enter into water more than superficially, kept the oceans and their communities safe from significant human interference or effects. As the human population of the world increased through the centuries, large land areas were significantly altered, but the oceans remained relatively untouched. The major usage of the sea during these centuries was as a source of food, but the primitive nature of gear and the limited range of vessels combined to make the effects of fishing insignificant.

In the twentieth century, all this has changed. The explosion of technology has enabled human beings to permeate all parts of the oceans. The rapidly increasing human population, coupled with sophisticated technology, in a few short decades has had significant effects on the ecology of the world's oceans. The purpose of this chapter is to discuss and evaluate some of these effects in light of our knowledge of the functioning of the oceans' ecosystems.

FISHERIES

The earliest use of the oceans by humans was probably for food. Early human populations living along the oceans captured various shore fishes and shellfishes for consumption. This is recorded in the various large "shell mounds" excavated by archaeologists in different parts of the world. With the advent of vessels to venture onto the sea and the development of more refined nets, fishing became an important source of food to people living near the oceans. However, the gear and the vessels were so inefficient through much of human history that, even as late as the 1880s, scientists such as T. H. Huxley believed that the major fisheries were inexhaustible. In the decades of this century, however, the old ships and gear have been replaced by much larger and more powerful vessels, more effective nets and traps, and electronic devices for detecting fish schools. The result has been a significant reduction in many fish populations and the disappearance or overexploitation of others at a time when increasing human populations are demanding more food. Since many believe the oceans are the major food reservoir for upcoming generations of humans, the impact of fishing and its potential to sustain future generations bear consideration here.

Major Fishing Areas

Geographically, the major fisheries are concentrated in the waters overlying the continental shelves around the world. The only major fisheries that operate in the open oceanic regions of the world are those for tuna and whales. There are several reasons for the concentration of fisheries in neritic waters. In the first place, the inshore waters have a much higher primary productivity than most open-ocean waters (see pp. 67–68) and, therefore, support larger populations of fishes at all trophic levels (Fig. 11.1). Second, the bottom is fairly shallow on the shelf and is accessible to the various nets and traps used by humans to capture fishes. The deep

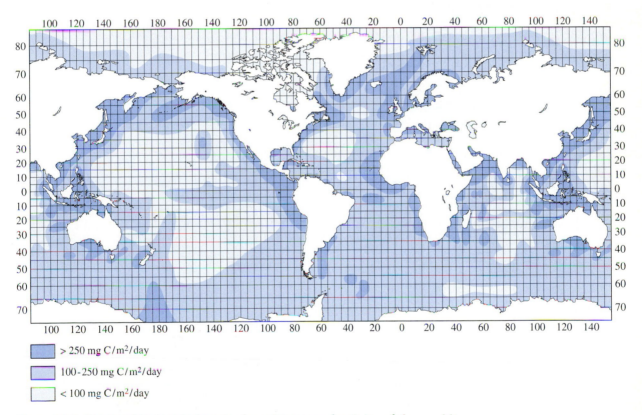

■ > 250 mg C/m²/day

▨ 100-250 mg C/m²/day

□ < 100 mg C/m²/day

Figure 11.1 Geographical variations in the primary productivity of the world's oceans. White areas are areas of lowest productivity (less than 100 mg C/m²/day); light stipple indicates medium productivity (100–250 mg C/m²/day); heavy stipple indicates areas of highest productivity (more than 250 mg C/m²/day). (After FAO atlas of the living resources of the sea, Food and Agriculture Organization of the United Nations.)

ocean floor, on the other hand, is so far removed from the surface that even with our current sophisticated capture mechanisms, we cannot commercially fish there. Finally, the great lack of food in abyssal depths precludes any large fish population from existing there. Any fishes there could likely not sustain a fishery for any length of time.

Although the fisheries are concentrated in the shelf regions, there is a very unequal distribution of the tonnage caught in various areas. The shelf areas of northwest Europe, along the upwelling coast of western South America, and off Japan produce the largest catches of fishes (Fig. 11.2). The southern oceans and the tropics, with the exception of the west coast of South America, contribute much less to world fisheries.

Major Commercial Species

The commercially important marine animals are drawn from four groups: bony and cartilaginous fishes, marine mammals, mollusks, and crustaceans. Of these groups, the various fishes constitute by far the greatest tonnage. Among the thousands of species of marine fishes, only a very few make up the majority of catches in fisheries throughout the world. These can be assembled into a few major groups (Table 11.1). The herrings, sardines, and anchovies account for the largest tonnage of fishes. Of these, a single species, the Peruvian anchoveta (*Engraulis ringens*) in the past has accounted for almost half the catch and was the basis of the largest single fishery in the world, off Peru (Fig. 11.3). Significant other fisheries in this group are for menhaden off the Atlantic and Gulf coasts of the United States and the herring fishery of northwestern Europe. Before its collapse in the 1940s and 1950s, the California sardine fishery was an important contributor to tonnage in this group. Herrings, anchovies, and anchovetas, commonly called clupeoid fishes, are small, pelagic, and feed on a low trophic level, either directly on phytoplankton (*Engraulis*) or on herbivores such as copepods. Most are not used as food for humans, but are reduced to protein meal for animal food.

Figure 11.2 The major marine fishing regions of the world and their respective 1987 landings. Each symbol represents 1 million metric tons. The circled numbers refer to the region in the FAO division of the oceans. (Modified from the FAO 1987 yearbook of fishery statistics, Food and Agriculture Organization of the United Nations.)

Table 11.1   WORLDWIDE CATCHES OF DIFFERENT GROUPS OF FISHES AND SHELLFISHES (IN MILLIONS OF METRIC TONS)

	1981	1982	1983	1984	1985	1986	1987
Herrings, sardines, anchovies	16.92	17.87	17.50	19.62	21.10	23.97	22.23
Cod, hake, haddock	10.65	10.98	11.22	12.26	12.45	13.54	13.70
Jacks, mullets, sauries	8.06	7.81	7.97	8.56	8.01	7.18	7.87
Miscellaneous marine fishes	8.11	8.00	8.19	8.43	8.77	9.51	9.62
Redfish, basses, congers	5.29	5.40	4.98	5.45	5.21	5.99	5.73
Mackerels, snocks, cutlass fishes	3.89	3.85	3.67	4.27	3.83	4.02	3.65
Tunas, bonitos, bill fishes	2.65	2.80	2.97	3.15	3.18	3.43	3.44
Shrimps, prawns	1.64	1.71	1.79	1.86	1.98	2.00	2.03
Squids, cuttlefishes, octopus	1.39	1.64	1.71	1.68	1.77	1.73	2.27
Clams, cockles, ark shells	1.32	1.28	1.33	1.51	1.64	1.90	2.12
Flounder, halibut, sole	1.09	1.13	1.12	1.19	1.35	1.31	1.28
Oysters	0.96	0.97	1.00	0.99	1.03	1.01	1.03
Crabs	0.76	0.80	0.82	0.84	0.89	1.00	1.07
Mussels	0.81	0.91	0.87	0.86	0.94	0.92	0.90
Sharks, rays, chimeras	0.63	0.63	0.58	0.60	0.62	0.64	0.66
Scallops, pectens, etc.	0.58	0.52	0.56	0.83	0.60	0.51	0.69
Lobsters, squat lobsters	0.18	0.17	0.19	0.20	0.21	0.22	0.23
Krill, planktonic crustacea	0.45	0.53	0.23	0.13	0.19	0.45	0.38

Source: From *United Nations Food and Agriculture Organization Yearbook of Fishery Statistics,* 1987.

The second largest group of fishes landed are the gadoids, comprising various cod, haddock, pollock, and hake. These are bottom-dwelling (demersal) fishes, and the major fisheries are on the shallow banks of the North Atlantic and North Pacific oceans (Fig. 11.3).

The third largest group are the scombroid fishes, generally known by the common name "mackerel." Fishes of this group are common in both temperate and tropical waters, where they are fast-moving pelagic carnivores in shallow water.

Closely related to mackerels are the various tunas. Tunas are among the largest fishes in commercial fisheries. They are also the basis of the only major open-ocean fishery. They are widespread, fast-swimming carnivores of tropical and warm temperate seas (see Chapter 3 for a discussion of their biology). Although the tonnage is not as great as for other fishes, they are significant because they command a high price as a canned fish product.

Redfishes, rockfishes, and sea basses are demersal, cold-water fishes that are primarily used as human food.

The flatfishes, such as halibut, sole, plaice, and flounder, are all well-known fishes that are of considerable economic importance because they are premium eating fishes. All are caught in fairly shallow waters, and many have been heavily exploited for years.

The sharks and other cartilaginous fishes have recently been more extensively exploited for food and some have been demonstrated to have excellent flesh.

One group not separated in Table 11.1 is the salmonids. They form a very minor part of the world landings, too small to merit separate listing, but are of great economic importance in North America and Europe because of their fine taste.

The major crustacean group in world fisheries is shrimp, various species being caught in both warm and cool waters in all oceans. The other significant crustaceans are various species of crabs and lobsters. All are economically valuable and command high prices for direct human consumption.

The major mollusk group in terms of tonnage is the squid. Significant numbers of squids are caught in Japan, Europe, and California. The other major mollusk group commercially harvested includes the bivalves, such as oysters and clams.

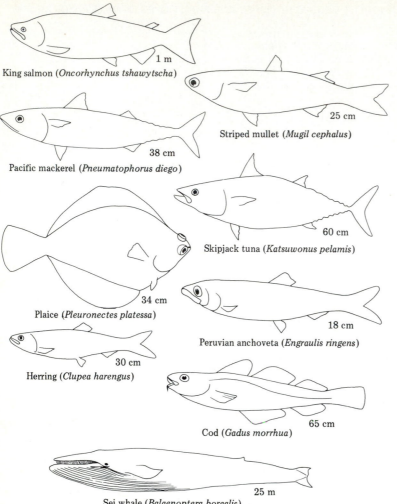

King salmon (*Oncorhynchus tshawytscha*) 1 m

Striped mullet (*Mugil cephalus*) 25 cm

Pacific mackerel (*Pneumatophorus diego*) 38 cm

Skipjack tuna (*Katsuwonus pelamis*) 60 cm

Plaice (*Pleuronectes platessa*) 34 cm

Peruvian anchoveta (*Engraulis ringens*) 18 cm

Herring (*Clupea harengus*) 30 cm

Cod (*Gadus morrhua*) 65 cm

Sei whale (*Balaenoptera borealis*) 25 m

Figure 11.3 Some of the major commercial fishes of the world. (Not to scale.)

The major fishery for marine mammals has been that for various species of whales. Persistent overexploitation of the stocks in this century has, however, virtually doomed this fishery. Other harvestable mammals are mainly seals and sea lions.

Other minor fisheries exist for various species of algae. Some are taken in Japan for human consumption, but most are collected to extract various products from them for use in industry (algin derivatives).

Sustainable Yield and the Future

In 1988, the total world fishery catch was 98.4 million metric tons. Of the 98.4 million metric tons, 84.6 million metric tons were from marine fisheries. The total fishes caught for human consumption represents perhaps 1 percent of all human food, but a significant 10 percent of the protein intake; hence its importance. The total marine tonnage in 1988 was the highest in history. In fact, the total marine landings have been increasing quite markedly since 1983 (Fig. 11.4). This is due primarily to the heavy fishing for certain unexploited fishes, such as pollock and hake, that are used mainly for fish meal, and the expansion of fisheries in the Southern Hemisphere oceans. Since the human population is still increasing and is expected to reach near 7 billion by the end of the century, and since many are anticipating that much of the food needed to feed these increased numbers will come from the sea, does this continued rise mean that we can count on continuing increases in fish protein to

Figure 11.4 Changes in the annual world fisheries catch in recent years.

meet the increased needs of a burgeoning human population? What are the facts with respect to current and potential amounts of food from the sea?

In order to answer these questions, it is first necessary to look briefly at the situation of current fisheries. Of the present fishing effort, 90 percent is concentrated on the 7 percent of the world ocean area represented by the continental shelves. These shallow areas are the most productive ocean areas.

The great growth of the world fisheries in the 1950s and 1960s was due to two factors. First was the discovery and exploitation of the anchoveta fishery of Peru, which sent Peru from having less than 1 percent of the world fisheries landings in 1950 to 17 percent in 1967 and to a position as the top fishing nation in the world. The second factor was the introduction of modern fish-capturing and -handling gear, including factory ships, larger nets, and more accurate methods of locating fishes. Most of this effort was expended on the continental shelf regions around the world with the result that, at the present time, perhaps most of this limited area has been fully exploited for its existing fish stocks and many have been overfished (see pp. 416–419). It would therefore appear that little increase can be expected from these areas.

What about the remaining 93 percent of the world's oceans? Currently, the major oceanic

fisheries of the world are for various tunas, swordfish, bill fish, and so forth. In 1987, the landings of this group were 3.44 million metric tons, or only 4 percent of total fishery landings for 1987. However, this 3.44 million tons is not a significant increase in the landings over previous years and suggests that we are approaching the limit in tunas that may be caught without depleting the stocks. Whereas 20 years ago the tuna stocks were considered nearly impossible to overfish, the advent of purse seines dramatically changed that fact in a few years. Today there is considerable concern about the continued ability of these fish stocks to sustain the current fishing pressure.

Bottom fishing on the abyssal plains under the oceanic area is not feasible, since the diminished amount of energy reaching the abyssal floor would not allow the existence of substantial fish stocks (see pp. 135–137).

Furthermore, as we noted in Chapter 1, the surface waters of these open-ocean areas are impoverished with respect to nutrients and phytoplankton. They cannot sustain the populations of fishes, even in the surface waters, that the continental shelf waters are able to. They are "watery deserts." This suggests that, even though this is a vast area, it cannot match the production of the shallow coastal waters. In fact, we are approaching the limit of what we can obtain from the sea.

What, then, is a reasonable estimate for the *maximum sustainable yield* of food that we can expect from the seas? By maximum sustainable yield we mean the largest number of fishes that can be harvested year after year without diminishing the stocks.

Perhaps the best estimate is based on total primary productivity of the oceans. As we have already discussed, most food chains and trophic levels are ultimately dependent on photosynthesis. The energy fixed in photosynthesis is finite in the world's oceans. By knowing the trophic level of the various fishes taken as food and the efficiency of energy transfer from level to level, it is thus possible to obtain a figure for maximum production. Such was the approach taken by Ryther (1969) to estimate potential yield.

In this estimate, he first divided the ocean into three areas, each representing waters with significantly different levels of primary productivity. The largest area (90 percent) was that of the open oceans of the world, but it had the lowest primary productivity (Table 11.2). The second area in size (9.9 percent) was the neritic zone, and the final—highest in productivity but smallest in size (0.1 percent)—was the restricted area where upwelling occurs (Table 11.2). Next, he estimated the efficiency of energy transfer among trophic levels in each division and, finally, the trophic level of the major fishes captured in each (Table 11.3). Combining the data, he arrived at a total world fish production of 24×10^7 tons (= 240 million tons).

This figure, however, does not represent the total amount that may be taken. A significant fraction of this 240 million tons must be left as breeding stock to sustain future yields. Second, another large fraction is consumed by other carnivores in the sea. Ryther estimated these latter fractions to account for somewhat more than half of the total production, thus leaving about 100 million tons as the sustained yield for the world's oceans.

Other scientists, such as Cushing (1969) and Schaefer (1965), have also estimated the total harvestable amounts of fishes in the sea. These estimates have been made on the basis of primary productivity, on the basis of extrapolating from current landings, and on the basis of trophic level and ecological efficiencies. Because scientists disagree on the proper figures to use for primary productivity, ecological efficiency, and average trophic level, these estimates have varied from a low of 60 million tons to a high of 290 million tons.

Given the current situation of world marine fisheries in which we are seeing continued increases in landings, at least of certain previously unexploited fishes, and at the same time are seeing declines in other stocks, coupled with a wide disparity in the estimates of the amount of sustainable yield, it is difficult to say at this time (1992) whether we are well under, near, or over the point of sustained yield for the world fisheries as a whole. However, that we may be reaching the limit on the number of fishes that may be taken is manifested in the decline of various fisheries.

Overexploitation

In recent years, there have been abundant examples of the decline of fish stocks of all types in all areas of the world. Some of these declines are

Table 11.2   DIVISION OF THE OCEANS INTO PROVINCES BASED ON LEVEL OF PRIMARY PRODUCTIVITY

Province	Percentage of the ocean	Area (km²)	Mean productivity (grams of carbon per m² per year)	Total productivity (10⁹ tons of carbon per year)
Open ocean	90.0	326.0×10^6	50	16.3
Coastal zone	9.9	36.0×10^6	100	3.6
Upwelling areas	0.1	3.6×10^5	300	0.1
Total				20.0

Source: After Ryther, 1969.

Table 11.3 ESTIMATED TOTAL FISH PRODUCTION OF THE WORLD'S OCEANS BASED ON THE THREE PROVINCES OF TABLE 11.2

Province	Primary production (tons of organic carbon)	Trophic levels	Efficiency (%)	Fish production (tons, fresh weight)
Oceanic	16.3×10^9	5	10	16×10^5
Coastal	3.6×10^9	3	15	12×10^7
Upwelling areas	0.1×10^9	1½	20	12×10^7
Total				24×10^7

Source: After Ryther, 1969.

clearly due to instances of overexploitation, but others have more complex origins. All indicate that we are approaching or exceeding the maximum amount of fishes we can take from the oceans, but also that we have often mismanaged many fisheries so that we have not achieved a maximum sustainable yield. The signs of overfishing are normally a decline in average size of fishes and an increase in effort needed to land the same amount of fishes. Age structure also changes. In some cases, overexploitation has been compounded by certain natural changes in the environment that also acted to reduce the stocks. In most situations, however, overexploitation is the direct result of human activity. Three examples are discussed here, one with environmental consequences affecting the fishery, one in which causes are uncertain, and one related directly to humans.

The small clupeid fish known as the Peruvian anchoveta (*Engraulis ringens*) is a small filter feeder that occurs in large schools in the upwelling water mass of the Peru Current off the west coast of South America, where it feeds on the abundant plankton. This small fish occurred in tremendous numbers in these cold waters before the advent of human fishing. The major predators were various species of marine birds, collectively called "guano birds," which existed by the millions along these shores and produced huge quantities of droppings or *guano,* which were harvested for fertilizer.

The anchoveta population depends on an abundant plankton, which, in turn, depends on the nutrient-rich upwelling water. In certain years, the upwelling is blocked in this area by an influx of warm surface water known as the *El Niño* Current (see Chapter 9, pp. 369–371). When

this occurs, plankton disappear and the population of anchoveta drops, as do the bird numbers. Hence, there is a natural abundance cycle.

Direct human exploitation of the anchoveta began in the 1950s and the fishery increased at a phenomenal rate until the late 1960s. The catch was 10 million metric tons or more than one-sixth of the entire world's catch!

This huge catch was accompanied by a decline in bird numbers, since they had fewer fishes to eat. The fishery experienced one drop in numbers in 1965 as a result of the occurrence of the El Niño, but this temporary setback was made up the following year.

Fishery experts had estimated a maximum sustainable anchoveta yield of 9.5 million metric tons. This was exceeded in 1967 when over 10 million tons were landed. By 1970, more than 12.3 million metric tons were landed, but the size of the fishes was decreasing, and it was apparent to experts that the fishery was in trouble and that too few fishes were escaping to provide the breeding stock. As could have been forecast, the crash came in 1972, aided by another El Niño, so that landings dropped to 4.7 million tons. This was followed by an even worse year in 1973, when only 1.7 million tons were landed. Subsequently, this fishery has not recovered, so that even in 1987 only 2.1 million tons were landed. Thus, a once huge resource has, due to overfishing, been reduced far below its sustainable yield, and the human population of the world has been denied a significant protein supply.

Much more mysterious is the demise of the California sardine fishery. The California sardine, *Sardinops Sagax,* is also a small, plankton-feeding clupeid fish with a range from Mexico to southern Alaska. In the 1930s, this species was the basis of the greatest fishery in North America. During the 1936–37 season, fishermen landed 650,000 tons in California alone, the major fishing region. The decline began in the 1940s, when the age structure of captured sardines began to change to older age groups, reflecting the decimation of the younger fishes and their failure to recruit to the fishery. Despite the establishment of the California Cooperative Oceanic Fisheries Research Program in 1949, the fishery was commercially dead by the early 1950s. In Monterey, California, a center of the industry, the famed Cannery Row of Steinbeck closed up. The fishery has not recovered in the nearly 40 years since then, though since 1981 there has been an encouraging increase in the numbers of fish. The population now appears to approach commercial levels and some limited fishing has begun.

Although most people attributed the demise of the fishery over the 10-year period from 1940 to 1950 to overfishing, it seems apparent now that, in this case, other factors were also at work. The decline of the fishery began with the onset of a 15-year period of unusually cold water in the sardine habitat. It is felt by Marr (1960) that this cold water affected the sardine population adversely, either directly through a reduction in breeding or through an effect on the food supply. The result was a catastrophic decline in numbers. If overfishing were the only reason for the demise, one would have expected that, in the more than 35 years during which there has been no fishing, the stocks would have recovered or at least returned to their known breeding grounds. But this has occurred only very slowly, and hence the strong suspicion that other factors are involved.

Finally, Soutar and Isaacs (1969) have presented evidence, from the examination of the scales of sardines preserved in layers of sediments in certain anoxic basins, that the California sardine naturally undergoes huge cyclic changes in population numbers, and that the cycles range from 500 to 1700 years. If this is correct, the California sardine would be present in numbers for a certain period of years, then vanish for a longer period before reappearing. During its absence, the niche is filled by another clupeid, the anchovy. Currently, the increase in anchovy numbers would seem to substantiate

this theory. If true, Cannery Row might again bustle to the activities of fish rendering in only 500 years or so!

The final example of the demise of a fishery is much more simple, for it is the direct result of human overfishing. Here, we refer to the whaling industry. The ignominious end of the world's whale fishery is now being observed.

In contrast to most animals taken in various fisheries, whales are warm-blooded mammals that have a long life span and a very low reproductive rate. This means that they are very slow to build up their numbers once reduced and are easily susceptible to extinction.

Large-scale whaling begain in the North Atlantic in the sixteenth century. At first, methods were primitive and refuges from whalers existed for most species. However, the advent of long-range vessels and, especially, the harpoon gun with explosive charge spelled the end of whale stocks in the North Atlantic and North Pacific by 1900.

Attention then turned to the large stocks of baleen whales in the Antarctic region, where most whaling of the twentieth century has been concentrated. Coincident with the exploitation of these southern stocks was the advent of huge factory ships, more speedy catcher boats, and airplanes and helicopters to find the whales. All these technological advancements led to a greater slaughter of whales in the first 60 years of this century than had been achieved by all whaling efforts in all preceding centuries. In approximately 60 years of Antarctic whaling, Rounsefell (1975) notes that perhaps 1.3 million whales were taken!

In an attempt to regulate this fishery, the International Whaling Commission (IWC) was set up in 1948. It was to obtain data on whale stocks and to set catch quotas for the various species so as to conserve the stocks. Unfortunately, because the Commission had no enforcement powers, because it had no independent means of obtaining data on the status of the various whale stocks outside of what the whalers reported, and because it was subject to intense political pressure by the whaling nations that it comprised, its history of attempts to conserve stocks has been a mockery. It has consistently set its quotas too high and refused to heed the advice of its own and outside experts. As a result, the stocks of whales have continued to decline to the point where, today, many species are on the verge of extinction, and others are so reduced in numbers that they are commercially extinct. Along with the demise of the whales has been the demise of the whaling industry. When the IWC was set up in 1948, its 20 member nations were whalers. By 1979, only two remained as international whalers, Russia and Japan.

The dramatic decline in this fishery can be understood from a few examples. In the 1930–1931 season in the Antarctic, 25,000 blue whales (*Balaenoptera musculus*) were taken, but by the 1963–64 season, only 112 were taken. In 1968, the IWC reluctantly gave complete protection to the species. Declining blue whale stocks forced whalers to change to fin whales (*Balaenoptera physalus*) after World War II, and from 1955 to 1961, more than 25,000 were taken each season. By 1965, this number had dropped to less than 2500 and by the 1974–75 season to less than 1000. Sei whales (*Balaenoptera borealis*) went from 25,453 in 1964–65 to 3859 in 1974–75.

Clearly, the unrealistically high quotas set by the IWC and the greed of the whaling nations have reduced the whale stocks to the point where it is, or soon will be, uneconomical to continue whaling. Even if all whaling is stopped now, as many countries have urged, it will take many species years, perhaps centuries, to recover because of their low fecundity and the long time necessary for them to reach reproductive size. That there is some hope for recovery of at least some of these leviathans lies in the recovery of the California gray whale (*Eschrichtius gibbosus*). This species was reduced to only a few individuals in the early decades of this century, but since complete protection in 1938, they have now increased in numbers to more than 8000.

Tragedy of the Commons

In 1968, Garrett Hardin published an article in *Science* magazine entitled "The Tragedy of the Commons," in which he pointed out that freedom of access to a common resource ultimately brings ruin to all who enter. The core of this argument is that whenever there is a resource held in common by a large number of individuals or political entities, such as countries, each will try to maximize its own gain from that resource because not to do so will allow another

to do so and because the negative aspects of acting selfishly do not devolve on the one acting alone, but on all who share the resource. Unfortunately, such an analysis is made by each and every entity sharing the resource. As a result, the resource is destroyed. Therein is the tragedy.

The plight of the whale fishery can certainly be likened to such a tragedy. What happened to it may also be applicable to other open-ocean fisheries, such as tuna, or potential fisheries, such as krill (see pp. 420–421).

Open-ocean fisheries lie in international waters—in other words, beyond the control of any one country—and presumably are held in "common" for the good of the world's people. Since they are owned by no one, they and their resources are open to access by all. Therefore, those countries with the means to exploit these areas do so because it serves their own interests. They do not have to answer to another nation or regulatory agency, and the tragedy is that, if they don't catch the fishes, another nation will. It is just this short-term greed that has destroyed the whale fishery and may next do so for tuna. Recent interest of some countries in mining of deep-ocean manganese nodules may also fall into this category.

Regulation

The decline of many world fisheries due to the tragedy of common access and the pressures of an increased demand for food by an ever-increasing human population have led to friction among fishing nations and various attempts at the regulation of fisheries resources. As long as the fishery exists on the continental shelf where individual nations have jurisdiction by international convention, the fishery can be controlled by a single nation. In the past, the area of an individual nation's jurisdiction was 12 miles (19.3 km). In recent years, however, several countries have unilaterally acted to extend these limits out to 50 or 200 miles. This has caused several confrontations such as the "cod war" between Iceland and Britain and the seizure and fining of many United States tuna boats by Ecuador and Peru. Most recently, through the actions of the United Nations, a standard 200-mile-wide fishing and economic zone was established for the waters of each coastal nation. This means that individual nations will now have full control of all fishery activity within 200 miles of their shores.

International agreements among nations for the control of fisheries have proven to be more difficult to bring about and also to make work effectively, as we witnessed in the IWC. However, there are some outstanding examples that have worked. One of the oldest is the International Fur Seal Treaty, established in 1911 among the United States, Russia, Japan, and Canada, to regulate exploitation of the northern fur seal (*Callorhinus ursinus*; Plate 15). It has been an outstanding success in regulating fur seal numbers to achieve a sustained yield. Another success has been the International Halibut Commission, established in 1924 between Canada and the United States to regulate the then declining Pacific halibut fishery. This was the first fish species to be conserved internationally. After establishment of the Commission, the stocks of halibut recovered (Fig. 11.5). Since these early attempts, other international regulatory agencies have been established, including the Atlantic and Baltic commissions, the International Tuna Commission, and the International North Pacific Fisheries Commission (INPFC). Many of these more recent regulatory agencies have had a beneficial effect on certain fisheries, but problems still remain. The major one is that the various commissions have limited jurisdiction geographically or are limited to only certain species. Thus, the waters around Japan are not in the jurisdiction of the INPFC and the Halibut Commission regulates only that species.

In order to effect a truly worldwide regulation of fisheries, the current commissions or others will have to expand their jurisdiction to include all species in their areas; and those areas, mainly in the Southern Hemisphere, where no regulatory agencies exist will have to be brought under control.

New Fisheries

Given the current approach of the world fisheries catch to the theoretical maximum sustained yield and the decline or plateauing in many long-established fisheries, are there any prospects for finding substantial new fisheries? The answer appears to be yes, but whether these new fisheries will be able to contribute significantly to the human food supply is much less certain.

Now that the stocks of baleen whales have been reduced to the point of extinction in Antarctic waters, a new fishery has developed for the "krill" (euphausiids) that formerly constituted the food of the whales. Krill exist in untold numbers in the seas around Antarctica, and they are large enough (5 cm) so they may be captured without resorting to very small-mesh plankton nets (Fig. 11.6). The most conservative estimates of stocks of krill are 50 million tons, but other

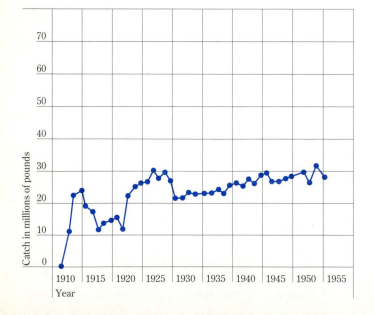

Figure 11.5 Pacific halibut fishery exploitation and recovery. (Modified from D. H. Cushing, Fisheries resources of the sea and their management. Copyright © 1975 by Oxford University Press.)

Figure 11.6 Krill, or *Euphausia superba,* of the southern ocean.

estimates range up to 100 million tons, according to Idyll (1970). Perhaps as many as 50 million tons could be harvested annually. Thus, the potential for this fishery is enormous, certainly the largest of any unexploited fishery in the world today. However, much remains to be learned about this fishery.

Another area of potential exploitation, until recently relatively untapped, is the area of the continental slopes in 300–1000 m. Recently, substantial stocks of blue whiting have been discovered in this zone off the British Isles. This area is now being exploited off the Pacific coast of North America for grenadiers (= Pacific roughy) and "rock cod" (*Sebastolobus*). The zone also contains substantial numbers of squids.

Finally, the entire Indian Ocean has a low level of fisheries development by current standards. It produces only about one-fifth the catch per unit area of the Atlantic and Pacific. It would appear that increased efforts in these waters would increase the catches without overexploiting the stocks. On the basis of primary production alone, Idyll (1970) estimated that this ocean can produce 10 million metric tons of fishes as opposed to a little over 5 million tons in 1987.

We can summarize this section by saying that, although the landings of marine fisheries have been increasing, it would appear from several aspects that many current world fisheries are approaching a maximum of exploitation, and some show signs of definite overexploitation due to unregulated access and overfishing. Furthermore, there appear to be only limited amounts of increases to be expected from fishing stocks and underexploited areas, with the possible exception of the Antarctic krill. The estimate of a total sustainable yield of 100 million metric tons or less for the world's oceans appears reasonable. With catches now reaching 84 million tons, it seems unlikely that significant increase in food can come from the sea. Therefore, the seas cannot and will not be the source of substantial amounts of food for an increasing human population.

MARICULTURE

As the limit is in sight for the harvesting of wild populations of marine animals and plants in fisheries, humans have begun work with culturing various marine and estuarine species in situations that are more analogous to our terrestrial-based agriculture, a field that is termed *aquaculture.* In aquaculture humans seek to rear selected aquatic plants and animals under controlled conditions to increase the amount of food available to humans with respect to that obtained through traditional fishing. We are concerned here only with marine aquaculture, or *mariculture.*

History and Extent

Despite evidence that culturing of marine organisms existed at least as far back as ancient Rome, aquaculture through the centuries has never become a significant contributor to human food. Food from the seas remained the province of the hunters and gatherers, the fishermen. Only in this century has aquaculture received wide recognition and undergone significant development. Partly, this development is in response to the pressure for more human food, especially protein, and partly from the declining wild stocks.

Mariculture encompasses marine organisms raised not only for human food, but also for other products, such as pearls, and terrestrial animal food.

Mariculture is practiced throughout the world in the tropical and temperate zones, but is more technologically advanced in the industrialized nations of the Northern Hemisphere. At present, it is confined to shallow coastal embayments or artificial ponds.

Species Cultivated

Few marine organisms are currently cultivated, and most are either the type that fetch a premium price on the market, which economically justifies the high cost of their culture, or else the type that produce large amounts of biomass under intensive culture. In the former category, we could include certain species of shrimps and abalone cultured in Japan. In the latter category are oysters and mussels. As might be anticipated, estuarine and bottom-dwelling species are more commonly cultivated than pelagic, open-ocean species.

Fishes that are successfully cultured commercially include several species of Pacific salmon (*Oncorhynchus*), Atlantic salmon (*Salmo*), mullets (*Mugil, Chanos*), yellowtail (*Seriola*), and flatfishes (*Pleuronectes, Solea*). Crustaceans cultured for market include various species of shrimps and prawns (*Penaeus, Metapenaeus, Leander*). Among the mollusks, oysters (*Crassostrea, Ostrea*) and mussels (*Mytilus*) are the subject of intensive commercial culturing (Fig. 11.7). Other mollusks raised include several clams (*Anadara, Tapes, Mactra, Mercenaria, Meretrix*) and the abalone (*Haliotis*). Compared to the total list of fishes, crustaceans, and mollusks taken for food in various fisheries, this is a very short list. Other species have been cultured experimentally, including lobsters, scallops, and crabs, but at present no commercial industry exists. This suggests that there is considerable room for future expansion.

Yields

Even though aquaculture has lagged far behind agriculture in development, its potential for production of human food even in its current low state is considerable. For example, Bardach (1968) has noted that the unmanaged common oyster grounds along the east coast of the United States yield only about 10 kg of oysters per hectare, while the intensive culture of oysters in hanging racks in Japan yields 58,000 kg/hectare! Even larger yields are possible with mussels. In Spain, an intensive culture on hanging racks produces 300,000 kg/hectare.

A similar situation occurs with fishes. In Indonesia, the milk fish (*Chanos chanos*) is cultured in brackish ponds, but is not subjected to such intensive care as in Taiwan. Yield in Indonesia is 400 kg/hectare, but 2000 kg/hectare in Taiwan.

Presently, the world yield from aquacultural practices is probably over 4 million tons, but most is from fresh water. Thus, mariculture

Figure 11.7 Oyster culture racks in a Pacific coast estuary. (Photo by the author.)

currently contributes little human food yield. However, the science of mariculture is as yet at a low level of development. If the above examples are any indication of what can be expected, it would appear that the potential future yield could be substantial; however, it is not great for reasons discussed below.

Problems and Restrictions to Mariculture Development

Mariculture is not a panacea for the problem of feeding the human race, nor is it likely to contribute significantly to food supplies in the immediate future. The reasons for this are numerous; space permits mentioning only the more significant. Marine animals require, first of all, that the water in which they live has the proper physical and chemical characteristics. This is difficult to maintain in ponds or tanks and requires complex filtering and water treatment to remove wastes and other toxic materials. Thus, attempts to raise marine organisms on a large scale mean an often economically crippling investment in equipment to maintain the water quality. Another drawback is that many marine species go through a series of larval stages, each requiring different conditions and food, before attaining adult size. To rear these forms successfully is often too costly or simply impossible in captivity. Still other species are pelagic in the open ocean and will not survive when they are confined in smaller spaces. Since we cannot control animals in the pelagic areas, we cannot culture there. Finally, there is the problem of various diseases and parasites, which tend to proliferate under captive or crowded conditions.

What this means, then, is that present and future mariculture appears most practical for those species with fewer or no larval stages, those that are nonpelagic, and those that live and reproduce well in crowded conditions. Furthermore, our inability to control conditions or organisms in the open ocean means that future mariculture and increased yields must come from sheltered inshore bays, lagoons, and estuaries. Since the areas suitable for such culturing are a very minor part of the total world oceans, even at maximum culturing intensity, we cannot expect really substantial quantities of food from mariculture in the near future.

POLLUTION

The seas have been considered as the ultimate appropriate dumping grounds for the wastes of human societies. In the same way that fishery stocks were once considered inexhaustible, so we have felt until recently that the immense volume of the world's oceans had an infinite capacity for absorbing all of our waste materials. Dilution, in other words, was the solution to pollution. In recent years, we have come to realize that however large, the oceans are not infinite in their capacity to absorb wastes and that certain of our wastes in very small amounts may have significant effects on communities and species. It is the purpose of this section to discuss briefly a few of the more significant sources of marine pollution and their effects or potential effects on marine ecosystems. For more extensive information on pollution, the student is referred to the bibliography at the end of this chapter.

Oil

Extensive media coverage over the last 15 years has certainly made the oil pollution problem in the seas widely known to the public. Oil pollution in the seas results primarily from the spillage of crude oil from offshore drilling platforms or from accidents involving tankers. What is less well known is that there are a few marine environments subject to natural oil contamination (Coal Oil Point in Santa Barbara, California, for example). One of the largest natural seeps was that reported by Harvey et al. (1979). They discovered a seep 1.5 km wide, extending for hundreds of kilometers off Trinidad. This seep was found at a depth of 100 m, had a thickness of 100 m, and contained in excess of 1 megaton of oil.

Crude oil released into the sea usually floats, although some components may sink, and after a time, evaporation of certain fractions may cause even more to sink. Away from land, the floating oil probably has little effect on the majority of planktonic and nektonic organisms. The exceptions are the marine birds and certain mammals such as sea otters. When birds contact the oil, they become coated, and their feathers lose their insulating qualities. As a result, most will die of exposure in the water or may be unable to feed, with the same result. Oil is therefore devastating to marine birds, causing significant losses.

Inshore, the oil may coat the shallow subtidal and intertidal and smother the communities. The devastation is nearly complete initially, but recovery usually occurs with time. More serious than the oil itself have been the various chemicals such as detergents used to break up or disperse the oil in the water. As Nelson-Smith (1973) notes, these were shown in the *Torrey Canyon* disaster in 1967 to have caused more mortality of marine organisms than the oil itself.

Oil disasters have become more numerous with the increased oil demand of the industrialized world, with the necessity of oil transport from distant sources, and with increased numbers of offshore drilling rigs.

The immensity of oil disasters involving the giant supertankers is revealed by the sinking of the *Amoco Cadiz* off the Brittany coast of France in March 1978. In this case, 220,000 tons of oil spilled and one-third reached shore to contaminate 320 km (192 miles) of shoreline. The spill contaminated or destroyed major commercial marine products, including oyster beds, mussel beds, and lobster holding pens. As of 1985, many of the areas impacted by the oil spill appear to the casual observer to be completely recovered from the effects of the spill. However, investigations such as that done by a joint United States–France commission (NOAA–CNEXO) have shown that some differences still exist between communities in the spill area and those that were present in the same area prior to the spill. This is particularly true for the bays and estuaries where studies such as those by Riaux-Gobin (1985) and Boucher (1985) suggest that the hydrocarbons have affected the long-term trends of community structure, particularly with respect to microalgae and nematodes.

The immensity of the area that can be impacted by a single oil spill from a modern large oil tanker was brought home to the world through the extensive media coverage of the disaster of the *Exxon Valdez*. Just past midnight on March 24, 1989, the tanker *Exxon Valdez* went aground in Prince William Sound, Alaska, spilling 260,000 barrels of Alaskan crude oil. This oil eventually was distributed by winds and currents over 30,000 km² of the coastal waters of Alaska (Fig. 11.8). According to Platt et al. (1990), at the time of the spill this area was occupied by about 1 million seabirds of which somewhere between 100,000 and 300,000 were killed—certainly a significant loss. Other organisms were undoubtedly impacted by this spill, including sea mammals and various commercial fisheries. However, at present (1991), it is virtually impossible to obtain any data on the impacts of this now well-studied spill because of a gag order placed on all researchers, according to Busch (1991), by the Alaska attorney general. When these data are finally released and published, we should have a much better understanding of the effects of oil spills on inshore marine communities as a result of the large number of studies that have been undertaken in the wake of this spill.

Until recently, we had experienced no such extensive spills from offshore wells, though they are potentially more devastating, since the amount of oil gushing out is not limited by the hold of a ship. Such a disaster happened in the Gulf of Mexico where, in 1979, the Ixtoc I well blew and spewed out from 450,000 to 1.4 million tons of oil in 295 days before being capped. The discrepancy in the amount spilled is due to the uncertainties of estimating the flow from the well. It was the world's greatest oil spill. It is also one of the least studied, so data on its effects are not available.

As if accidental oil spills were not enough of a problem, we were witness in 1991 to a massive deliberate oil spill, one in the northern Persian Gulf, that was the result of the Iraqi army releasing stored Kuwaiti crude oil from tanker-loading piers. The extent and damage of this spill remain unknown at present (1992) because scientific study was impossible in what was at the time of occurrence a war zone.

Only a few of these larger disasters per year could add up to significant destruction, particularly of shallow-water communities. We can expect more such disasters in future years with increased oil shipping and increased offshore drilling. A more optimistic view comes from a recent study. According to a comprehensive

Figure 11.8 Observed distribution of oil from the *Exxon Valdez* oil spill between March 24 and May 20, 1989. (From J. F. Platt et al., Immediate impact of the "Exxon Valdez" oil spill on marine birds, Auk 107:387–397, 1990.)

review by the National Academy of Sciences (1985) of the fate of oil in the sea, biological impacts vary considerably. Low-energy environments such as estuaries, salt marshes, and intertidal and subtidal sedimentary areas appear most vulnerable and also show the slowest rates of recovery. There appears to be little detectable long-term effect on pelagic communities, save for birds, and in all cases examined there appear to be mechanisms for oil degradation that should give the potential for eventual recovery.

Sewage and Garbage

The discharge of human sewage and garbage into coastal waters is practiced throughout the world. The sewage may or may not have had some treatment before discharge. Sewage adds a large volume of small particles to the water and also

large amounts of nutrients. In small volumes and with adequate diffusing pipes, it is difficult to detect any long-term effect on the communities of the open coast. In large volumes and in semienclosed embayments, the effect can be devastating. Two examples should suffice.

In southern California, the Los Angeles area discharges in excess of 330 million gallons of sewage per day at the White's Point outfall off the Palos Verdes Peninsula. Extensive studies in and around this outfall and others in the area over a period of years by personnel of the Southern California Coastal Water Research Project (Bascom, 1978) have revealed that sewage has caused significant degradation in benthic invertebrate communities in areas near the outfall when compared to similar areas some distance away. In addition, the kelp beds in the vicinity of the outfall have disappeared; sea urchins have markedly increased; and diseased fishes are more prevalent. About 4.6 percent or 168 km^2 of the 3640-km^2 southern California mainland shelf has been changed or degraded as a result of sewage discharge from four major outfalls (Fig. 11.9).

Kaneohe Bay on the Island of Oahu is a large, shallow embayment with restricted water circulation that was once known for its flourishing coral reefs. Since World War II, Banner (1974) reports that the area around the bay was subjected to a tenfold increase in population. As a direct result of this urbanization, the bay was subjected to massive domestic sewage discharges along with significant siltation from runoff during storms. The result was the total destruction of the coral reef communities in two-thirds of the bay and their replacement by a massive, smothering growth of the green alga *Dictyosphaeria carvernosa* (Fig. 11.10). By 1978, the sewage discharges were eliminated from the bay, and Maragos et al. (1985), in a resurvey of the area in 1983, have reported remarkable recovery

of corals and water clarity, suggesting an early recovery of the reefs in this bay.

In addition to sewage, large amounts of garbage are dumped into the sea each year from shore and from ships. What effect this material may have on the marine communities is as yet unknown. Evidence of the extent of the garbage pollution is found in the occurrence of small bits of plastic in plankton tows taken in many parts of the North Atlantic and Pacific oceans and in the digestive tracts of fishes, diving seabirds, and marine mammals. As yet, there is no evidence that these plastic bits cause mortality, though it has been suggested that they cause intestinal blockage. More ominous is that they do not decompose, and increasing use of plastics by modern human societies probably means increasing amounts in the oceans.

The whole spectrum of the offal of civilization characterizes the material dumped into the waters off New York City. It includes dredge spoils, sewage, chemicals, garbage, and various construction materials. The amounts of these

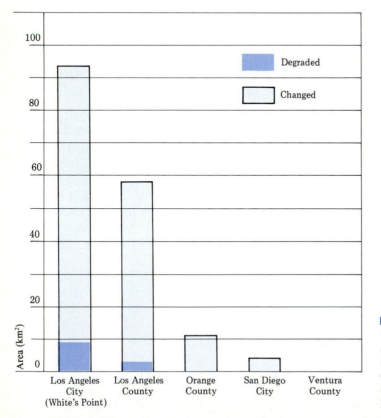

Figure 11.9 "Changed" or "degraded" areas near southern California's major sewer outfalls. (After W. Bascom [ed.], Coastal water research project annual report, Los Angeles, 1978.)

Figure 11.10 Kaneohe Bay, Hawaii, showing the extent of destruction of coral and overgrowth by the alga *Dictyosphaeria* at its maximum extent before cleanup was initiated. The darker blue portion indicates the area of destruction. (Modified from A. H. Banner, Kaneohe Bay, Hawaii: Urban pollution and a coral reef ecosystem, pp. 685–702. *In:* Proceedings of the second international coral reef symposium, Volume 2, Great Barrier Reef Committee, Brisbane, 1974. Used by permission of the Great Barrier Reef Committee.)

materials poured into these open coastal waters is of such volume as to be visible even from satellites, according to Gunnerson (1981). For sewage alone, 127 municipal discharges contribute 2.6×10^9 gallons per day, or more than 100 times that of the White's Point outfall in the Los Angeles area. The results of 35 years of dumping have been significant. Oxygen levels near zero have occurred over extensive bottom areas off New Jersey and have led, on numerous occasions, to mass fish and shellfish mortalities, as well as to chemical and bacteriological contamination of other fish and shellfish. Some fisheries have even had to be abandoned, according to Gunnerson (1981). Furthermore, even though

most of the waste is dumped many miles offshore, certain materials, such as plastic, tar, and grease balls, and even medical wastes such as hypodermic needles, have come ashore to contaminate bathing beaches throughout the area. Clearly, human activities have had and will continue to have a significant effect on the marine ecosystem.

Chemicals

More insidious than oil or sewage, which are at least visible, are the various invisible toxic chemicals produced by the industrialized nations that find their way into the oceans' ecosystems. These chemicals are often transferred through the food chains in the sea and exert their effects in animals and places removed in time and in space from the source. Because of this, it is difficult to pin down the effects of a given chemical, especially if the effects turn up years later. Certain marine organisms also enhance the toxic effects of many chemicals because they have the ability to accumulate the substances in their bodies far above that found in the surrounding water. Another factor that tends to increase the effects of chemicals on living systems is *biological magnification,* in which the chemical increases in concentration in the bodies of organisms with succeeding trophic level. This results from the fact that these chemicals are not metabolized in an organism, and therefore the amount accumulated in the tissues remains there. When several such individuals are consumed by a carnivore of the next trophic level, the carnivore, in turn, gains the chemicals from all the individuals, which, in turn, increase the total concentration in its body. Continuation of this process can lead to rather significant levels in the top carnivore, if the food chains are long. As we saw in Chapter 1, marine food chains are long, and so top carnivores in marine systems often have heavy loads. Humans also tend to consume marine organisms mainly from the higher trophic levels. This has already had an effect on certain of our fisheries such as swordfish and tuna, in which the levels of the element mercury have often been higher than safe levels established by the FDA for humans, causing fishes to be rejected for consumption. More tragic than that is the case of Minimata disease, documented by Goldberg (1974).

In the late 1930s, the Chisso Corporation of Japan established a factory on the shores of Minimata Bay to produce vinyl chloride and formaldehyde. By-products of this plant contained mercury and were discharged into the bay. Through biomagnification, the marine fishes and shellfish accumulated high concentrations of the toxic compound methyl-mercury chloride. The fishes and shellfish were, in turn, consumed by the inhabitants of the area. About 15 years after the dumping of mercury into the bay began, a strange, permanently disabling neurological disorder began to appear among the inhabitants, especially among the children. This disorder was called *Minimata disease.* The cause was diagnosed as due to mercury poisoning in 1959, but it was not until the early 1960s that the active mercury compound was identified and the link to the factory discharge established. It took even longer, until the 1970s, for Japan to halt the discharge of mercury into the ocean and for other nations to blacklist mercury dumping and establish standards for acceptable levels in food.

A second, well-documented example is the case of the pesticide DDT. Dichlorodiphenyltrichloroethane (DDT) was the first of the synthetic pesticides of a class known as *chlorinated hydrocarbons.* It was put into worldwide use in 1945 and was hailed as a boon to humankind for its effectiveness in destroying a wide range of insect pests while remaining relatively nontoxic to humans. It has, however, a few attributes that were ultimately to bring about serious environmental consequences. First, it is a remarkably stable compound in natural systems. It, or its first breakdown product, DDE, persists for years; just how long is unknown. It is relatively insoluble in water, but soluble in fats or lipids, and it adheres strongly to particles.

Although DDT is not used in the marine environment, it has entered the marine food webs through runoff from land, precipitation, and dumping. As a result, by the 1960s, DDT had been found in marine organisms as remote as Antarctica. The insidious effects of DDT in marine food webs began to be apparent in the 1960s, when certain top carnivores were found to have high levels in their tissues. DDT enters the marine food chains at the level of plankton, where the chemical absorbs on the surfaces of the plankton organisms or is dissolved in the lipid. Since it is not metabolized, biomagnification concentrates it through succeeding trophic levels to the top carnivore. Concentrations of DDT in certain marine fish such as mackerel were found to exceed the permissible 5 ppm established as safe for human consumption by the FDA, and many catches had to be destroyed.

The most serious effects, however, occurred among marine birds. In birds, once DDT reached certain high levels, it had a direct effect in interfering with the calcium deposition in the eggshell. As a result, thin eggs were laid, which broke open when the birds incubated them. The greatest amount of attention was focused on the brown pelican (*Pelicanus occidentalis*), in which successful breeding virtually disappeared in the main rookeries on the islands off southern California in the late 1960s and early 1970s. As Goldberg (1976) notes, this was traced directly to the high levels of DDT in their tissues. What was curious here was that most of southern California is urbanized and had not used much DDT, certainly not as much as that used in agricultural areas. Yet the waters around Los Angeles had concentrations of DDT averaging 370 ppm as opposed to 1 ppm off Baja California. Where did it come from? The source was discovered to be the White's Point sewer outfall. It turned out that a single chemical plant that produced most of the world's DDT was located in Los Angeles and was dumping its wastes into this outfall at the rate of 100 tons a year, a figure ten times as high as the total amount of DDT carried per year by the Mississippi River into the Gulf of Mexico. Thus, the decline of the brown pelican was linked through the marine food web to a toxic chemical discharged from a single source. Subsequent cessation of DDT dumping in Los Angeles has been followed by the slow recovery of the pelican.

These are only two examples of the problems of toxic chemicals. They emphasize the insidious nature of the often much-delayed effects and the great difficulty of tracing the source of a problem through complex food webs, about which we know relatively little. Since modern industrial states now produce thousands of actually or potentially toxic chemicals each year and many will enter the oceans, how many more of these disasters must we look forward to in the future?

Radioactive Wastes

Since 1944, the world's oceans have been receiving an input of radioactive wastes generated from production of nuclear weapons and electric power. Major sources of input have been fallout from nuclear weapons tests, releases or dumping of wastes from nuclear fuel cycle systems, mainly power or reprocessing plants, and certain accidents. In the latter category, one of the most significant contributors was the 1964 incident involving an aerospace nuclear generator that reentered the atmosphere when the launch mal-

functioned and deposited into the oceans a quantity of plutonium 238 equal to half the total oceanic deposits of plutonium 238.

There are two main categories of radioactive wastes. *Low-level wastes* contain less than 1×10^{-8} curies of radioactivity per gram. *High-level wastes* contain more than 1×10^{-8} curies per gram.

Presently, the dumping of high-level radioactive wastes in the oceans is prohibited by international agreement, but dumping of low-level wastes is permitted. Low-level ocean dumpsites have been designated and used at one site in the northeast Atlantic, off Europe at three sites, off the United States East Coast at one site, and at one off the Pacific coast. From 1946 to 1970, the United States dumped approximately 107,000 containers of low-level radioactive wastes at the three sites off our coasts, according to Park et al. (1983). Currently the only active site is off the East Coast. Whereas low-level radioactive waste seems manageable, there is considerable concern about high-level wastes such as those that are generated from discarded fuel assemblies from nuclear generators or arise from various reprocessing operations. These wastes usually have very long half-lives and powerful ionizing radiation. Such wastes must be safely isolated from the biosphere for periods of time that are geological in scale.

If a radioactive waste can be contained for 10 times the half-life of the radionuclide, the radioactivity will be reduced by 1024 times. Plutonium 239, a common radionuclide of spent nuclear fuel, has a half-life of 24,100 years. This means that it must be isolated from the biosphere for 241,000 years.

Currently, there is considerable debate over what to do about disposal of these high-level nuclear wastes. This debate and its resolution have been heightened by the increased proliferation of nuclear power plants and weapons, generating more wastes, and by the fact that the United States also needs to decommission and safely dispose of three or four nuclear submarines per year for the next 25 or more years. This problem may become more acute with the demise of the Cold War and the decomissioning of more weapon systems.

Since the high-level radioactive wastes remain very toxic to humans for many thousands of years, it is necessary to find a disposal site that will be stable for an equivalent time period, and that preferably also will be beyond the reach of human tampering for the same length of time. Given the ingenuity of humans and our present technological prowess, there are few places on the planet that will meet these requirements. One such place is the deep ocean, well away from the edges of the tectonic plates. The abyssal plains of the deep ocean lie under 4 to 5 km of water. This is beyond any regular current human activity in the ocean. Furthermore, the bottom is covered with a thick layer of sediment that has accumulated over a long time and has remained stable. If the radioactive waste could be emplaced in this sediment, there is a good chance that it would remain safely away from the realm of human activities for the required time. Even if wastes should leak from the containers after some years, the nuclides would adhere to the sediment and probably remain in place (Fig. 11.11).

Because of the foregoing advantages, the deep-sea bed has been suggested as the one area

of the planet where we could safely isolate these toxic wastes for thousands of years. However, in order to evaluate the deep sea for feasibility of high-level waste disposal, we need to consider the probable results of leakage or failure of the containment vessels. Therefore, it is necessary to understand the fate of escaped radionuclides once they are out of the containment vessel. This requires a considerable knowledge of both the physical and biological mechanisms that will influence radionuclide movement. Since, as we saw in Chapter 4, the deep sea is the least understood of all the ocean areas of the planet, there is obviously a great deal of information that we need. As we grapple with the problem of safe disposal of radioactive wastes, we should encourage the research necessary to answer the important questions lest we again fail to make an intelligent choice and resort to expediency.

Miscellaneous Pollution Problems

In order to provide docking facilities for commercial and pleasure vessels, humans have en-

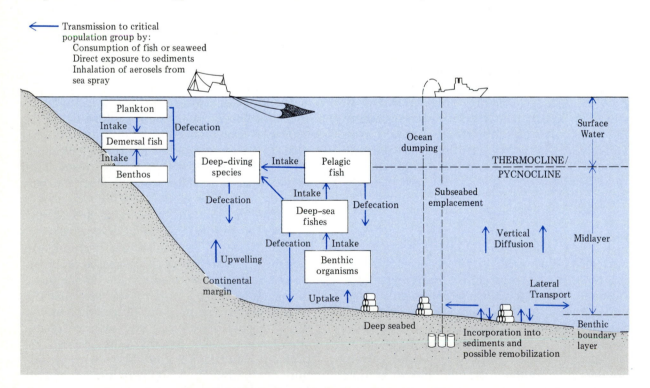

Figure 11.11 Physicochemical and biological transmission of waste radionuclides from deep-sea disposal sites. (Modified from P. K. Park et al., Wastes in the ocean, Vol. 3, Radioactive wastes and the ocean, Wiley, 1983.)

gaged in a great deal of dredging of estuaries and bays. Such activities destroy large areas of productive habitat for marine organisms and may have effects beyond the immediate area if nursery grounds for commercial species caught offshore are destroyed. Not only are communities destroyed by physically being removed, but the increased load of silt suspended in the water reduces light, and hence photosynthesis, and clogs the feeding and respiratory surfaces of many invertebrates, leading to their deaths. Such activity has in the past destroyed many productive areas worldwide, and the expanding human population suggests that we can look forward to even more destruction in the future.

The siting of plants along the seacoasts to take advantage of the marine waters for cooling purposes has led to the perturbation of marine communities by heated water discharge. This *thermal pollution* has been little investigated for the marine environment; indeed, thus far, the effects would appear to be restricted to the communities immediately adjacent to the discharge. The deleterious effects of thermal discharges are most pronounced in the tropics, where the organisms are living naturally at water temperatures that are close to their thermal maximum. It takes little increase in temperature to stress such communities. An example of the effects of a thermal discharge on tropical communities is described by Banus and Kolehmainen (1976) in Guayanilla on the south coast of Puerto Rico, where a power-generating station discharges heated water into a lagoon. The discharged water is 8–10°C higher than the ambient seawater and produces water temperatures in the lagoon of 39°C in the summer. This is high enough to affect reproduction of the dominant mangrove trees in the bay, and no young rooted and growing mangrove seedlings have been found in the lagoon.

Eutrophication refers to the release of excess amounts of nutrients into coastal waters. These nutrients have their origin in the fertilizers that are spread on agricultural land and that make their way to the sea via rivers and streams. The results of such increases are the red tides, various yellow and green foams, slimes and slicks— all of which indicate excess phytoplankton growth. Although on a small scale certain of these blooms are natural, they have become more common and extensive in recent years. For example, in 1988 a yellow brown slick off the Scandanavian countries threatened the entire Norwegian salmon mariculture industry and led to the most massive peacetime maritime rescue operation in the history of Norway, according to Cherfas (1990). In another incident, in 1978, the German North Sea coast beaches were inundated by a meter-deep, foul-smelling foam created by surf beating on a massive outbreak of the gelatinous encapsulated alga of the genus *Phyaeocystis.* Aside from aesthetic considerations, many of these massive outbreaks are also increasingly toxic. This further contributes to the disaster.

How do we know that these blooms are not natural events and their apparent increases are not just the result of better observation and quick communication common to modern times? Studies in Japan and Europe, areas with good historical records, show that such phenomena have increased markedly in the past 20 years. The good side to this is that the increased numbers of algal cells may absorb more carbon dioxide and thus retard the greenhouse effect (see subsequent section). However, at present, no one knows whether this is true or not. Another unanswered problem concerns dimethyl sulfide (DMS), a natural compound of phytoplankton. When the phytoplankton die, DMS is released into the atmosphere where it may be changed to sulfuric acid and contribute to acid rain.

It may seem strange to suggest that dams built on freshwater rivers and streams can affect the oceans, but by preventing the flow of fresh water into the oceans, these structures have had effects in certain areas. Probably the best example is the Aswan Dam on the Nile River in Egypt. Prior to its construction in 1965, the Nile carried into the Mediterranean Sea 43 billion cubic meters of water. It now carries only 3 billion. The result of this dam has been the erosion of the Nile Delta and the loss of the fisheries of the eastern Mediterranean.

A final consideration concerns pollution by the introduction of *alien animals.* The devastating consequences on native fauna and flora through the introduction of alien animals and plants are well known and documented for the terrestrial environment. We know, for example, of the loss of the American chestnut through a disease imported from Europe, and more recently, the demise of our native elm trees through Dutch elm disease. Equally dramatic have been the devastations of European rabbits on the flora and fauna of Australia. What about introductions in marine waters?

Our knowledge of the effects of introducing a marine species is limited. The major intentional introductions of marine species have been those of commercial or sports fishing value. Thus, the eastern (*Crassostrea virginica*) and Japanese (*Crassostrea gigas*) oysters were introduced into the bays and estuaries of the Pacific coast of North America to provide the basis of an oyster industry. Similarly, the striped bass (*Roccus saxatilis*) was introduced to the Pacific coast from the Atlantic and has become the basis of a valuable sports fishery.

Along with these purposeful introductions, however, have come other, unintentional species. Other inadvertent introductions have occurred from the fouling organisms on ships and in ballast and through commercial bait and seafood importations. The number of these species and their cumulative effects on indigenous communities are very little known, but potentially great. That they may have profound effects is witnessed by recent studies of introduced species on the central California coast. Carlton (1975, 1989) has shown that between 150 and 200 marine invertebrates from various parts of the world have become established in the bays and estuaries of the Pacific coast. In the case of San Francisco Bay, the majority of the invertebrate fauna has been introduced, and it is now nearly impossible to know what the indigenous communities of this great bay were like.

How serious these introductions are as a threat to marine communities around the world remains to be seen, but if the results we have observed in terrestrial communities are any guide, the threat could be considerable. We have an example in the marine environment in the case of the carnivorous snail *Rapana,* which was introduced into the Black Sea from Japan. The result was the devastation of the shellfish beds in the Black Sea. The potential for such introductions in the future is great. If, for example, a sea-level canal is put through in Central America, allowing free interchange of the tropical Atlantic and Pacific faunas, massive changes may occur. The giant offshore oil-drilling rigs are another source. They accumulate a whole marine fauna in one area during drilling. Then, when they are moved, often to an entirely different area, the whole fauna is brought along, transferred intact.

DRUGS FROM THE SEA

Many marine organisms produce chemicals that are not found, or are rare, in terrestrial organisms. These chemicals are used mainly by the marine organisms to defend against predators. Certain of these compounds have been found to have biomedical properties that may potentially be useful in combating human disease. Organisms especially productive of such biomedical compounds include sponges, cnidarians, tunicates, and opisthobranch mollusks. Among the useful chemicals are a shark-repelling saponin, ichthyotoxin, from the fish *Pardachirus marmoratus;* antitumor chemicals from the sponges *Haliclona* and *Halichondria;* and a potent nonprotein toxin, palytoxin, from the soft coral *Palythoa.* More than 1700 compounds with some biomedical property have been reported in the last eight years, and more are certain to come with additional testing of organisms. The main stumbling block thus far to commercial pharmaceutical development has been the inability to raise most of the marine organisms in large numbers in captivity.

GLOBAL WARMING AND SEA LEVEL CHANGE

Certainly, one of the most talked about subjects in science today, not only among scientists but among the public, politicians, and governments, is that of *global warming* or the *greenhouse effect* and how it will affect the climate of the planet. Global warming refers to the increase in the temperature of the planet primarily as a result of the increase in the atmosphere of gases such as carbon dioxide, methane, nitrous oxide, and chlorofluorocarbons. These gases have the ability to "trap" the radiant energy coming from the sun, thus leading to increases in atmospheric temperatures. Increased temperatures, in turn, may cause changes in the climate, alter rainfall and storm patterns, and increase the rate of sea-level rise.

There seems little doubt that the "greenhouse" gases have increased over the last 100 years, and much of this increase is due to the activities of humans in burning fossil fuels and wood. Certainly, the increase in chlorofluorocarbons is directly attributable to humans since this gas is a human invention and used in refrigeration systems. The other gases are derived from natural sources. Exacerbating the problem has been the wholesale cutting of the world's forests, which would otherwise serve as a major sink to absorb excess carbon dioxide. The other major sink for carbon dioxide is the world's oceans (see the carbon cycle in Chapter 1).

It is both the increase in sea-level rise and the potential of the ocean to absorb greenhouse gases that directly concern us here. As global temperatures rise, the oceans of the earth will expand because of the melting of glaciers and ice caps, the most significant of which are those of Greenland and Antarctica, and because of the thermal expansion of the ocean water. Depending on the amount of warming, sea levels could rise as much as 5–7 ft over the next century. This is opposed to the rise of 0.5 ft in the last 100 years. If the "worst case" rise were to occur, there would be massive inundations of low-lying coastal areas, such as Louisiana, Cape Cod, Florida, Egypt, and Bangladesh, and many low-lying island archipelagos, such as the Maldives and Marshall islands, might be completely eliminated.

If these unprecedented disasters could occur, why then is there such controversy concerning global warming? The answer stems from our own lack of knowledge. Whereas we can chart with relative accuracy the trends in atmospheric gases, sea-level rise (or fall), and the advance or retreat of glaciers over the last 100 to 200 years, this time frame is geologically insignificant and subject to so much variation that it cannot serve as a basis to predict really long-term changes. Besides, we know that the planet has undergone various sea-level and climatic changes in the past, but that these have usually proceeded very slowly, over thousands or millions of years, not within a hundred or so years, as now seems to be the case. Thus, the real question, and the source of the controversy, is whether or not the changes we are now observing are simply natural variations in an otherwise stable climate since the end of the last ice age, or whether they are the forewarnings of a long-term change. In spite of our current lack of the necessary long-term data, there now seems to be a gathering consensus among scientists, according to Kerr (1990), that global warming is a reality.

Given the fact that greenhouse gases are continuing to increase, and the known fact that if this trend continues there will be considerable sea-level rise, is there anything that humans can do to forestall this? The answer is yes, but it would require massive action on the part of all nations on the planet. Forests must be prevented from being cut (indeed, more need to be planted to absorb the carbon dioxide), chlorofluorocarbons must be taken out of use, and fossil fuel usage must be reduced. There is, finally, the intriguing suggestion by Martin et al. (1990) that much of the excess carbon dioxide could be removed from the atmosphere to the ocean sink if a few hundred thousand tons of iron could be made available to the phytoplankton of the Antarctic seas. If Martin is correct, these phytoplankton are limited in growth and photosynthesis by the lack of trace amounts of iron in the water. If iron were added, the increase in photosynthetic activity would be great enough to draw down the carbon dioxide from the atmosphere. This hypothesis, however, remains to be tested.

CONCLUDING REMARKS

For too long, the seas of the world have been considered an inexhaustible source of food, as having an infinite capacity to absorb and purify our wastes, and as a source of all the raw materials needed to maintain an industrial society. It is now apparent that none of these assumptions is true and that the human population at the current level of technological development has the ability to inflict massive destruction on the seas, just as we have done on land. At present, the seas remain in good condition relative to the land, and we cannot afford to permit them to be degraded in the same way if we wish to continue our tenure as a species on this planet. We must employ ecological principles, such as those outlined in this text, in order to ensure that the potential of the oceans is realized—without degradation or reliving yet another tragedy of the commons.

BIBLIOGRAPHY

Banner, A. H. 1974. Kaneohe Bay, Hawaii: Urban pollution and a coral reef ecosystem, pp. 685–702. *In:* Proceedings of the Second International Coral Reef Symposium, Vol. 2. Great Barrier Reef Committee, Brisbane.

Banus, M. D., and S. E. Kolehmainen. 1976. Rooting and growth of red mangrove seedlings from thermally stressed trees, pp. 46–53. *In:* Esch, G. W., and R. W. McFarlen (eds.). Thermal ecology, II. Washington Technical Information Center of the Energy Research and Development Administration.

Bardach, J. E. 1968. Aquaculture. Science 161:1098–1106.

Bardach, J. E., J. H. Ryther, and W. O. McLarney. 1972. Aquaculture. Wiley, N.Y. 868 pp.

Bascom, W. (ed.). 1978. Coastal water research project annual report. Los Angeles. 253 pp.

Boucher, G. 1985. Long term monitoring of meiofauna densities after the *Amoco Cadiz* oil spill. Mar. Pollution Bull. 16:328–333.

Busch, L. 1991. Science under wraps in Prince William Sound. Science 252:772–773.

Carlton, J. 1975. Introduced intertidal invertebrates, pp. 17–25. *In:* Smith, R. I., and J. Carlton (eds.). Light's manual: Intertidal invertebrates of the central California coast, 3rd ed. University of California Press, Berkeley.

Carlton, J. 1989. Man's role in changing the face of the ocean: Biological invasions and implications for conservation of near-shore environments. Cons. Biol. 3(3):265–273.

Cherfas, J. 1990. The fringe of the ocean—under siege from the land. Science 248:163–165.

Culley, M. 1971. The pilchard, biology and exploitation. III, The California sardine. Pergamon, N.Y. pp. 143–176.

Cushing, D. H. 1969. Upwelling and fish production. FAO Fish. Tech. Paper 84:1–40.

Cushing, D. H. 1975. Fisheries resources of the sea and their management. Oxford University Press, Oxford. 87 pp.

Cushing, D. H. 1977. Science and the fisheries. Studies in biology no. 85. Edward Arnold. 60 pp.

FAO. 1978. 1977 Yearbook of fishery statistics, Vol. 44. Food and Agriculture Organization of the United Nations, Rome.

FAO. 1984. 1983 Yearbook of fishery statistics, Vol. 56. Food and Agriculture Organization of the United Nations, Rome.

FAO. 1989. 1987 Yearbook of fisheries statistics, Vol. 64. Food and Agriculture Organization of the United Nations, Rome.

Fautin, D. (ed.). 1987. Biomedical importance of marine organisms. Memoirs of the California Academy of Sciences no. 13. 159 pp.

Goldberg, E. D. 1974. Marine pollution: Action and reaction times. Oceanus 18(1):6–18.

Goldberg, E. D. 1976. The health of the oceans. UNESCO Press, Paris. 172 pp.

Gunnerson, C. 1981. The New York Bight ecosystem, pp. 313–378. *In:* Geyer, R. A. (ed.). Marine environmental pollution. 2, Dumping and mining. Elsevier, N.Y.

Hardin, G. 1968. The tragedy of the commons. Science 162(3859):1243–1248.

Harvey, G. R., A. G. Requejo, P. A. McGillivary, and J. M. Tokar. 1979. Observations of a subsurface oil-rich layer in the open ocean. Science 205:999–1001.

Hess, W. 1978. The AMOCO CADIZ oil spill. A preliminary scientific report. NOAA/EPA. 354 pp.

Idyll, C. P. 1970. The sea against hunger. Crowell, N.Y. 221 pp.

Kerr, R. A. 1990. New greenhouse report puts down dissenters. Science 249:481–482.

Maragos, J., C. Evans, and P. Holthus. 1985. Reef corals in Kaneohe Bay six years before and six years after termination of sewage discharges. Proc. 5th Int. Coral Reef Symp. 2:236.

Marine pollution. 1974. A series of articles in *Oceanus* 18(1).

Marr, J. C. 1960. The causes of major variations in the catch of the Pacific sardine, *Sardinops caerulea.* Spec. Sci. Rept., U.S. Fish & Wildlife Ser. 208:108–125.

Martin, J. H., S. E. Fitzwater, and R. M. Gordon. 1990. Iron deficiency limits phytoplankton growth in Antarctic waters. Global Biogeochem. Cycl. 4:5–12.

McVay, S. 1966. The last of the great whales. Sci. Amer. 215:13–21.

Nelson-Smith, A. 1973. Oil pollution and marine ecology. Plenum, N.Y. 260 pp.

National Academy of Sciences. 1985. Oil in the sea, inputs, fates and effects. National Academy Press, Washington, D.C. 601 pp.

Park, P. K., D. R. Kester, I. W. Duedall, and B. Ketchum. 1983. Wastes in the ocean. Vol. 3, Radioactive wastes and the ocean. Wiley, N.Y. 522 pp.

Platt, J. F., C. J. Lensink, W. Butler, M. Kendziorek, and D. R. Nysewander. 1990. Immediate impact of the "Exxon Valdez" oil spill on marine birds. Auk 107:387–397.

Riaux-Gobin, C. 1985. Long-term change in microphytobenthos in a Brittany estuary after the "Amoco Cadiz" oil spill. Mar. Ecol. Progr. Ser. 24:51–56.

Rounsefell, B. A. 1975. Ecology, utilization and management of marine fisheries. Mosby, St. Louis. 516 pp.

Russell-Hunter, A. D. 1970. Aquatic productivity. Macmillan, N.Y. 306 pp.

Ryther, J. H. 1969. Photosynthesis and fish production in the sea. Science 166:72–76.

Schaefer, M. B. 1965. The potential harvest of the sea. Trans. Amer. Fish. Soc. 94:123–128.

Scheuer, P. J. 1990. Some marine ecological phenomena: Chemical basis and biochemical potential. Science 248:173–177.

Soutar, A., and J. D. Isaacs. 1969. History of fish populations inferred from fish scales in anaerobic sediments off California. Calif. Mar. Res. Comm., CalCOFI Rept. 13:63–70.

Glossary

abyssal The bottom zone of the oceans at depths between 4000 and 6000 m.

abyssal gigantism Phenomenon observed among several crustacean groups in which general size increases with increasing depth.

abyssal plain That area of the deep ocean floor lying between 4000 and 6000 m.

acidity A measure of the concentration of hydrogen ions in a solution.

adenosine triphosphate The energy storage molecule of most living systems.

advection Movement of seawater horizontally or vertically, as in a current.

aerenchyma Tissue of thin-walled cells and spaces found in stems and roots of certain marsh plants serving to transfer oxygen.

aerobic Condition in which oxygen is present.

ahermatypic coral A non-reef-producing coral without the symbiotic zooxanthellae in the tissues.

algal ridge Coral-free ridge of encrusting coralline algae lying immediately behind the buttress zone.

alien animal Animal that is not indigenous to the area.

alkalinity A measure of the concentration of hydroxyl ions in a solution.

allochthonous Of foreign origin; transported into the area from outside.

alternate stable states The occurrence in a single habitat of several different persistent communities in which the structure of each is the result of historical events.

anaerobic Condition where oxygen is absent.

anchor ice Ice that forms around any convenient nucleus in the area below the permanent pack ice in Antarctica; it tends to carry organisms out of the area to be incorporated in the sea ice above.

anchoveta Common name for *Engraulis ringens,* the Peruvian anchovy.

anoxic Without oxygen.

anticyclonic Moving in a clockwise direction.

aphotic Without light; that area of the oceans without light.

apogee That point during the orbit of the moon around the earth when the moon is farthest from the earth.

aquaculture The culture of aquatic organisms.

aquatic Living or existing within or on water.

asexual In reproduction, without involving sex.

atoll A type of coral reef that is a modified ring-shaped reef arising out of deep water far from continental landmasses and enclosing a shallow lagoon.

ATP *See* adenosine triphosphate.

autochthonous Formed or occurring in the place where it is found.

autotrophic Living organisms capable of producing their own energy resources.

auxospore In diatoms, a reproductive cell that reestablishes the initial size of the species.

bacterioplankton Bacteria that live in the plankton, usually of a size range of 0.2 to 2.0 μm (picoplankton).

baleen Horny material growing in comblike, fringed units from the upper jaws of whales of the order Mysticeti.

barrier reef Reef adjacent to landmasses and separated from them by a lagoon or channel of variable extent.

bathyscape A free-moving, deep submersible designed to carry human observers into the deep sea.

bay A partially enclosed inlet of the ocean.

"bends" Name of disease occurring in humans who live breathing compressed air; results from a rapid decrease in pressure, which causes nitrogen bubbles to form in the internal tissues and blood vessels.

"bet hedging" A theory of life-history evolution in which the life-history traits displayed by the organisms result from the fluctuations in juvenile survivorship probabilities relative to those of the adult; opposite of "r-K" selection theory.

binocular vision Type of vision providing depth-of-field focus due to overlap of the field of vision of two closely set eyes.

biochemistry A scientific discipline concerned with the study of the chemistry of living organisms and their products.

bioerosion The dissolution or breakdown of substrates, usually calcium carbonate, by a variety of living organisms.

biogeochemical cycle Cyclical movement of an element or compound through living organisms and the nonliving environment.

biological interaction In ecology, a general term in which organisms have a mutual or reciprocal action or influence, including predation, competition, and grazing.

biological magnification Increase in concentration of chemicals in the bodies of animals with increasing trophic level.

bioluminescence The production of light by living organisms.

bioturbator Term given to an organism that is capable of modifying the local environment to enhance or exclude other organisms.

blade In a kelp plant, the flattened part of the plant that terminates the stipe.

blubber The layer of lipid that serves as an insulating layer under the skin of whales and other marine mammals.

blue-green algae Group of prokaryotic photosynthetic organisms of the kingdom Monera.

boiling point Temperature at which a liquid changes to a gas.

bradycardia Slowing of the heart rate.

buffer A chemical solution that resists or dampens changes in pH with the addition of acids or bases.

carbohydrate Group of biochemical compounds composed of carbon, hydrogen, and oxygen, often in the ratio of 1 carbon to 2 hydrogen and 1 oxygen.

Carnivora An order of mammals adapted to feed on other animals.

carnivore An animal that consumes other animals as food.

caudal The tail or posterior end of an organism.

caudal peduncle In fishes, that part of the body immediately in front of the tail.

cellulose A long-chain carbohydrate composed of repeated units of the sugar glucose.

centrifugal force Force tending to pull objects out from the center of rotation.

Cetacea The order of mammals that contains the whales and porpoises.

chemosynthetic bacteria Those bacteria able to obtain energy and therefore synthesize organic material through oxidation of reduced sulfur compounds.

chitin Biochemically, a polymer of the carbohydrate glucosamine that forms the hard outer integument of crustaceans and other marine invertebrates.

chlorophyll maximum Depth in the ocean where the concentration of chlorophyll reaches its highest level on a per-volume basis.

ciguatera Disease of humans caused by consumption of tropical fishes with toxins in the flesh and organs.

cleaning behavior Special category of symbiosis in which large animals, usually fishes, permit themselves to be cleaned of various parasites by smaller fishes or invertebrates.

climax In ecology, the final stage of a successional sequence that is able to persist in the absence of environmental change.

clone A group of genetically identical individuals of a plant or animal species produced by asexual reproduction from a single sexually produced individual.

coastal plain estuary Estuary formed by flooding of a low-lying coastal valley due to rising sea level.

coccolithophore Small, unicellular, flagellated algae with an external covering of small pieces of calcium carbonate.

cohesion Mutual attraction of similar molecules, which resists external forces that would break them.

commensalism A symbiotic relationship between two different species in which one benefits and the other is neither benefited nor harmed.

compensation depth Depth at which the processes of photosynthesis and respiration are equal.

compensation intensity In the water column, that point at which light intensity is equal to 1 percent of surface intensity.

competition The interaction among organisms for a necessary resource that is in short supply.

competitive exclusion The ecological principal that states that complete competitors cannot coexist.

competitive interference Exclusion of one species by another species through interruption of its normal activities.

continental shelf The shallow underwater extension of a continent; usually limited in depth to 200 m.

continental slope The steeply descending bottom between the edge of the continental shelf and the abyssal plain; the ocean bottom between the depths of 200 and 4000 m.

copepod Small crustacean of the order Copepoda; the dominant planktonic herbivore.

copepodid The larval stages of a copepod, which follow the nauplius and metanauplius stages and precede the adult.

coral bleaching Expelling of zooxanthellae from the tissues of reef-building corals.

coral reef Massive limestone structure built up through the constructional cementing and depositional activities of anthozoans of the order Madreporaria and certain other invertebrate and algal species.

corallite The cup into which a coral polyp fits.

corallivore An animal that consumes corals.

Coriolis effect The deflection imparted to moving water masses due to spinning of the earth on its axis; the deflection is to the right in the Northern Hemisphere and to the left in the Southern Hemisphere.

cosmopolitan In biogeography, an organism that is distributed throughout the world in suitable habitats.

crinotoxin A poisonous material secreted onto the surface of an organism.

critical depth The depth at which photosynthesis for the water column is equal to respiration for the water column; the depth to which phytoplankton cells may be mixed and still spend sufficient time above the compensation depth so that respiration and photosynthesis are equal.

critical tide level Points on the shore where a small change in vertical distance results in a disproportionate change in exposure time.

cropper In the disturbance theory of Dayton and Hessler, the term applies to deep-sea animals that consume living and dead animals smaller than themselves.

cryptic coloration Color hue or pattern that mimics the background.

cryptic symbiosis Symbiotic association between an algal cell and a nonphotosynthetic organism in which the algal species' cellular integrity is so reduced that it appears as an integral part of the host.

cryptofauna General term referring to the fauna living on or in coral substrates.

current Water movements that result in the horizontal transport of water masses.

cyanellae Symbiotic blue-green algae.

Cyanophyceae A class of blue-green algae.

cyclonic Moving in a counterclockwise direction.

decomposer An organism that breaks down dead protoplasm, freeing simple chemical substances for use by other organisms.

deep scattering layer Concentration of midwater organisms that reflect sound waves produced by sonar devices.

deep sea A general term referring to that area of the ocean beyond the continental shelf and below the lighted zone.

demersal Living close to the bottom of the sea.

density The mass per unit volume of a substance (physics); the number of individuals per unit area (biology).

deposit feeder An animal that feeds by consuming particles on or in the substrate.

desiccation The process of losing water.

diatom Microscopic autotrophic organism of the algae class Bacillariophyceae characterized by being enclosed in a two-part siliceous capsule.

diel Daily or once every 24 hours.

dinoflagellate Microscopic organism of the order Dinoflagellata possessing two locomotory flagellae.

disphotic The area of the ocean between the euphotic and aphotic zones; it has insufficient light for photosynthesis but sufficient light for detection and response by animals.

dissipative beach A flat beach, maximally eroded, with fine sediments and with wave energy consumed in a surf zone away from the beach proper.

disturbance theory A hypothesis for explaining the high diversity in the deep sea that suggests that extreme predation pressure prevents any species from becoming numerous enough to eliminate others by competition, therefore perpetuating a large number of species and high diversity.

diurnal Occurring daily or relating to daytime.

diurnal tide Tide with a single high and low each day.

divergence Surface horizontal flow of water away from a central area, permitting water to upwell.

diversity *See* species diversity.

DOM Abbreviation for dissolved organic matter.

drag Resistance to movement through a medium.

dysphotic *See* disphotic.

echolocation *See* sonar.

ecotone An area of intergradation between two biological communities or associations.

ecothermic Cold-blooded; unable to regulate body temperature.

eddy A circular movement of water.

electrical conductivity Relative ability of a material to allow the passage of electricity.

electron A negatively charged elementary particle in all matter.

element A substance composed of atoms having the same atomic number.

elutriation Means of separating interstitial fauna from the sand through constant stirring with water and subsequent filtering on a fine mesh.

endobenthic Meiofaunal-sized organisms that move within the sediments displacing particles.

endothermic Warm-blooded; able to regulate internal body temperature.

endozoite An animal living symbiotically inside another animal.

energy The capacity to do work.

epibenthic Organisms living at the sediment-water interface.

epicercal tail A caudal fin of fishes with the dorsal lobe larger than the ventral lobe.

epifauna Benthic organisms that live on or move over the substrate surface.

epiphytic Plants that are not parasitic, but that are attached to other plants.

epizoite An animal that is not parasitic, but that lives attached to another animal.

equilibrium species A species that has a life history characterized by long life, long development time to reach maturity, low death rates, and few reproductive periods per year.

estuary A partially enclosed coastal embayment where fresh water and seawater meet and mix.

euphotic *See* photic.

euryhaline Able to tolerate wide fluctuations in salinity.

eutrophic Containing abundant nutrient material.

evaporite estuary *See* negative estuary.

exploitative competition Competition among different species or members of the same species for a necessary resource that is in short supply.

extinction coefficient Ratio between the intensity of light at a given depth and the intensity at the surface.

femtoplankton A size class of plankton ranging from 0.02 to 0.2 μ m.

fin A flattened appendage of an aquatic animal used in locomotion or maneuvering in the water.

fishery The place or act of taking fishes or other sea organisms for human use.

fjord Estuaries occurring in deep, drowned valleys, originally cut by glacial action; characterized by the presence of a shallow sill at the mouth, restricting water interchange between ocean water and deeper fjord waters.

flipper The pectoral or pelvic appendage of a marine mammal or reptile.

flushing time Time interval required for a given mass of fresh water to be discharged from the estuary.

form resistance Condition in which drag is proportional to the cross-sectional area of the object in contact with the water.

frictional resistance Condition in which drag is proportional to the amount of surface area in contact with the water.

fringing reef A coral reef that forms immediately adjacent to a landmass.

fugitive species *See* opportunistic species.

fundamental niche The sum total of all the ranges of all the biological and physical factors within which a species can exist; a multidimensional or *n*-dimensional hyperspace.

gametophyte In algae, the haploid plant that produces gametes.

gas bladder Structure on dorsal side of body cavity of bony fishes that contains gas and is used by the fishes to regulate buoyancy.

geomorphology Discipline dealing with the form and configuration of the surface of the earth.

global warming Increase in average global temperature due to build-up of gases such as CO_2, NO_2, and chlorofluorocarbons in the atmosphere.

gorgonians Anthozoans of the subclass Octocorallia that have spiculate internal skeletons; commonly called sea fans or sea whips.

grab A collecting device that picks up a volume of the bottom; usually deployed from a vessel on a cable.

gravity The attraction of terrestrial bodies toward the center of the earth (= the earth's gravitational attraction).

grazer An animal that feeds on plants or other sessile animals.

gross photosynthesis The total amount of photosynthesis before subtracting losses due to respiration.

guild Species that make their living in the same way.

guyot A submerged, isolated, flat-topped mountain.

gyre Circular motion of water in the major ocean basins.

hadal Parts of the ocean below 6000 m.

halophyte A plant that is adapted to grow in soils with high salt concentrations.

harem In mammals, a group of females controlled by a single male during the breeding season.

heat Energy that moves from a higher to a lower temperature system.

heat capacity Amount of energy required to raise 1 g of material 1°C.

heat of vaporization Amount of energy required to evaporate a unit mass of a substance at constant temperature and pressure.

herbivore An animal that consumes plants.

hermatypic coral A reef-building coral with symbiotic zooxanthellae in the tissues.

hetertrophic Organisms that require an external source of food.

high-level radioactive wastes Those wastes generating greater than 1×10^8 curies of ionizing radiation.

holdfast In kelp, that part of the plant that fixes it to the substrate.

holoepipelagic Nektonic animals that spend their entire lives in the open ocean.

holoplankton Plankton organisms that spend their entire lives in the plankton.

homogeneous estuary An estuary with complete mixing such that the salinity is similar from surface to bottom.

homoiothermic *See* endothermic.

host The major provisioning partner of a symbiotic or commensal relationship.

hydrodynamic The physical features of water motion.

hydrogen bond Weak bond formed between a hydrogen atom in one molecule and an electronegative atom of the same or another molecule.

hydrographic The physical and chemical features of the oceans.

infauna Benthic organisms that live in or burrow through the bottom sediment.

infralittoral fringe In the Stephenson Universal Zonation scheme, the lowest zone on the shore, bounded above by the upper limit of laminarians and below by the lowermost tide level.

inquilinism Subcategory of commensalism in which the animal lives in the home or digestive tract of another without being parasitic.

insolation Being exposed directly to the sun's rays.

interstitial The space between adjacent particles in a soft bottom.

interzonal fauna Mesopelagic animals that migrate vertically into other zones.

ion A positively or negatively charged atom or molecule produced through loss or gain of one or more electrons.

isohaline A line joining points of equal salinity concentration.

isotherm A line joining points of equal temperature.

iteroparous Referring to organisms that reproduce more than once during their life spans.

keel A median longitudinal ventral ridge in certain nektonic fishes.

kelp bed A community dominated by large brown algae in which the plants do not form a surface canopy.

kelp forest A community dominated by large brown algae in which the plants form a surface canopy.

keystone species A species that is disproportionately important in the maintenance of community integrity, and without which drastic alterations of the community would occur.

krill Colloquial name for crustaceans of the order Euphausiacea.

lagoon (a) Type of estuary formed through the cutting off of inshore waters by the buildup of sandbars parallel to the shore; (b) a shallow stretch of water separated from the open ocean by a coral reef or island.

Langmuir convection cell A localized area of water a few meters wide and hundreds of meters long created by the wind in which there is a local circulation pattern causing water to upwell and diverge from the center and converge and sink at the boundaries.

latent heat of fusion Increase in heat content that accompanies converting a unit mass of substance from a solid to a liquid state.

law of the minimum Law stating that of those essential substances that are necessary to the survival of an organism, the one that is present in the smallest quantities or is reduced to a minimum first will limit the growth and survival of the organisms, even if all others are plentiful.

lecithotrophic larvae Planktonic larvae that do not feed in the plankton.

leeward The side of an island or reef protected from winds and waves.

light Electromagnetic radiation with wavelengths between 400 and 700 μ m.

lithothamnion ridge Synonym for algal ridge.

low-level radioactive wastes Those wastes that generate less than 1×10^8 curies of ionizing radiation.

luciferase The enzyme that catalyzes the production of bioluminescence in the presence of luciferin and ATP.

luciferin The protein that produces bioluminescence in the presence of ATP and luciferase.

mackerel Common name for several species of economically important fishes of the family Scombridae.

macrofauna A general term referring to benthic organisms more than 1 mm in size.

macroplankton Plankton with a size range from 2 to 20 cm.

Madreporaria The order of the class Anthozoa that contains the corals.

mangal *See* mangrove forest.

mangrove Common name for any of several species of inshore tropical trees or shrubs that dominate the mangal associations.

mangrove forest A variety of tropical inshore communities dominated by several species of shrubs or trees that have the ability to grow in salt water.

mariculture The culture of marine organisms.

marine-dominated estuary *See* homogeneous estuary.

marine snow Name given to the various types of amorphous particulate matter found in the water column and derived from living organisms.

mass strandings When large numbers of whales beach themselves.

maximum sustainable yield The largest number of fishes that can be harvested year after year without diminishing the stock.

megaplankton Plankton with a size range above 20 cm.

meiobenthos A synonym for meiofauna.

meiofauna Benthic organisms within the size range between 0.5 mm and 62 μ m, often used synonymously with interstitial fauna.

melon The large lipid-filled body in the head of whales that serves as an acoustical lens.

meroepipelagic Open-ocean nekton that spend part of their lives in the open ocean and part in other areas.

meroplankton Planktonic organisms that spend only part of their life cycles in the plankton.

mesobenthic Meiofaunal-sized organisms living and moving within the interstitial space.

mesoplankton Plankton with a size range between 0.2 and 20 mm.

mesopsammon Organisms living between sand grains in fresh water.

metabolism A general term referring to all physical and chemical processes that occur in a living organism.

metabolite A biochemical produced by an organism and secreted into the surrounding medium.

metamorphosis The process of structural transformation in an animal, which changes the organism from larva to adult.

microbial loop Name given to that part of the planktonic food web made up of bacteria and various nanoplankton organisms that recovers and cycles dissolved and particulate carbon back into the larger planktonic food web.

microfauna Benthic organisms below the size of 0.1 mm.

microplankton Plankton with a size range from 20 to 200 μ m.

midlittoral zone In the Stephenson Scheme of Zonation, that area between the upper limits of laminarians and the upper limits of barnacles.

migration Periodic movement of animals from one place to another.

mimic An organism that assumes the shape, pattern, or behavior of a different species, usually for protection.

Minimata disease Crippling neurological disorder caused by mercury poisoning.

mixed tide Tides consisting of a mixture of diurnal and semidiurnal tides.

multivore *See* omnivore.

mutualism That form of symbiosis in which two species associate to their mutual benefit.

myoglobin An iron-containing protein compound in muscle, which serves as an oxygen carrier and reserve oxygen supply.

nanoplankton Plankton with a size range from 2 to 20 μm.

nauplius First larval form of most crustaceans; characterized by possessing three pairs of appendages.

negative estuary Estuary in which evaporation is high and freshwater input low, so that salt water enters at the surface, evaporates, becomes hypersaline, and sinks, forming an outflowing bottom current.

nekton Pelagic animals that are powerful enough swimmers to move at will in the water column.

net photosynthesis The amount of photosynthesis in excess of respiration.

net plankton Organisms caught in a standard zooplankton net (above 0.2 mm).

net production Amount of total production left after losses from respiration (or that amount left to support other trophic levels).

neutral In chemistry, having equal numbers of OH^- and H^+ ions; a pH of 7.

Niño, El An episodic increased warming of the equatorial Pacific Ocean due to atmospheric changes; it can cause large-scale changes in oceanographic and atmospheric conditions that may lead to catastrophic mortality of many marine organisms.

nonpelagic development Development without a planktonic period.

nucleus The central, dense, positively charged part of an atom.

oligotrophic Containing little nutrient material.

omnivore An organism that consumes both plant and animal material (synonym = multivore).

ooze A deep-sea sediment composed, at least partially, of the skeletal remains of certain pelagic organisms.

opportunistic species A species that has a life history characterized by short life span, short development time to reach maturity, high death rate, and many reproductive periods per year.

osmoconformer An organism unable to regulate its internal fluid and salt balance, and therefore one with a varying internal salt concentration.

osmoregulation Physiological activity within an organism that serves to maintain the internal salt and fluid balance within a narrow acceptable range.

osmosis The movement of water across a semipermeable membrane separating two solutions of differing solute concentrations, movement being from the more dilute solution to the more concentrated.

oxygen minimum zone An area in the ocean, usually between 500 and 1000 m, where oxygen values approach zero.

parallel bottom community Concept that similar sediment types at similar depths around the world contain similar communities of organisms in which the dominant animals are closely comparable taxonomically and ecologically.

parasitism Association in which one species lives on, or in, another, drawing nourishment from that species at the expense of, or to the detriment of, the other.

pectoral Pertaining to the chest region of the body.

pelvic Pertaining to the hip region of the body.

perigee That point during the orbit of the moon around the earth when the moon is closest to the earth.

Petersen grab A quantitative grab designed by C. G. Joh. Petersen to take a fixed area of soft bottom, usually a tenth or one-half of a square meter.

pheromone A chemical secreted by a species that influences the behavior of others of the same species.

photic The lighted area of the ocean; the epipelagic zone.

photophore A complex organ of certain deep-sea animals that produces light and regulates its emission and use.

photosynthesis Process occurring in green plants, whereby they use the energy of sunlight to synthesize energy-rich organic compounds from carbon dioxide and water.

phototaxis Movement of an organism in response to light.

phototrophic Name applied to organisms that utilize light as a source of energy.

physoclist In bony fishes, those fishes with no open duct between the gas bladder and the digestive tract.

phytoplankton Planktonic plants.

picoplankton Plankton in the size range of 0.2 to 2.0 μm.

Pinnipedia An order of mammals containing the seals, sea lions, and walruses.

plankton Those organisms free-floating or drifting in the open water of the oceans having their lateral and vertical movements determined by water motion.

plankton net Fine-mesh conical nets dragged in the water to collect plankton.

planktotrophic larvae Larvae that feed in the plankton.

planula The sexually produced larvae of many cnidarians, including corals.

pneumatocyst The gas-filled floats on certain kelps.

POC Abbreviation for particulate organic carbon.

poikilothermic *See* ectothermic.

polar The Arctic or Antarctic zones.

pollution Degradation of the natural environment.

polyp A sessile or sedentary cnidarian individual with a cylindrical body and, usually, tentacles surrounding a mouth at the free end.

positive estuary Estuary with substantial freshwater input and reduced evaporation to the extent that fresh water flows out at the surface and salt water flows in at the bottom; also known as a salt wedge estuary.

preadaptation Possession by a species or group of species of features that favor easy occupation of a new habitat.

predator An animal that feeds on another animal.

protein Complex biochemical compounds of large molecular weight composed of joined amino acids.

psammon An obsolete term referring to organisms living interstitially.

pycnocline That area in the water column where the highest rate of change in density occurs for a given change in depth.

realized niche That portion of the fundamental niche that the species actually occupies or uses within its habitat.

redox potential discontinuity (RDP) layer Zone in muddy shores characterized by a rapid change from positive to negative redox potential; the boundary layer between aerobic and anaerobic sediments.

red tide The name given to massive blooms of dinoflagellates in which the concentrations of the plants are such that the water becomes a red-brown in color, toxins often being secreted in sufficient quantities to affect other marine organisms.

reflective beach Beach with no surf zone, a steep slope, and waves that break directly on the beach and that are then reflected back, producing large swashes on the beach.

refractive index For a medium, the ratio of the speed of light in a vacuum to that in the medium.

rete mirabile A large network of anastomosing small blood vessels serving a number of functions in marine vertebrates.

reverse migration Opposite of vertical migration; animals migrate down at night and toward the surface during the day.

rhodopsin A light-sensitive pigment found in the rods of the eye.

rings *See* eddy.

river-dominated estuary *See* salt wedge estuary.

rod The sensory body of the retina that is most responsive to low light levels.

salinity The total amount of dissolved material (salts) in seawater.

salt marsh Communities of emergent vegetation rooted in soils that are alternately inundated and drained by tidal action.

salt wedge estuary Estuary with a pronounced difference in salinity between surface and deeper water, the deeper waters being more saline (*see also* positive estuary).

Scleractinia Synonym for Madreporaria.

sclerosepta The calcium carbonate septa in the skeletons of corals.

seagrass Collective name for marine flowering plants of the families Potamogetonaceae and Hydrocharitaceae.

sea mount *See* guyot.

seasonal estuary Estuary having freshwater inflow and mixing with ocean water only for part of a year; the rest of the year it is a stagnant lagoon.

seasonal succession The change of dominant species in the phytoplankton with succeeding seasons of the year.

sedentary In reference to organisms, those that remain for long periods in one place or have limited movement.

semelparous Breeding or reproducing only once during a lifetime.

semidiurnal tide A tide with two highs and two lows per lunar day.

sere One of the stages in a successional sequence leading to a climax community.

sessile An organism that is fixed in place.

silicoflagellate Flagellated photosynthetic organism of the class Chrysophyta with an internal skeleton of silicon dioxide.

Sirenia The order of the class Mammalia that contains the sea cows (dugongs and manatees).

slough *See* lagoon.

sonar A method employing sound waves by which objects may be located in the water column.

species diversity In ecology, a numerical measure combining the number of species in an area with their relative abundance.

species richness The number of species in a given area.

spermaceti organ The melon of the sperm whale.

spermatophore A special packet that contains sperm.

sporophyte In algae, the diploid plant that produces spores.

stability-time hypothesis Theory seeking to explain the occurrence of high diversity among communities of the marine environment; the hypothesis states that high diversity occurs because highly stable environmental conditions have persisted over long periods of time, allowing numerous, highly specialized species to evolve.

standing crop The biomass of an organism or group of organisms present per unit volume or per unit area at a given point in time.

stenohaline Able to tolerate only a narrow range of salinity changes.

stipe That part of a kelp plant between blade and holdfast.

stratified estuary *See* salt wedge estuary.

subsidence theory Darwin's theory as to the origin of coral atolls whereby atolls begin as fringing reefs around volcanic islands; subsequent subsidence of the island, coupled with upward reef growth, then produces first a barrier reef and finally, with the disappearance of the volcano, an atoll.

succession The process of community change through time. The facilitation model of succession is the classical definition of succession, whereby there is an orderly sequence of seral stages, the organisms of each modifying the environment making it less suitable for it and more suitable for the next. The end point or climax perpetuates itself. The inhibition model of succession has no organism as competitively superior and those that first occupy the area remain until a disturbance removes them. A third model, the tolerance model, is intermediate between the other two.

supralittoral fringe In the Stephenson Universal Zonation scheme, the highest zone on the shore, bounded below by the upper limit of barnacles and above by the upper limit of *Littorina*.

surface tension With respect to water, the attraction of the molecules of water at the surface (air–water interface) so that the surface resists penetration by small objects.

surge channel Deep channels in the windward face of a coral reef through which water moves in and out of the reef.

suspension feeding Feeding by filtering particles out of the water column.

swim bladder *See* gas bladder.

symbiont Name given to the partners in a mutualistic relationship.

symbiosis The interrelationship between two different species.

tectonic estuary Estuary formed when the sea reinvades the land due to subsidence of the land as a result of earthquake activity.

terrigenous Ocean sediments derived from the land.

territory An area occupied and protected by a species against others of its own species, usually for breeding or feeding purposes.

thalassopsammon The interstitial organisms in marine waters.

thermal pollution Degradation of the environment through increases in temperature.

thermocline That portion of the water column where temperature changes most rapidly with each unit change in depth.

thigmotaxis Behavioral or movement response triggered by contact with a solid body.

tide The periodic rise and fall in the surface water of the oceans due to gravitational attraction of the sun and moon and the rotation of the earth.

tidepool Area in the rocky intertidal that retains some volume of water at low tide.

toxin Any of a variety of chemical substances produced by organisms that are, in turn, poisonous to other organisms.

trenches The narrow, steep-sided depressions in the ocean floor usually lying between 6,000 and 10,000 m in depth.

trophic group amensalism Exclusion of one group of organisms by modification of the environment by another.

trophic level An ecological system of classification of organisms according to their means of obtaining nutrition; the basic level is that of the autotrophs, the second is that of the herbivores, and the succeeding levels are carnivores.

trophogenic zone The area in the ocean above the compensation point.

tropholytic zone The area in the ocean below the compensation point.

tsunami A very long period wave generated by an underwater earthquake, landslide, or volcanic eruption; also known as a tidal wave.

tuna The common name for any of several large pelagic fishes of the family Thunnidae.

turbidity Condition of reduced visibility in water due to the presence of suspended particles.

ultraabyssal *See* hadal.

ultraplankton Plankton less than 2 μ m in size.

vernal The spring of the year.

vertical migration Diurnal vertical movement of pelagic organisms in the water column toward the surface at night and down to depth during the day.

viscosity That property of a liquid to resist movement through it.

vitamin An organic compound essential to certain metabolic processes in organisms.

wave With respect to the oceans, a disturbance that moves through the water but does not cause particles of the water to advance with it.

whalebone *See* baleen.

windward The side of an island or reef that faces the prevailing wind.

zoochlorellae Symbiotic algae that are green in color.

zooplankton Planktonic animals.

zooxanthellae Dinoflagellates symbiotic in the tissues of various marine invertebrates; symbionts that are brownish in color.

Index

Note: Boldface type indicates pages on which organisms are illustrated; "t" indicates a table; "p" indicates a color plate.